INTRODUCTION TO PHYSICAL GEOGRAPHY

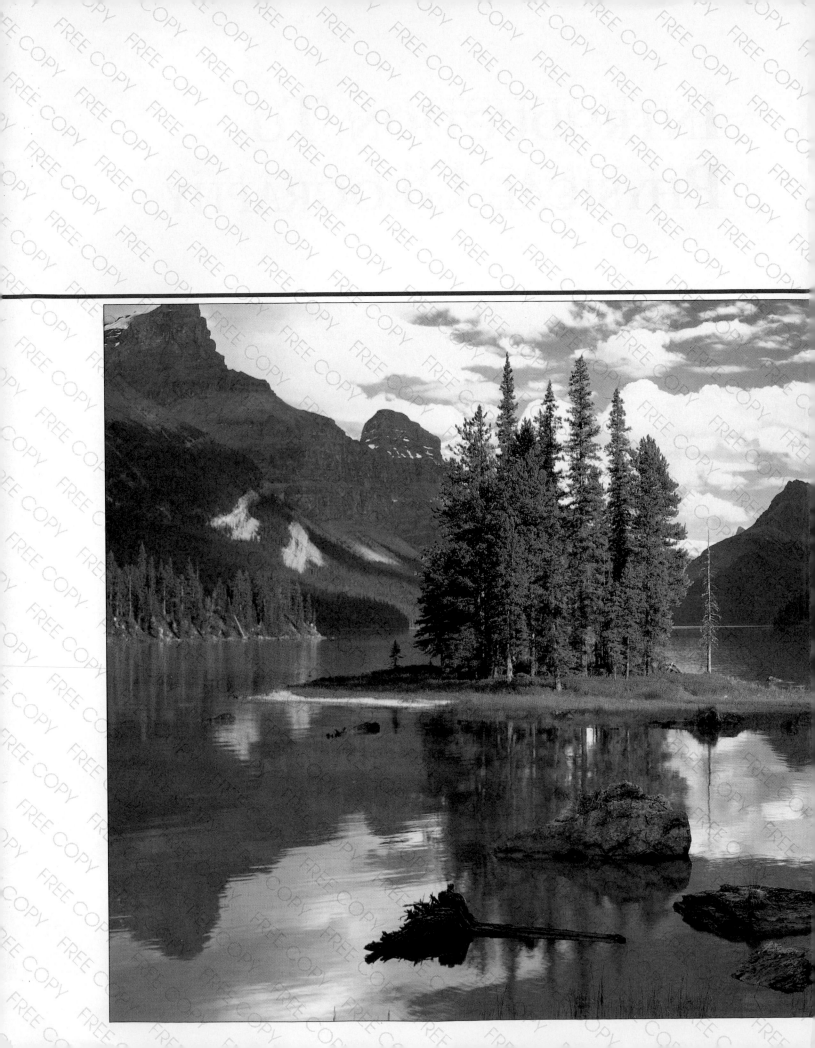

INTRODUCTION TO PHYSICAL GEOGRAPHY

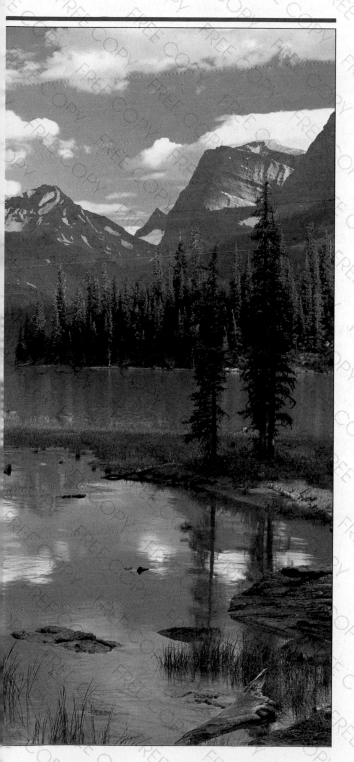

RALPH C. SCOTT

TOWSON STATE UNIVERSITY

WEST PUBLISHING COMPANY

ST PAUL/MINNEAPOLIS

NEW YORK

LOS ANGELES

SAN FRANCISCO

DEDICATION

This book is dedicated to my wife Judi and to my children Scottie, Kelli, and Jill. They make it all worthwhile.

WEST'S COMMITMENT TO THE ENVIRONMENT

In 1906, West Publishing Company began recycling materials left over from the production of books. This began a tradition of efficient and responsible use of resources. Today, 100% of our legal bound volumes are printed on acid-free, recycled paper consisting of 50% new paper pulp and 50% paper that has undergone a de-inking process. We also use vegetable-based inks to print all of our books. West recycles nearly 22,650,000 pounds of scrap paper annually—the equivalent of 187,500 trees. Since the 1960s, West has devised ways to capture and recycle waste inks, solvents, oils, and vapors created in the printing process. We also recycle plastics of all kinds, wood, glass, corrugated cardboard, and batteries, and have eliminated the use of polystyrene book packaging. We at West are proud of the longevity and the scope of our commitment to our environment.

West pocket parts and advance sheets are printed on recyclable paper and can be collected and recycled with newspapers. Staples do not have to be removed. Bound volumes can be recycled after removing the covers.

Printed with **Printwise** ∞
Environmentally Advanced Water Washable Ink

British Library Cataloguing-in-Publication Data. A catalogue record for this book is available from the British Library.

COPYRIGHT ©1996   By WEST PUBLISHING COMPANY
610 Opperman Drive
P.O. Box 64526
St. Paul, MN 55164-0526

Printed in the United States of America

03 02 01 00 99 98 98 97 96   8 7 6 5 4 3 2 1 0

Library of Congress Cataloging-in-Publication Data

Scott, Ralph C. (Ralph Carter), 1944–
    Introduction to physical geography / Ralph C. Scott.
        p.   cm.
    Includes index.
    ISBN 0-314-06260-2 (soft : alk. paper)
    1. Physical geography.   I. Title.
GB54.5.S373   1996
550--dc20   95–45242
                                     CIP

CREDITS

COPYEDIT: Pat Lewis
TEXT DESIGN
   AND LAYOUT: Metier, Inc.
COVER DESIGN: John Edeen
COVER PHOTO: Maligne Lake, Canada (Dietrich Stock Photos)
COMPOSITION: Carlisle Communications
ARTWORK: B. Suter Graphics & Design; Carlyn Iverson
MAPS: Carto-Graphics
INDEXING: Catalyst Communication Arts

PRODUCTION, PRINTING AND BINDING BY WEST PUBLISHING COMPANY.

CHAPTER OPENING PHOTOS

CHAPTER 1   M. Thonig/H. Armstrong Roberts. Alpine meadow vegetation in the French Alps near Dauphine.

CHAPTER 2   NASA. Apollo 17 view of the Earth. Africa and the Arabian Peninsula appear in the upper portion of the photo, while ice-covered Antarctica is partially visible beneath the clouds near the bottom of the photo.

CHAPTER 3   Tony Stone Images. A rainbow and clearing skies follow a storm in California.

CHAPTER 4   Bud Lehnhausen/Photo Researchers. Desert vegetation on Isla San Esteban, in the Sea of Cortez, Baja California, Mexico.

CHAPTER 5   © Ken Fraser/FPG International Corp. Yachts sailing with the wind in the Caribbean Sea.

CHAPTER 6   Ralph Scott. Summer cumulonimbus clouds forming over South Carolina are being observed by the author.

CHAPTER 7   Richard Kaylin/Tony Stone Images. An approaching severe thunderstorm at sunset.

CHAPTER 8   Nicholas DeVore/Tony Stone Images. Moroccan town in the foothills of the Atlas Mountains.

CHAPTER 9   R. Matassa/H. Armstrong Roberts. Loch Duich, near Kintail in northwest Scotland.

CHAPTER 10   © Travelpix/FPG International Corp. A variety of desert vegetation in Arizona's Saguaro National Monument.

CHAPTER 11   M. Thonig/H. Armstrong Roberts. Dense tropical forest in Sri Lanka.

CHAPTER 12   J. Erwin/H. Armstrong Roberts. Wheat fields follow the contours of the land in Palouse County, Washington. Alternating strips of land are left unplanted to conserve soil moisture.

CHAPTER 13   G. L. French/H. Armstrong Roberts. Nevada Falls and adjacent granitic exfoliation dome, Yosemite Valley, California.

CHAPTER 14   EOSAT. Satellite view of the Bogda Shan Mountains, east of the city of Urumqui in the Gobi Desert of western China. Healthy vegetation appears green and snow cover near the crest of the mountains is pale blue. The steep slopes near the top of the photo give way to a series of alluvial fans. The large orange-red patch at the lower right is an area of sand dunes.

CHAPTER 15   Paul Johnson. Differential weathering and erosion of weak sedimentary rock accounts for the spectacular rock formations of Bryce Canyon National Park in Utah.

CHAPTER 16   Joura/H. Armstrong Roberts. Arizona's Grand Canyon as viewed from Hermit's Rest.

CHAPTER 17   V. Clevenger/H. Armstrong Roberts. The spectacular south face of Alaska's Mt. McKinley, North America's highest peak.

CHAPTER 18   EOSAT. Fault movements are largely responsible for the topography of this portion of southern California just north of Los Angeles.

CHAPTER 19   EOSAT. The east-central portion of the Florida Peninsula in the vicinity of Cape Canaveral.

CONTENTS

6 Atmospheric Moisture 102

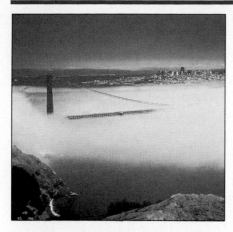

7 Weather Systems 134

8 Climates of the World 171

9   WATER ON AND BENEATH THE EARTH'S SURFACE   210

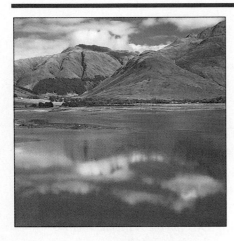

10   LIFE-FORMS AND ECOSYSTEMS   240

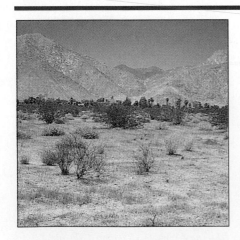

11 Distribution and Characteristics of Terrestrial Biomes   266

12 Soils   292

13 Landforms and Earth Materials   323

14 Tectonic Processes and Landforms　350

15 Weathering, Mass Wasting, and Karst Topography　390

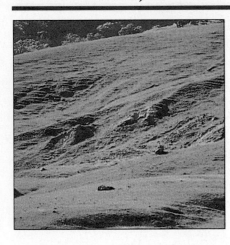

16 Fluvial Processes and Landforms　415

17 GLACIAL AND PERIGLACIAL LANDFORMS   443

18 EOLIAN PROCESSES AND DESERT LANDSCAPES   473

19 COASTAL LANDFORMS AND SEAFLOOR TOPOGRAPHY   496

PREFACE

My primary goal in writing this book is to provide its readers with a greater understanding and appreciation of the world around them. The book is intended chiefly for college students with limited physical science and geography backgrounds who are taking the course either to meet general university requirements or as an introductory course in a geography major program. The book provides an intermediate depth of material coverage that, I believe, is appropriate for the majority of college and university undergraduate students. I hope that students and instructors will also find the material to be clearly written and well organized.

Over the past two decades, geographic education in North America, and particularly within the United States, has been experiencing a long-overdue revival. Students, educators, and the general public once again are coming to realize that other places, and the events that occur in them, have an important bearing on our own lives.

We are also discovering that our planet is a fragile one, and it has become clear that all its components are interconnected to form a single life-support system of enormous complexity. Humanity, with its explosive growth in numbers and technological capabilities, now has a greater potential than ever before to permanently alter the Earth's physical characteristics on a global scale. It is essential that today's students—the next generation of decision-makers—display a greater understanding and appreciation of the environment than did the leaders of the past.

APPROACH AND CONTENT OF THE BOOK

Introduction to Physical Geography has the same sequence of major topics as its predecessors—*Physical Geography* and *Essentials of Physical Geography*. Following a brief introductory chapter discussing the field of geography as an academic and intellectual discipline, nearly all material coverage is devoted to the traditional "big five" subjects of physical geography. These include the Earth in space (planetary dimensions, motions, and depictions), weather and climate, natural vegetation and wildlife, soils, and landforms. Although a discussion of the optimal order for these subjects can stimulate endless debate, I believe that the chapters within these sections are sufficiently self-contained that they can be treated in a variety of differing sequences.

The approach I use for covering the book's major topics might be called the process/distribution approach. By this I mean that a certain amount of general science background appears first to explain the origin and characteristics of each topic. This is followed by a discussion of the topic's geographical distribution, and finally by the reasons for its distribution. Throughout the book, I have tried to stress the causative and geographical linkages among the various Earth phenomena discussed.

I decided not to base my coverage of world climates on the Köppen system, so as to avoid forcing it upon instructors and students. I have always been somewhat partial to this system, however, and my climate regions correspond closely to Köppen's. I have also included a Focus Box within the chapter on Climate (Chapter 8) that explains this system, including its advantages and shortcomings, in some detail.

A variety of important environmental concerns are examined in the Case Studies at the end of the chapters and appearing in many of the Focus Boxes throughout the text. I hope that these studies will promote student interest, will illustrate the relevance of the material to contemporary issues, and will foster an increased environmental consciousness.

Several additional learning aids have been provided in the text. At the beginning of each chapter is a subject outline as well as several Focus Questions that broadly address the core subjects of the chapter material. Within the body of the chapters, important terms appear in boldface type for easy recognition, and again in an extensive glossary at the end of the book. A number of Geographical Technique boxes highlighting methods used by geographers to organize and analyze geographic data also can be found within the chapters. At the end of each chapter is a summary that contains answers to the chapter-opening Focus Questions. Also appearing at the end of each chapter is a set of Review Questions and a list of Key Terms; some chapters also have a Problems section containing mathematical problems. Finally, at the end of the text are appendices that contain information on scale conversions, weather map interpretation, a world political map, and answers to the end-of-chapter problems.

DIFFERENCES BETWEEN THIS BOOK AND ITS PREDECESSORS

This book draws heavily upon the best aspects of its predecessors. Depth of material coverage is slightly less than that of the *Physical Geography* textbook in most areas, but greater than that of the *Essentials* text. All chapters, of course, have been thoroughly reviewed, revised, and updated.

One change of particular significance is the expansion of the Biogeography material to two chapters. Important new material has been added to this section that deals with the forms and functions of plants and animals and with the makeup and activities of biological communities. I have also added a number of new Focus Boxes, several of them by guest authors. Among the new topics these boxes cover are Geographic Information Systems, Hurricane Andrew, water pollution, and National Park management. Because several Canadian universities have used this book's predecessors, I have also increased the emphasis on North America as a single region, rather than focusing largely on the United States. I have also added material on mapping to the chapter dealing with the Earth as a planet ("The Planetary Setting") because the reviewers and I felt that basic information about the Earth's surface should be followed closely by a discussion of methods used to depict that information.

Perhaps the most striking way in which this book differs from that of my earlier texts is in the number and quality of the illustrations. Improved processing techniques, a better grade of paper, and a larger selection of photos and diagrams have helped produce a marked improvement in this important aspect of the book.

AVAILABLE ANCILLARIES

West has provided a comprehensive package of ancillary materials that I believe will greatly aid adopters of this book. These include an instructor's manual, a laboratory manual, a student study guide, and a computerized test bank for IBM, Macintosh, and Apple computers. The choice of a set of approximately 100 color slides or overhead transparencies of the more important maps and diagrams in the text is also available to teachers using the book. A great number of additional illustrations are available in a complementary video disk that contains thousands of still frames as well as short animations and video segments. Finally, West now maintains a Video Library of longer videos, available to qualified adopters. Please contact the publisher or your local sales representative for further information on any of these materials.

ACKNOWLEDGEMENTS

The preparation of a book of this magnitude involves a great many individuals, all of whom are essential to its successful

completion. I wish once again to thank the West editorial, production, and sales staffs, whose efforts enabled the book to become a reality and who have been chiefly responsible for the success of its predecessors. Foremost among these individuals is my Acquiring Editor, Clark Baxter, who has overseen this project through all four textbook publications and has provided constant encouragement and support. Linda Poirier, our Editorial Assistant, has provided perceptive analyses of the numerous manuscript reviews. Pat Lewis spent innumerable hours reading, correcting, and improving the manuscript of the text, and, in my opinion, did an excellent job. Stephanie Gard Buss, our Promotion Manager, has been responsible for the development of advertising brochures and for making sure that they reached potential users of the book. Lastly, the person whose efforts have made it possible to turn my manuscript into a book is my Production Editor, Holly Henjum.

Another group to whom I owe a debt of gratitude are the eight individuals who were kind enough to serve as "guest authors" for Focus Boxes appearing in the text. These include Masamu Aniya, University of Tsukuba (Japan); Robert Balling, Arizona State University; Margaret F. Boorstein, Long Island University; Gregory Elmes, University of West Virginia; Bryon Middlekauf, Plymouth State College; John Morgan III, Towson State University; David A. Padgett, Austin Peay State University; and Laurel Phoenix, of Lafayette, Colorado.

Another group whose long hours of work with the manuscript have been vital to both its content and quality have been the reviewers. The work of the reviewers of my earlier books *Physical Geography* and *Essentials of Physical Geography* is also reflected in the content of this book, and I want to thank them again for their generous assistance.

I wish particularly, however, to thank the reviewers of this current edition. I found myself in agreement with most of their suggestions, and I believe they will see them incorporated in the book. These individuals are Thomas D. Anderson, Bowling Green State University; Bernard Bauer, University of Southern California; Gerald Brothen, El Camino College; William R. Brown, Prairie View A & M University; David R. Butler, University of North Carolina at Chapel Hill; Robert Champlin, Fitchburg State College; Anthony O. Clarke, University of Louisville; Don W. Duckson, Jr., Frostburg State University; Richard W. Foley, Cumberland College; Robert J. French, University of Southern Maine; Roland L. Grant, Eastern Montana College; Clark G. Hilden, Blue Mountain Community College; David A. Howarth, University of Louisville; Jim Hull, Towson State University; Rudi Kiefer, University of North Carolina at Wilmington; Paul A. Knapp, Georgia State University; Theodore W. Kury, Buffalo State College; Steve LaDochy, California State University at Los Angeles; Bernard J. McGonigle, Community College of Philadelphia; David R. Privette, Central Piedmont Community College; Michael B. Sady, Western Nevada Community College; Tony B. Shaw, Brock University; Danny M. Vaughn, Weber State

University; Thomas B. Williams, Western Illinois University; and Brent Yarnal, Pennsylvania State University.

Still another group of people who played a key role in this project were the artists and photo suppliers. I wish especially to thank Alice Thiede, who drew most of the maps appearing in the book, and Carlyn Iverson, who drew a number of the new figures. I owe a special thanks to Gerald Brothen, Bernard McGonigle, Rudi Kiefer, and Don Duckson, who reviewed these maps and figures and many of the photographs in the text. Thanks also go to the photo suppliers, including my wife Judi and my Towson State University colleague Wayne McKim, who each supplied several photographs used in the book.

Ralph C. Scott
December, 1995

1 THE SCOPE OF GEOGRAPHY

FOCUS QUESTIONS

1. What is geography? How does the geographic approach differ from other approaches to information?
2. What are the major subdivisions of the field of geography, and what topics do they cover?
3. What are regions, and why are they commonly employed in geographic studies?
4. What is a system? Why can the Earth be considered an open system?

THE DISCIPLINE OF GEOGRAPHY

When we look at news reports in the paper or on television, we are struck by the fact that the Earth undergoes constant change. Evidence for the dynamic nature of our planet includes major natural disasters, such as the earthquakes that struck California in 1991 and 1994; volcanic eruptions, such as the eruptions of Mt. St. Helens in Washington State in 1980 and Mt. Pinatubo in the Philippines in 1991; floods, such as those that devastated the Upper Mississippi Valley in the summer of 1993; and hurricanes, such as Hugo in 1989 and Andrew in 1992.

We may also be struck by the fact that the physical characteristics of the Earth's surface vary greatly from place to place. We may be well aware that due to higher Sun angles, the southern United States from California to Florida has a much warmer climate than the northern United States and Canada. But the natural conditions on our continent also exhibit numerous other major differences. Why, for instance, does much of western North America have a high elevation with extensive mountain ranges and deserts, while eastern North America is generally lower, flatter, and wetter? What effect do these differences in climate and topography have on the types and distributions of plants and animals that occupy these regions? And why are portions of western North America especially prone to earthquakes and volcanic eruptions, while the East Coast often experiences major hurricanes?

We may also be aware that the surface of our planet has experienced major changes over time. For instance, less than 15,000 years ago, nearly all of Canada and much of the northern United States were buried under massive ice sheets. A warmer climate caused the ice to melt, but now researchers are concerned that continued global warming may melt additional ice, raising sea levels and altering existing patterns of climate and human activities.

Over much greater time spans, even the positions of the continents themselves are not fixed. At one time North America was connected to Europe and Africa; eventually, they separated, and our continent drifted westward, opening the basin now occupied by the Atlantic Ocean. Meanwhile, the continued northward movement of India, which at one time was connected to Antarctica, has caused it to collide with southern Asia and, in so doing, to thrust up the Himalayas—the world's loftiest mountains.

If you are curious about subjects such as these, you are enrolled in the right course; the only academic discipline that examines collectively the patterns of natural phenomena on the face of the Earth is physical geography.

The Geographical Approach

Geographers are concerned with understanding the locational aspects of all Earth phenomena, both those that are natural and those that are produced by human activities. In its emphasis on location, geography's approach to its subject matter is unique among all the academic disciplines. The German philosopher Immanuel Kant (1724–1804) provided the philosophic basis for geography in the late eighteenth century when he identified three approaches that can be used to study any subject. The first, the **systematic approach,** is organized by the type of entity to be studied. Today most fields of knowledge are systematically defined. They study everything about one topic. An example of a topically oriented discipline is biology, which is the study of living organisms. All life-forms as well as the environmental factors that directly influence them are considered to fall within the bounds of biological studies.

Kant's second approach is the **historical approach,** which involves examining how a topic changes over time. Because everything changes with time, the historical approach is not topically restricted.

The final approach in Kant's conceptual framework is the **geographical approach,** which studies topics by their differences in location or, to put it another way, by their spatial characteristics. Just as everything changes with time, so too does everything vary in location. This includes not only physical objects such as people, rivers, and rock types, but also less tangible things such as religious beliefs, military alliances, education levels, and tastes in music. Because geography is defined by its method of approach rather than by the topic being studied, it can be described as a field that examines one aspect of every topic, rather than examining every aspect of one topic, as the systematic disciplines do.

Earthly phenomena are not randomly distributed; everything has a reason for being located where it is. In nearly all cases, the distribution of one type of entity is influenced by the presence or distribution of others. For example, human population distributions are influenced by factors such as climate, landforms, patterns of transportation routes, and the availability of housing and employment. World climates form patterns controlled by factors such as the availability of solar energy, proximity to water bodies, and elevation. In examining the causes for the distributions of Earth phenomena, geographers cannot avoid noting that these phenomena are interrelated. The systematic study of interrelationships among Earth phenomena has therefore also become a focus of geographic inquiry.

Geographers feel that most disciplines, especially those with rigidly defined topical boundaries, do not adequately emphasize the broader relationships that exist between their topics and other phenomena. The familiar expression of being "unable to see the forest for the trees" is particularly applicable in this context. Whereas many fields of study, by analogy, provide their practitioners with detailed knowledge of individual trees, geography is more concerned with the characteristics of the forest as a whole. Geography, then, is an integrative discipline that has a wide breadth of coverage

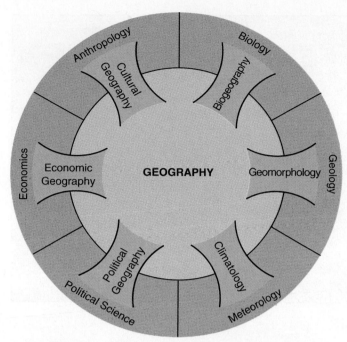

◆ **FIGURE 1.1** Geography overlaps with most topically defined fields of knowledge, because each contains spatial components.

SOURCE: Reprinted with permission from Julia A. Tuason, "Reconciling the Unity and Diversity of Geography," *Journal of Geography*, vol. 86, no. 5, p. 193

(Figure 1.1). It overlaps with most topically defined fields of knowledge because each contains spatial, or geographical, components. Like a person trying to assemble a jigsaw puzzle, the geographer's goal is to put enough pieces of knowledge about the Earth in their correct locational settings that the "big picture" begins to emerge.

In summary, *geography is the study of the distributions and interrelationships of Earth phenomena*. The geographer therefore is concerned both with the locations of phenomena and with the causes and consequences of those locations. A basic objective of geographic research is to achieve a better understanding of the nature of places, and especially how distributions of terrestrial phenomena interact to cause similarities and differences between places. The ultimate goal of geography is to develop a sufficient understanding of spatial interaction to allow for a comprehension of the entire Earth as a functioning system.

Subdivisions of Geography

As an examination of a college textbook or course syllabus will show, the major academic disciplines typically are subdivided into separate units of material. This is done largely to permit individuals to specialize in a body of related information of manageable proportions. The field of geography is no exception. Although geography *as a whole* is defined by its method of approach rather than by topic, its *subdivisions* are topically defined. While the basic spatial emphasis remains, each branch of geography deals with a different category of Earth phenomena.

Geography's subdivisions can be arranged in a hierarchy. In general, the more advanced a study, the more narrowly its particular borders are defined. At the most general level, the field of geography is partitioned into two major branches: physical geography and human geography.

Physical geography is concerned with locational aspects of natural Earth phenomena—that is, with phenomena not produced or primarily controlled by human beings. The four basic divisions of physical geography are weather and climate, surface features (including landforms and water bodies), natural vegetation, and soils. Most of this textbook is devoted to the geographical analysis of these four subjects.

Human geography, on the other hand, deals with subjects in which distributional patterns are largely or entirely controlled by people. It includes such topics as patterns of population, agricultural and industrial activities, urban areas, religions, transportation routes, political regions, and recreational facilities.

In reality, the interaction of terrestrial phenomena, both physical and human, produces a great deal of overlap between the different topical specializations, both in geography and in other disciplines. Geographers, because of the breadth of their interests and training, are especially able to appreciate these interrelationships. The wide range of subjects covered in geography courses often makes it difficult to group this discipline with others that are topically defined. (Colleges and universities have experienced problems in this regard, because they typically contain separate academic divisions or schools of physical and social sciences.) Geography as a whole, then, is both a physical and a social science, and one of the geographer's chief goals is to emphasize the interrelationships between the two.

The Importance of Geography

What advantage can a college student gain from a knowledge of geography and, in particular, of physical geography? The answer is that a knowledge of the locations and nature of places enables us to gain a fuller appreciation of the global, national, and local events that continually shape our lives. For example, a true understanding of the causes and likely consequences of recent events in the Middle East can be achieved only with a basic knowledge of the nations, peoples, customs, religions, climates, land and water bodies, and natural resources of that part of the world.

A knowledge of physical geography is also needed to appreciate fully the delicate complexity of the Earth's surface environment as the only known abode of life in the universe. As a species, human beings have now developed the technological capability of irreparably damaging the natural environment on a global scale. We will need to be far more environmentally sensitive in the future than we have been in the past. A number of case studies and focus boxes highlighting important contemporary environmental issues appear throughout this book. They illustrate the range of impacts that human beings have had on the Earth's natural systems.

◆ FIGURE 1.2 Some natural landscape changes, such as this Wisconsin stream's shift in course after it abandoned a meander, can occur over short periods of time. Most large-scale landscape alterations, however, occur slowly over long time spans.

SOURCE: JLM Visuals

FUNDAMENTAL GEOGRAPHICAL CONCEPTS

As already noted, the geographic distribution of the Earth's physical features is not the result of chance, but exists in response to natural processes that provide reasons for the characteristics and locations of terrestrial phenomena. A basic understanding of these processes makes the study of physical geography—and of any physical science—much more logical and intellectually satisfying. Explanations for the distribution of Earth features will comprise an essential component of our geographic study throughout this book. In this section, we examine a few fundamental concepts related to the distributions and interactions of Earth surface phenomena.

Regions

Although no two places on the Earth's surface are identical, different places may have some characteristics that are quite similar. For example, the coastal region of southern California much more closely resembles the coast of Spain than it does the nearby Mojave Desert, even though California and Spain are some 9,600 kilometers (6,000 mi) apart. This resemblance is largely due to similarities in climate, landforms, and vegetation between the two sites.

An important method of geographical analysis is the formulation of **regions,** areas that are similar in selected attributes. Regionalization simplifies and organizes existing patterns of Earth phenomena. This process, in turn, helps geographers to discover and understand the factors responsible for the existence of these patterns. Throughout this book, maps and diagrams will display regional patterns of physical Earth phenomena, and the text will discuss their causes and characteristics.

Regions—and the illustrations and text used in this book to describe them—have limitations, however. First, descriptions of broad areas are necessarily general. Nearly every rule has its exceptions, and often these exceptions are so localized or limited in importance that they cannot be shown on a map or covered in the text. A second limitation of regionalization is that spatial changes in the characteristics of Earth phenomena generally occur in a gradual manner. As a result, regional boundary lines on maps should not be viewed as indicating sudden changes, but rather as representing zones of transition. Similarly, conditions throughout all portions of a given region should not be regarded as homogeneous. The climate, for example, normally does not change suddenly when one crosses a line on a climate map, nor is the climate identical in all places lying within a single climatic region.

Temporal Changes

Places change through time as well as over distance. Some of these changes, such as a shift in the course of a stream or the passing of the seasons, are small or temporary (Figure 1.2). Others, like the uplift of a mountain range or the onset of an ice age, are much longer in duration and of greater regional significance. Although small-scale characteristics often change over short periods of time, broad regional patterns normally change very slowly, at least from a human perspective. Human activities are proving particularly disruptive to environmental conditions in many areas and are causing rapid changes, often of an undesirable nature.

The Earth as a System

A **system** is a set of interrelated components through which energy flows to produce orderly changes (Figure 1.3). We commonly use many types of mechanical systems; two examples are clocks and automobiles. Because all terrestrial phenomena are interrelated, any action that affects one component will ultimately influence everything else. In its association of parts, then, the Earth can be viewed as a single system of enormous complexity.

Systems can be divided into two basic types. A **closed system** has a limited energy supply confined within it. Such a system can operate only as long as the energy supply lasts; then it will "run down." Familiar examples of closed systems include a wind-up clock and an automobile with one tank of gasoline.

An **open system,** on the other hand, has access to an unlimited supply of energy from an external source. Changing the previous examples of closed systems to an electric clock and to an automobile that can obtain refills at a gas station whenever needed would make them open systems. Of course, no source of energy is truly infinite; any open system—and even the universe itself—will probably eventually exhaust its energy supply. For practical purposes, though, an open system has access to such a large energy supply that the exhaustion of this energy is generally not considered an issue.

The Earth system can be thought of as consisting of two major subsystems—a materials system and an energy system. The materials system consists of all the matter that comprises our planet and its atmosphere. This system is essentially a closed system because the material of the Earth (with the minor exceptions of incoming meteorites and the gradual leakage of atmospheric gases into outer space) remains almost constant. Earth scientists, including physical geographers, frequently regard the Earth materials system as being composed of four interconnected subsystems, often

referred to as "environmental spheres" (Figure 1.4). These are the **atmosphere,** the ocean of air that overlies the Earth's surface; the **hydrosphere,** the water of the surface and near-surface regions of the Earth; the **lithosphere,** the massive accumulation of rock material and metal that forms the solid body of the planet; and the **biosphere,** the distribution of living organisms of which we are a part.

The Earth's energy system, in contrast, is an open system, since energy is constantly arriving from the Sun and is being continuously generated by atomic decay processes in our planet's interior. At the same time, approximately equal amounts of energy are escaping from the Earth to outer space, producing a global energy balance. The terrestrial energy supply is distributed about our planet largely by the actions of a mobile subsystem of the atmosphere—the winds, which are discussed in Chapter 5, and a mobile subsystem of the hydrosphere—the ocean currents, discussed in Chapters 4 and 9.

Most terrestrial subsystems exist in a state of relative balance or **equilibrium,** at least under natural conditions, because a nearly constant stream of energy and materials passes through them. For example, total global precipitation is relatively constant from year to year, even though precipitation amounts vary greatly from place to place. This consistency occurs because global precipitation is closely tied to the amount of water vapor made available through evaporation, which, in turn, is largely controlled by the relatively steady receipt of solar energy by the Earth.

In most cases, minor fluctuations in terrestrial subsystem inputs are dampened by internal self-regulating processes so

◆ **FIGURE 1.3** The Earth's climate system is a complex open system driven chiefly by solar energy. It involves components of the atmosphere, hydrosphere, lithosphere, and biosphere.

a

b

c

d

◆ **FIGURE 1.4** The four major terrestrial subsystems, or "environmental spheres," are the (a) atmosphere (cumulonimbus cloud over South Carolina), (b) hydrosphere (Big Sur coast of California), (c) lithosphere (view from Dead Horse Point State Park, Utah), and (d) biosphere (tree ferns in Costa Rica).

SOURCE: a: Jerry Scott; b: D. Muench/H. Armstrong Roberts; c: VU/Doug Sokell; d: M. Thonig/H. Armstrong Roberts

that the equilibrium conditions tend to be maintained. On occasion, however, a crucial input to a system may change sufficiently to cause the system to adjust, perhaps rapidly, to a new equilibrium condition that will in turn affect other Earth subsystems. For example, many researchers are concerned that a variety of human activities may be causing the Earth's atmosphere to retain more solar energy. Computer models indicate that this will cause global temperatures to increase, which, in turn, will significantly alter planetary precipitation patterns, sea levels, and plant and animal distributions. This topic is discussed in the Case Study at the end of Chapter 4.

The various terrestrial subsystems respond in complex ways to the flow of energy and materials through the Earth system as a whole. The distributive patterns and movements resulting from these subsystems form the basis for the material content of this book.

▌SUMMARY

This chapter has provided an overview of the geographic approach to the study of Earth phenomena and the material content of the field of geography. It has also examined a few basic concepts that are central to geography.

As a means of summarizing the main points discussed in the chapter, we return to the Focus Questions posed on the chapter title page:

1. *What is geography? How does the geographic approach differ from other approaches to information?*

Geography is the study of the distributions and interrelationships of Earth phenomena. The distinctiveness of geography results not from the kinds of phenomena studied, but rather from the *way* they are studied, with the emphasis being placed on spatial or locational components. A geographer generally asks three questions when studying any topic: "Where is it?" "Why is it there?" "What is the significance of it being there?"

2. *What are the major subdivisions of the field of geography, and what topics do they cover?*

The field of geography generally is divided into two major branches. The emphasis of **physical geography** is on Earth phenomena that are natural and have not been created or extensively modified by human actions. The chief topical subdivisions of physical geography are the atmosphere, hydrosphere, lithosphere, and biosphere. **Human geography** deals with Earth phenomena that are created, controlled, or modified by humans. Examples of human geography topics include economic, political, urban, and religious patterns.

3. *What are regions, and why are they commonly employed in geographic studies?*

Regions are portions of the Earth's surface that display similarity in one or more selected physical or human attributes. Regions are employed as a means of simplifying and organizing the geographic study of large areas by emphasizing selected spatial patterns.

4. *What is a system? Why can the Earth be considered an open system?*

A system is an organized set of interrelated parts through which energy flows to produce orderly changes. Closed systems contain a finite supply of materials and energy, while open systems are powered by an essentially limitless supply of energy from an external source. The Earth can be regarded as an enormously complex open system, powered by a continuous flow of energy both from our planet's interior and from the Sun. One of the physical geographer's basic goals is to gain a greater understanding of the workings of the Earth system.

▌QUESTIONS FOR THOUGHT AND REVIEW

1. What is geography? How does geography differ from the other physical and social sciences in its method of approach to the study of a subject?
2. What are the two major subdivisions of the field of geography? Explain the basis of the difference in their material content.
3. What is a geographic region? What are the major advantages and disadvantages of using regions to describe global patterns of physical phenomena?
4. Explain why the Earth can be described as a system. In what respects is it a closed system? In what ways is it an open system?

▌KEY TERMS

geography	open system
physical geography	atmosphere
region	hydrosphere
system	lithosphere
closed system	biosphere

▌SELECTED REFERENCES FOR CHAPTER 1

The following books are recommended for those seeking a more comprehensive discussion of geographic methodology, philosophy, and history:

Broek, Jan O. M. *Geography: Its Scope and Spirit.* Columbus, Ohio: Merrill Publishing Co., 1965.

Dunn, Margery G., and Donald J. Crump. *Exploring Your World—The Adventure of Geography.* Washington, D.C.: National Geographic Society, 1989.

Gaile, Gary, and Cort Wilmott, eds. *Geography in America.* Columbus, Ohio: Merrill Publishing Co., 1989.

Holt-Jenson, Arild. *Geography: Its History and Concepts.* Totowa, N.J.: Barnes & Noble Books, 1980.

James, Preston E., and Geoffrey J. Martin. *All Possible Worlds: A History of Geographical Ideas.* 2nd ed. New York: John Wiley & Sons, 1981.

2 THE PLANETARY SETTING

OUTLINE

FOCUS QUESTIONS

1. What is the geographical setting of the Earth in space, and what movements is the Earth making through space?
2. What basic directional and locational systems are employed on the surface of the Earth?
3. How is the global system of time zones organized?
4. What causes the seasons?
5. What is a map? With what basic map information must the user become familiar before a map can be employed to full advantage?
6. How is map information collected and compiled?

The Earth is the third of nine known planets orbiting a star we call the Sun (Figure 2.1). Sharing our **solar system** with the Sun and planets are a number of planetary satellites (moons), as well as asteroids, comets, dust, gases, and a great deal of empty space. Stability is maintained in the solar system by two opposing factors: (1) the outward thrust of centrifugal force caused by the inertia of the planets as they orbit the Sun and (2) the inward pull of gravity. The revolution of the planets around the Sun, and of the moons around the planets, is made possible by the precise balance between these forces.

The Sun is only one relatively undistinguished member of a giant assemblage of stars called the **Milky Way galaxy.** Galaxies are large groups of stars held together by their mutual gravitational attraction. They vary tremendously in size, shape, and density and contain anywhere from a few thousand to hundreds of billions of stars (Figure 2.2). Most galaxies also contain extensive clouds of gases and dust, out of which new stars constantly form.

The Milky Way is a large, discus-shaped galaxy consisting of more than 100 billion (or 10^{11}) stars. Despite its enormous stellar population, the great preponderance of the Galaxy consists of virtually empty space containing only the most rarefied concentrations of gas and dust. To illustrate, light takes 8 minutes and 20 seconds to travel the 150 million kilometers (93 million miles) from the Sun to the Earth. The next nearest star is so distant, however, that its light takes 4.22 *years* to reach the Earth.

The Milky Way galaxy revolves about its center of gravity. This motion has generated the centrifugal force responsible for the Galaxy's discus-like shape. The Galaxy's peripheral disk is not homogeneous; rather it is organized into spiral arms of stars, giving the Milky Way something of a pinwheel appearance. Our solar system is located on a spiral arm an estimated 30,000 light-years from the center of the Milky Way.[1] This places us about two-thirds of the way from the galactic center to its outer fringe. The center of the Galaxy is rendered invisible to us by vast clouds of interstellar gas and dust. The stars that so liberally sprinkle the sky on a clear night are our neighbors, occupying nearby portions of our arm of the Milky Way. The origins of the universe, Milky Way galaxy, Sun, and Earth are discussed in greater detail in the Focus Box beginning on page 12.

PLANETARY DIMENSIONS AND MOTIONS

The Earth is engaged in several motions through space, each at a differing level of magnitude. These movements can be divided into two categories—large-scale movements and small-scale movements. The large-scale movements are of limited direct significance to the Earth because they require millions or even billions of years to produce major changes. The Earth's small-scale movements, however, critically affect nearly every aspect of our environment and will be discussed in greater detail. For clarity, the Earth's various motions through space will be discussed in order of decreasing magnitude.

---

[1] A light-year is the distance that light travels in one Earth year. It equals 9.5×10^{12} kilometers, or 5.9×10^{12} miles.

◆ **FIGURE 2.1** The planets of our solar system all orbit the Sun in the same direction and, except for Pluto, on nearly the same orbital plane.

◆ **FIGURE 2.2** The Andromeda galaxy, pictured here, is a spiral galaxy similar to the Milky Way.
SOURCE: Palomar Observatory, California Institute of Technology

Large-Scale Motions

The largest-scale motion of all is the movement of the Earth, along with the rest of the solar system and Galaxy, away from the center of the universe—that is, from the site of the Big Bang (see the Focus Box on p. 12). Astronomical observations indicate that the universe is still expanding at great speed.

The second large-scale motion is the revolution of our solar system around the center of the Milky Way galaxy. An estimated 230 million years are required for one full revolution; this period is a function of our distance from the center of the Galaxy.

A third motion of the Earth is associated with the drift of our solar system among the neighboring stars in our arm of the Milky Way. The positions of the various stars relative to one another are gradually changing. Over thousands of years, this movement will radically alter the present constellations of the night sky.

Small-Scale Motions

The small-scale motions of rotation and revolution cause the Earth to change its position constantly with respect to the Sun. These movements are extremely crucial to the Earth and its life-forms because not only is the Sun our only major source of planetary energy but it also powers external terrestrial processes such as the weather, plant growth, and erosion. These motions are also responsible for the seasons and for the alternation of day and night, and as such, they form the basis for our system of keeping time.

Revolution

One of the two primary small-scale motions of the Earth is its **revolution** around the Sun. This motion was apparently initiated by the rotary motion that developed during the initial contraction of the **solar nebula**—the gas cloud that formed our solar system. The motion has since been main-

tained by the combined effect of the centrifugal force generated by the inertia of the Earth as it orbits the Sun, and the inward pull of the Sun's gravitational force.

The Earth's revolution is the basis of our calendar year. The precise period of time needed for one complete revolution around the Sun is 365 days, 5 hours, and 49 minutes. For convenience, though, calendar years are defined in whole-day increments. Most years are therefore 365 days long. Because the extra 5 hours and 49 minutes required for a full revolution is very close to an extra quarter of a day, every fourth year has an added day (February 29) and is termed a **leap year.**

The Earth orbits the Sun on a plane termed the **plane of the ecliptic** in the same direction as all the other planets (Figure 2.1). This direction appears as counterclockwise when Earth is viewed from a point in space above the northern hemisphere. (The vantage point in space is crucial, because the direction of revolution is clockwise when Earth is viewed from above the southern hemisphere.)

The Earth's orbit is roughly circular, but more precisely, it forms a broad ellipse (Figure 2.3). The mean radius of this ellipse—the average distance between the Earth and Sun—is approximately 150 million kilometers (93 million miles). At its closest approach to the Sun, the Earth is said to be at **perihelion** (from the Greek *peri*, or "near," and *helios*, or "Sun"). At this point in its orbit, which occurs about January 3 each year, the Earth is only 147.5 million kilometers (91.5 million miles) from the Sun. Six months later, on or about July 4, the Earth is at its farthest point from the Sun, or at **aphelion** (from the Greek *apo*, or "away from"). Our distance from the Sun is then approximately 152.5 million kilometers (94.5 million miles). Therefore, the Earth's distance from the Sun varies by about 3 percent during the course of a year.

Rotation

Rotation, the smallest-scale motion of the Earth, involves the spinning of our planet on its **axis.** The Earth completes

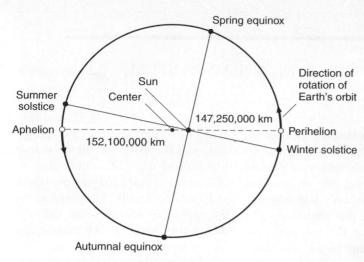

◆ FIGURE 2.3 The Earth's orbit is slightly elliptical. As a consequence, we are approximately three million miles nearer the Sun in early January than in early July.

one full rotational cycle in a 24-hour period. The rotation of the Earth on its axis is similar to the rotation of a model globe on its axis bar except that, of course, the Earth's axis is not a physical entity and provides no support.

The rotation of the Earth allows us to define three important and very basic surface locations. The **North Pole**

and **South Pole,** located on opposite sides of the planet, are the two points where the axis intersects the Earth's surface. Midway between the poles, and equidistant from both along its entire length, lies the **equator.** The equator extends due east-west and is the line of fastest rotational speed on the Earth.

Rotation is the basis of our calendar day and is the primary reason for the apparent westward motion of the Sun, Moon, and stars through the sky. (The situation is analogous to riding eastward in an automobile and watching the outside scenery apparently rushing by to the west.) The Earth rotates in the same direction as it revolves; this rotational direction can be described as counterclockwise when viewed from above the North Pole, clockwise when viewed from above the South Pole, or eastward if the Earth is viewed from a point in space above the equator.

Because all places on Earth complete one 360° rotation in a 24-hour period, they rotate through an angle of 1° every 4 minutes. The linear speed of rotation is highly variable, however, and depends upon distance from the equator. The speed of rotation is at its maximum value of approximately 465 meters per second (1,040 mph) at the equator, but diminishes to 0 meters per second at the poles.

The Earth's axis is not perpendicular to the plane of the ecliptic, but is inclined from it at an angle of approximately 23½° (Figure 2.4). Because this value does not change significantly over the short term as the Earth revolves around the Sun, the Earth's axis is said to exhibit **parallelism,** and the poles remain directed toward approximately the same

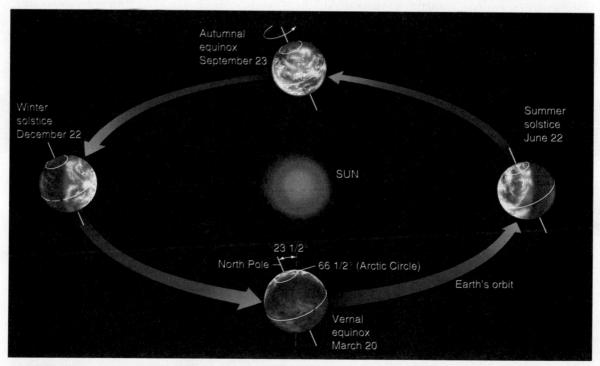

◆ FIGURE 2.4
The Earth's axis is tilted at a constant angle with respect to the plane of the ecliptic as our planet revolves around the Sun during the course of the year.

FORMATION OF THE UNIVERSE, SOLAR SYSTEM, AND EARTH

Astronomical observations have revealed that the universe is composed of vast expanses of almost perfectly empty space, as well as the widely separated clouds of gases, dust, and stars we call galaxies. Within the known universe are many billions of galaxies, each containing, on the average, tens of billions of stars (Figure 2.2). The number of stars in the universe is for all practical purposes limitless. It has been said that there are more stars in the universe than grains of sand on all the beaches of the world, and this is apparently a great understatement. Moreover, a large proportion of these stars may have planetary systems similar to our own solar system.

The origin of the universe is one of the great mysteries facing humanity, and numerous theories concerning its formation have been developed over the centuries. Any modern scientific theory, however, must take into account a startling fact that has been revealed through the spectrographic analysis of the light from distant galaxies. This fact is that the galaxies are rushing away from one another. In general, the farther apart two galaxies are, the faster their rate of recession seems to be. The only way astronomers can explain this phenomenon is by theorizing that the occupied universe is expanding in all directions, much like an inflating balloon, as the result of a titanic explosion.

THE BIG BANG THEORY

Over the past few decades, a theory of the origin of the universe—the so-called Big Bang Theory—has been developed and refined. It seems to fit most known facts and is now generally accepted by the scientific community. According to this theory, some 13 or so billion years ago, all matter in the known universe was concentrated into one tiny and inconceivably dense mass sometimes nicknamed the "cosmic egg." How it formed, and what events occurred prior to this time, nobody knows. The universe as we know it was initiated by the explosion of this mass. The explosion, the "Big Bang," destroyed the mass, and all the matter it contained was hurled into space in all directions, much like shrapnel from a detonating bomb. The violence of the explosion was so great that the superdense matter of the cosmic egg was reduced to subatomic particles, which soon reassembled to form only two elements—hydrogen and helium. Shortly after the explosion, then, the universe consisted of an expanding cloud of gas composed of approximately 73 percent hydrogen and 27 percent helium.

Gravity, however, began to alter the characteristics of the cloud. The expanding cloud gradually lost its homogeneity as it was drawn into many smaller gas clouds, each held together by its own internal gravitational attraction. Within these gas clouds—the future galaxies—gravity produced much denser and more localized concentrations of gas. As these smaller gas clouds contracted, their central cores were gradually heated by compression. When a temperature of approximately

points in space for extended periods of time. This is why the North Star, or Polaris, occupies a fixed position above the North Pole.

Size and Shape of the Earth

It is common knowledge today that the Earth is roughly spherical. The Earth's shape is not readily apparent from observation, however, and was the subject of speculation and debate for centuries. Most early scientists and philosophers believed that the Earth was flat. This view was based on the physical appearance of the Earth's surface, just as the early belief that the rest of the universe revolved around the Earth was based on the apparent motion of the Sun, Moon, and stars.

The Earth's spherical shape is produced by gravity. This force causes all matter to have an attraction for, and to be attracted by, all other matter. The mutual gravitational attraction of the Earth's matter has caused our planet to assume a spherical form—the most compact form that exists. Gravity has also caused Earth materials to be layered by density, with lighter materials tending to overlie heavier ones. (Fortunately for us, this layering is far from perfect.) Almost all other known astronomical objects of large size are also essentially spherical, indicating that gravity is apparently

11 million degrees Celsius (20 million degrees Fahrenheit) was reached, thermonuclear fusion was spontaneously initiated, and stars were born.

The basic thermonuclear reaction that powers the stars, including the Sun, involves the conversion of hydrogen to helium. A series of other stellar fusion reactions are believed to have manufactured all the elements heavier than helium.

The earliest stars formed under highly favorable conditions and were large, extremely hot, and short-lived. These massive stars consumed their fuel at a prodigious rate and soon became unstable as their fuel supplies were exhausted. They eventually collapsed and then exploded as supernovas, spewing a portion of the heavy elements they had produced throughout the Galaxy. Succeeding generations of stars, including our Sun, thus formed from concentrations of gas containing not only hydrogen and helium, but the heavier elements as well. It is believed that the elements that constitute most of the Earth, and even our own bodies, originated in the cores of these supernova stars!

FORMATION OF THE SOLAR SYSTEM AND EARTH

Our Sun formed approximately 4.6 billion years ago from a cloud of interstellar gas consisting mostly of hydrogen and helium, but also containing a mix of heavier elements from prior supernova explosions. During the contraction of a cloud of gas to form a star, any initial rotational motion is magnified because of the conservation of angular momentum. (This is the same principle that causes ice skaters to spin faster when they pull in their arms.) The centrifugal force that is generated draws the contracting gases into a disk shape. The major concentration of gases at the center of the disk is under sufficient gravitational pressure to initiate thermonuclear reactions and become a star, while it is believed that smaller nodes of gases and solid particles develop to form bodies, termed *planetesimals,* on the rotating rim of the disk. As these planetesimals grow, primarily through collisions with smaller bodies, they gradually become planets.

The rotational origin of the solar system explains why the planets all orbit the Sun on roughly the same angular plane and travel in the same direction. If this system of star formation operates in similar fashion elsewhere in the universe, most stars should have planetary systems, and, of course, the chances of earthlike planets and the existence of life elsewhere in the universe would be greatly enhanced.

In the case of our solar system, it is believed that the weaker gravity of the Earth and the other inner planets, coupled with intense radiation emanating from the nearby Sun (the so-called solar wind), drove off the low-mass hydrogen and helium gases that formed the initial bulk of the masses of these planets. The inner planets therefore consist primarily of the heavier residual elements that originated in supernova stars. The massive outer planets, though, were able to retain most of their light gases. They still consist largely of hydrogen and helium, which probably surround small solid cores of heavier elements. ∎

a universal force that operates wherever a sufficient mass of matter has assembled.

The three basic measurements of the size of a sphere are its **radius, diameter,** and **circumference.** For a given sphere, if any one of these measurements is known, the other two can be readily calculated. The radius of the Earth, or the distance from its surface to its center, is approximately 6,350 kilometers (3,950 mi). Earth's diameter, the linear distance from any point on the surface through the center of the Earth to its opposite (**antipodal**) point, is equal to two radii, or 12,700 kilometers (7,900 mi). Finally, Earth's circumference, the greatest distance around the Earth's surface, is 40,000 kilometers (24,900 mi).

Departures from Sphericity

Although gravity causes the Earth to be nearly spherical, it is not precisely so because other forces also influence the shape of our planet. The two most important of these forces are the outward thrust of centrifugal force generated by the Earth's rotation on its axis and the tectonic forces produced in the planet's deep interior.

The centrifugal force of rotation is much too weak to overcome the binding force of gravity and cause the Earth to fly apart. It is, however, sufficient to distort our planet's shape slightly. In the vicinity of the equator, where the speed of rotation—and therefore the centrifugal force—is greatest, the Earth is bulged outward slightly. In contrast, the slowly

(a) True sphere   (b) Oblate ellipsoid

◆ **FIGURE 2.5** The Earth's shape is somewhat distorted from that of a true sphere (a) because of the centrifugal force of the Earth's rotation on its axis. This motion causes our planet to assume the shape of an oblate ellipsoid (b), with its equatorial diameter slightly larger than its polar diameter. The amount of polar flattening in (b) is exaggerated for effect.

rotating polar areas are somewhat flattened as the Earth's rotational inertia displaces their mass toward the equator (Figure 2.5). The Earth thus assumes the shape of an **oblate ellipsoid**, a three-dimensional ellipse that is flattened at the poles. The amount of bulging and flattening is too small to be visible, and from space the Earth looks perfectly spherical. The departure from exact sphericity, however, affects the precise calculations required for surveying and mapping and also influences the force of gravity on different parts of the Earth.

The second, more important departure from sphericity (discussed at length later in this book) results from the Earth's variable surface features. If our planet were a true sphere, or even a true oblate ellipsoid, its surface would be smooth and would appear flat to an observer at ground level. Instead, the surface is highly wrinkled and contorted in places, exhibiting a great variety of landforms of many sizes and shapes. Total surface irregularity increases if we consider the sea floor, rather than the ocean surface, which comprises approximately 71 percent of our planet's surface. Thus, the Earth, when examined in sufficient detail, has a unique shape and complex surface features.

How great a departure from true sphericity do Earth's surface features represent? One way of judging this is by determining the total vertical distance between the world's highest mountain and its greatest ocean depth. Mt. Everest in the Himalayas is the highest peak, reaching an elevation of 8,847 meters (29,028 ft) above sea level. Challenger Deep in the western Pacific is the deepest known point of any ocean, with a maximum depth of 11,033 meters (36,198 ft) below sea level. The total vertical difference between these two extreme points is 19,880 meters (65,226 ft), or about 20 kilometers (12.3 mi). This value is less than 0.3 percent of the Earth's radius. Indeed, if the Earth were the size of an apple, its surface features would cover a vertical extent only as great as an apple peel.

Although the Earth's surface features are diminutive compared to the size of the entire planet, they are extremely important to humankind because we happen to live precisely where these features are located—at the surface. As a result, the Earth's surface features have strongly influenced most major aspects of human society, including population distributions, transportation routes, and agricultural patterns. In addition, as we shall see, Earth's landforms profoundly influence global patterns of climate, vegetation, and soils.

DIRECTION

The concept of direction is vital to the field of geography because of the discipline's emphasis on location. Unfortunately, no natural "universal" directions exist in space due to a lack of reference lines or surfaces. In outer space, in fact, even the words "up" and "down" are meaningless.

The basis for the horizontal directional systems used on Earth is our planet's rotational orientation. Rotation has established the positions of the North and South Poles and the equator. These locations enable us to construct reference lines for stating horizontal directions. Vertical directions are based on the orientation of the Earth's gravitational force, which pulls directly downward toward the center of the Earth.

North, often considered the most basic of all directions, is defined as the straight-line direction from any point toward the North Pole. Likewise, **south** is the direction from any point toward the South Pole. These two directions are always 180° apart with respect to a given starting point. **East** and **west** may be defined either as the directions perpendicular to north and south, or as the directions parallel to the equator. The Earth rotates in an eastward direction, and when an observer stands facing north, east is to the right and west to the left. Any other direction can be determined by reference to these four cardinal directions.

It is important to note that the directions north, south, east, and west are all equally horizontal. Expressions such as "up north" and "down south" are strictly figures of speech and probably arose because many maps are displayed vertically, as on classroom walls.

A number of systems have been devised for stating direction on the Earth's surface. The ideal directional system is one that is uncomplicated, allows for easy directional computations, and is capable of accurately describing any direction. At present, two systems are widely employed in the United States. One, the **azimuth system,** has the advantages just noted; the other, sometimes called the **Mariner's Compass,** has serious shortcomings but has long been in use and is familiar to the general public. These two systems are illustrated in Figures 2.6 and 2.7.

Magnetic North and South

The exact determination of direction is difficult without an accurate map or surveying equipment, although estimates can be made from the position of the Sun, the North Star, or other celestial objects (see the Focus Box on p. 17). The

◆ **FIGURE 2.6** The azimuth system states directions in degrees, progressing in a clockwise direction from due north (0°).

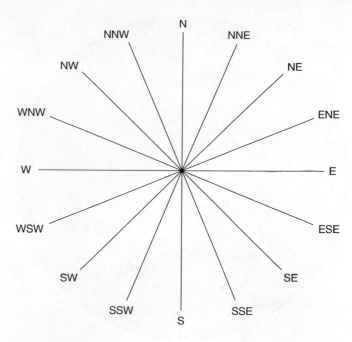

◆ **FIGURE 2.7** The 16 basic points of the Mariner's Compass.

most commonly employed instrument for determining direction in the field is the magnetic compass. A compass does not point toward the geographic (rotational) North Pole, however, but aligns itself with the magnetic field extending outward from the **magnetic North and South Poles.**

Unlike the geographic poles, the magnetic North and South Poles are not fixed in position, but slowly drift about. The magnetic North Pole is currently located in the Canadian Arctic, north of Hudson Bay, at a point some 1,600 kilometers (1,000 mi) south of the geographic North Pole. The magnetic South Pole is located near the coast of Antarctica south of Sydney, Australia, at a point about 2,900 kilometers (1,800 mi) from the geographic South Pole.

The origin of the Earth's magnetic field is incompletely understood. It is apparently produced by the flow of liquid iron in the Earth's core. Because the magnetic poles are at considerable distances from the geographic poles, compass directions can differ greatly from geographical directions depending on one's location, and corrections are generally needed.

LATITUDE AND LONGITUDE

The analysis of spatial data necessitates the development of an orderly system of reference points, distances, and directions. Geographers are concerned with understanding spatial phenomena and interactions over the entire Earth. They must therefore employ locational systems that can accurately describe the geographic position of any point on the Earth's surface.

A number of global locational systems have been devised, but the most widely used, by far, is the system of **latitude** and **longitude,** often referred to collectively as the **geographic grid.** This system is based on the locations of the North and South Poles and the equator. Related to these locations are the directions established by the Earth's rotational orientation. All lines in the geographic grid are oriented either due north-south or due east-west.

Latitude

Lines of latitude have the following characteristics, as illustrated in Figure 2.8:

1. They extend in an east-west direction but are used to state locations in a north-south reference frame.
2. Lines of latitude are really circles of latitude if looked at in their entirety, because they form full circles on the globe.
3. Lines of latitude are parallel, or equidistant to one another, along their entire lengths. For this reason, they are commonly termed **parallels.**
4. Lines (circles) of latitude vary greatly in length (circumference), but do so in an orderly fashion. The longest circle of latitude is the equator. Progressing poleward, the circumferences of the circles of latitude become progressively shorter, until at the poles they have shrunk to a point.
5. An unlimited number of parallels exist, but on a map they are usually spaced at set intervals, established by the **cartographer** (mapmaker) in view of the map's intended purpose and the space available.

The equator is used as the starting line for the determination of latitude values. The equator is unique in that it is the only individual line in the system that is a full circumference of the Earth. A circle that is a circumference on a sphere bisects the sphere and is termed a **great circle.** Any

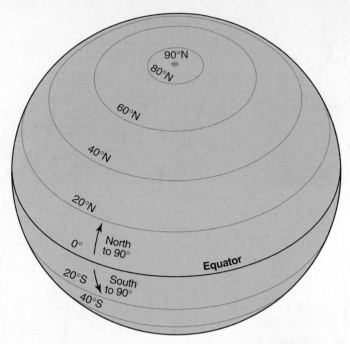

◆ **FIGURE 2.8** The globe with selected lines of latitude. These lines, termed parallels, actually take the form of concentric circles, numbered from the equator.

circle drawn on a sphere that is not a circumference is termed a **small circle.** All circles of latitude other than the equator are small circles.

Parallels of latitude are identified by their angular distance, in degrees, north or south of the equator. As the starting line, the equator is assigned a value of 0° latitude. The half of the Earth's surface located north of the equator is termed the **northern hemisphere;** the numerical values of all latitudinal positions here are followed by an *N.* Likewise, the half of the Earth's surface located south of the equator is termed the **southern hemisphere,** and all latitude values are followed by an *S.* If we took a large globe and drew a parallel for each degree of curvature north of the equator, we would find that the fortieth line we drew would pass through Philadelphia, Pennsylvania. Philadelphia therefore has a latitude of 40°N. So do Columbus, Ohio; Boulder, Colorado; Ankara, Turkey; Beijing, China; and many other cities and towns. It is evident, then, that a statement of latitude is insufficient by itself to pinpoint a location.

The higher the numerical value of a line of latitude, the farther it is from the equator. The most northerly point on Earth is the North Pole, which is 90° north of any point on the equator. Its latitude is 90°N. Similarly, the latitude of the South Pole is 90°S. Because all low-numbered parallels are near the equator, areas within about 30° of the equator are frequently referred to as the **low latitudes.** Areas between about 30° and 60°N and S latitude have intermediate values and are referred to as the **middle latitudes.** Most of the United States, southern Canada, and Europe lie within this zone. Areas poleward of about 60°N and S are considered to be in the **high latitudes.**

If the Earth is considered to be a sphere with a circumference of 40,000 kilometers (24,900 mi), the length of one degree of latitude can be calculated readily. A circle contains 360° of arc. The length of one degree of latitude therefore is 40,000/360 or 111 kilometers (69 mi). Actually, since the Earth is slightly flattened at the poles, a degree of latitude is slightly longer at the poles (111.7 kilometers or 69.4 mi) than it is at the equator (110.6 kilometers or 68.7 mi).

For greater precision, a degree of either latitude or longitude can be subdivided into sixtieths, termed **minutes** ('), and each minute, in turn, can be divided into sixtieths, termed **seconds** ("). Minutes and seconds of curvature therefore subdivide degrees as minutes and seconds of time subdivide hours. One minute of latitude is approximately 2.85 kilometers (1.15 mi) long, and one second of latitude is about 31 meters (100 ft) long. A latitudinal reading is given from left to right in progressively smaller units, for example, 39°07′15″S.

Longitude

Lines of longitude, or **meridians,** are similar in some respects to lines of latitude, but they also have a number of unique characteristics (see Figure 2.10 on p.18). The term *meridian* means "noon" or "midday," and, as we shall see, longitude is a basis of our system of time.

Lines of longitude have the following characteristics:

1. They extend in a north-south direction but are used to state locations in an east-west reference frame.
2. Each meridian is exactly half a great circle with its end points at the North and South Poles and its midpoint at the equator.
3. Meridians are *not* parallel to one another; they are farthest apart at the equator and converge northward and southward to meet at the poles. Because meridians converge, the surface length of a degree, minute, or second of longitude depends on its latitude. The length of one degree of longitude at the equator is approximately the same as that of a degree of latitude—111.2 kilometers (69.2 mi). This length decreases to zero at the poles.
4. Meridians intersect parallels at 90° angles.
5. Any number of meridians may be placed on a map, but, like parallels, they usually are spaced at set intervals.

Meridians are numbered in the same way as parallels—by choosing a 0° line and then measuring subsequent lines in degrees of surface curvature, this time to the east and west. Choosing an appropriate starting meridian long posed a problem, though, because no one meridian differs from the rest in any geometrical manner. As a result, many countries that made maps established 0° meridians through their own capital cities or other points of national importance. This practice caused a great deal of confusion because longitude values for the entire world depended upon the country of origin of the map or book being consulted. It was finally decided at an international meeting in 1884 to adopt the English system, in which the starting or **prime meridian** passed through Greenwich Observatory on the outskirts of London.

GEOGRAPHIC POSITIONING

The ability to accurately determine one's position on the Earth's surface has for centuries been crucially important to individuals in a variety of occupations. Only in recent years, however, has modern technology provided highly accurate solutions to this perplexing problem. Determining the position of a ship at sea—and out of sight of any identifiable terrestrial landmarks—has been particularly challenging, and for this reason, the navigator is one of the most essential members of a ship's crew. Over the centuries, navigation techniques have evolved from rough "guesstimates," based on a combination of luck, intuition, and visual observations of such phenomena as the Sun and stars, waves, and clouds, into a highly sophisticated and accurate science.

The basic problem in oceanic navigation has been to determine both the ship's north-south (latitudinal) and east-west (longitudinal) position in the absence of visual landmarks. In ancient times, the latitudinal position was the easier of the two to determine with reasonable accuracy. This was done by observing the altitude of various celestial objects (Figure 2.9). Polaris, the North Star, was best for this purpose because its position in the sky does not change appreciably with time at any one location. In addition, the altitude of Polaris above the horizon corresponds to the ship's angular distance, or latitude, north of the equator. For example, at a latitude of 30°N, Polaris will appear at an altitude of 30° above the northern horizon. If the latitude of one's destination was known, it was a fairly easy matter to use Polaris as a guide to reach this latitude and then to

◆ **FIGURE 2.9** The astrolabe, an ancient instrument for determining geographic location. The altitude of the Sun or stars above the horizon was determined by sighting through pinholes on the alidade, or pointer bar (crossing the center of the astrolabe). From this, one's latitude could be determined.
Source: The Bettmann Archive

sail due east or west along the parallel to the destination.

When sailing in the southern hemisphere where Polaris is not visible, navigators used the position of southern hemisphere stars, such as those of the Southern Cross.

The accurate determination of longitudinal positions remained a problem into the late eighteenth century, when the development of the **chronometer**—a highly precise clock set to Greenwich

mean time—provided the eventual solution. The Earth's rotation, as we have seen, causes the Sun's position in the sky to shift westward at a rate of exactly 15° each hour. By comparing the east-west position of the Sun at one's own location to the position of the Sun at Greenwich as calculated from the chronometer, the longitude could be readily determined.

In recent decades, navigational instruments have been developed that provide highly accurate locational data. Among these is **RDF** (radio direction finder), which employs radio beams from land-based transmitters to allow the determination of lines of position at sea. **Loran** (long-range navigation) uses synchronized, intersecting radio signals that provide precise positions.

An even more recent development—one that allows positions at sea or on land to be quickly determined to within 5 meters (15 ft)—is the *Global Positioning System (GPS)*. The system is based on a constellation of 21 satellites that orbit the Earth twice a day at an altitude of 32,000 kilometers (19,900 mi). At least four satellites are over North America at any one time. These satellites, whose orbits are known with precision, send radio signals that are picked up by GPS receivers on the ground. Since the speed of light (the speed of the radio signal) is also precisely known, the receivers are able to calculate their distance to each satellite by multiplying the speed of light by the time it takes the signal to arrive. With distance measurements known to four satellites, the receiver can compute its own exact latitude, longitude, and altitude. ■

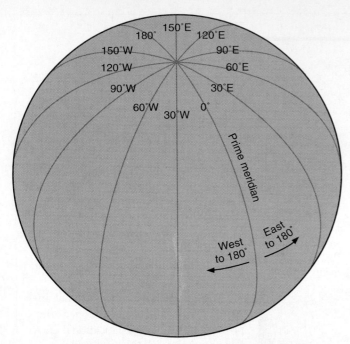

◆ **FIGURE 2.10** Lines of longitude, or meridians, are half-circles with their end points at the poles. They are spaced farthest apart at the equator and are numbered in degrees east and west of the prime meridian.

From the prime meridian (also called the **Greenwich meridian**) westward, west longitude values increase from 0° to a maximum of 180° halfway around the world. Similarly, going eastward, values of east longitude increase from 0° to 180°. The numerical values of meridians therefore go twice as high as those of parallels. Longitude values are followed by an *E* or *W*.

Longitude may be used to divide the world into the **eastern hemisphere** and the **western hemisphere.** These terms are not as frequently employed as are the terms northern hemisphere and southern hemisphere, but such longitudinally derived expressions as the Western World and the Far East are in common usage.

Determining Geographical Coordinates

Taken together, the latitude and longitude of any location describe its geographical position. A given parallel intersects a given meridian at only one point on the Earth, so if the correct value of each of these lines is provided, the position of any place can be pinpointed. The accuracy of placement of any point depends on the accuracy with which its position is stated. Locations stated to the nearest whole degree contain a potential margin of error of many miles, while those stated to the nearest second of latitude and longitude are accurate to within 15 meters (50 ft).

When the geographical coordinates of a site are given, latitude is always stated first. In the following example, a location is stated to the nearest second:

39°22′18″N 76°38′05″W.

It must be remembered that, with the exception of the equator, the prime meridian, and the 180th meridian, the numerical value of each parallel or meridian is followed by the letter *N, S, E,* or *W.* It should also be kept in mind that both minutes and seconds are sixtieths, not hundredths. Most maps have selected lines of latitude and longitude either drawn on them or indicated by tick marks on their margins. The numerical spacing of a map's lines of latitude is not necessarily the same as that for its lines of longitude.

THE SEASONS

Over most of the Earth's land surface, and particularly for those areas lying outside the tropics, the seasonal changes in climate are among the most important characteristics of the natural environment. Outside the tropics, all life-forms must adjust to the seasonal rhythm of the climate, and most species have evolved biological cycles tied to the orderly seasonal progression of climatic changes.

Cause of the Seasons

The seasonal climatic changes are caused by the Earth's axial inclination of $23^1/_2$° from perpendicular to the plane of the ecliptic as our planet revolves around the Sun (Figure 2.11). Due to this axial tilt, all places on Earth experience a yearly cycle of change in both the angle and the duration of sunlight. The resulting changes in solar energy produce annual cycles of weather conditions over most of our planet.

Before examining Earth-Sun relationships associated with each of the four seasons, we must discuss the key factor responsible for the relationship between the Sun's angle and its heating ability. The angle of the Sun in degrees above the horizon is termed its **altitude.** In all but the high latitudes, the altitude of the Sun is the most important factor controlling the amount of solar heating received at the Earth's surface. When the Sun is directly overhead, or at an altitude of 90°, its heating power is greatest; as its altitude decreases, its rays gradually lose intensity. This decline in intensity occurs because a decrease in the Sun's altitude causes an equivalent amount of solar energy to strike a larger Earth surface area, spreading and reducing the energy received per unit area (Figure 2.12). The solar energy reaching the Earth's surface is further reduced when the Sun is at a low altitude because the sunlight's oblique passage through the atmosphere results in increased scattering and absorption.

The tilt of the Earth's axis causes the noon Sun to vary in altitude at any point on the Earth's surface during the course of the year. This, in turn, causes the concentration and heating ability of the solar energy received to vary considerably. As the Earth revolves around the Sun, the North and South Poles constantly point toward fixed locations in space. The inclination of the poles, or of any other location on Earth, is *not* constant, though, with respect to the Sun. In fact, the combination of rotation and revolution causes a constant shift in the position of the point on the Earth's

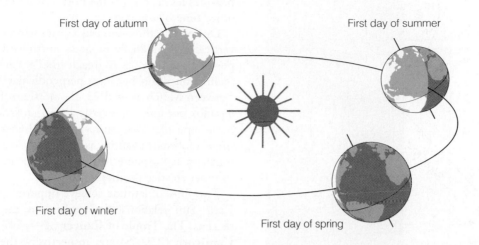

First day of autumn

First day of summer

First day of winter

First day of spring

◆ **FIGURE 2.11** The four seasonal positions of the Earth. Each season officially begins when the Earth reaches the solstice or equinox positions shown here.

surface that has the Sun directly overhead (Figure 2.13). Through the course of a year, this point shifts alternately northward and southward because of the parallelism of the Earth's axial inclination and also moves constantly westward because of the Earth's rotation.

The seasons are based on the changing orientation of the Earth's axis with respect to the Sun during the course of a year. The official beginning of each season marks the instant that a key Earth-Sun positional relationship occurs. The decision to begin each of the four seasons at the four "quarter positions" of the Earth's orbit around the Sun was an arbitrary one; no natural law mandates this. As a consequence, the change of seasons does not lead immediately to the changes in the weather that some people seem to expect. In fact, in large parts of the tropics, day-to-day weather conditions remain relatively constant throughout the year.

Summer begins in the northern hemisphere when our hemisphere reaches its maximum inclination of 23$1/2$° (actually 23°27′) toward the Sun (Figure 2.14a). When this occurs, the Sun is at its **summer solstice** position and, at solar noon, is directly overhead at a point 23$1/2$° north of the equator. The northern hemisphere at this time receives a larger proportion of the Earth's total allotment of solar energy than at any other time of the year. Because the southern hemisphere is on the opposite side of the Earth, it is receiving a smaller amount of solar energy than at any other time during the year. Therefore, the northern hemisphere's summer solstice also marks the beginning of winter in the southern hemisphere, where the Sun is at its **winter solstice** position. All the seasons are exactly reversed be-

tween the two hemispheres, so that when any seasonal event occurs in one, the opposite event occurs in the other.

In the northern hemisphere, the summer solstice occurs on, or within two days of, June 21. In fact, each of the four seasonal positions occurs between the nineteenth and twenty-third day of its respective month. The exact time and date change from year to year because of leap years as well as

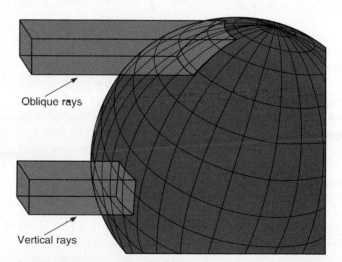

Oblique rays

Vertical rays

◆ **FIGURE 2.12** The intensity and heating power of the Sun's rays are directly related to the Sun's altitude. The same amount of energy is contained in both "boxes" of incoming radiation, but the oblique rays striking the higher latitudes are spread out over a larger area, resulting in lower temperatures.

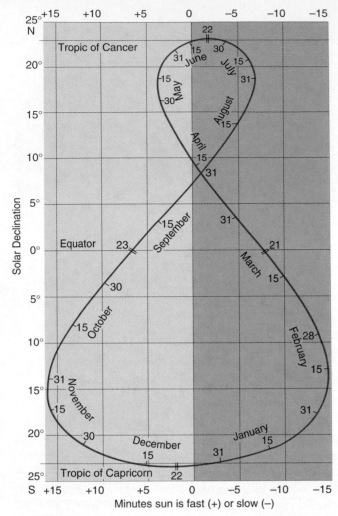

◆ **FIGURE 2.13** The analemma indicates both latitudinal and longitudinal changes in the Sun's position with respect to the Earth at different times of the year. The analemma is asymmetrical because the Earth is nearer the Sun in winter, causing it to orbit the Sun at a faster speed at that time.

slight yearly variations in the Earth's orbit, caused by external gravitational forces.

During the three months of summer that follow the northern hemisphere summer solstice, the revolution of the Earth causes the inclination of the North Pole toward the Sun to be progressively reduced. Consequently, the Sun's vertical rays gradually shift southward from 23½°N until they reach the equator. When this occurs, between September 19 and 23, the Sun is at its **autumnal equinox** position for the northern hemisphere, and the fall season begins. In the southern hemisphere, where spring is beginning, the Sun is in its **spring** (or **vernal) equinox** position. At this time the axial tilt of the Earth is perpendicular to the direction of the Sun's vertical rays, and both hemispheres receive an equal amount of solar energy.

As the fall season progresses, the continued revolution of the Earth causes the northern hemisphere to be inclined more and more away from the Sun. Finally, between December 19 and 23, the entire 23½° tilt of the axis is directed

away from the Sun in the northern hemisphere (and toward the Sun in the southern hemisphere), and the northern hemisphere winter begins (Figure 2.14b). At this time, the Sun follows its lowest course through the southern sky and provides less energy to the northern hemisphere than at any other time of the year.

During the three months of the winter season, the Sun's vertical rays gradually progress northward from their initial position at 23½°S to reach the equator as the Earth's inclination again becomes perpendicular to the Sun's rays. Between March 19 and 23, the northern hemisphere spring equinox occurs as the vertical Sun crosses the equator, and both hemispheres again receive equal energy. The following three months of spring witness the increasing tilt of the northern hemisphere toward the Sun until once again the summer solstice position is reached.

Two lines of latitude achieve importance as a result of the Earth-Sun relationships involved in the changing of the seasons. The **Tropic of Cancer** (23½°N) and the **Tropic of Capricorn** (23½°S) are, respectively, the highest latitudes north and south that the Sun's vertical rays ever reach. Stated differently, the Sun is always directly overhead somewhere between 23½°N and 23½°S. Places between these two parallels have the Sun directly overhead on two occasions during the year—once while the Sun is shifting northward toward the Tropic of Cancer, and once while it is shifting southward toward the Tropic of Capricorn. More than half the world is poleward of these lines (or outside the **tropics**) and never has the Sun directly overhead. For nontropical locations, the Sun reaches its highest and lowest altitudes at solar noon on the dates of the summer and winter solstices, respectively.

Lengths of Day and Night

The changing of the seasons involves not only constant changes in the Sun's noon altitude, but also variations in the lengths of day and night for all places on Earth except the equator. All locations in the northern hemisphere experience longer days than nights during the spring and summer, when the northern hemisphere is inclined toward the Sun. Days are shorter than nights during the fall and winter, when the northern hemisphere is tilted away from the Sun. The lengths of day and night at equivalent latitudes in the southern hemisphere are exactly reversed from those in the northern hemisphere because the seasons are reversed. Only the equator, the line separating the two hemispheres, experiences equal-length (12-hour) days and nights throughout the year.

The length of daylight is not evenly distributed by latitude; it becomes progressively more unequal poleward, particularly at the times of the solstices. Figure 2.14a shows that more of each line of latitude progressing northward from the equator is on the day side of the Earth than on the night side. At the time of the summer solstice, all parallels north of the **Arctic Circle** (66½°N) are entirely in daylight. The Sun therefore does not set in this area (sometimes called the "Land of the Midnight Sun") at this time. Conversely,

(a)   (b)

◆ **FIGURE 2.14** The angle of incidence of the Sun's rays at the times of the solstices. At the time of the northern hemisphere summer solstice (a), the Sun is directly overhead at the Tropic of Cancer. Days are longer than nights in the northern hemisphere, while nights are longer than days in the southern hemisphere. The situation is reversed during the winter solstice (b), at which time the Sun is directly overhead at the Tropic of Capricorn.

increasing proportions of parallels progressing southward from the equator (and at their winter solstice) are in darkness until, south of the **Antarctic Circle** (66½°S), continuous night prevails. This situation is exactly reversed six months later (Figure 2.14b) so that for both hemispheres and indeed all places on Earth, the Sun is above the horizon for a total of six months over the course of a year and below the horizon for the other six months.

The Arctic and Antarctic Circles therefore mark the equatorward limits of the two portions of our planet in which day and night exceed 24 hours at the times of the summer and winter solstices, respectively. The situation is most extreme at the North and South Poles, where day and night each last for six months. At these two locations, the Sun remains constantly below the horizon during the fall and winter and constantly above the horizon during the spring and summer. On the dates of the equinoxes, all places on Earth experience 12 hours of day and 12 hours of night.

Path of the Sun in the Sky

Contrary to popular opinion, the Sun does not rise and set vertically except at the equator. It actually describes a curved path through the sky. The horizontal component of the Sun's apparent motion increases directly with latitude. For example, at 43° N or S latitude, the Sun rises and sets at an angle of 43° from the vertical. At the poles, the Sun's motion is essentially horizontal as it circles the sky in a clockwise direction at the North Pole and a counterclockwise direction at the South Pole.

In the higher latitudes, the horizontal component of the Sun's motion causes a substantial proportion of the night hours to experience **twilight** rather than complete darkness. Twilight is the familiar condition of semidarkness that occurs before sunrise and after sunset as sunlight is scattered by air molecules, clouds, and dust particles in the upper atmosphere. It occurs when the Sun is less than 18° below the horizon. Just as the Sun never rises more than 23½° above the horizon at the poles, it is never more than 23½° below the horizon. The polar regions therefore experience up to 14 weeks of twilight during the year, mostly in early autumn and late winter. On the other hand, the vertical motion of the Sun at the equator is responsible for the often expressed observation that "night falls swiftly in the tropics," because here the Sun moves directly and rapidly past the 18° mark below the horizon.

TIME

While the seasons are based on the inclination of the Earth's axis to the plane of the ecliptic, our system for establishing the time of day is based on the east-west (or longitudinal) position of the Sun. The Sun's position and apparent motion during the course of a day are associated primarily with the Earth's rotation but are also slightly affected by the Earth's revolution around the Sun. The basic unit of time is the **solar day,** which is the period needed for the Sun to make one apparent 360° circuit of the sky. The solar day is subdivided into the familiar units of hours, minutes, and seconds.

Local Time

The event historically used for establishing the time of day is the occurrence of **solar noon.** Solar noon occurs the instant the Sun reaches its daily zenith (high point in the sky). For centuries, until the late 1800s, communities made the time of the mean occurrence of solar noon the official instant of noon and then simply subdivided the periods between solar noons into the appropriate hours, minutes, and seconds. This system of **local time** was highly accurate with respect to the position of the Sun because the time was individually adjusted to each locality.

During the nineteenth century, however, a major short-coming in the local time system became increasingly evident and eventually led to its discontinuance over most of the world. The problem was that different places, even those located near one another, inevitably had somewhat different times. Because the local time system used the east-west position of the Sun in the sky, only places on precisely the same meridian had exactly the same time at any one instant. For example, at 40°N, the Earth's rotation causes the position of the noon Sun to shift westward over the Earth's surface at a speed of approximately 300 meters (1,000 ft) each second. This means that a location only 300 meters

(1,000 ft) to the east of a reference point would, when strictly observing local time, be a second later, and a place 300 meters (1,000 ft) to its west would be a second earlier than at the reference point.

Prior to the nineteenth century, time differences between nearby localities using the local time system were only a minor inconvenience. Timepieces were rare and rather inaccurate, appointments were rarely made for exact times, and means of transport were so slow that it took a long period to reach a destination far enough away to have a significantly different time. In the nineteenth century, however, the advent of the railroads provided a rapid and long-distance means of transportation, while the telegraph offered a means of instantaneous long-distance communication. The railroad and telegraph companies both experienced major difficulties because of the multiplicity of times existing throughout the world. As a result, they brought increasing pressure to bear on government agencies to standardize time over large areas.

Standard Time

Our present **standard time system** was officially established for the United Sates in November 1883 and soon was adopted by most other countries. The standard time system,

◆ **FIGURE 2.15** The global system of time zones. Numbers indicate the number of hours that must be added to or subtracted from the local time to bring it to Greenwich Time. (The Alaskan time zones have recently changed. For a correct depiction, please see Figure 2.16.)

Standard time zones of North America.

like the local time system, is based on the position of the noon Sun, but only at selected meridians of longitude rather than at each specific site. The areas surrounding these standard meridians all have the same official time. Because the Earth rotates 15° in one hour, the standard time zones differ by one-hour intervals. Thus, 24 time zones cover the Earth (Figure 2.15). In changing from local time to standard time, therefore, accuracy of time determination at each particular place was sacrificed for much greater convenience.

The meridians used as the centers of their respective time zones are multiples of 15°. The meridian that provides the basis for all the time zones is the Greenwich meridian, and the time zone established by this meridian is called Greenwich mean time (GMT). Many people employed in fields such as transportation and communications base their activities directly on Greenwich mean time.

If the standard time system were rigorously followed, each time zone would be centered exactly on its own standard meridian and would extend 7°30′ east and west from the meridian to the borders of the adjacent time zones. For example, the 75°W meridian is used as the basis for the eastern time zone of the United States. In theory, eastern time should extend westward to 82°30′W, changing to central time from that line westward to 97°30′W. In reality, though, considerable liberties have been taken in placing the time zone boundaries; they are generally drawn along state or county lines and are sometimes quite distant from the theoretical 7°30′ boundaries (Figure 2.16). The rationale for this is to avoid the problems that arise when time lines cut through the middle of a county, town, or other political unit.

Internationally, time zone boundaries also have been adjusted considerably, as is evident from Figure 2.14. In many cases entire countries use the same time, and the time changes at the international boundary. Most large countries, especially those of great longitudinal extent, have multiple time zones. The United States, for example, has six different time zones (four in the coterminous United States), and Russia has ten! Because time zones are based on 15° increments of longitude, they are wedge-shaped and taper poleward. Over the open ocean, the official time boundaries generally are used because no practical advantage is gained from shifting them.

matter of the map. The title should be clear and concise, because many people obtain maps based on their titles.

DATE. The date of the map is important because most types of map information change over time, and the information depicted will likely become outdated. In many cases the year is sufficient, but in some instances, as with weather maps, information changes so rapidly that the map should be dated to the hour or even to the minute.

LOCATION. Most maps depict only a small portion of the Earth, and the map user will need to be able to determine the specific locations that the map displays. Many maps indicate locations by the use of numbered parallels and meridians, although a variety of other locational systems may also be employed.

DIRECTIONAL ORIENTATION. The compass orientation of the area shown on a map is commonly indicated by the use of parallels and meridians. Not all maps maintain constant directions everywhere, so these lines, although actually straight in a two-dimensional sense, may curve on the map. A north directional arrow also is frequently used. Most maps are oriented so that north is toward the top, south toward the bottom, east toward the right, and west toward the left.

LEGEND. The legend is the information section of a map. The primary purpose of the legend is to explain the major map properties and symbols. **Symbols** are the devices employed to represent the phenomena displayed on the map. They may be **point symbols,** such as dots to represent towns; **line symbols,** such as those used to represent the roads on a highway map; or **area symbols,** such as colors that indicate elevations or shading patterns that represent types of land use.

Map Scale

The **map scale** is the ratio between distances on a map and corresponding distances on the Earth's surface. All maps are reductions of reality, and the scale indicates the extent of this reduction. Scale information is especially vital when one of the chief purposes of the map is to indicate distances between places, as on a road map. Unfortunately, for reasons to be explained shortly, the scale of a map showing a large area cannot be made consistent over its entire surface. The scales of maps depicting restricted areas such as states or cities may, for practical purposes, be considered constant, however.

Three basic methods of stating map scale are commonly employed, and one or more of these should appear in the map legend. Maps designed for use by people with limited map experience often have **verbal scales.** They frequently appear, for example, on road maps. In this case, the scale is stated in a sentence format, as in the following example:

1 inch equals 8 miles.

◆ FIGURE 2.18 The map scale information section in a United States Geological Survey (USGS) topographic map legend. The scale appears both in representative fraction and in multiple graphic scale formats.

By convention, the first figure given (1 inch) is the map distance, and the second (8 miles) is the corresponding Earth distance.

A **graphic scale** generally takes the form of a bar or line graph showing the map distances corresponding to real Earth units. On some maps, multiple bars may be used to show either different units of measure (Figure 2.18) or the scale on different parts of the map. Graphic scales have the advantage of providing a visual indication of the scale and of allowing for direct measurements to be taken. Unlike the other two methods of stating scale, they remain correct if the map is enlarged or reduced.

A **representative fraction (RF) scale** is stated as a direct ratio between map distances and corresponding Earth distances. It may be written in either of the following fashions:

1:63,360 or 1/63,360
(values, of course, vary)

This fraction is a dimensionless quantity that can be thought of as the "degree of reduction" used to display the portion of the Earth's surface depicted on the map. In both cases the first number (the fraction numerator) is the map distance, which is always set at a value of one. The second figure (the fraction denominator) is the corresponding Earth distance. A major advantage of the RF scale is that any unit of measure can be employed, as long as the same unit is used for both map distances and Earth distances. In the preceding example, one inch on the map equals 63,360 inches on the Earth, one centimeter on the map equal 63,360 centimeters on the Earth, and so forth. Because there are 63,360 inches in one statute mile, the RF scale of 1:63,360 is equivalent to the verbal scale "one inch equals one mile."

The cartographer (mapmaker) chooses the scale of a map based largely on the intended use of the map, the amount of information to be shown, and the amount of space available on the map sheet. An unlimited number of scales are possible, but, for general descriptive purposes, maps are often described as being small-scale or large-scale.

A **small-scale map** is greatly reduced from reality. This large reduction enables the map to show a large surface area, but has the disadvantage of leaving little room for surface detail. Most classroom wall maps are small-scale maps, even though the maps themselves are made large in size for easy viewing from a distance. **Large-scale maps** are less reduced than are small-scale maps. They cannot cover as much Earth surface area, but are capable of showing more detail. An example would be a road map of a city or county.

Map Properties and Distortion

As we have seen, maps are representations of the spherical surface of the Earth. The great majority of maps, however, are drawn on plane (flat) surfaces, usually sheets of paper. The planar nature of these maps is responsible for the major problem in cartographic depiction—that of **map distortion.** Distortion in this context refers to inconsistencies in the spatial properties of different portions of the map. The crucial map properties of size, shape, distance, and direction may differ from one place to another, and it is not possible to eliminate this problem completely.

Map distortion exists because it is geometrically impossible to accurately depict a spherical surface, or any substantial part of it, on a plane surface. This dilemma can be illustrated by painting a map of the Earth on a large rubber ball, such as a basketball, and then attempting to flatten it out. This obviously cannot be done without considerable stretching, cutting, tearing, or compression. Even pieces of the ball (map) will be curved and cannot be flattened without stretching or tearing. Yet, this transformation from a spherical surface to a planar one is exactly what the cartographer must accomplish in producing a flat map. In the process, some of the basic map properties of size, shape, distance, and direction must be distorted, and on many maps all of these properties are distorted.

Because map distortion results from the flattening of the Earth's curved surface, maps depicting larger surface areas are more distorted than those depicting limited areas. Maps of the entire Earth, therefore, are subject to the greatest distortion. Conversely, large-scale maps generally show such a small geographical area that distortion can be considered negligible.

The only way to map the Earth without distortion is to use a globe, which duplicates the near-spherical shape of our planet. The geometrical accuracy of globes is their greatest advantage over flat maps. Because globes are not distorted, relative sizes, shapes, distances, and directions can all be shown as they actually exist on the Earth.

If flat maps are distorted, and globes are not, why are flat maps so much more commonly used? The reason is that the handling and use of globes presents numerous practical difficulties. Undoubtedly, the biggest problem is their very small scale and resulting lack of detail. There is no way, for example, that a globe can be used as a practical road map. Other major shortcomings are the great bulkiness and high cost of globes, the difficulty of storing them (see how many you can fit in your car's glove compartment!), the difficulty of making measurements on a curved surface, and the impossibility of seeing more than half the Earth at any one time. Because of these problems, globes are used primarily for display and general reference purposes. They provide an accurate depiction of the Earth's surface, but once the correct geographical relationships are mastered, practical work then normally proceeds by use of flat maps. On the other hand, it is undoubtedly true that globes are not used enough for educational purposes, and as a result, many students have developed erroneous impressions about the geography of our planet by taking literally the views of the Earth appearing on flat maps. This problem is explored further in the Focus Box beginning on p. 28.

Although flat map distortion cannot be eliminated, it can be manipulated. Maps can be made to have their greatest distortion in what are generally considered the less important parts of the Earth's surface, such as the polar regions or oceans. It is also possible to construct maps in which the distortion is confined to certain spatial characteristics, while other characteristics remain undistorted. The types and amounts of distortion on an unfamiliar map can be estimated by comparing it to a globe or by examining the behavior of the map's parallels and meridians.

All maps, regardless of distortion, should show the absolute locations of mapped features correctly with respect to latitude and longitude (or other reference system). Two additional basic map properties are their method of depicting an area's size and shape. These two properties are strongly influenced by the problem of distortion; in fact, it is impossible for a single flat map to display both an area's size and its shape correctly. It is possible, however, to keep either one or the other of these map properties from being distorted.

A map that has the property of consistency in relative size is called an **equal-area map.** All places on such a map are shown in their proper proportional sizes—that is, subject to the same total areal reduction. Proportional sizes can be maintained only by distorting lengths, and therefore shapes; lengthening in one direction is compensated for by shortening in another. Shape is therefore badly represented on an equal-area map, particularly near its margins.

A map that displays shapes correctly is termed a **conformal map.** Actually, only shapes of small areas can be shown without distortion on conformal maps; over large distances, shapes and directions remain distorted. Conformality can be achieved only by manipulating sizes, and high-latitude regions usually appear at a much larger scale than low-latitude regions. Conformal maps are particularly useful in determining directions, making them valuable as navigational charts. The parallels and meridians on a conformal map cross at right angles, just as they do on a globe.

On conformal maps, directions from any point start out correctly, but gradually become distorted with distance. It is impossible to devise a flat map on which all directions remain correct over the entire map surface. However, a third type of map, the **azimuthal map,** retains true directions radially outward from one central point, regardless of distance.

MAP MISINFORMATION

We use maps in order to gain an accurate understanding of the sizes, shapes, and locations of places around the world. Because flat maps of the Earth are necessarily distorted, however, the geographical concepts we derive from them may be distorted as well. Of particular concern to geographic educators are the erroneous ideas about our planet's geography that elementary school students get from using rectangular wall maps. When students who are learning world geography for the first time study these highly distorted maps, they can gain lasting impressions that are difficult to correct later.

The worst offender of all is the popular Mercator projection, which excessively enlarges the sizes of high latitude regions at the expense of the tropics and cannot show the poles (Figure 2.19a). The lines of latitude on a Mercator map are all the same length, although these lines actually become shorter poleward, and they are spaced progressively farther apart toward the poles, even though they are virtually equidistant on the globe. In addition, lines of longitude are parallel, while in reality they converge poleward. Finally, many Mercator maps display more of the northern hemisphere than the southern hemisphere, thus placing the equator below the center of the map.

In 1989, six major professional geographical and cartographic organizations urged book and map publishers, the media, and government agencies to stop using rectangular world maps. Recommended instead were some of the more than 100 nonrectangular world maps—many of them oval—created to overcome the problem of rectangular map distortion. Of course, these map types are also distorted, but the relative sizes of mapped features are generally not nearly as distorted as on most rectangular maps. In December 1988, the prestigious National Geographic Society adopted the Robinson projection for its world maps, and many publishing companies followed suit (Figure 2.21b). The Robinson projection is an oval projection, but is flattened in the polar regions. Its developer is Arthur Robinson, Professor Emeritus at the University of Wisconsin—Madison, who is sometimes called the "dean" of American cartographers.

Another possible but highly controversial replacement for the Mercator projection is the Peters projection, developed in the late 1970s by the German historian Arno Peters (Figure 2.21c). Billed as "the most honest map of the world yet devised" when it first appeared, it is an equal-area rectangular map. In order to avoid the high-latitude distortion typical of rectangular projections, longitudes on the Peters projection are compressed in the tropics, and latitudes are greatly compressed in the polar regions. The result is the preservation of correct relative sizes at the expense of a great distortion in shapes.

Compared to most other rectangular projections, the Peters projection makes Africa and southern Asia look very large and greatly reduces the size of high-latitude countries such as Canada and Russia. Its use was promoted enthusiastically by many tropical developing nations, especially in Africa, and it has been used on United Nations publications dealing with international development.

An important point to be made about the controversy on the most appropriate world projection is that people with particular viewpoints may choose maps that serve to advance those views. To some extent, then, the mapmaker or map user may not only be trying to display the world, but also may be making a statement about the world. ■

Many maps are neither equal-area, conformal, nor azimuthal, Instead, they represent some degree of compromise between these basic types. Their distortion is distributed among the various map properties, usually producing a better appearing map. Such maps are commonly used for general reference or classroom purposes.

Map Projections

A geometric figure such as a sphere that cannot be flattened without distortion is said to be **undevelopable.** The basic goal in mapmaking is to transfer the Earth's undevelopable surface onto a **developable** surface—one that can be flattened. A **map projection** of the Earth is produced by

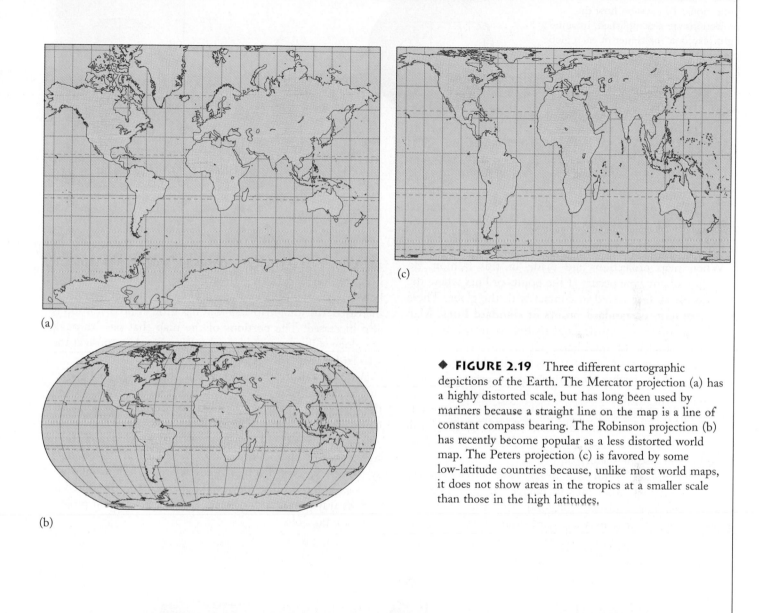

(a)

(b)

(c)

◆ **FIGURE 2.19** Three different cartographic depictions of the Earth. The Mercator projection (a) has a highly distorted scale, but has long been used by mariners because a straight line on the map is a line of constant compass bearing. The Robinson projection (b) has recently become popular as a less distorted world map. The Peters projection (c) is favored by some low-latitude countries because, unlike most world maps, it does not show areas in the tropics at a smaller scale than those in the high latitudes.

mathematically transferring or projecting the image of all or part of the Earth's surface onto a developable surface, which is then "developed," or flattened. Three particular developable surfaces—the plane, the cone, and the cylinder—are commonly used for map projections (Figure 2.20).

In order to visualize the map projection principle, imagine that a lightbulb has been placed in the center of a transpar-

ent globe that has drawn on it parallels, meridians, and the outlines of the continents. Place in turn a plane, a cone, and a cylinder over the globe as shown in Figure 2.20. The shadows of the features on the globe will then be projected onto these surfaces to produce maps that, in the cases of the cone and cylinder, can then be cut and flattened, as shown in Figure 2.21.

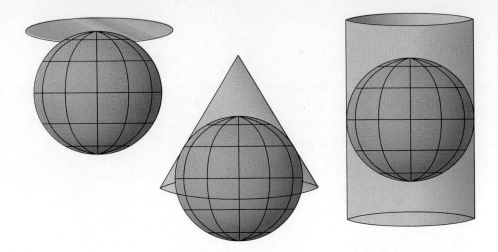

◆ **FIGURE 2.20** The principles for plane, conic, and cylindrical map projections. To envision how the projections are accomplished, imagine a light placed at the center of each globe projecting the outline of the Earth's features onto the geometrical surface.

When map projections are made in this fashion, no distortion of any type occurs at the points or lines where the developable surface rested in contact with the globe. These places are termed **standard points** or **standard lines.** Map projections often are constructed so that standard lines are particular parallels or meridians. Distortion occurs everywhere else on the map, increasing as the developable surface lies progressively farther from the globe's surface. Because the projected images expand outward in all directions after they leave the globe, locations on portions of the developable surfaces that are situated well away from the globe will appear at a larger scale, and angular relationships in these locations will also be distorted.

Another option is for the developable surface to cut through the globe, rather than lie tangent to it, as shown in Figure 2.22. For a plane projection, this will produce a standard line rather than a standard point. For both conic and cylindrical projections, the result will be two standard lines rather than one. Because the amount of contact between the globe and the developable surface has been increased, the total map area that has little or no distortion is also increased. The portions of the map that pass through the globe will now appear at a smaller scale than those at the standard points or lines.

Although plane, conic, and cylindrical projections may still be constructed according to the principles described in the preceding paragraphs, many modern projections are constructed by computer and often are altered so as to achieve some desired property. Whatever the projection, however, the map information must be transferred in a systematic fashion. Because many types of developable surfaces exist and because they can be mathematically altered in various ways, an unlimited number of types of map projections are possible.

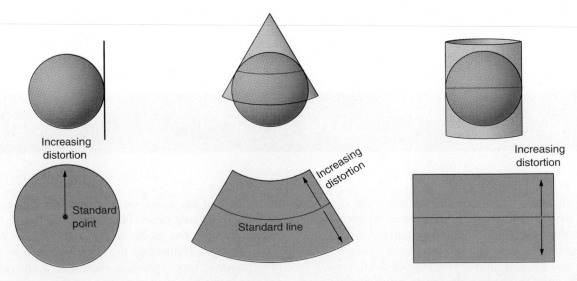

◆ **FIGURE 2.21** A map projection formed by a plane, cone, or cylinder lying tangent to the Earth's surface has only a single standard point or standard line, with distortion increasing outward from it.

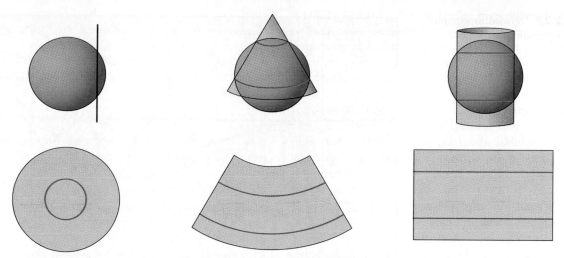

◆ **FIGURE 2.22** A map projection formed by a plane passing through the Earth's surface has a circular standard line of no distortion. A map projection formed by a cone or cylinder passing through the Earth's surface has two standard lines, allowing a larger map area to display little or no distortion.

Characteristics of Major Map Projections

Most commonly used map projections are members of the plane, conic, and cylindrical projection families. Each family has characteristic strengths and weaknesses. Because maps have such varied uses, no single projection group or type can be considered best; in choosing a type of map projection, the user should select the one whose properties will allow the map to display the desired information most clearly and accurately.

PLANE PROJECTIONS. A plane projection is a direct projection of a portion of the Earth's surface onto a plane surface. The resulting maps are circular and are capable of showing half of the Earth at most. Most plane projections are constructed so that they have a single point of contact with the globe. If so, they are azimuthal (correct in direction) from that point, with distortion increasing symmetrically outward.

Plane projections are typically centered on the North or South Pole; these are termed **polar projections.** They may, however, be centered on any point, such as a city, airport, or park, to show directions and distances to other locations. Any straight line drawn from the standard point to any other point will represent the line of shortest Earth distance between the two. Such lines are termed **great circle routes** because they form arcs of great circles on the globe. Because of the ease of plotting great circle routes on them, plane projections are commonly used in preparing navigational charts for ships and aircraft.

CONIC PROJECTIONS. Conic projections are constructed by transferring the Earth's features onto a cone. Most commonly, the apex of the cone is located directly above the North or South Pole. This produces a standard parallel with no distortion in the middle latitudes. Conic projections are therefore well suited for depicting middle latitude areas with long east-west dimensions, such as the United States, Canada, Europe, and Australia. If the cone cuts through the globe so as to produce two standard parallels, the width of the zone of little distortion can be expanded. On conic projections, parallels appear as arcs of circles, while meridians appear as straight lines that gradually converge poleward.

CYLINDRICAL PROJECTIONS. Cylindrical projections are produced by transferring the Earth's features onto a cylinder. The resulting maps are almost always oriented so that the equator forms the standard parallel. As a result, they display the low latitudes most accurately, with scale—and distortion—increasing poleward. Because only the poles cannot be shown, such projections are frequently used for world maps. On a cylindrical projection, both parallels and meridians appear as straight lines, giving the map a rectangular shape (see Figures 4.9 and 4.10, for example). Unless they have been mathematically adjusted, the spacing of the meridians remains constant in a poleward direction, while that of the parallels increases.

The best-known and probably the oldest cylindrical projection is the **Mercator projection,** developed by the Flemish cartographer Gerhardus Mercator in 1569 for use in the construction of navigational charts (Figure 2.19a). The unique attribute that has made this projection so valuable and durable is the fact that any straight line drawn on a Mercator map is a **rhumb line,** or line of constant compass bearing. In plotting a course on a Mercator map, navigators simply draw a straight line on the map between their current position and their destination and measure the compass orientation of the line. Even though such a course normally does not follow the shortest (or great circle) route, the ability to easily determine an accurate compass heading to one's destination was a major breakthrough at a time when navigational techniques were very imprecise.

1. Each contour line has a specific elevation. The elevation value of a given contour line may be written on it, or it may be necessary to determine the value from the elevations of nearby contour lines.

2. A contour line separates areas with elevations higher than its value from areas with elevations lower than its value.

3. The map distance between contour lines indicates slope steepness. Widely spaced contour lines indicate gentle slopes, while closely spaced contour lines indicate steep slopes.

4. Contour lines do not meet or cross one another, except in the rare cases of vertical or overhanging topographic features.

5. The bends in contour lines point uphill when crossing streams or other linear lowlands and point downhill when crossing ridges.

6. A closed contour line normally indicates the presence of a high point of land, such as a hilltop. Closed depressions, which are less common, are indicated by special **depression contours,** which are drawn with inward-facing (i.e., downhill-extending) line segments.

The cartographer determines the number and elevation values of the contour lines used on a given topographic map. The vertical interval used, called the **contour interval,** is normally constant over the entire map. Some topographic maps, such as those prepared by the chief American topo-graphic mapping agency, the U.S. Geological Survey (USGS), also use thicker **index contours** with a larger contour interval for faster elevation determination. For example, on a map with a 20-foot contour interval, every fifth line might be thickened to produce a 100-foot index contour interval.

USGS topographic maps are available for purchase by the public and frequently are used for fieldwork and lab exercises in Earth science and engineering courses. In addition to elevations, which are indicated by brown contour lines, these maps contain a great deal of other information (Figure 2.25). The larger-scale maps in particular contain a wealth of surface detail pertaining to both natural and cultural features, and more than 100 standardized symbols have been developed to identify these features (Figure 2.26).

Remote Sensing and Computer Mapping

As we have seen, sophisticated topographic maps such as those prepared by the USGS contain a tremendous quantity of spatial data. How is this information collected, and how is it located on maps? In contrast to the laborious manual ground surveying and cartographic techniques of the past, most map information currently is gathered by remote sensing instruments, and the maps themselves are largely computer generated.

◆ **FIGURE 2.25** A portion of a modern USGS topographic map. Contour lines are colored brown.

TOPOGRAPHIC MAP SYMBOLS

Hard surface, heavy-duty road ....

Hard surface, medium-duty road ....

Improved light-duty road ....

Unimproved dirt road ....

Trail ....

Railroad: single track ....

Railroad: multiple track ....

Bridge ....

Drawbridge ....

Tunnel ....

Footbridge ....

Overpass—Underpass ....

Power transmission line with located tower ....

Landmark line (labeled as to type) .... TELEPHONE

Dam with lock ....

Canal with lock ....

Large dam ....

Small dam: masonry — earth ....

Buildings (dwelling, place of employment, etc.) ....

School—Church—Cemeteries .... Cem

Buildings (barn, warehouse, etc.) ....

Tanks; oil, water, etc. (labeled only if water) .... Water Tank

Wells other than water (labeled as to type) .... Oil Gas

U.S. mineral or location monument — Prospect ....

Quarry — Gravel pit ....

Mine shaft—Tunnel or cave entrance ....

Campsite — Picnic area ....

Located or landmark object—Windmill ....

Exposed wreck ....

Rock or coral reef ....

Foreshore flat ....

Rock: bare or awash ....

Horizontal control station ....

Vertical control station .... BM ×671 × ×672

Road fork — Section corner with elevation .... 429 +58

Checked spot elevation .... × 5970

Unchecked spot elevation .... × 5970

Boundary: national ....

State ....

county, parish, municipio ....

civil township, precinct, town, barrio ....

incorporated city, village, town, hamlet ....

reservation, national or state ....

small park, cemetery, airport, etc. ....

land grant ....

Township or range line, U.S. land survey ....

Section line, U.S. land survey ....

Township line, not U.S. land survey ....

Section line, not U.S. land survey ....

Fence line or field line ....

Section corner: found—indicated .... + +

Boundary monument: land grant—other ....

Index contour .... Intermediate contour ....

Supplementary cont. .... Depression contours ....

Cut — Fill .... Levee ....

Mine dump .... Large wash ....

Dune area .... Tailings pond ....

Sand area .... Distorted surface ....

Tailings .... Gravel beach ....

Glacier .... Intermittent streams ....

Perennial streams .... Aqueduct tunnel ....

Water well—Spring .... Falls ....

Rapids .... Intermittent lake ....

Channel .... Small wash ....

Sounding—Depth curve ...... Marsh (swamp) ....

Dry lake bed .... Inundated area ....

Woodland .... Mangrove ....

Submerged marsh .... Scrub ....

Orchard .... Wooded marsh ....

Vineyard .... Bldg. omission area ....

◆ **FIGURE 2.26** USGS standardized topographic map symbols.

GEOGRAPHIC INFORMATION SYSTEMS: A NEW TOOL FOR MAPPING AND DECISION MAKING

From a historical perspective, maps were prepared to help people navigate through unfamiliar terrain. Today, many government agencies and businesses are using maps to help make important decisions regarding the Earth. This revolution in mapping has been made possible by the development of geographic information systems, or GIS as they are commonly called.

A GIS is a computer-based system that provides for the storage, retrieval, manipulation, and display of geographic (physical and human) data. Although the first efforts to develop such systems began in the late 1960s, it was not until the advent of powerful desktop computers and workstations in the mid-1980s that GIS gained prominence as a computer-mapping and decision-making tool. A typical GIS today consists of a microcomputer or workstation, GIS software, specialized hardware for the input and display of map data (digitizers, scanners, and plotters), and a database of map and related data that have been converted to digital form.

A GIS stores data about physical and human features in the form of coordinates for points, lines, and areas. For example, a meteorological station can be identified by a "single" latitude/longitude coordinate (point), a stream as a "string" of latitude/longitude coordinates (line), and a soil type as a "closed loop" of latitude/longitude coordinates (area). In addition to locating features on the Earth's surface, a GIS also allows us to store data about those features in a computer database. Temperature and precipitation records can be stored for meteorological stations; data on depth, volume, and velocity can be stored for streams; and data on physical and chemical properties can be stored for soils. Data in the database can be related to the location of any point, line, or area feature on the Earth by means of its coordinate(s). Many government agencies and businesses have already begun to convert map and related data from paper to digital form for use with GIS. The U.S. Geological Survey, for example, has converted the elevation, roads and other transportation features, and place-name data on many of its maps to digital form.

Although GIS are being used to produce many of today's maps, the real power of this new tool is its ability to provide information for decision-making purposes quickly and cost-effectively. The manipulation and analysis capabilities of GIS allow geographers, Earth scientists, resource managers, planners, and others to readily "overlay" different maps of physical and human data. For example, a GIS can be used to combine data on climatic characteristics, vegetation, soils, and land use to map areas that are ideally suited for agriculture. Maps portraying a combination of these data could be used to identify areas that could be brought into agricultural production through irrigation, or to identify agricultural land close to an urban area that should be protected from development. Maps like these would have been difficult, if not impossible, to produce using traditional cartographic techniques.

Figure 2.27 shows how GIS can be useful in helping to convey information about natural disasters quickly and accurately. The United States has experienced an unusual number of natural disasters since 1993, including hurricanes, major floods, and earthquakes. GIS software was used to monitor the impact caused by a 6.8 magnitude earthquake in Northridge,

Every object or substance on the Earth's surface absorbs, reflects, and emits energy in the form of electromagnetic radiation. The quantity and wavelengths of this radiation depend on the nature and temperature of the surface materials. In the rapidly evolving field of **remote sensing,** information is acquired from a distance by mechanical devices that receive portions of this electromagnetic energy.

This information enables us to gain more knowledge of the nature or characteristics of the radiation sources.

A large number of remote sensing devices are used at the Earth's surface for a great variety of purposes. Some familiar examples are radios, cameras, binoculars, telescopes, and radar. Of greatest importance to physical geographers, however, are remote sensing devices that are sent aloft to gather

California, which occurred in January 1994. Within days, maps were prepared using GIS to show earthquake seismology and patterns of epicenters for earthquake aftershocks. GIS was also used for emergency response to the earthquake, such as tracking highway closures, locating aid centers, and providing other information necessary to assist earthquake victims.

As more and more data are converted from analog (paper) to digital form, GIS will very likely change the way we use maps. Folded paper maps that we use solely for navigation will give way to computer-generated map displays. GIS will allow us to use maps to help make decisions on a broad array of subjects concerning the Earth's physical and human features. In doing so, we will be able to ask questions of maps and related data and quickly get information on which to base decisions. ■

This focus box was written by Dr. John Morgan III, Department of Geography, Towson State University.

◆ **FIGURE 2.27**   This map, which was produced using the Arc/Info GIS, shows the location of the initial Northridge, California, earthquake epicenter, along with the location of aftershock epicenters. The earthquake data have been overlaid on a block diagram that shows the terrain in the vicinity of Northridge.

SOURCE: Graphic image provided courtesy of Environmental Systems Research Institute, Inc., California Institute of Technology Seismological Laboratory, Pasadena, and Thomas Bros. Maps

data on the nature of the Earth's surface and lower atmosphere. These devices, which operate on a variety of electromagnetic wavelengths, have enabled Earth scientists to accumulate vast quantities of spatial data.

One of the oldest remote sensing tools is black-and-white photography. Surface photography was developed in the mid-nineteenth century, and the earliest aerial photographs were taken in the late 1800s by cameras carried aloft by balloons. With the development of the airplane, it became possible to control the area being photographed, and systematic air photo coverage began in the United States and portions of Europe after World War I. This led to the growth of the field of **photogrammetry**—the making of accurate measurements from photographs.

Following World War II, the use of color film, which more realistically portrays the surface and provides a greater variety of surface information, gradually increased. At about the same time, remote sensing devices that used radiation wavelengths other than visible light were developed. Of particular importance were infrared sensors, which recorded patterns of heat rather than light. These sensors record emissions that readily penetrate clouds, fog, and darkness.

In the 1960s, the deployment of orbiting satellites started a revolution in remote sensing technology equivalent to that resulting from the development of the camera and the airplane. From their high altitudes, satellites have greatly expanded the field of view of the Earth. Unlike aircraft, satellites can remain aloft and continuously collect and transmit surface data for months and even years. Satellites have been able to acquire data from all parts of the world because they are affected neither by adverse weather conditions nor by international boundaries. They therefore are critically important in gathering military intelligence, and many of their data-collecting capabilities remain classified.

Weather satellites, which were among the first satellites developed, have been operating since 1960 when the United States launched the first TIROS satellite. The current generation of GOES (Geostationary Operational Environmental Satellite) satellites remain at fixed positions 36,000 kilometers (22,300 mi) above the Earth's surface, permitting continuous monitoring of weather systems affecting North America. Infrared sensors in these satellites allow cloud patterns to be photographed both during the day and at night.

Other satellites, including the Landsat series, provide high-resolution color imagery of surface features. These satellites have permitted the accurate mapping of remote areas, such as mountainous and desert regions, and have also been extensively employed in monitoring air and water pollution patterns, identifying areas of diseased crops and forests, and exploring for minerals.

At present, literally hundreds of satellites serving a great variety of purposes orbit the Earth, and additional satellites of progressively increasing sophistication are constantly being deployed. The U.S. government makes aircraft and satellite imagery of all parts of the Earth's surface available to the public, and some of this imagery appears in this book.

Another recent and ongoing revolution with major applications to both cartography and remote sensing is occurring in the areas of **computer mapping.** Computers can be used to store, organize, and display huge quantities of data in a tiny fraction of the time that such tasks could be accomplished manually. Modern cartography and remote sensing technology depend upon the use of computers for preparing maps. For example, plotting machines have been developed that can produce accurate topographic maps directly from overlapping series of aerial photographs (Figure 2.28).

In addition to assisting in map preparation, some computers store, retrieve, manipulate, and display a variety of geographic data for portions of the Earth's surface. These computer systems, or **geographic information systems,** are

◆ **FIGURE 2.28**   Pattern of overlapping photographs (individual squares) taken from an aircraft flying back and forth over an area. Because surface features are viewed from different angles on the overlapping portions of adjacent photos, it is possible to determine differing surface elevations directly from the photographs. In similar fashion, our brains construct a single three-dimensional image from the separate two-dimensional views gathered by each of our eyes.

revolutionizing the way geographers work with mapped data. Geographic information systems can be considered automated atlases of mapped data. They allow geographers to synthesize geographical information by superimposing several maps in the computer in order to identify spatial patterns of Earth surface phenomena. The Geographic Techniques feature on the previous pages examines this important geographic decision-making tool in greater detail.

❙ SUMMARY

This chapter has examined the Earth's origin, position, and movements in space, as well as the size and shape of our planet. These factors have enabled us to devise systems for describing locations and determining times on the Earth's surface; they also are responsible for the changing of the seasons. The methods by which information about the Earth is gathered and depicted on maps have also been discussed. We return to the Focus Questions posed at the beginning of the chapter to summarize the most important concepts presented:

1. *What is the geographical setting of the Earth in space, and what movements is the Earth making through space?*

The Earth is the third planet from the Sun in a solar system that contains nine known planets. Our planet is roughly spherical because of the inward pull of gravity and has a diameter of approximately 12,700 kilometers (7,900 mi). The solar system is located in the outer portion of a spiral arm of the Milky Way galaxy—a rotating collection of many billions of stars. The Earth is involved in several movements through space, occurring at greatly differing orders of magnitude. At the largest scale, we, along with the entire galaxy, are moving away from the center of the universe as an apparent consequence of the Big Bang. We are also revolving around the center of the Milky Way galaxy. On a smaller scale, the Earth revolves around the Sun once each year and rotates on its axis once each day.

2. *What basic directional and locational systems are employed on the surface of the Earth?*

Most directional and locational systems in common use are based on the positions of the North and South Poles and of the equator. Two directional systems widely used in the United States are the Mariner's Compass and the azimuth system.

The system of latitude and longitude is the most important locational system currently in use. Degrees of latitude are measured north and south of the equator, and degrees of longitude are measured east and west of the prime meridian.

3. *How is the global system of time zones organized?*

The standard time zones ideally form 15°-wide swaths of longitude, each centered on a meridian divisible by 15. The time in each of the 24 time zones differs by one hour from adjacent zones. Over land areas, the zones usually are adjusted to coincide with political boundaries. The International Date Line, located in the central Pacific, is a line over which a 24-hour time difference exists.

4. *What causes the seasons?*

The seasons occur because the Earth's axis is inclined at 23$^1/2$° from a perpendicular angle to the plane of the ecliptic. As the Earth revolves around the Sun during the course of the year, the northern and southern hemispheres alternately are inclined at varying angles toward and away from the Sun. Spring and summer occur when a hemisphere is inclined toward the Sun; at this time of year, the Sun is relatively high in the sky, periods of daylight are longer than nights, and total energy receipts are relatively large. Conversely, fall and winter occur when the hemisphere is inclined away from the Sun.

5. *What is a map? With what basic map information must the user become familiar before a map can be employed to full advantage?*

A map is a pictorial representation of the geographic locations of selected surface features at a reduced scale. Maps are essential geographical tools and are extensively employed to display the distributions of phenomena of all types. The following categories of basic map information should be examined before a map is selected or used: (1) the map title, which should clearly indicate the purpose of the map; (2) the date of the map or of the information contained on the map; (3) the locations covered by the map; (4) the directional orientation of the map with respect to the map sheet; (5) the symbols used on the map, as explained in the map legend;

and (6) the map scale, indicating the degree of reduction of surface features appearing on the map.

6. *How is map information collected and compiled?*

A variety of remote sensing techniques now enable map data to be collected and assembled relatively quickly and easily. Most surface information is currently obtained by aircraft and satellites, which often use electromagnetic wavelengths other than visible light to obtain their imagery.

Computers can generate maps directly from aerial photographs. Sophisticated computer programs termed *geographic information systems* are revolutionizing the geographical analysis of mapped data. These programs permit researchers to superimpose several different categories of map data so as to identify complex spatial patterns resulting from the interaction of multiple factors.

❙ QUESTIONS FOR THOUGHT AND REVIEW

1. Briefly describe the Big Bang theory of the origin of the universe. Include an explanation of why the universe is expanding and how the galaxies and stars were formed.
2. Why is the Earth spherical? What two major departures from a precise spherical shape exist? Why do they occur?
3. What is the purpose of the system of latitude and longitude? What similarities and differences exist between parallels and meridians?
4. What problems would you encounter in your schedule of daily activities if the local time system were still in use?
5. What is the justification for using daylight saving time? Why is it used in most of the United States during only part of the year?
6. Draw a single diagram illustrating the parallelism of the Earth's axis and the position of the Earth with respect to the Sun at the times of the solstices and the equinoxes. Label each position.
7. What is a map? Why are maps particularly associated with the field of geography?
8. Why are all flat maps distorted? Is the problem of map distortion greater with small-scale maps or with large-scale maps? Why? Why aren't globes more widely employed as a means of avoiding the problem of map distortion?
9. Explain the basic principles of constructing a map projection. What are the three basic types of map projections? Why are conic projections especially well suited for depicting the United States?
10. Why are contour lines better suited for the making of precise elevation measurements than the other elevation depiction methods discussed in the text?
11. What is remote sensing? What remote sensing devices have been most important in the acquisition of information about the Earth's surface during the twentieth century?

PROBLEMS

1. If you travel 60° south of 45°N, at what latitude will you be located?
2. If you travel 50° west of 140°W, on what meridian will you be located?
3. If it is 11 P.M. Monday at 45°W, what are the time and day at the same instant at 135°W?
4. If it is 9 A.M. Thursday at 120°W, what are the time and day at the same instant at 165°E?
5. Is the Sun's noon altitude higher in Maine or in Florida during the summer? Does Maine or Florida have the longer daylight period during the summer? Which U.S. state experiences the most hours of twilight during the year?
6. Change the verbal map scale of "1 inch equals 8 miles" to the equivalent RF scale. How many kilometers would one centimeter represent at this scale?
7. Change the RF map scale of 1:1,000,000 to the equivalent miles-per-inch scale. Change it to the equivalent kilometers-per-centimeter scale.
8. If a map with an RF scale of 1:126,720 measures 50 centimeters (20 in) from side to side and 20 centimeters (10 in) from top to bottom, how many square miles of Earth surface area does it depict? How many square kilometers?
9. If the mapped area described in the previous problem were remapped at an RF scale of 1:63,360, how many times as much paper would be needed? What would be the dimensions of the new map?
10. If a topographic map had a contour interval of 25 feet, how many contour lines would encircle a mountainous oceanic island with a summit elevation of 673 feet? What would be the elevation of the uppermost contour line?

KEY TERMS

solar system	longitude
Milky Way galaxy	parallel
revolution	meridian
rotation	great circle
geographic North and South Poles	small circle
equator	prime meridian
parallelism	solstice
oblate ellipsoid	equinox
latitude	Tropics of Cancer and Capricorn
Arctic and Antarctic Circles	map scale
local time	map projection
standard time	topographic map
International Date Line	contour lines
daylight saving time	remote sensing
map	geographic information systems

3 EARTH'S ATMOSPHERIC ENVELOPE

OUTLINE

FOCUS QUESTIONS

1. What basic influences does the atmosphere have on the Earth's surface
　environment?
2. What is the composition of the atmosphere?
3. How do the characteristics of the atmosphere change with altitude?
4. How are human activities influencing the composition of the atmosphere?

The **atmosphere** is the gaseous envelope that surrounds the Earth. Although nearly transparent and seemingly insubstantial, the atmosphere nonetheless profoundly affects nearly every aspect of our physical environment. Life on Earth could not exist without it.

The atmosphere is the outermost of the four environmental spheres that interface at the surface of the Earth. In essence, it forms a canopy over the hydrosphere and the lithosphere and plays a crucial role in supporting the biosphere. The cycles of development and change that characterize the four environmental spheres require tremendous amounts of energy. Nearly all this energy is derived from the Sun, although a small quantity comes from the interior of the Earth. Because the atmosphere is the outermost of the four spheres, it is the first to receive the solar energy on which it and the others depend. As solar energy passes downward through the atmosphere, it is partially absorbed and redirected before it reaches the Earth's surface and is converted to the thermal and chemical energy that largely powers the other spheres.

◆ **FIGURE 3.1** The gray, pockmarked surface of the Moon stands in stark contrast to the brilliance of the distant Earth in this photo taken by the *Apollo* astronauts. If the Earth had no atmosphere, its surface would closely resemble the Moon's surface.

SOURCE: Department of the Interior, U.S. Geological Survey

ATMOSPHERIC ORIGIN AND FUNCTIONS

Earth's atmosphere is believed to have been derived primarily from volcanic gases released from our planet's interior. Volcanic eruptions not only produce large quantities of lava and ash, they also liberate great volumes of gases. Scientists believe that more than four billion years of volcanic eruptions have released not only the gases that have formed our atmosphere, but also the water vapor that has formed all the Earth's water bodies and glaciers. Present-day eruptions continue to liberate gases, originally dissolved under great pressure in rock deep beneath the surface. Evidence for the volcanic origin of the atmosphere is provided by measuring the composition of gases currently being released from active volcanoes. These gases consist of approximately 85 percent water vapor, 10 percent carbon dioxide, and 1 or 2 percent nitrogen. These three gases, and their derivatives (including oxygen), comprise about 99 percent of our atmosphere.

The atmosphere performs a number of vital functions that profoundly affect the nature of the Earth's surface and its ability to support life. The most obvious and vital function is the supplying of oxygen to animals and carbon dioxide to plants. These gases are essential to their biological processes.

Earth's atmosphere also protects our planet from the impact of small meteors, which are incinerated by friction before reaching the surface. Large **meteorites** do reach the surface and produce craters similar to those on the Moon (Figure 3.1). Unlike lunar craters, meteorite craters on Earth have relatively short spans of existence because the forces of erosion associated with atmospheric phenomena, especially precipitation and wind, gradually wear them away. In fact, all the Earth's landforms exist for limited periods of geologic time because of erosional and depositional processes related to the atmosphere.

The blanketing influence of the atmosphere also protects us from most cosmic radiation. **Cosmic radiation** consists of charged particles, mostly hydrogen and helium nuclei, that enter Earth's atmosphere at nearly the speed of light. Because of their great kinetic energy, these particles are capable of destroying or mutating plant and animal cells and would be lethal to Earth's life-forms in the concentrations that bathe the Moon's surface. Fortunately, the upper atmosphere acts to greatly reduce the influx of cosmic radiation.

The resistance of the atmosphere to the passage of energy greatly reduces the range of Earth surface temperatures. On the Moon's surface, daytime temperatures can exceed 94°C (200°F), only to fall to –128°C (–200°F) at night. Earth's atmosphere reflects some solar energy and distributes the remainder through a vertical distance of many kilometers, so temperature ranges are much less extreme. Earth's atmosphere also has a substantial capacity to absorb solar radiation (see Chapter 4), permitting it to escape only slowly into space. This causes mean temperatures at the Earth's surface to be much higher than they would be in the absence of an atmosphere.

Finally, the atmosphere makes possible the cycle of evaporation and precipitation that provides water to the land areas of the Earth. Without this cycle (discussed in Chapter 6), there would be no clouds, no rain or snow, no rivers or lakes, little or no land vegetation or soil, and little animal life on the land or in the sea. In addition, many important minerals could not form, and most landforms with which we are familiar would not develop.

COMPOSITION OF THE ATMOSPHERE

The atmosphere is a mixture of gaseous elements and compounds. Most of these gases exist in constant proportions in all but the atmosphere's outermost layers (above 80 kilometers—50 mi—altitude). The reason for the homogeneity of the atmosphere is the mixing action of the wind. Without the turbulent transfer of air molecules from one level to another by wind currents, the various atmospheric gases would become partially stratified into horizontal layers on the basis of density. A few gases, however, have characteristics that prevent the wind from mixing them evenly, and consequently their atmospheric concentrations vary.

In addition, small quantities of liquids and solids are suspended in the atmosphere. Atmospheric liquids and solids vary greatly in quantity at different times and in different places because they are much denser than the atmospheric gases and tend to settle to the Earth's surface. The distribution of these substances depends on the size and location of their sources, the rate at which they enter the atmosphere, and the speed and direction of the wind currents.

◆ **TABLE 3.1** **Composition of the Atmosphere**

NAME	SYMBOL	PERCENT BY VOLUME*
Nonvariable Gases		
Nitrogen	N_2	78.08
Oxygen	O_2	20.95
Argon	Ar	0.93
Neon	Ne	0.002
Others		0.001
Variable Gases and Particulates		
Water Vapor	H_2O	0.1 to 4.0
Carbon Dioxide	CO_2	0.036
Ozone	O_3	0.00006
Other Gases		trace
Particulates		normally trace

*Percentage figures, except for water vapor, are for dry air.

The components of the atmosphere can be divided into three categories: nonvariable gases, variable gases, and particulates (liquids and solids).

Nonvariable Gases

The **nonvariable gases** comprise over 98 percent of the total atmosphere by volume (Figure 3.2). These gases are sufficiently stable, both physically and chemically, to remain in the atmosphere long enough to be dispersed evenly by wind currents.

As Table 3.1 indicates, the most abundant gas by far is **nitrogen,** which by volume comprises 78 per cent of the atmospheric total of nonvariable gases. Although it is only the third most abundant gas released by the volcanic eruptions that formed the atmosphere, nitrogen has achieved its position of dominance because of its physical and chemical stability. Gaseous nitrogen is diatomic; that is, two atoms of nitrogen combine to form one nitrogen molecule (N_2). In this form, nitrogen is chemically inactive and does not participate directly in the respiratory processes of either plants or animals. Small amounts of atomic nitrogen released by lightning or bacterial action, however, are incorporated into compounds that serve as essential plant nutrients.

Most of the remainder of the atmosphere consists of **oxygen** (O_2), another diatomic molecule, which is essential to the respiratory processes of plants and animals. Oxygen also enters into many chemical reactions with rock materials at or near the Earth's surface to form a variety of products, including soil. Free oxygen is not liberated in quantity by volcanic eruptions. Instead, these eruptions release carbon dioxide (CO_2), which plants later separate into its constituent carbon and oxygen during the process of photosynthesis. Plants therefore have been responsible for placing nearly all the free oxygen in the atmosphere. Nitrogen and oxygen

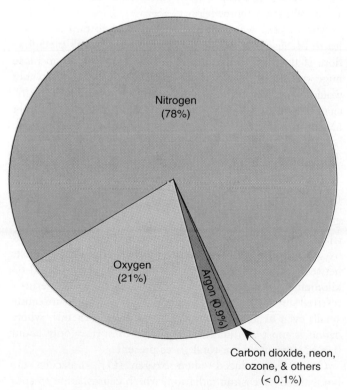

Nitrogen (78%)

Oxygen (21%)

Argon (0.9%)

Carbon dioxide, neon, ozone, & others (< 0.1%)

◆ **FIGURE 3.2** The relative proportions of the major atmospheric gases.

together comprise slightly over 99 percent, by volume, of dry air.

One other nonvariable atmospheric gas exists in more than a trace amount. This gas is **argon** (from the Greek *argos,* or "inactive"), which comprises 0.93 percent of dry air. It is an inert gas and is virtually incapable of entering into any chemical reactions. Argon is produced chiefly by the radioactive decay of potassium-40 in the Earth's crust and enters the atmosphere by slow diffusion from near-surface rocks.

Variable Gases

Several gases exist in variable quantities in the atmosphere. Although they collectively comprise only a very small proportion of the atmosphere's total mass and volume, three of these **variable gases**—carbon dioxide, water vapor, and ozone—are crucial to the existence of life.

Carbon Dioxide

Carbon dioxide is essential to the photosynthetic processes of plants and is the source of the carbon that is the basic element of Earth's plant and animal life. The amount of carbon dioxide in the atmosphere varies slightly during the year because of variations in its rate of absorption by plants. Its quantity declines slightly each northern hemisphere summer, when the vegetation of the vast northern hemisphere landmasses is growing most vigorously, and increases again during the winter as it is released by organic decay. The atmosphere's carbon dioxide content also has been slowly increasing, especially during the twentieth century, as a result of human activities. This phenomenon and its potential environmental consequences are discussed in the Case Study at the end of Chapter 4.

Carbon dioxide and oxygen are involved in a constant cycle of interchange. Because the dissociation of a carbon dioxide molecule into its component carbon and oxygen is an endothermic (energy-absorbing) reaction, plants require sunlight to supply the energy for this process. Conversely, when animals recombine carbon from the food they have ingested with atmospheric oxygen, an exothermic (energy-producing) reaction occurs. This reaction enables animals to perform all their activities. In a sense, then, we are powered by stored sunlight, which has been "fixed" by plants in the carbohydrate molecules we digest.

Although approximately 10 percent of the gas emitted by volcanic eruptions is carbon dioxide, only a very small percentage (0.036 percent) of the atmosphere currently is made up of this gas. This invites the question of where the remainder has gone. A small portion of the carbon dioxide forms the bodies of plants, especially trees, that are alive or have recently died. A considerably larger quantity is stored beneath the Earth's surface in the form of peat, coal, oil, and gas deposits. These deposits represent the remains of plants and animals, most of which lived more than one hundred million years ago. The bulk of the carbon dioxide, though, has dissolved into the ocean waters. Carbon dioxide is highly soluble in water (a fact that the carbonated beverage industry uses to its advantage), and over the course of geologic time, most atmospheric carbon dioxide has dissolved into the oceans, later to be precipitated in the form of carbonate rocks such as limestone and dolomite.

Water Vapor

Water vapor (H_2O), or water in the invisible gaseous state, is one of the four most abundant gases in the atmosphere, but is highly variable in amount. Cold, dry air may contain less than 0.1 percent water vapor by volume, while very warm, moist air in the tropics may contain as much as 4 percent. Worldwide, air just above the Earth's surface contains an estimated average of 1.4 percent water vapor. This proportion declines rapidly with altitude, so that approximately half the atmosphere's water vapor content exists within 1.6 kilometers (1 mi) of sea level.

The major reason that the atmosphere's water vapor content varies so greatly is that the water vapor holding capacity of the air is highly dependent upon its temperature (Figure 6.5). Warm air can hold much more water vapor than can cold air. In addition, the availability of water vapor differs greatly in different areas because of the nature of the Earth's surface. For example, air over land areas is generally drier than air over large bodies of water. As a result of the continuous temperature and locational changes to which the air is subjected, its water vapor content alternately increases through the process of evaporation and decreases through the processes of condensation (conversion to liquid water) and sublimation (conversion to ice).

Atmospheric water vapor is critically important to life on Earth because it is the source of all of the Earth's precipitation. If the atmosphere could not alternately gain and lose huge quantities of this gas, the land areas of our planet would be sterile deserts. Atmospheric moisture is of sufficient importance to be treated in its own chapter (Chapter 6), and its effects also are dealt with in subsequent chapters covering the subjects of climate, the Earth's surface waters, fluvial landforms, and glacial landforms.

Ozone

The third important variable gas is ozone. **Ozone** (O_3), a triatomic form of oxygen, has three chemically bonded oxygen atoms instead of the normal two. Most ozone exists between an altitude of about 15 kilometers (10 mi) and 65 kilometers (40 mi), a portion of the atmosphere sometimes referred to as the **ozone layer.** Concentrations are quite small; even at an altitude of 24 kilometers (15 mi), where ozone is most abundant, it typically comprises only about 0.0012 percent of the total gases present.

Ozone is produced when oxygen (O_2) molecules are bombarded by cosmic radiation, which causes some to split and to recombine as ozone (Figure 3.3). The ozone molecule

High-energy UV radiation — Oxygen molecule → 2 Oxygen atoms

Oxygen atom + Oxygen molecule — UV radiation → Ozone molecule

◆ **FIGURE 3.3** Ozone forms when a free oxygen atom, released by the splitting of an oxygen molecule by cosmic radiation, combines chemically with another oxygen molecule.

is not particularly stable chemically and has a tendency to revert to the diatomic oxygen (O_2) form.

The minute amounts of ozone produced in the upper atmosphere are important because of their ability to absorb the potentially lethal quantities of cosmic and ultraviolet radiation that enter the atmosphere. The absorption of this short-wave radiation releases heat, producing a zone of relatively warm temperatures centered at an altitude of about 48 kilometers (30 mi). Considerable concern has developed in recent years that human activities may be depleting the ozone layer. This subject is explored further in the Focus Box on pages 46–47.

The atmosphere also contains trace amounts of several other variable gases. Among the more significant in quantity are carbon monoxide (CO), methane (CH_4), sulfur dioxide (SO_2), and nitrogen dioxide (NO_2). A substantial proportion of these gases is produced by human activities, although they are formed by natural processes as well.

Particulates

Atmospheric particulates consist of liquids and solids (with the exception of water droplets or ice crystals) that are suspended in the atmosphere because of their small size. The smallest are capable of remaining aloft almost indefinitely. Airborne particulates are often considered to be atmospheric "impurities," because the atmosphere is essentially a mixture of gases, and particulates represent temporary additions derived largely from the Earth's surface. Nonetheless, these particulates play a vital role in atmospheric processes. Most particulates are solids, rather than liquids, and are collectively referred to as **dust.**

Dust derives from a variety of sources including both natural processes and, increasingly, human activities (see Figure 3.5 on page 48). Natural sources may be divided into six categories. First is the meteoric dust that is continuously sifting through the atmosphere. A second source, volcanic dust and ash, comes from the opposite direction—the interior of the Earth. Although major volcanic eruptions occur only sporadically, they have the short-term capacity to release more dust into the atmosphere than any other source. A third source consists of windblown surface materials, usually soil, derived from areas not protected by vegetation. A fourth source is smoke from fires, which may attain impressive dimensions, particularly in the lower latitudes. Fifth are minute salt crystals that are released from the evaporation of sea spray and affect primarily oceanic and coastal areas. Finally, many "dust" particles of biological origin exist, including pollen, spores, seeds, and bacteria.

Particulates released by human activities are increasing rapidly in many areas, especially in the less-developed countries. They originate primarily from urban sources such as factory and automobile exhausts and smoke from heating and cooking, as well as from agricultural areas unprotected by vegetation (see the Case Study at the end of this chapter). The well-known hazy conditions that prevail in summer over much of North America and Europe result largely from the combination of dust-generating human activities and the occurrence of relatively light winds at this time of year.

Excessive quantities of airborne particulates can produce unpleasant and occasionally unhealthful atmospheric pollution. Conversely, dust particles play a vital role in the formation of clouds and precipitation by serving as **condensation nuclei.** Water vapor normally requires solid or liquid surfaces on which to condense or sublime in order to produce water droplets or ice crystals, and airborne particulates provide these surfaces. Every raindrop is composed of many thousands of cloud droplets, each of which has formed around a mote of dust. If no atmospheric dust particles were present, the air near the surface would become saturated with water vapor, but there would be no way for this moisture to be precipitated.

VERTICAL CHARACTERISTICS OF THE ATMOSPHERE

In the next few chapters, we examine the horizontal distribution of such atmospheric phenomena as temperature, air pressure, winds, and moisture. The horizontal (or at least relatively horizontal) level in which we are most interested is the base of the atmosphere; that is, the interface between the atmosphere and the Earth's surface. It is in the vertical direction, however, that the greatest variations in the atmosphere occur. These variations are produced largely by the force of gravity, which causes the air to differ both in pressure and in composition at different altitudes. An additional major influence on the characteristics of the lowest portion of the atmosphere is the surface of the Earth itself.

Weight, Density, and Vertical Extent

The air may appear weightless, but in reality it has mass and weight, just as any other substance does. Although the Earth's atmosphere contains less than one-millionth of our planet's total mass, the atmosphere's total weight is enormous. It is estimated to be 5.44 quintillion kilograms,

ARE HUMAN ACTIVITIES DEPLETING THE OZONE LAYER?

Considerable evidence has accumulated in recent years that human activities are seriously depleting Earth's ozone layer. Researchers have found that the average global concentration of stratospheric ozone declined by about 3 percent between 1969 and 1989. A study released by the U.S. Environmental Protection Agency in April 1991 reported ozone depletion of 4 to 5 percent over the United States. These statistics are of concern because ozone is the only gas in the atmosphere capable of limiting the amount of ultraviolet radiation reaching the surface. Because this radiation, in sufficient quantity, can destroy or mutate the cells of exposed plants and animals, continued ozone layer depletion could eventually have a devastating impact on Earth's life-forms. For example, the yields of many important agricultural crops would be greatly diminished, and the incidence of skin cancer among humans would likely increase significantly.

The chief culprit responsible for the loss of ozone is the release of chlorofluorocarbon gases (CFCs), which currently are widely employed as refrigerants and aerosol propellants. When CFCs break down, they release chlorine atoms that act as catalysts to trigger the conversion of ozone molecules into normal oxygen molecules. The chlorine atoms are not affected by this transformation and act over and over again, eventually destroying great quantities of ozone.

CFCs are relatively stable gases, with chemical half-lives of 75 to 100 years. Once released, they are spread around the Earth by the wind currents and gradually diffuse upward into the ozone layer over a period of years. Because there is a considerable period between the time CFCs are produced and the time they begin to affect the ozone layer, even a worldwide ban on their use would be slow to show results. Currently, about 5 percent more ultraviolet radiation reaches the Earth's surface than in 1969, and recent computer models indicate that the continued erosion of the ozone layer will likely cause a further increase in ultraviolet radiation of from 5 to 20 percent within the next 40 years.

The scientific community's concern about possible human effects on the ozone shield made headlines in 1985, when the British Antarctic Survey discovered a hole in the ozone layer over Antarctica (Figure 3.4). This hole apparently first appeared in 1975 and has been returning each Antarctic spring. The ozone-depleted zone extends from 12 to 20 kilometers (7 to 12 mi) in altitude.

The hole in the ozone layer over Antarctica is apparently caused by the extremely low air temperatures that develop over the continent during the sunless winter season. This leads to the formation of ice crystal clouds in the normally clear lower stratosphere. The ice crystals react with CFC gases to release chlorine molecules. When sunlight returns to Antarctica during August and September, these molecules are photochemically separated into individual chlorine atoms, which then react with the ozone molecules and convert them to normal oxygen.

The strength of the ozone hole in any given year appears to depend on temperatures in the Antarctic stratosphere, but the hole has tended to increase gradually in size and intensity. In 1993, the ozone hole was the largest to date—during the period from late August through October, 70 percent of the ozone disappeared over a large portion of Antarctica, and the hole extended over 23 million square kilometers (9 million sq mi), an area almost the size of North America. The hole disappears each summer as temperatures in the stratosphere rise.

Within the past few years, a second, weaker ozone hole has begun to form over the North Pole. Early in 1989, a scientific team measured 25 percent reductions in ozone concentrations at altitudes between 22 and 26 kilometers (13 and 16 mi) over the Arctic Ocean. During the winter of 1992–1993, however, this ozone hole expanded greatly, and ozone levels over much of the higher latitudes of the northern hemisphere fell to values 10 to 15 percent below average levels recorded from 1979 to 1990. Scientists believe this record drop likely resulted from the 1991 volcanic eruption of Mt. Pinatubo in the Philippines and is therefore a short-term phenomenon. The volcano released tremendous volumes of sulfur dioxide gas that were

continued on next page

converted in the atmosphere to sulfuric acid droplets. These droplets, it is believed, assisted in the release of ozone-destroying chlorine atoms from CFC gases.

Aerosol cans currently are the world's largest single source of ozone-destroying CFCs. Although both the United States and Canada banned the nonessential use of CFCs as aerosol propellants in 1978, the resulting decline in emissions has been largely offset by an increase in their use in foreign countries. A significant international agreement to limit future releases of CFCs was reached in September 1987, when 47 nations signed the Montreal Protocol, pledging to reduce CFC production by 50 percent over a 10-year period. In June 1990, more than 70 countries agreed to discontinue CFC use totally by the year 2000. As ozone losses continued to accelerate, the Protocol was further strengthened in 1992, when 92 countries agreed to ban CFC production and use by January 1, 1996. A monetary fund was also established to assist less-developed nations in phasing out their use of CFCs. ■

◆ **FIGURE 3.4**   Color-coded satellite view of the ozone hole over Antarctica on October 5, 1989, as measured by the Total Ozone Mapping Spectrometer (TOMS) aboard the NIMBUS–7 satellite. The areas of black and lavender shading centered over Antarctica have the greatest ozone depletion.
SOURCE: NASA

(b)

(a)   (c)

◆ **FIGURE 3.5** Among the chief sources of atmospheric dust particles are (a) erupting volcanoes (in this case, Washington's Mt. St. Helens), (b) urban and industrial activities (here, a copper refinery in Nevada), and (c) forest fires like this one in Paulina, Oregon.
SOURCE: a: JLM Visuals; b: © Paolo Koch/Photo Researchers; c: ©Lois Moulton/Tony Stone Images

normally written as 5.44×10^{18} kg (or as 12×10^{18} lb). The weight of the overlying air subjects each square inch of the Earth's surface at sea level to an average pressure of 6.7 kilograms (14.7 lb). Because the mean elevation of the Earth's land and water surface, taken as a whole, is about 273 meters (900 ft) above sea level, the surface air pressure for the entire Earth averages approximately 1 kilogram per square centimeter (14.2 lb/in^2).

This weight is very unevenly distributed throughout the vertical extent of the atmosphere because the air is highly compacted by gravity. The great vertical extent of the atmosphere and the imperceptible way it merges with the scattered molecules of interplanetary space make it impossible to locate the top of the atmosphere precisely, but it extends several thousand miles above the Earth's surface. Half the atmosphere's total mass, however, lies below an altitude of only 5.6 kilometers (3.5 mi, or 18,500 ft). At this altitude, the air pressure averages just half its value at sea level, as indicated on the left side of Figure 3.6. Many of the world's higher mountain peaks extend well beyond this altitude; Mt. Everest, for example, has a summit elevation of 8.8 kilometers (5.5 mi). Because the air pressure is so low, many mountain climbers employ oxygen masks when climbing above 6,200 meters (20,000 ft).

Compositional Layers

The lowest 80 kilometers (50 mi) of the atmosphere, which contain more than 99.9 percent of the total air, are kept mixed by wind currents so that the proportions of the nonvariable gases remain constant. Because of its relatively homogeneous composition, this portion of the atmosphere is sometimes termed the **homosphere.** The ability of the homosphere's gases to remain mixed actually results from both the mixing action of vertical wind currents and the numerous collisions between air molecules in this dense portion of the atmosphere. When air molecules collide, they tend to bounce in random directions. This action thoroughly mixes the various gases, regardless of their masses.

Above an altitude of 80 kilometers (50 mi), the atmosphere's density is so low that the number of air molecule collisions is greatly reduced. The wind flow at these higher altitudes is also predominantly horizontal, so the air currents are subjected to very little vertical mixing. As a result, even though strong winds are common in this low-friction environment, no effective mechanisms exist to mix gases of differing densities. At higher altitudes, therefore, the various gases exhibit an increasing tendency to be layered by density. Scientists have named this zone the **heterosphere** because its composition is vertically heterogeneous.

Thermal Layers

The atmosphere not only contains layers of varying composition, it also is stratified with respect to temperature (Figure 3.6). The lowest and most important of these thermal layers is the **troposphere,** which contains approximately 80 percent of the atmosphere's total mass. This is the zone in which most observable weather phenomena occur and in which we live. The troposphere contains most of the atmosphere's water vapor supply. Consequently, it contains nearly all clouds and is the origin of all precipitation reaching the surface. Winds in the troposphere are highly variable in both speed and direction, and these variations are directly respon-

sible for many of the changing weather conditions at the surface. In fact, the term *troposphere* was derived from the Greek word *tropos,* which means "turning" or "changing."

The dominant vertical characteristic of the troposphere is a rather rapid decrease in temperature with increasing altitude. This thermal pattern exists because the troposphere is heated primarily from below—from the warm surface of the Earth, which has absorbed energy from the Sun. The actual rate of temperature change at any given time and place is quite variable, but over long periods the mean rate of decline at all levels in the troposphere is approximately 6.4 C° per 1,000 meters (3.5 F° per 1,000 ft).

The sharply defined upper boundary of the troposphere, the **tropopause,** is the level at which the temperature stops falling with increasing altitude. The tropopause bulges outward above the tropics because of their high surface temperatures. Over the equator, it is located at an altitude of about 18 kilometers (11 mi); it gradually lowers poleward, until at the North and South Poles it is only about 8 kilometers (5 mi) above sea level. The tropopause also varies somewhat in altitude with the season of the year, being about 1.5 kilometers (1 mi) higher in summer (due to surface heating) than in winter over the middle and high

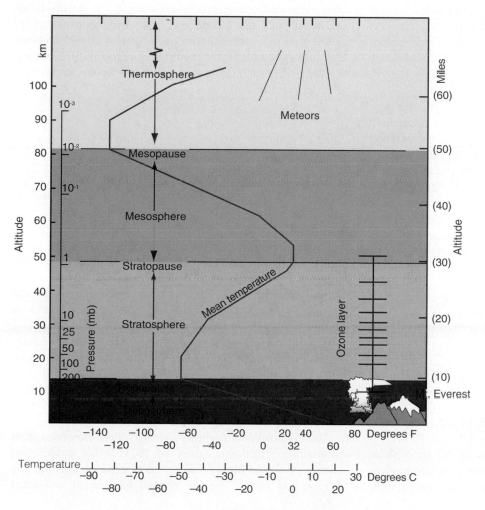

◆ **FIGURE 3.6** The average vertical temperature profile of the atmosphere, and its relationship to the four thermal atmospheric layers.

latitudes. Temperatures at the tropopause are usually between –50°C and –80°C (–60°F and –110°F).

The tropopause can be visualized as a boundary surface separating the troposphere from the **stratosphere**. This second thermal layer extends from the tropopause to an altitude of 50 kilometers (30 mi). The temperature remains fairly constant through the lower stratosphere, up to about 32 kilometers (20 mi) in altitude, but above this level, it gradually rises. This increase in temperature is caused by the absorption of solar ultraviolet energy by ozone, which converts this energy into heat. Temperature readings near the top of the stratosphere are quite variable, depending on the quantity of ultraviolet emissions from the Sun. Normally, they are near or slightly below freezing, but on occasion they can reach 7°C (45°F). Because temperatures increase with altitude in the stratosphere, there is much less tendency for the vertical air movements that form clouds to develop. The tropopause thus acts as a "lid" for cloud development, so that the stratosphere is generally characterized by cloudless, thin, dry air, often subject to strong horizontal winds.

The upper edge of the stratosphere is marked by the **stratopause**, another well-defined thermal boundary. Crossing the stratopause, the temperature stops rising and begins falling as the atmosphere's third thermal layer, the **mesosphere**, is entered. The mesosphere is located between 50 and 80 kilometers (30 and 50 mi) above the Earth's surface. Because little solar energy is absorbed at this level, temperatures within the upper mesosphere fall to the lowest values found anywhere in the atmosphere. At the upper boundary of the mesosphere, termed the **mesopause**, readings are in the vicinity of –90°C (–130°F).

Above the mesosphere, a final reversal in temperature occurs as the **thermosphere** is entered. In this zone, located at altitudes above 80 kilometers (50 mi), the temperature gradually rises to eventually reach the highest readings occurring anywhere in the atmosphere. In the upper thermosphere, temperatures may exceed 1,100°C (2,000°F) because of the absorption of ultraviolet radiation from the Sun. Although the temperature is very high, very little thermal energy is actually represented because the air is so rarefied at these great heights that few air molecules are present to contain the energy.

THE ATMOSPHERE, WEATHER, AND CLIMATE

The variability of the atmosphere is made evident by the constantly changing array of weather conditions with which we are all familiar. This chapter, in fact, serves as a preface to Chapters 4 through 8, which deal with global weather and climate.

Weather is the short-term condition of the atmosphere. Variations in the weather occur both over time at any given location and between different locations at any one time. As a result, the weather becomes more complex when examined over progressively larger segments of time and space.

When discussing long-term weather conditions, some generalizations are necessary. These long-term conditions constitute an area's **climate.** The climate of a region is sometimes defined as its typical weather because mean (or statistically average) conditions are usually stressed. Also of great importance, though, are extreme conditions, because they can have disastrous consequences. People therefore should expect that weather conditions will frequently be near the climatic means, but also should be prepared to cope occasionally with the extremes.

The branch of science having weather as its subject is termed **meteorology** (from the Greek *meteora*, or "atmospheric phenomena," and *logos* or "description"). Today the field of meteorology is considered a branch of applied physics. Forecasting meteorology involves the acquisition, rapid transmittal, and computer analysis of large volumes of surface and upper air data. **Climatology**, the science that deals with the subject of climate, is concerned with the causes and global distributions of climatic characteristics.

The atmospheric phenomena studied in weather and climate can be divided into six discrete segments or "elements": temperature, air pressure, wind, humidity, clouds, and precipitation. Each element varies almost continuously over time and space, but when analyzed in terms of its permanent or average condition, each displays an organized and logical geographical pattern. The next several chapters examine each weather element in turn to provide an understanding of what it is, how and why it operates, and how it is distributed around the world.

▌SUMMARY

This chapter has examined the physical and chemical properties of Earth's atmosphere and has briefly discussed its influence on the surface environment. The atmosphere originated from gases released from our planet's interior by volcanic eruptions over the course of several billion years. The mixture of atmospheric gases has been altered gradually by physical, chemical, and biological activities. Earth's atmosphere is unique in its properties among the atmospheres of the planets of our solar system and is the only one capable of supporting life as we know it.

To organize our summary of the significance and characteristics of the atmosphere, we return to the chapter-opening Focus Questions:

1. *What basic influences does the atmosphere have on the Earth's surface environment?*

The atmosphere performs several vital functions that profoundly affect the nature of the Earth's surface and enable plant and animal life to exist. First, it provides the oxygen and carbon dioxide needed for animal and plant

Air Pollution: A Global Environmental Problem

Humans have faced the problem of air pollution for thousands of years. Within the past century, though, air pollution has increased in magnitude from a local source of annoyance into an important global environmental concern. While air pollution is produced by both natural and human sources, the focus here is on the human sources.

Two general categories of human activities are major producers of polluted air: urban-related activities and agricultural activities. Urban air pollution is associated primarily with motor vehicles, factory emissions, and home heating. The leading agricultural air pollutant is windblown soil resulting from the removal of protective vegetation, plowing, and other soil-loosening activities. Airborne pesticides and the burning of grass to clear land for crops also create problems in some agricultural areas.

URBAN AIR POLLUTION

A combination of gaseous, liquid, and solid pollutants lowers the quality of life in many large urban areas today. These substances produce unsightly deposits, reduce visibilities, irritate eyes and respiratory systems, and under some conditions can cause serious illness and even loss of life. As indicated in Table 3.2, most urban air pollution results from the burning of fossil fuels. Transportation sources, especially automobiles, are the most important single producers of urban air pollution. Vehicle exhausts contribute most of the carbon monoxide, half the hydrocarbons, and about a third of the nitrogen oxide gases. An additional 30 percent of urban air pollution results from the burning of fuel at stationary sources, notably power plants and industries. Industrial processing operations also release a variety of gases, liquids, and solids.

Whether these pollutants readily disperse into the atmosphere or accumulate to potentially harmful levels is determined largely by meteorological conditions. The conditions most commonly associated with severe urban air pollution are a lack of precipitation, light winds, and a near-surface temperature inversion. A **temperature inversion** occurs when a shallow surface layer of dense, cool air is trapped beneath warmer air, as illustrated in Figure 3.7. It is termed an inversion because it represents a localized reversal of the normal tropospheric characteristic of declining temperatures with increasing altitude. Short-lived temperature inversions are common occurrences. They frequently form in the late night and early morning as a result of the nightly radiation of surface heat into space, only to dissipate by midday because of solar heating. Occasionally, though, foggy weather within a stagnant air mass helps maintain a temperature inversion for several days. This enables pollutants to accumulate to potentially dangerous levels in the trapped layer of cool air at the surface.

Conditions favoring air pollution occur most frequently in late summer and autumn in much of North America and western Europe. A prime site for severe air pollution in the United States is the Los Angeles Basin (Figure 3.8). Its eastern rim of mountains helps trap the pollutants produced by this major urban area,

continued on next page

◆ **TABLE 3.2** **Major Urban Air Pollutants and Pollution Sources in the United States, 1990**

AIR POLLUTANTS	PERCENTAGE BY WEIGHT
Carbon Monoxide	47
Sulfur Oxides	17
Nitrogen Oxides	16
Hydrocarbons	15
Particulates	5
POLLUTION SOURCE CATEGORY	**PERCENTAGE BY WEIGHT**
Transportation	43
Stationary Fuel Combustion	30
Industrial Processes	15
Solid Waste	2
Miscellaneous	10

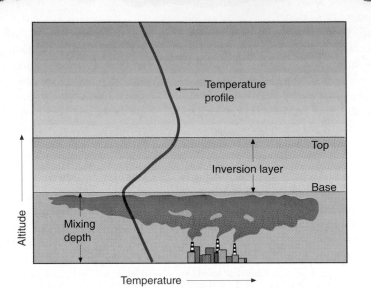

◆ **FIGURE 3.7** A low-level temperature inversion exists when a shallow layer of cool, relatively dense air lies just above the ground, with warmer air aloft. In urbanized or industrialized areas, they can trap pollutants near the surface, as shown here.

while the cold waters of the offshore California Current generate low-level temperature inversions and frequent fog. As a consequence, Los Angeles experiences more days with moderate to severe air pollution than any other large city in the nation. Fortunately, a combination of increased concern and technological advances has brought about an improvement in the urban air quality of most developed countries during the past decade. Both the United States and Canada have recorded marked improvements recently. In the United States, the emission of particulates has declined by approximately two-thirds since 1970, while sulfur dioxide emissions have declined by 30 percent. Factors contributing to these improvements include the desulfurization of coal and oil, the increased use of solar and nuclear power, the trend toward smaller cars with better-designed engines, the mandatory use of vehicle emissions control systems, and a substantial increase in the quantity and quality of industrial gaseous effluent treatment equipment. Antipollution legislation, enacted to meet the growing public concern about urban air pollution, has been responsible for many of these changes (Figure 3.9).

Improved air pollution control technology, however, has been largely offset by the rapid population growth and industrial development of many developing countries, particularly those in Africa, Latin America, and Southeast

◆ **FIGURE 3.8** Smog in the Los Angeles Basin sometimes gives the sky and setting Sun an orange cast.
SOURCE: R. Krubner/H. Armstrong Roberts

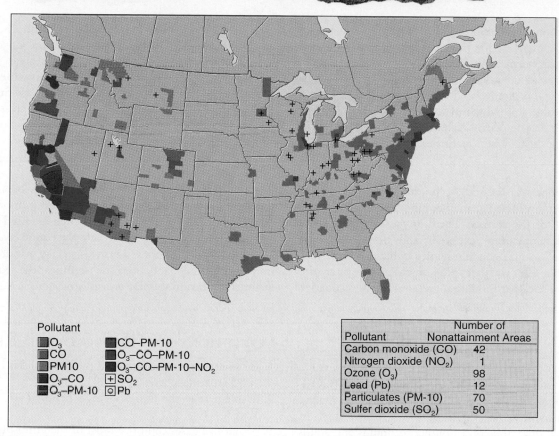

Pollutant

- ▮ O_3
- ▮ CO
- ▮ PM10
- ▮ O_3–CO
- ▮ O_3–PM-10
- ▮ CO–PM-10
- ▮ O_3–CO–PM-10
- ▮ O_3–CO–PM-10–NO_2
- ⊞ SO_2
- ⊡ Pb

Pollutant	Number of Nonattainment Areas
Carbon monoxide (CO)	42
Nitrogen dioxide (NO_2)	1
Ozone (O_3)	98
Lead (Pb)	12
Particulates (PM-10)	70
Sulfer dioxide (SO_2)	50

◆ **FIGURE 3.9** Shaded areas are counties that, as of October, 1991, failed to meet minimal clean air quality standards as specified by the Clean Air Act of 1990. Colors indicate the major air pollutants. These counties collectively contain about one third of the nation's population.

SOURCE: Data courtesy of United States Environmental Protection Agency

Asia, which are for the first time producing significant quantities of air pollution. Large-scale industrialization of the countries of Eastern Europe since World War II also has been carried out with little regard to air pollution controls. The geographic consequence of these events is that urban air pollution is increasingly a global phenomenon and is no longer primarily associated with established industrial areas such as North America, Europe, and Japan.

AGRICULTURAL AIR POLLUTION

While the severity of urban air pollution for the world as a whole probably is remaining nearly steady or even declining slightly, agricultural air pollu-

tion continues to increase. Although smoke from fires used to clear land for agriculture causes serious air pollution in many areas, the chief air pollutant of agricultural origin is windblown dust from unprotected soil. Dust storms occur in many parts of the world, especially where semiarid conditions prevail. These areas, despite their dryness, contain some of the world's most fertile soils, and extensive areas have been cleared for agriculture.

As in the case of urban air pollution, the erosion of soil by the wind is gradually being reduced in more technologically advanced countries such as Canada and the United States. During the Dust Bowl era of the 1930s, tremendous dust storms carried off millions of tons of fertile soil from the

American Great Plains (see Figure 18.1), reducing visibilities in the cities of the East Coast and creating spectacular sunsets in Europe. Today improved agricultural techniques and changes in land use patterns have reduced the frequency and severity of dust storms in the Great Plains, although soil erosion continues to be a serious problem (see the Case Study at the end of Chapter 12).

In contrast, high population growth rates in many of the developing countries of the subtropics, especially in Africa and Asia, have necessitated a rapid expansion in agricultural land use. Much of this land is marginal in quality, and the removal of protective vegetation, plowing operations, and

continued on next page

frequent droughts combine to augment the potential for major dust storms. Dust storms not only temporarily add great quantities of dust to the air, they also remove vast amounts of irreplaceable soil. Much of this soil is deposited in the oceans, where it is irretrievably lost. The quantities of dust involved are so large that some scientists theorize they may have reduced the influx of solar energy sufficiently to account for the estimated worldwide drop in temperature of about 0.3C° (0.5F°) that occurred over the northern hemisphere between 1940 and the mid-1980s. Vegetation removal and soil erosion also are leading to the formation of desert-like conditions over large areas. This process, called **desertification,** is discussed further in the Case Study at the end of Chapter 18. It represents one of the more serious long-range environmental problems faced by humanity.

respiratory processes. Second, it produces frictional resistance to incoming meteors, incinerating all but the largest before they reach the surface. Third, it protects the Earth's surface from temperature extremes and from potentially lethal doses of cosmic and ultraviolet radiation. Finally, it makes possible the weather phenomena that water the Earth's surface and that produce a great variety of distinctive landforms and surface materials.

2. *What is the composition of the atmosphere?*

The atmosphere is composed of a physical mixture of gases. The most abundant are nitrogen (N_2), oxygen (O_2), argon (Ar), water vapor (H_2O), and carbon dioxide (CO_2). The proportions of the first three of these gases remain relatively constant, while those of the last two vary. In addition, small quantities of a number of liquids and solids exist within the atmosphere.

3. *How do the characteristics of the atmosphere change with altitude?*

The most important vertical changes in the atmosphere are in density and temperature. The atmosphere rapidly becomes less dense with altitude, and half of the atmospheric gases are concentrated within 5.6 kilometers (3.5 mi) of sea level. Four distinct thermal layers exist, each with vertical temperature characteristics reversed from adjacent layers. From the Earth's surface to the top of the atmosphere, these layers are the troposphere, the stratosphere, the mesosphere, and the thermosphere. The composition of the atmosphere also varies with altitude. Water vapor is concentrated in the troposphere, ozone (O_3) is most abundant in the stratosphere, and hydrogen and helium are most abundant in the thermosphere.

4. *How are human activities influencing the composition of the atmosphere?*

Human activities add a variety of substances to the air, especially in urban areas. The three substances of greatest environmental concern are carbon dioxide, chlorofluorocarbons, and dust. Carbon dioxide, which is released in large quantities by the burning of fossil fuels, may be altering the Earth's climate. (This situation is explored in more detail in Chapter 5.) Chlorofluorocarbons, which are used as propellants and refrigerants, act to deplete the ozone layer, particularly in the polar regions. Dust, especially windblown dust from agricultural land, may be significantly reducing the quantity of incoming solar energy.

▌QUESTIONS FOR THOUGHT AND REVIEW

1. In what major respects would the Earth differ if it did not have an atmosphere?

2. Explain how the atmosphere is believed to have originated.

3. What five gases comprise most of the atmosphere? What is the direct source of atmospheric oxygen?

4. What is ozone, and how is it produced? What essential function does it serve with respect to life on Earth?

5. What are the major natural and human sources of atmospheric dust? What vital function do atmospheric dust particles perform?

6. List the four thermal layers of the atmosphere, and describe the altitude and temperature characteristics of each. Why is the troposphere the only "livable" layer of the atmosphere?

7. What causes temperatures to vary with altitude in the atmosphere? What are the two warmest portions of the atmosphere, and what are their respective sources of heat?

▌KEY TERMS

nonvariable gases	**mesosphere**
variable gases	**thermosphere**
ozone	**weather**
condensation nuclei	**climate**
troposphere	**meteorology**
tropopause	**climatology**
stratosphere	

4 ENERGY FLOW AND AIR TEMPERATURE

OUTLINE

FOCUS QUESTIONS

1. How does solar energy get to the Earth, and what happens to it once it reaches the Earth?
2. What are the major factors controlling air temperatures at the Earth's surface?
3. What is the geographic pattern of temperatures that results from the action of these controls?

In late January 1994, a record-breaking Arctic air mass sent temperatures tumbling throughout the central and eastern United States. Among the dozens of cities with all-time lows were Pittsburgh, which recorded –30°C (–22°F); Akron, with –32°C (–25°F); and Indianapolis, with –33°C (–27°F). Throughout the Midwest and East, schools and businesses shut down, while people stayed in their homes and shivered.

Temperatures at the Earth's surface, however, do much more than influence human comfort and activities. The global temperature pattern largely controls the world's distribution of life-forms—both plants and animals. In addition, such vital Earth surface processes as storm formation, soil production, and the sculpting of landforms by erosional forces are controlled chiefly by the global pattern of thermal energy derived from the Sun.

ENERGY AND TEMPERATURE

The constant cycles of change that occur in the Earth system,[1] including its various subsystems—the atmosphere, hydrosphere, lithosphere, and biosphere—are driven by energy derived from the Sun and from the Earth's interior. By definition, **energy** is the capacity to do work on some form of matter.

When we measure the **temperature** of any object or substance, we are actually making a quantitative measure of its thermal energy. **Thermal energy** is a form of energy in which the temperature of the substance is proportional to the rate of motion (average speed) of its molecules. As energy is absorbed (gained from another energy source and converted to molecular kinetic energy), the temperature rises. If enough heating takes place, the substance involved may undergo phase changes from a solid, to a liquid, and finally to a gaseous (vapor) form. Because no limiting maximum temperature exists, the temperature attained by any substance is restricted only by the limits of its ability to continue to absorb more energy than it loses. The Earth's air temperature therefore is a measure of the rate of motion of the air molecules that results from the absorption of energy from all sources.

Of the several energy sources that heat the atmosphere, the Sun is overwhelmingly dominant. It is estimated to supply at least 99.97 percent of the atmosphere's total heat. A second source, heat from the interior of the Earth, provides nearly all the remainder.[2] The Sun's energy is produced by atomic fusion: within the Sun's core, atoms of hydrogen are fused into helium under conditions of tremen-

◆ **FIGURE 4.1** During hydrogen fusion, four hydrogen atoms fuse into one helium atom. The attendant release of energy powers the Sun and most other stars.

dous heat and pressure (Figure 4.1). This process produces helium at a rate of approximately 3.6×10^{11} kilograms (or 400 million tons) per second. The mass of the helium produced is slightly less than that of the original hydrogen, and this lost mass is converted to the energy that is radiated in all directions from the Sun. The Earth is such a small target and is so far away from the Sun that it intercepts only about .0000000005 (or one/two billionth) of the Sun's total energy output. This small proportion of energy, however, is the dominant power source for the Earth system.

ENERGY TRANSFER PROCESSES

How does solar energy cross the great void separating the Earth from the Sun? Once it has reached the Earth system, how is this energy distributed to all parts of our planet? Three energy transport processes—radiation, conduction, and convection—collectively perform these tasks.

Radiation is the process of energy transmission by electromagnetic waves. These waves, so-named because they display both electric and magnetic properties, have the ability to pass through a vacuum such as the 1.5×10^8 kilometers (93 million miles) of space separating the Earth from the Sun. Radiation travels through space at the speed of light (light itself being a form of electromagnetic radiation), or 300,000 kilometers per second (186,000 mi/sec). Radiation is divided into several major categories on the basis of wavelength to form an **electromagnetic spectrum** (Figure 4.2). As explained in the Focus Box beginning on page 58, the Sun produces shortwave energy that is absorbed after reaching the Earth system, converting it to the heat energy that warms the Earth's surface and atmosphere.

Once solar radiation reaches the Earth system, the processes of conduction and convection play major roles in its subsequent distribution. **Conduction** involves heat transfer by physical contact between objects or substances of different temperatures. Heat always flows from the warmer body to the colder one and, if contact is maintained, will continue to do so until their temperatures have equalized. Conduction

---

1. The Earth and its atmosphere, taken together, will be referred to as the **Earth system.**

---

2. Heat from the Earth's interior, termed **geothermal heat,** sometimes manifests itself spectacularly in the form of earthquakes, geysers, and volcanic eruptions. This energy source, although unimportant in providing atmospheric heat, plays a crucial role in the development of the Earth's surface features and will be examined in detail later in this book.

Wavelength

- 10^5 m
- 10^4 m
- 10^3 m
- 10^2 m
- 10 m
- 1 m
- 0.1 m — 100 mm
- 0.01 — 10 mm
- 1 mm
- 0.1 mm — 100 μm
- 0.01 mm — 10 μm
- 1 μm — 10,000 Å
- 0.1 μm — 1,000 Å
- 100 Å
- 10 Å
- 1 Å

Type of radiation

Long radio waves

Standard AM radio broadcasts

Short radio waves

FM radio broadcasts
TV broadcasts

Microwaves

Infrared radiation

Visible light — 0.7 μm
— 0.4 μm
Ultraviolet radiation

X rays

Gamma rays

0.710 μm — Red light
0.647 μm — Orange light
0.585 μm — Yellow light
0.575 μm
0.491 μm — Green light
0.424 μm — Blue light
0.400 μm — Violet light

◆ **FIGURE 4.2** The electromagnetic spectrum. Electromagnetic radiation is divided into categories based on wavelength, or distance between successive electromagnetic waves. Those wavelengths range from distances of hundreds of kilometers down to subatomic dimensions.

is the basis of our ability to sense heat and cold. For example, when you hold an ice cube, heat flows from your hand to the much colder ice. The loss of heat makes your hand cold, while the ice may gain enough energy from your hand to melt.

Most conduction within the Earth system occurs between the Earth's surface (both land and water) and the atmosphere. Because air, water, and land are all rather poor conductors, this energy transfer process is restricted largely to the lowest few meters of the atmosphere.

Convection is the process of energy transfer through the physical movement of a fluid (a liquid or gas) from one place to another. At the Earth's surface, winds and ocean currents are the major convectors. (If the motions of these two fluids are predominantly horizontal, the energy transfer process is termed **advection**.) A given mass of air has a lower capacity for transporting heat than does the same mass of water, but wind typically travels at a much greater speed than do ocean currents. In addition, wind currents exist over both land and water surfaces. As a result, approximately twice as much energy is convected by the atmosphere as by the ocean waters.

On a global basis, convection produces a poleward net transport of heat (see Figure 4.5 on page 60). If it were not

for this exchange of energy between the lower and higher latitudes, the tropics would be significantly hotter, and the polar regions substantially colder, than they already are. Such a situation would greatly reduce the habitability of our planet. As we shall see, the latitudinal imbalance in solar energy reaching the Earth system is also largely responsible for driving our planet's atmospheric and hydrologic (water) subsystems.

The Global Energy Balance

Solar shortwave radiation passes through the 1.5×10^8 kilometers (93 million miles) of space between the Sun and the Earth system without being appreciably altered in any way. During its comparatively short passage through the atmosphere to the Earth's surface, however, solar radiation is subjected to a number of redirections and transformations of great importance to life on Earth.

Solar energy can be visualized as reaching the Earth system in a steady stream of shortwave radiation. As this **insolation** (a shortening of the term *incoming solar radiation*) enters the atmosphere, it is effectively divided into three segments, as illustrated in Figure 4.6 on page 60. Approximately 30 percent of the energy is redirected back to

CHARACTERISTICS OF ELECTROMAGNETIC RADIATION

Although electromagnetic radiation constantly surrounds us and all objects or substances absorb and emit radiation, we still have a very imperfect understanding of this phenomenon. The radiation spectrum is commonly divided into categories based on wavelength, or distance between consecutive waves, but all forms of electromagnetic radiation are essentially similar in most respects.

The quantity and wavelengths of radiation emitted by any object depend on the object's temperature and **emissivity,** which is a measure of its efficiency as an emitter of radiation. A perfect emitter is termed a "black body" and has an emissivity of 100 percent. Good emitters of radiation are good absorbers of the same radiation wavelength, and vice versa. The Sun, the Earth (both land and water), and even clouds approach black body charac-

teristics with respect to emissivity, but the variation in the emissivity of the atmosphere is quite complex, depending on the wavelength of the radiation reaching it (Figure 4.3). At some wavelengths, especially between 0.3 and 1.0 micrometers, the air's emissivity is near zero. Consequently, virtually no radiation of these wavelengths passing through the atmosphere is absorbed. Most of the Sun's radiation falls within this wavelength range; it there-

◆ **FIGURE 4.3** The solar radiation spectrum. Values indicate the proportion of solar radiation within each segment of the electromagnetic spectrum. Most energy is visible light and near-infrared radiation.

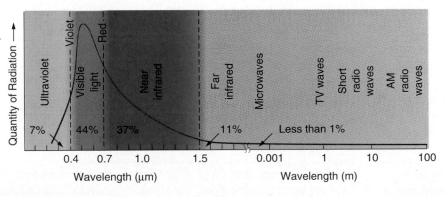

space. While most of this redirected energy is reflected from the upper surfaces of clouds, some is scattered by air molecules and dust particles, and some passes through the atmosphere to be reflected from the Earth's surface.

Energy that is scattered is broken up and reflected in all directions by air molecules and dust particles. Without the diffuse light produced by scattering, objects not exposed to a direct light source would be totally black, and therefore invisible. Because the wavelengths that produce the color blue are scattered more effectively than are those of the longer visual wavelengths, a clear sky in the presence of sunlight takes on a bright blue color. The sky as viewed from the Moon, which has no atmosphere, is black.

The ability of different materials to reflect or scatter solar energy varies greatly; this ability is indicated by their visual

brightness in sunlight. For example, the almost blinding brightness of ice and snow or of a cloud deck when viewed from above is produced by reflected sunlight and shows that these objects are excellent reflectors. In contrast, the dark green color of a dense forest or the black color of an asphalt road indicates that they are poor reflectors, and therefore good absorbers, of insolation. The reflection rate, or **albedo,** of the Earth system as a whole is 30 percent, but local albedos display considerable variation, both temporally and spatially (Table 4.1). In general, albedos are higher in the high latitudes than in the lower latitudes because of the presence of ice and snow and the oblique passage of sunlight through the atmosphere in the high latitudes. For the same reasons, albedos are higher, on average, in winter than in summer outside the tropics.

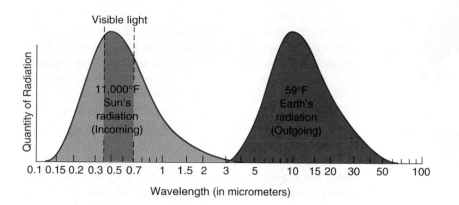

◆ **FIGURE 4.4** Comparison of solar and terrestrial radiation spectra. Although the total quantities of incoming and outgoing energy are equivalent, the peak radiation wavelength for outgoing terrestrial radiation is about 20 times longer than that of incoming solar radiation (insolation).

fore readily penetrates the Earth's atmosphere to reach and warm the surface. At other wavelengths, such as those shorter than 0.3 micrometers, the atmosphere's absorption rate is nearly 100 percent, so that this energy is effectively blocked from reaching the surface.

The hotter an object is, the greater the quantity of radiation it emits per unit area, and the shorter the wavelength of the emitted energy. The Sun, with a surface temperature of approximately 6,000°C (11,000°F), radiates

maximum energy in the vicinity of 0.5 micrometers, which is in the visible light portion of the electromagnetic spectrum (Figure 4.3). Because the Sun is so much hotter than the Earth, a given area on the surface of the Sun emits about 160,000 times the radiation of an area of similar size on the Earth's surface. The maximum radiation for the Earth occurs at a wavelength of about 10 micrometers, well within the portion of the infrared range that we sense as heat (Figure 4.4). Because wavelengths in the zone

of maximum terrestrial radiation are roughly 20 times longer than wavelengths in the zone of maximum solar radiation, terrestrial radiation is generally referred to as **long-wave radiation** and solar radiation as **short-wave radiation.** ■

◆ **TABLE 4.1** Typical Albedos of Terrestrial Materials

MATERIAL	ALBEDO (%)
Earth System	30
Sand	25
Dense Forest	8
Field Crops	25
Water (Sun Near Horizon)	40–75
Water (Overhead Sun)	6
Fresh Snow	85–95
Concrete	20–30
Dense Low Cloud Cover	90

The 30 percent of insolation that is reflected or scattered upward is not used in terrestrial energy processes because it is redirected back into space. The remaining 70 percent is absorbed to heat the Earth's surface and atmosphere. The actual absorption of this energy occurs in two places—the atmosphere and the Earth's surface. Approximately 19 percent of the total insolation does not reach the surface, but is absorbed in the atmosphere. The remaining 51 percent passes through the atmosphere to be absorbed at the surface (Figure 4.6).

Although energy is neither gained nor lost during the absorption process, a major shift in the wavelengths of the radiation takes place. Most insolation arrives as visible light and as near-infrared radiation, with an average wavelength of about 0.5 micrometers (μm) (see the Focus Box on

◆ **FIGURE 4.9** Mean January air temperatures at sea level in degrees Fahrenheit. Celsius equivalents are in parentheses.

offset radiation losses. Cooling is especially slow when the air is moist or a cloud cover is present to radiate energy back to the surface more efficiently.

The global radiation balance, illustrated diagrammatically in Figure 4.8, can be written as the following equation:

$$Q = K\downarrow - K\uparrow + L\downarrow - L\uparrow$$

In this equation, Q represents the net quantity of radiative energy at the Earth's surface. It is the amount of solar energy that is available to do work, including heating the air, melting snow, evaporating water, and powering plant photosynthesis. $K\downarrow$ is incoming solar radiation (insolation); $K\uparrow$ is outgoing (reflected and scattered) short-wave solar radiation; $L\downarrow$ is incoming (reradiated) long-wave terrestrial radiation; and $L\uparrow$ is outgoing long-wave terrestrial radiation.

WORLD TEMPERATURE PATTERN AND CONTROLS

Geographers are interested in terrestrial energy sources and energy transfer processes primarily because they help us to better understand the global temperature pattern. This pattern, in turn, strongly influences vegetative patterns, soil types, the distribution of human and animal populations, and many other factors. In this section we examine the influence of the physical factors primarily responsible for the Earth's surface air temperature pattern. Before proceeding, a few basic terms used in geographic temperature analysis must be introduced.

Temperature patterns usually are mapped by the use of **isotherms,** lines connecting places with the same temperature. Figures 4.9 and 4.10 depict the mean temperatures for the world in January and July, the months that represent the extremes of summer heat and winter cold for most land areas.

Weather reporting stations record a variety of temperature data, but the daily high and low temperatures provide the basis for calculating several basic temperature statistics. The **daily mean temperature** is calculated by adding the high and low temperatures and dividing the total by two. The **monthly mean temperature** is the average of the daily means for the entire month. The **daily temperature range** is the number of degrees between the day's high and low temperatures.

◆ **FIGURE 4.10** Mean July air temperatures at sea level in degrees Fahrenheit. Celsius equivalents are in parentheses.

Last, and probably most important in terms of world temperature analysis, is the **annual temperature range,** defined as the number of degrees between the warmest and coldest monthly mean temperatures. For example, the warmest month in Baltimore, Maryland, is July, with a long-term mean of 25°C (77°F). Baltimore's coldest month is January, with a monthly mean of 1°C (33°F). Its annual temperature range is thus 24 Celsius degrees (44 Fahrenheit degrees). This figure gives a good indication of the "season-ality" of an area's climate, with higher values indicating a climate with a more pronounced summer/winter temperature variation. The global distribution of annual temperature ranges is displayed in Figure 4.12.

Primary Temperature Controls

Although a number of factors influence the Earth's tempera-ture pattern, five stand out as the primary controls: the altitude of the Sun in the sky, the duration of daylight, the distribution of land and water, the elevation, and the pattern of ocean currents. After examining each of these primary controls, we will briefly discuss a number of secondary temperature influences.

Altitude of the Sun and Duration of Daylight

We have already noted that nearly all atmospheric heating is derived directly or indirectly from the Sun. When the Sun is high in the sky, the radiant energy received per unit area of surface is maximized (see Figure 2.12). When the Sun's altitude is low, radiation received per unit area of surface is greatly reduced, and more of the energy that does arrive is likely to be reflected and scattered by the atmosphere before reaching the surface. For these reasons, *the altitude of the Sun is the most important factor controlling air temperatures at the Earth's surface.* This control affects air temperature patterns over three differing time spans—long term, annual, and daily.

The long-term pattern, the most basic and important, can be seen on both the January and July isotherm maps (Figures 4.9 and 4.10). The most fundamental long-term character-istic of the global temperature distribution is that tempera-tures are warmest in the tropics and decrease poleward. Put another way, as a general rule, temperatures decline with increasing latitude because of decreasing mean solar alti-tudes. Because isotherms are lines of equal temperature, they display an overall zonal, or east-west trend, paralleling lines of latitude.

TEMPERATURE SCALES

Two temperature scales are in common use in the United States. Both have values based on the boiling and freezing points of water under standard atmospheric pressure conditions. The **Fahrenheit** scale, developed in 1714 by the German physicist Gabriel Daniel Fahrenheit, has been losing popularity internationally but is still the most used scale in the United States. This scale has values of 32°F and 212°F, respectively, for the freezing and boiling points of water. The metric system temperature scale is the **Celsius** (or centigrade) scale, named in honor of Anders Celsius, a Swedish astronomer who devised it in 1742. The freezing and boiling points for water on the Celsius scale are 0°C and 100°C, respectively.

Where both temperature scales are commonly employed, as in the United States, it is often necessary to convert from one to the other. The formulas used are as follows:

$$°C = \frac{°F - 32}{1.8}$$

$$°F = 1.8°C + 32$$

A third temperature scale, used primarily in the laboratory sciences, is the **Kelvin** (or absolute) scale. Like the Celsius scale, it is calibrated to have 100 equal temperature increments between the freezing and boiling points of water. Numerical values, however, begin at absolute zero, the temperature at which all molecular vibration ceases. The freezing point of water is 273 K, and the boiling point is 373 K. Any

Kelvin temperature has a numerical value 273 degrees higher than its Celsius equivalent. Stated as a formula, this relationship is as follows:

$$K = °C + 273$$

The Fahrenheit, Celsius, and Kelvin scales are illustrated in Figure 4.11. ■

◆ **FIGURE 4.11** Comparison of the Fahrenheit, Celsius, and Kelvin temperature scales.

◆ **FIGURE 4.12** World annual temperature ranges in Fahrenheit degrees, with Celsius equivalents in parentheses. Values indicate the differences between the warmest and coldest monthly mean temperatures. The largest ranges occur over continental interiors in the high latitudes.

Although the Sun's average altitude at solar noon becomes progressively lower with increasing latitude, it is subject to seasonal variations in all parts of the world because of the changing inclination of the Earth's axis with respect to the Sun. For all locations from the Tropics of Cancer and Capricorn poleward, the Sun gets progressively higher in the noon sky during the six-month period from the winter solstice to the summer solstice. It then decreases in altitude during the following six months. Locations within the tropics have two periods of increasing and decreasing solar noon altitudes during the course of a year, as indicated by the analemma on page 20. At all locations except the equator itself, however, the Sun's average noon altitude is higher in the spring and summer than in the fall and winter. In addition, all places on Earth (again excepting the equator) have longer days than nights in spring and summer and longer nights than days in fall and winter. These two related factors are responsible for the seasonal variations in global temperature distributions that can be observed between the January and July isotherm maps. For example, the combina-

tion of vertical Sun angles, limited cloud cover, dry surface conditions, and 13- to 14-hour periods of sunlight contribute to the fact that land areas in the subtropics, rather than those in the equatorial regions, experience the hottest summer temperatures in the world.

The greater intensity and duration of insolation during the spring and summer typically cause these two seasons to be warmest. This tendency becomes progressively stronger poleward, because seasonal differences in the receipt of insolation increase with latitude. In the high latitudes, poleward of the Arctic and Antarctic Circles, seasonal differences in the daily duration of sunlight reach extremes, with 24 hours of sunlight at the time of the summer solstice and zero hours at the time of the winter solstice. Annual temperature ranges therefore are very large.

Because the Earth's surface heats and cools rather slowly, seasonal temperature lags exist. The hottest and coldest times of the year actually occur a month or more after the dates of the summer and winter solstices. For example, July and August are the two warmest months in most of North

America, even though the greatest insolation receipts occur during the summer solstice month of June. Likewise, January and February typically are colder than the winter solstice month of December. As a result of the seasonal temperature lag, summer is warmer than spring, and winter is colder than fall in most places.

The familiar day/night cycle of air temperature is also caused by differing Sun altitudes due to the rotation of the Earth on its axis. The resulting continuous change in the Sun's altitude rarely permits temperatures to remain steady for very long, especially when skies are clear. Although the atmosphere restricts the magnitude of daily temperature ranges, variations of from 3C° to 22C° (5F° to 40F°) are common.

Daily heating and cooling lags also occur. Although the altitude of the Sun begins to decrease after solar noon, the Sun normally remains high enough in the sky for total insolation amounts to exceed outgoing terrestrial radiation totals until mid-afternoon. This causes maximum temperatures to typically occur between 2 P.M. and 4 P.M. By the same token, minimum temperatures more commonly occur near sunrise than near midnight (see Figure 6.7).

Land and Water Distribution

The complex distribution of land and water over the Earth's surface acts as a highly important world temperature control. While the *mean* temperatures of different parts of the world are most affected by the altitude of the Sun, temperature *ranges*, both on a daily and an annual basis, are more greatly influenced by the distribution of land and water.

This factor's importance stems from the ability of land to change temperature far more readily than water. The following factors are responsible for the differences in the thermal behavior of land and water:

1. Water has a specific heat five times greater than that of dry land. The **specific heat** of a substance is the quantity of heat energy that must be absorbed or released in order to produce a given temperature change. Thus, a given mass of water must absorb or release five times as much heat in order to undergo the same temperature change as a similar mass of dry land.
2. Water is transparent while land is opaque. Insolation striking a water surface can therefore penetrate to a considerable depth before being completely absorbed; consequently, solar energy is distributed through a large volume of water. This is particularly true when the Sun is high in the sky. On land, insolation is converted to heat at the immediate surface; as a result, the surface is heated much more intensely.
3. Unlike land, water is horizontally and vertically mobile. Heat absorbed near the water surface is distributed in all directions by waves and currents.
4. Water loses large amounts of energy as latent heat through the process of evaporation, because considerable

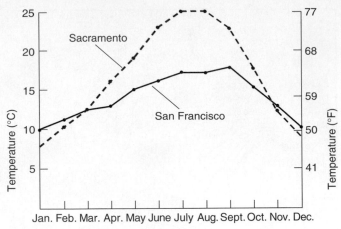

◆ **FIGURE 4.13** The differences between the annual temperature graphs of the coastal city of San Francisco and of Sacramento, located about 145 kilometers (90 mi) inland, illustrate the influence of continentality.

energy is required to convert water from the liquid to the vapor state. Water, of course, will also evaporate from a damp land surface, but a dry surface cannot lose heat to evaporation, so more energy is converted to sensible heat instead. This factor is largely responsible for the fact that the world's highest temperatures occur in arid desert regions.

The net effect of these factors is that land areas tend to have larger daily and annual temperature ranges and shorter seasonal and daily heating and cooling lags than do water bodies at similar latitudes. In fact, the influence of the surrounding land or water surface is so marked that climates often are described as being either continental or maritime (Figures 4.13 and 4.14).

◆ **FIGURE 4.14** Eureka, California, and Lincoln, Nebraska, are at the same latitude and have identical annual mean temperatures. The difference in their continentality, however, causes their annual temperature ranges to differ greatly.

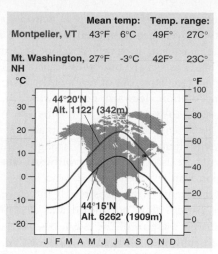

	Mean temp:		Temp. range:	
Montpelier, VT	43°F	6°C	49F°	27C°
Mt. Washington, NH	27°F	-3°C	42F°	23C°

◆ **FIGURE 4.15** Temperature graphs for Montpelier, Vermont, and Mt. Washington, New Hampshire, illustrate the influence of elevation on air temperature.

Continental climate characteristics include large daily and annual temperature ranges, one-month seasonal heating and cooling lags, and a tendency toward relatively low amounts of precipitation, fog, cloudiness, and humidity. Continental climates are best developed in the interiors of large continents. The interiors of both Canada and the United States, for example, have highly continental climates. In addition, the east coasts of both countries experience moderately continental climates despite their proximity to the Atlantic. This is because the prevailing winds blow predominantly from west to east, transporting continental air masses from the North American interior to the coast.

Maritime climate characteristics, conversely, include relatively small daily and annual temperature ranges, two-month seasonal heating and cooling lags, and a greater quantity of atmospheric moisture. Maritime climates are best developed over the oceans, but also occur in many coastal areas that have a predominantly onshore wind flow, such as the west coast of North America. They even develop to some extent along the shores of large lakes such as the Great Lakes. In reality, describing the climate of any location as just continental or just maritime is an oversimplification because all degrees of continentality exist between, for example, the supercontinental conditions of the central Sahara and the ultramaritime conditions of Easter Island in the South Pacific.

The influence of the Earth's land and water distribution on temperature ranges is clearly apparent from the world map of annual temperature ranges (Figure 4.12). A striking variation can be seen between the temperature ranges of large land and water masses at similar latitudes. Land areas have much larger ranges than do adjacent water bodies, especially in the higher latitudes. The world's largest annual temperature ranges occur on large continents in the high latitudes, where the influences of large seasonal variations of insolation and continentality are combined. In northeastern

Asia, the monthly means of January and July differ by as much as 64C° (115F°). Ranges as large as 50C° (90F°) occur in northwestern Canada and northeastern Alaska on the somewhat less massive continent of North America.

Elevation

Another important influence on the global temperature pattern is elevation. This factor primarily affects temperature means (Figure 4.15). For locations above approximately 2,000 meters (6,560 ft), elevation becomes a dominant temperature influence. Peaks above 5,500 meters (18,000 ft) are perpetually covered with snow and ice even at the equator. Only a small proportion of the world's surface area, however, is situated at such high elevations. The importance of this control is therefore reduced by the limited areal extent of the Earth's highland surfaces.

Elevation is capable of producing substantial differences in both temperature and precipitation means over smaller distances than is any other major climatic control. Climatic characteristics in mountainous areas therefore vary greatly; a hike or a short drive often brings travelers to places with climatic conditions very different from those they left. In lowland areas, in contrast, it often is necessary to travel hundreds of kilometers to reach locations with substantially different climates. Travel to both higher elevations and higher latitudes results in colder temperatures. Depending on the location and season of the year, however, temperatures decrease some 500 to 1,000 times faster with elevation than with latitude, so an increase in elevation of 300 meters (1,000 ft) may be equivalent, in thermal terms, to traveling 160 to 320 kilometers (100 to 200 mi) poleward.

The decline in temperature with altitude in the troposphere is a function of distance from a direct source of heat. High elevation locations are simply farther away from sea level, the level at which most insolation is converted to heat. In addition, the air is less dense at high elevations and has a reduced ability to absorb or conduct heat energy from either above or below. As a consequence, the Sun's rays are very intense, but temperatures remain colder, and readings fall quickly after sunset during clear weather.

As mentioned in Chapter 3, the worldwide average rate of vertical temperature decrease is 6.4C° per 1,000 meters (3.5F°/1,000 ft) in the troposphere. This value is termed the **average environmental lapse rate.**[4] It should be stressed that this is a mean value, and that the existing lapse rate for a location at any given time will probably be somewhat different. For a few applications of the average environmental lapse rate, refer to the Geographic Techniques Box on page 69.

Ocean Currents

The surface ocean currents constitute the final major world temperature control (Figure 4.16). We have already seen that

---

4. This value also is often referred to as the **normal lapse rate.**

	Mean temp:		Temp. range:	
Santos, Brazil 23°58'S	71°F	21°C	12F°	7C°
Antofagasta, Chile 23°32'S	62°F	17°C	13F°	8C°

◆ **FIGURE 4.16** Ocean current effects on temperature. Santos, Brazil, on the east coast of South America, is warmed by the Brazil Current. Antofagasta, Chile, on the west coast, is chilled by the Humboldt (Peru) Current.

water can absorb and transport tremendous quantities of heat, and that ocean currents are responsible for about a third of the net poleward transport of thermal energy over the Earth's surface.

Ocean currents usually are classified as warm or cold. Warm currents form when water that has absorbed heat in the low latitudes is transported poleward. Cold currents, conversely, have spent an extended period in the higher latitudes before flowing equatorward.

Ocean currents can exert a strong effect on air temperatures, but their overall geographical significance is reduced by the limited areal extent of well-developed warm and cold currents. This is especially true with respect to land areas. Many areas are far inland, out of reach of the effects of ocean currents. In addition, **leeward coasts,** which have a predominantly land-to-water wind flow, often do not experience the inland penetration of maritime air masses. Those land areas with temperatures most strongly influenced by ocean currents are associated with **windward coasts,** which have a predominantly water-to-land wind flow.

The ocean currents that most strongly influence the world temperature pattern flow northward and southward along the east and west margins of the ocean basins in the middle latitudes and subtropics. In general, the east coasts of the continents are paralleled by warm ocean currents, while the west coasts are paralleled by cold currents (see Figure 9.19). As might be expected, the thermal effects of these currents are most pronounced along the immediate coasts and diminish inland. The distance to which oceanic influences penetrate inland depends primarily on the direction and strength of the prevailing winds and the presence or absence of mountain barriers.

The January and July isotherm maps (Figures 4.9 and 4.10) illustrate, in places, the thermal effects of warm and cold currents. As a general rule, the influence of warm currents is most marked in winter, when the water is warmest relative to the land. For example, the pronounced poleward bulge of isotherms over the North Atlantic on the January temperature map (Figure 4.9) is produced by the

◆ **FIGURE 4.17** The Gulf Stream is clearly shown in this satellite mosaic depicting sea surface temperatures of the western North Atlantic during June, 1984. The warmest water (about 27°C or 80°F) is indicated in red, followed by orange, yellow, green, blue, and violet for progesssively colder water.

SOURCE: O. Brown, R. Evans, and M. Carle, University of Miami, Rosentiel School of Marine and Atmospheric Science

APPLYING THE LAPSE RATE

It is easy to estimate expected temperatures at different elevations in mountainous regions if you keep in mind the fact that the air temperature can be expected to fall approximately 6.4C° for each 1,000 meters of elevation. (The equivalent value is 3.5F° for each 1,000 feet.)

It should be stressed that this value, termed the average environmental lapse rate (or normal lapse rate), is strictly an average value; the existing lapse rate varies constantly and also generally changes with altitude. In addition, because of intense surface heating, average lapse rates tend to be slightly higher in summer than in winter in the midlatitudes and higher year-round in the tropics than elsewhere.

Nevertheless, the average environmental lapse rate value gives us some idea of the temperatures to expect when changing elevations. Let's try a couple of problems applying this value:

1. If the temperature near the equator averages 27°C at sea level, what should be the approximate mean freezing level in the Andes Mountains of Ecuador? (These mountains reach elevations exceeding 6,100 meters, or 20,000 feet.)

   Solution: We know that the temperature of 27°C is expected to fall by 6.4C° for each 1,000 meters of elevation. Dividing, we find that 6.4 goes into 27 approximately 4.2 times. Therefore, the freezing level should be at an elevation of about 4,200 meters.

2. If the temperature in Denver, Colorado, at 5,200 feet elevation is 60°F, what should be the temperature atop nearby Mt. Evans at an elevation of 14,200 feet?

   Solution: We know that the average environmental lapse rate on the Fahrenheit scale is 3.5F° for each 1,000 feet of elevation. In traveling from Denver to the top of Mt. Evans, we need to ascend a total of 14,200 – 5,200 or 9,000 feet. The temperature should therefore decline by 9 × 3.5, or by 31.5F°. Subtracting 31.5F° from 60°F gives an estimated temperature of 28.5°F for the summit of Mt. Evans. ■

warm North Atlantic Drift, a continuation of the Gulf Stream, which flows along the southeastern coast of the United States (see Figure 4.17). Near Iceland, this current keeps January air temperatures more than 22C° (40F°) warmer than the world average for that latitude at that time of year. Mild air from the North Atlantic makes the climate of Europe much more comfortable than would be expected on the basis of latitude alone. The climate of London, for example, is often considered to be rather similar to that of Boston, Massachusetts. London, however, has a latitude of 52°N, 10 degrees farther north than Boston. In fact, London's latitude is the same as that of southern Labrador, where snow lies on the ground from September through May.

The effects of cold ocean currents are most apparent on summer isotherm maps. The pronounced equatorward bulge of isotherms off the California coast on the July map, and off the west coasts of South America, South Africa, and Australia on the January map, are quite evident.

Secondary Temperature Controls

In addition to the five primary controls, a number of secondary controls affect the global distribution of temperature. The influence of these factors often is not obvious on world temperature maps, but collectively, they significantly modify the global temperature pattern. The most important secondary controls are the seasonally changing distance between the Earth and the Sun, geographical variations in cloud cover, variations in wind patterns, topographic influences, and, increasingly, the effects of human activities, which are discussed in the Case Study at the end of the chapter.

Changing Distance between the Earth and Sun

The Earth's orbit around the Sun is somewhat elliptical (see Chapter 2), causing the distance between the Earth and the Sun to vary by about 4.8 million kilometers (3 million miles)

WORLD RECORDS OF HEAT AND COLD

World weather records of many kinds have been established, and this type of information is so popular that several books on the subject have been published. Among the most basic weather records are extremes of heat and cold.

The world's highest generally accepted air temperature was 58°C (136.4°F), recorded in El Azizia, Libya, in the northern Sahara, on September 13, 1922. The hottest annual mean temperature reported from a regular weather station is 34.4°C (94°F), recorded in Dallol, Ethiopia.

The hottest temperature ever recorded in North America was 57°C (134°F) in Death Valley, California, on July 10, 1913. Canada's highest recorded temperature was 45°C (113°F), registered at Midale, Saskatchewan, on July 5, 1937.

The world's coldest temperatures occur in Antarctica. At Vostok, a Russian research station located on the Antarctic Plateau at 78°S 96°E and at an elevation of 3,420 meters (11,220 ft), annual temperatures average -57.8°C (-72°F), and a world record low temperature of -89.2°C (-128.6°F) was set on July 21, 1983. The coldest tempera-

ture ever officially recorded in any permanently inhabited town was -71°C (-96°F) in Oymyakon, Russian Siberia, in 1964.

The lowest temperature recorded in North America was -63°C (-81°F) at Snag, Yukon, on February 3, 1947. The coldest reading for the United States was -62°C (-80°F), registered on January 23, 1971, at Prospect Creek, Alaska. For the lower 48 states, the coldest temperature so far recorded was -57°C (-70°F) at Rogers Pass, Montana, on January 20, 1954. ■

---

during the course of a year. When the Earth is at perihelion in January, it intercepts approximately 7 percent more insolation than it does at aphelion in July.

Because of its varying distance from the Sun, the Earth might be expected to be somewhat warmer as a whole near the time of perihelion than it is near the time of aphelion. With some time lag factor, this may be true; assessing the overall significance of this influence is difficult because it is masked by two much more powerful temperature controls—seasonal differences in Sun angle and the effect of land and water distribution.

Geographical Variations in Cloud Cover

Changing amounts of cloudiness and dust cause atmospheric transparency to vary both spatially and temporally. As a consequence, the proportion of total insolation that is absorbed often varies greatly on a short-term basis. Even when the sky is clear and the air appears transparent, insolation must penetrate many miles of atmosphere to reach the Earth's surface, and a significant proportion is scattered by air molecules and dust particles. As we have seen, about half the total insolation entering the outer fringes of the atmosphere does not reach the surface because of reflection and scattering. When the Sun is low in the sky,

its oblique rays must penetrate a much greater thickness of atmosphere before reaching the ground. The receipt of energy is therefore reduced substantially in the early morning, late afternoon, and near the time of the winter solstice. The greatest overall reduction occurs in the polar regions, where Sun angles are always low.

Clouds are even more effective than dust at reducing insolation by scattering and reflection. Satellite photographs have shown that, on the average, slightly more than half the Earth's surface is covered by clouds at any given time, and that some areas are much cloudier than others. Because clouds are excellent reflectors of insolation, it might be expected that the temperature means of these cloudy zones would be substantially lowered. A cloud cover, however, is nearly as effective at keeping terrestrial radiation ($L\uparrow$) from escaping the atmosphere as it is at keeping solar radiation ($K\downarrow$) from entering. As a consequence, the daily and annual temperature ranges, much more than the means, are lowered by cloudy conditions. For this reason, not only are daytime temperatures held down by a cloud cover, but nighttime temperatures are raised. The abundant cloudiness associated with many maritime locations strengthens the effect of the nearby water in maintaining relatively constant temperatures, while the clear skies of desert regions help produce large daily and annual temperature ranges.

◆ **TABLE 4.2** Windchill Equivalent Temperatures

WIND SPEED (M/SEC)	AIR TEMPERATURE (°C)															
	6	3	0	−3	−6	−9	−12	−15	−18	−21	−24	−27	−30	−33	−36	−39
3	3	−1	−4	−7	−11	−14	−18	−21	−24	−28	−31	−34	−38	−41	−45	−48
6	−2	−6	−10	−14	−18	−22	−26	−30	−34	−38	−42	−46	−50	−54	−58	−62
9	−6	−10	−14	−18	−23	−27	−31	−35	−40	−44	−48	−53	−57	−61	−65	−70
12	−8	−12	−17	−21	−26	−30	−35	−39	−44	−48	−53	−57	−62	−66	−71	−75
15	−9	−14	−18	−23	−27	−32	−37	−41	−46	−51	−55	−60	−65	−69	−74	−79
18	−10	−14	−19	−24	−29	−33	−38	−43	−48	−52	−57	−62	−67	−71	−76	−81
21	−10	−15	−20	−25	−29	−34	−39	−44	−49	−53	−58	−63	−68	−73	−77	−82
24	−10	−15	−20	−25	−30	−35	−39	−44	−49	−54	−59	−63	−68	−73	−78	−83

WIND SPEED (MPH)	AIR TEMPERATURE (°F)																		
	45	40	35	30	25	20	15	10	5	0	−5	−10	−15	−20	−25	−30	−35	−40	−45
5	43	37	32	27	22	16	11	6	1	−5	−10	−15	−20	−26	−31	−36	−41	−47	−52
10	34	28	22	16	10	4	−3	−9	−15	−21	−27	−33	−40	−46	−52	−58	−64	−70	−76
15	29	22	16	9	2	−5	−11	−18	−25	−32	−38	−45	−52	−58	−65	−72	−79	−85	−92
20	25	18	11	4	−3	−10	−17	−25	−32	−39	−46	−53	−60	−67	−74	−82	−89	−96	−103
25	23	15	8	0	−7	−15	−22	−29	−37	−44	−52	−59	−66	−74	−81	−89	−96	−104	−111
30	21	13	5	−2	−10	−18	−25	−33	−41	−48	−56	−63	−71	−79	−86	−94	−102	−109	−117
35	19	11	3	−4	−12	−20	−28	−35	−43	−51	−59	−67	−74	−82	−90	−98	−106	−113	−121
40	18	10	2	−6	−14	−22	−29	−37	−45	−53	−61	−69	−77	−85	−93	−101	−108	−116	−124
45	17	9	1	−7	−15	−23	−31	−39	−47	−55	−62	−70	−78	−86	−94	−102	−110	−118	−126

Variations in Wind Patterns

Wind currents are another secondary temperature influence. The origin of the wind largely determines the day-to-day temperatures experienced throughout most of the middle and higher latitudes. Daily temperature variations can be quite striking, especially in the winter half of the year. The United States, for example, is frequently invaded by cold air masses from Canada and by warm air masses from Mexico and the Gulf of Mexico. In general, though, the effects of warm and cold air masses largely offset one another when it comes to computing long-term temperature means. This is especially true since the prevailing global wind belts are **zonal** (west to east, or east to west) in direction.

Many people believe that the speed of the wind, as well as its direction, strongly influences temperature. It is well known that the air normally feels much colder on a windy day than on a calm day. Windchill values, which indicate just how cold it feels due to the combined effect of air temperature and wind, have become popular statistics with radio and television weather broadcasters (Table 4.2). In reality, the wind does not lower the air temperature, but merely blows away the insulating layer of warm air next to our bodies and replaces it with colder outside air. With sufficiently thick or impermeable clothing, this insulating layer stays intact, and we are not chilled by a strong wind. However, an important influence of wind on air temperatures is to reduce daily temperature ranges by mixing air at different levels.

Topographic Influences

The *topography*, or the geographic distribution of the land surface features, also has an effect on temperatures. Like elevation, the topography most strongly influences temperatures in mountainous regions, because mountain barriers act to block or redirect air masses. Just as high mountain ranges restrict the movement of people, so too do they impede the passage of air masses. Especially affected are cold, dense air masses, which resist the lifting necessary to cross mountain barriers, and air masses that are moving slowly because of the presence of weak upper-level winds. Mountains that are exceptionally high and continuous make the most effective barriers.

Probably the best example of how topography influences temperatures is provided by the Himalayas along the India-Tibet border in south-central Asia. The January isotherm map (Figure 4.9) shows that the isotherms are compressed in this area, indicating a very large or "steep" temperature gradient. Part of this gradient results from the ability of the Himalayas to block the southward penetration of cold central Asian air masses into India. Similarly, the west coast of the United States, especially California, normally is spared incursions of cold Canadian air masses in winter by the blocking effects of the Rockies and the Sierra Nevada. The mountains of California also are quite effective in restricting the inland penetration of cool, damp air masses from the Pacific, especially during the summer months (Figure 4.18). This causes interior southern California to

	Mean temp:		Temp. range:	
San Diego, CA 32°43'N	62°F	17°C	14F°	8C°
Calexico, CA 32°41'N	71°F	22°C	37F°	20C°

◆ **FIGURE 4.18** Mountain barrier effect on temperature. San Diego, California, is situated on the coast and frequently experiences onshore winds from the cool Pacific. Calexico is located about 145 kilometers (90 mi) inland behind several low mountain ranges.

experience much hotter conditions than do the state's coastal regions.

Additional topographic influences on temperature result from the effects of slope angle and slope **aspect,** or compass orientation, on the receipt of insolation. In the northern hemisphere, particularly in the higher latitudes, steep south-facing slopes are inclined toward the Sun and therefore have substantially warmer daytime temperatures than do slopes facing in other directions, especially north. Slope aspect can produce marked differences in vegetation types and soil moisture characteristics over small distances and is an important microclimatic influence in mountainous regions. In the southern hemisphere, of course, the situation is reversed; north-facing slopes receive the most insolation and have the warmest temperatures.

▌SUMMARY

This chapter has discussed both the causes and the geographic distribution of the world temperature pattern. Temperature is the quantitative measurement of heat energy; the heat energy that drives physical processes at the Earth's surface comes primarily from the Sun, which is a giant hydrogen fusion reactor. Most terrestrial systems are ultimately powered by this energy source.

We now return to the Focus Questions at the beginning of the chapter:

1. *How does solar energy get to the Earth, and what happens to it once it reaches the Earth?*

Solar energy is radiated to the Earth in the form of electromagnetic waves with wavelengths mostly in the visible light and near-infrared portions of the electromagnetic spectrum. Once this energy reaches the Earth system, much

is converted to long-wave terrestrial radiation, which we sense as heat. The heat energy is distributed around the Earth's surface and atmosphere by the combined processes of radiation, conduction, and convection before finally being reradiated to space. The distributional pattern of this energy controls the global temperature pattern.

2. *What are the major factors controlling air temperatures at the Earth's surface?*

Numerous factors help determine the global pattern of air temperatures. The most important are the altitude of the noon Sun, the duration of daylight, the distribution of land and water, surface elevation, and ocean currents. Additional factors of more limited importance are the varying distance between the Earth and Sun, the amount of atmospheric cloudiness and dust, the wind direction, topographic influences, and the influence of human activities.

3. *What is the geographic pattern of temperatures that results from the action of these controls?*

Because each of the world temperature controls operates with differing intensities in different regions, the global pattern of temperatures is complex. The Sun's altitude, the most dominant of all the controls, causes the overall temperature pattern to be latitudinal in nature, with mean temperatures generally decreasing as latitude increases. The duration of daylight is a seasonally varying factor that tends to cause progressively greater seasonal temperature contrasts with increasing latitude. Much of the thermal influence of the Earth's land and water distribution also is seasonal. Areas with continental climates, which are dominated by winds from the land, tend to have large temperature ranges on both a daily and an annual basis. Maritime climate areas, which receive the majority of their air from large water bodies, tend to have small daily and annual temperature ranges. Elevation reduces mean temperature values at an average worldwide rate of 6.4C° per 1,000 meters (3.5F° per 1,000 ft), so that high mountain and plateau regions are noted for their cold temperatures. Ocean currents may be either warm or cold, depending on their origins, and correspondingly act to warm or cool regions in their vicinities.

▌QUESTIONS FOR THOUGHT AND REVIEW

1. Describe the three energy transfer processes of radiation, conduction, and convection. In which portions of the Earth system does each primarily operate?
2. Explain what happens to the insolation reaching the Earth system. What proportion is absorbed, and where? What happens to the energy when it is absorbed? What proportion is reflected and scattered, and from where?
3. What is the greenhouse effect, and how does it influence world temperatures? Explain the origin of the term.
4. What are the five primary world temperature controls? Briefly explain how each control influences the global temperature pattern.

The Uncertain Greenhouse Effect

Interest in climate change has increased tremendously in recent years due to the highly popularized apocalyptic predictions associated with the greenhouse effect. According to this popular interpretation, the continued buildup of greenhouse gases can lead to a substantial increase in global temperatures during the next century. This projected rise in temperature may result in (1) the melting of large ice caps, (2) rising sea levels due both to melting of ice and to thermal expansion of seawater, (3) more frequent and destructive severe storms, (4) increasing droughtiness in continental interiors, and (5) shifts in modern-day climatic zones. Such rapid changes in climate would be associated with disruptions to physical, political, economic, and social systems at all geographic scales. Very simply, the greenhouse effect has become the single most important environmental issue of our time.

The physical principles underlying the greenhouse scare are relatively straightforward. The planetary temperature should reflect a balance between incoming solar energy absorbed and the amount of energy returned to space. In the absence of any greenhouse gases, a planetary temperature of approximately $-20°C$ ($-4°F$) should be established. However, the average global temperature is near $15°C$ ($59°F$). The reason for the difference is that greenhouse gases delay the escape of the heat energy back into space. The naturally occurring "greenhouse effect" thus produces an approximate $35C°$ ($63F°$) increase in the global temperature.

Following the Industrial Revolution, human activities resulted in the emission of large quantities of greenhouse gases, and the atmospheric concentration of these gases has increased steadily (Figure 4.19). For example, the atmospheric concentration of carbon dioxide (CO_2) in 1800

continued on next page

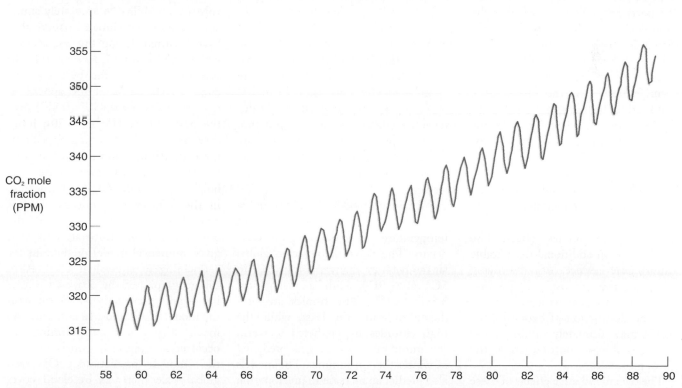

◆ **FIGURE 4.19** The gradual rise in global concentrations of atmospheric carbon dioxide is well illustrated by this graph of CO_2 levels monitored at the isolated measuring station atop Mauna Loa, Hawaii. CO_2 levels fall each northern hemisphere summer due to vegetative absorption, but then rise each winter to new record highs.

was approximately 280 parts per million (ppm), but by 1990, the atmospheric concentration of CO_2 had risen to approximately 350 ppm. Although CO_2 represents the most important greenhouse gas related to human activities (water vapor is the overall most important greenhouse gas), many other gases, including methane (CH_4), contribute to the greenhouse effect. Methane concentrations were near 0.75 ppm in 1800;however, recent measurements show methane levels to be near 1.70 ppm, and the increase is related largely to various agricultural activities. Nitrous oxide (N_2O), another naturally occurring greenhouse gas, has increased in atmospheric concentration due to deforestation, fossil fuel burning, and the use of some fertilizers. Atmospheric concentrations of N_2O have risen from about 285 parts per billion (ppb) before the Industrial Revolution to approximately 310 ppb in 1990.

Carbon dioxide, methane, and nitrous oxide are greenhouse gases that occur naturally in the atmosphere. Unlike these other greenhouse gases, the pre–Industrial Revolution atmospheric concentrations of the chlorofluorocarbons (CFCs) were essentially zero. CFCs are very powerful greenhouse gases, and despite having concentrations that are measured in parts per trillion, they add significantly to the overall greenhouse effect. The CFCs have an additional undesirable atmospheric effect of destroying ozone in the stratosphere. Because ozone also operates as a greenhouse gas, the destruction of ozone by the CFCs may ultimately minimize the total greenhouse contribution of the CFC molecules.

The overall radiative effects of these multiple greenhouse gases may be approximated by their "equivalent carbon dioxide" values. The resultant value gives an indication of how much CO_2 would be required to produce the same greenhouse effect as other trace gases found in the atmosphere. Equivalent CO_2 levels were approximately 290 ppm at the beginning of the Industrial Revolution, 310 ppm in 1900, and 430 ppm in 1990. By the middle of the next century, the equivalent CO_2 value should pass the 600 ppm level, double the values before the Industrial Revolution.

In an attempt to estimate climate changes associated with this expected doubling of equivalent CO_2, climatologists use numerical models that are representations of the climate system based on the underlying physics governing the atmosphere-ocean-Earth system. To conduct a doubled CO_2 experiment, the value of CO_2 in the model is varied from 300 ppm to 600 ppm. In the equilibrium experiments, the model is run first for 300 ppm, then for 600 ppm, and the differences in the simulated climate are determined. In more difficult, but more realistic time-dependent experiments, the model's equivalent CO_2 levels are raised gradually from 300 ppm to 600 ppm, and the resultant climate is determined along the time continuum. These approaches to the problem are somewhat different, but in gross terms, they tend to produce similar results.

All climate models predict a global increase in annual average surface air temperatures as a doubling of CO_2 occurs. The range of projected global air temperature changes is from $1C°$ to $5C°$ ($1.8–9F°$), with a mean near $3.5C°$ ($6.3F°$). The tropics are predicted to warm the least, while the high latitudes are predicted to warm the most; more warming is predicted in winter than in summer. The models also predict an increase in precipitation due to an intensification of the hydrologic cycle and the ability of the warmer atmosphere to hold more water. Irrespective of the changes in precipitation, the warming should act to increase the rate at which water can be evaporated into the atmosphere. The rise in evaporation is predicted to overwhelm any increases in precipitation, and soil moisture levels are expected to decrease over many continental areas.

These models are far from perfect representations of reality, however. The role of the ocean in absorbing CO_2 and storing and transporting heat is not adequately represented in the existing models. Cloud representations are particularly questionable. Other widely acknowledged problems in the models are related to the representation and treatment of sea ice, snow cover, localized storms, biological systems, and the role of other pollutants that may act to cool, rather than warm, the Earth.

One obvious test of the models involves their ability to accurately simulate the observed climate changes that have occurred during the last century when equivalent CO_2 levels rose by 40 percent. Based on the best available data, global temperatures appear to have increased by $0.45C°$ ($0.8F°$) over the past century (Figure 4.20); however, most of the warming appears in the early decades and not in the most recent period of greenhouse gas buildup. Given the 40 percent increase in the atmospheric concentration of equivalent CO_2 over this period, the observed rise in temperature is low, given numerical model predictions for a doubling of equivalent CO_2. Also, the geography of the observed warming is not particularly consistent with the model predictions. In addition, too much of the observed warming has occurred at night to be consistent with the model predictions. Observed global-scale increases in cloud cover and precipitation are in reasonable agreement with the model predictions, however.

(a)

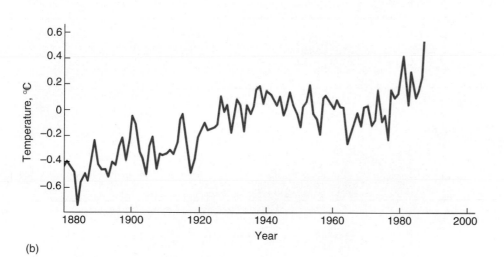

(b)

◆ **FIGURE 4.20** Atmospheric CO_2 and temperature changes since 1880. (*a*) Changes in concentration of CO_2 obtained from ice cores (smooth curve) and from measurements at Mauna Loa, Hawaii (annual oscillations). (*b*) Average annual surface air temperature plotted as a deviation from the average.

Some of the apparent mismatches between the $2 \times CO_2$ model predictions and observed temperature changes may be resolved when the cooling effects of other gases are more fully included in the model simulations. Along with greenhouse gases, humans are emitting other gases and particulates that are known to have a cooling effect. Sulfur dioxide, for example, is a pollutant that causes acid rain but also has an overall cooling effect on the atmosphere. Numerical models are currently being constructed to simulate climate conditions given realistic estimates of atmospheric chemistry in the future.

The great interest in the greenhouse effect has stimulated climatic research on many fronts. The numerical models are improving, analyses of the observational climate record are continuing, and the climate responses to changing CO_2 levels are being evaluated vigorously. Nevertheless, policymakers are now confronted with critical decisions that must be made in an environment of significant scientific uncertainties. The greenhouse issue is likely to be with us for years to come, and physical geographers will find an important role in resolving critical questions that surround the global warming debate.

SOURCE: This Case Study was written by Robert C. Balling, Jr., Assistant Professor of Climatology at Arizona State University.

5. What are the basic characteristics of continental climates and of maritime climates? Which coast of North America (the east or west) has the most maritime climate? Why?

6. Why are mean temperatures lower in the Rocky Mountain region than in most other areas at similar latitudes? Explain the mechanism responsible for these reduced temperatures.

7. Briefly discuss the influence of each of the world temperature controls in your hometown or campus area. How does your geographic location influence the effects or importance of each control?

8. Explain how human activities might produce a global warming trend in the near future.

▌PROBLEMS

1. Convert 50°F to the equivalent Celsius reading. Convert it to the equivalent reading in degrees Kelvin.

2. What is the Fahrenheit equivalent of absolute zero (0 K)?

3. Convert 32°C to the equivalent temperature in degrees Fahrenheit.

4. If the average environmental lapse rate happens to exist and the temperature at sea level is 67°F, what is the temperature at an elevation of 8,000 feet?

5. If the temperature at an elevation of 6,500 feet is 37°F, and the existing lapse rate is 2.5F° per thousand feet, what is the temperature at an elevation of 2,500 feet?

6. If the temperature at an elevation of 1,500 meters is 12°C and the average environmental lapse rate happens to exist, what is the temperature at an elevation of 4,500 meters?

▌KEY TERMS

energy
radiation
conduction
convection
insolation
greenhouse effect

continental climate
maritime climate
environmental lapse rate
leeward coast
windward coast

5 AIR PRESSURE AND WIND

OUTLINE

FOCUS QUESTIONS

1. What causes variations in air pressure, and how do these variations affect the weather?
2. What is the basic pattern of global air pressure belts?
3. What factors control wind speed and direction?
4. What is the basic pattern of global wind belts?
5. What are the jet streams, and how do they influence surface weather conditions?

A portion of the solar energy the Earth receives is used to produce the atmospheric air currents, or winds. For centuries humans have harnessed the power of the wind for energy and for transportation. For example, the use of wind currents by sailing ships made possible the European discovery and settlement of North America. Although wind power is less important as a means of transportation than it once was, the wind is increasingly used at present as a generator of electricity. The force of the wind can also produce great destruction and loss of life, particularly in association with major storms like hurricanes and tornadoes.

Winds blow because of differences in air pressure. Air pressure differences, in turn, are caused by the geographical variations in thermal energy discussed in Chapter 4. Because they are so closely related, the two weather elements of air pressure and wind are discussed together in this chapter. Although day-to-day patterns of air pressure and wind, like those of temperature, are complex and subject to constant change, both elements display relatively simple patterns when examined at the global scale.

AIR PRESSURE

Air pressure is the force that air exerts on its surroundings because of its weight. The amount of pressure exerted on any point at the surface or in the atmosphere is proportional to the weight of the overlying air from that point directly upward to the outer edge of the atmosphere. Because the atmosphere is gaseous, this pressure is applied equally in all directions. The air pressure at sea level averages 1,013.2 millibars[1] (14.7 lb/in^2). The pressure at any given site, however, fluctuates constantly and rarely equals this exact value.

Analyzed at the molecular level, air pressure is produced by collisions of air molecules. At room temperature, air molecules travel in random directions at speeds of about 450 meters per second (1,000 mph); near sea level, they collide elastically with one another or with surrounding objects billions of times each second. The air pressure is actually determined by the number and force of these molecular collisions on a pressure-sensing device (a barometer).

Air pressure varies both horizontally and vertically. Because of the dominating influence of gravity on the vertical distribution of the atmosphere, however, the vertical **pressure gradient,** or the vertical variation in air pressure per unit of distance, is typically several thousand times greater than is the horizontal pressure gradient.

Measured through time at any one place, air pressure variations generally are very small and can be neither seen nor felt. The practical importance of this weather element results largely from the fact that air pressure variations greatly influence, directly or indirectly, the global distributions of each of the other five weather elements. The air pressure pattern controls both the speed and direction of the wind; the wind, in turn, strongly influences the global patterns of temperature and moisture.

Causes of Air Pressure Variations

Why do changes in air pressure occur? The constant variation in the atmospheric pressure pattern is largely related to two factors: temperature changes and the rotation of the Earth. As discussed in Chapter 4, temperatures at the Earth's surface and in the lower atmosphere vary continuously because of the interplay of a variety of factors. When the air is heated, the air molecules increase their rate of vibration, resulting in more energetic collisions that push the molecules farther apart. This causes the heated air to expand in all directions, reducing the total mass of the air molecules within any given area and causing a reduction of air pressure (Figure 5.1). Similarly, when air cools, its molecular activity decreases. This allows the air molecules to crowd more closely together, increasing the surface air pressure. In summary, heating causes the air pressure to decrease, while cooling causes it to increase.

◆ **FIGURE 5.1** Effect of temperature on air pressure. At higher temperatures, air molecules collide more energetically, causing the air to expand outward and upward. This reduces air pressure values near the surface, but results in higher pressures in the upper atmosphere.

---

1. The millibar is the metric system measure of air pressure commonly employed by meteorologists. For a mathematical definition of this unit of pressure, see the Focus Box on pages 80–81.

The temperature/air pressure relationship produces a minor daily cycle of air pressure change in many areas, especially over land, with somewhat lower pressures during the warmer daytime hours and higher pressures at night. Over a longer time span, cold surfaces, such as high-latitude continental interiors during the winter, tend to develop shallow, thermally induced areas of high air pressure known as **thermal highs.** Similarly, the summertime heating of some land surfaces tends to foster the development of **thermal lows.** These thermal pressure centers strongly influence seasonal weather conditions in large parts of the world.

The second major cause of air pressure change is variations in air density resulting from the deflection of air into or out of certain latitudinal zones because of the Earth's rotation. The causes of this deflection are rather complex and are related to the Coriolis effect and pressure gradient, concepts explained later in this chapter. At this point in our discussion, it is sufficient to note that a combination of factors cause air to accumulate in certain areas, forming high pressure systems, and to be deflected away from other areas, producing low pressure systems.

Global Distribution of Air Pressure

The Earth's rotation, coupled with the influence of warmer and colder surface temperatures, produces constantly changing patterns of air pressure in the lower troposphere. The resulting high and low pressure systems, often referred to simply as "highs" and "lows," are each associated with

◆ **FIGURE 5.3** Three-dimensional wind-flow pattern in the northern hemisphere associated with a low pressure center.

characteristic weather conditions. The high pressure in a "high" is produced by descending air that is compressed as it approaches the Earth's surface (Figure 5.2). The compression of the air makes it warmer and relatively drier. Highs therefore are normally associated with fair weather, and a rising barometer is viewed as a sign of good weather to come. Lows operate in precisely the opposite manner (Figure 5.3). They are areas of rising air, which expands and cools, thereby reducing its ability to hold moisture (see Chapter 6). Lows therefore are normally associated with clouds and precipitation. By signaling the approach of a low, a falling barometer often indicates that inclement weather is on the way. The alternating passage of highs and lows at intervals of a few days is largely responsible for the changing weather of locations within the middle and high latitudes.

Global Surface Pressure Belts

Although the precise distribution of pressure systems around the world varies daily, the global pattern of air pressure is not entirely random. On a long-term basis, some parts of the world are dominated by high pressure; others are dominated by low pressure. In addition, some regions typically experience marked seasonal differences in air pressure patterns, while others do not. The whole situation is somewhat analogous to the pattern of traffic on a busy highway. Although the potential for variability in terms of the number, types, and locations of the vehicles is limitless, an underlying logic and order govern the pattern. Likewise, pressure systems—and weather systems in general—tend to

◆ **FIGURE 5.2** Three-dimensional wind-flow pattern in the northern hemisphere associated with a high pressure center.

AIR PRESSURE MEASUREMENT

The most frequently measured weather element other than temperature is air pressure, and the instrument used to measure it is the **barometer.** Barometers may be grouped into two basic categories: mercurial barometers and aneroid barometers.

The *mercurial barometer* was invented by the Italian physicist Evangelista Torricelli in 1643. Its operating principle is quite simple: changes in air pressure control the height of the mercury column in a graduated glass tube from which the air has been removed (Figure 5.4).

The second type, the *aneroid barometer*, is much more popular and is found in many homes. It generally takes the form of a dial-type instrument that is compact, sturdy, adjustable, and accurate (Figure 5.5). The heart of an aneroid barometer is the sylphon cell—a collapsible, accordion-shaped box from which the air has been partially evacuated. An increase in air pressure causes the sylphon cell to be compressed, while a decrease in air pressure allows it to expand. It is connected through a series of chains, levers, and springs to a pointer on a dial.

The two standard scales used for barometric pressure readings in the United States are the inches of mercury and the millibar scales. The older **inches of mercury** scale simply measures the height to which the mercury column in a mercurial barometer rises. Standard sea level air pressure of 1 kilogram per square centimeter (14.7 lb/in^2) is equivalent to 29.92 inches (76 centimeters) of mercury. Both an-

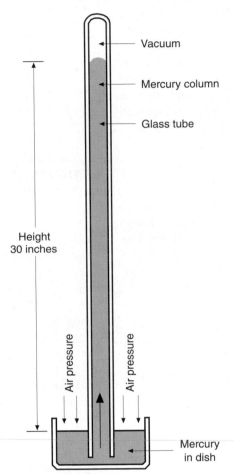

Height 30 inches

Vacuum

Mercury column

Glass tube

Air pressure

Air pressure

Mercury in dish

◆ **FIGURE 5.4** Operating principle of the mercurial barometer. Air pressure exerted on the mercury in the bowl is transmitted to the mercury inside the glass tube. The mercury in the tube rises to the level at which its weight exactly counterbalances the air pressure.

◆ **FIGURE 5.5** The aneroid barometer (pictured here) contains a sealed box that expands or contracts in response to changes in air pressure.
SOURCE: Courtesy of Qualimetrics, Inc., Sacramento, California

eroid barometers and mercurial barometers often use this scale. The second scale, which is currently employed by meteorologists throughout the world, uses units of pressure termed **millibars.**[1] Standard sea level air pressure is 1,013.2 millibars.

All surface air pressure measurements must be adjusted to eliminate the influence of elevation. Because of the weight of the atmosphere, air pressure values are greatly influenced by the elevation of the measuring station.

[1]By definition, a millibar is 0.001 bar. A bar is a force of one million dynes per square centimeter. A dyne is the force required to accelerate a one-gram object at a rate of 1 centimeter per second each second.

◆ **FIGURE 5.6** Surface weather maps prepared by the U.S. National Weather Service depict air pressure values by the use of isobars (lines of equal pressure) drawn at 4-millibar intervals. The air pressure at each reporting station is also given to the nearest tenth of a millibar. Additional information on the data contained on this map appears in Appendix B.

To eliminate the influence of elevation, barometric readings are normally "corrected" to sea level. In other words, the value given is not the actual air pressure at a station, but rather the air pressure that the station would have if it were situated at sea level.

On weather maps prepared by the U.S. National Weather Service (Figure 5.6), air pressure patterns are depicted by the use of **isobars.** These are lines connecting points with the same air pressure; they are typically drawn at 4-millibar intervals. ■

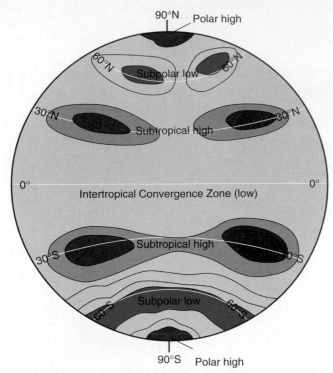

◆ FIGURE 5.7 Idealized latitudinal distribution of the global air pressure belts.

follow certain principles that control their development and movement.

Computing long-term averages of air pressure reveals a belted pattern of high and low pressure systems. These pressure belts encircle the Earth in an east-west direction. In a general sense, the arrangement of the global air pressure belts can be considered symmetrical with respect to the equator, with the northern hemisphere pattern being a mirror image of the southern hemisphere pattern. Each hemisphere contains three complete pressure belts, while a fourth straddles the equator and is shared by both hemispheres (Figure 5.7). There are thus seven belts in all. In reality, the shapes of the belts are far from symmetrical or beltlike, especially in the northern hemisphere, and they vary considerably in strength and location from summer to winter in many areas. The reasons for this will be explained shortly.

The most equatorward of the global pressure belts dominates the region between about 5°N and 5°S. It is a broad but rather weak trough of low pressure termed the **Intertropical Convergence Zone (ITCZ).** This low pressure zone is thermally induced by the high Sun angles and resulting year-round warmth of the equatorial region. Because warm, moist tropical air from both hemispheres converges and rises within this zone, it is associated with some of the world's rainiest climates.

Proceeding poleward in either direction, air pressures gradually rise until the centers of the **subtropical highs** are reached at latitudes between 25° and 40°. Unlike the ITCZ, the two subtropical highs are produced largely by air at high altitudes that is deflected into the subtropics by the Earth's rotation and subsequently sinks to the surface. These highs therefore are dynamically, or rotationally, induced. Sinking air is heated by compression, causing it to become relatively drier. The subtropical highs therefore act to suppress precipitation, and in many regions, they cause arid or semiarid conditions. As a result, most of the world's major deserts are located in the subtropics.

Air pressures typically decline poleward through the middle latitudes, until the **subpolar lows** are reached at latitudes of approximately 55° to 70°. These lower pressures are generated both by the Earth's rotation, which deflects air away from this latitudinal zone, and by the lifting of warm air from the lower latitudes as it encounters denser cold air from the polar regions along weather boundaries, or **fronts.** Although often described as pressure belts, the subpolar lows are less beltlike than either the ITCZ or the subtropical highs. Instead, especially in the northern hemisphere, the subpolar lows tend to form a zone in which storms occur frequently. These storms can become very powerful, especially over ocean areas in the winter half of the year. A semipermanent storm center that develops over the subpolar North Atlantic is termed the **Icelandic Low,** while its counterpart in the North Pacific is the **Aleutian Low** (Figure 5.8).

The most poleward (and typically the most weakly developed) of the four global pressure belts are the **polar highs.** In actuality, they are not belts at all, but ideally form roughly circular caps over the polar regions. They are thermally induced by the cold temperatures that prevail in this energy-deficient region and are quite shallow. The South Polar high is much the better developed of the two because the continent of Antarctica is considerably colder than is the Arctic Ocean location of the North Polar high.

It should be stressed that the world pressure belts merely represent average pressure conditions. They display considerable daily and seasonal variation in location and strength. In addition, numerous traveling high and low pressure centers exist independent of the pressure belts. These traveling pressure systems are especially common in the middle latitudes, where they strongly control daily weather conditions. Their characteristics and influences on the weather will be explored more fully in Chapter 7.

Seasonal Pressure Variations

As we have seen, the combination of latitudinal temperature differences and the Earth's rotation produces a beltlike world pressure pattern. If these two factors were the only controlling influences on air pressure, the seven pressure belts would be strongly developed and would probably appear much as they do in Figure 5.7. Two additional factors, however, distort the global pressure pattern and produce marked seasonal variations in air pressures over much of the Earth's surface. These two factors are the seasonal shift in latitude of

◆ **FIGURE 5.8**   January global mean sea level air pressure values in millibars.

the Sun's rays and the uneven distribution of the Earth's land and water surfaces.

The entire pattern of world pressure belts experiences a seasonal shift in latitude that mirrors, on a reduced scale, the annual latitudinal shift of the Sun's vertical rays between the Tropics of Cancer and Capricorn (see Chapter 2). The belts shift north or south from their mean positions only a few degrees over mid-ocean areas, but up to 10 degrees or more over large continents.

The shift of the pressure belts is thermally generated by the seasonal movement of Earth's surface temperature pattern with the Sun. As already noted, the uneven distribution of land and water bodies produces marked seasonal differences in temperature over areas at similar latitudes. The greater pressure belt displacement over land is caused by the ability of land to heat and cool more readily than water. These temperature differences strongly influence air pressure patterns, resulting in the formation of thermal highs and lows. Because land areas heat and cool more readily than do water bodies, land areas typically experience greater seasonal pressure fluctuations. Specifically, land areas usually are hotter than water bodies during the summer months and therefore tend to have lower surface pressures; during the winter months, land areas are colder and tend to have higher pressures. Seasonal heating and cooling variations are the dominant factor responsible for disrupting the orderly pattern of global pressure belts. Because the southern hemisphere has relatively little land, it heats and cools much more evenly than does the northern hemisphere. As a result, the air pressure belts of the southern hemisphere are better defined than those of the northern hemisphere. The January and July world pressure maps (Figures 5.8 and 5.9) portray the global pressure patterns during the midsummer and midwinter months, when the temperature contrasts between land and water are most pronounced (see Figures 4.9 and 4.10 for comparison).

WIND

Wind is simply air in motion. This motion may be in any direction, but in most instances the horizontal component of wind flow far surpasses the vertical. The wind is tremendously important as a global distributor of both moisture and thermal energy. It carries water evaporated from the oceans

◆ **FIGURE 5.9** July global mean sea level air pressure values in millibars.

to the land, where much of this moisture falls as precipitation. The wind also transports tremendous quantities of both sensible and latent heat from the tropics to the higher latitudes, moderating the temperatures of both the low and the high latitudes. The uneven solar heating of the Earth system is largely responsible for generating the air pressure imbalances that cause the winds. The relationship between air temperature and wind therefore is complex and reciprocal: spatial differences in temperatures create pressure differences that cause winds to blow, and the resulting winds transport heat and cold over the Earth's surface, thereby influencing world temperatures.

Factors Controlling Wind Speed and Direction

The wind has two distinct components: speed and direction. These components not only are measured and stated separately, but also are produced by differing combinations of controlling forces.

Wind Speed Controls

The wind blows because of differences in air pressure. Wind results from the tendency of the atmosphere, in the absence of offsetting forces, to achieve a pressure equilibrium between different points at the same altitude. The wind always blows from areas of high pressure to areas of low pressure, and if other factors remain constant, its speed is determined by the rate of air pressure change per unit of distance between the pressure systems. The situation is analogous to the flow of water down a hill. Just as water flows from higher to lower elevations at a speed that increases with the steepness of the slope, air flows from areas of higher to lower atmospheric pressure at a speed that increases with the steepness of the pressure slope, or pressure gradient. On an air pressure map, the steepness of the pressure gradient is indicated by the spacing of **isobars,** lines of equal air pressure. In areas where the isobars are closely spaced, the pressure gradient is "steep" and strong winds can be expected. In areas of widely spaced isobars, the pressure gradient is gentle and winds are typically light.

◆ FIGURE 5.10 Wind directions in response to a hypothetical air pressure pattern if the wind were controlled solely by the orientation of the pressure gradient. The wind would blow perpendicularly across the isobars from high to low pressure. The Earth's rotation prevents this from actually occurring.

A second force that influences wind speed and that always serves to partially offset the pressure gradient force is friction. Some molecular friction occurs within an air stream, but friction between the atmosphere and the Earth's surface is more significant. The amount of friction generated by the surface is partly a function of wind speed. With increasing wind speeds, friction between the air and the nonmoving surface increases rapidly. For this reason, a steepening of the pressure gradient will produce a less-than-proportional increase in the speed of the wind.

The frictional resistance provided by the Earth's surface is also influenced by its roughness. Land areas vary greatly in this respect. At one extreme, a smooth, ice-covered surface such as the interior of Antarctica offers very little frictional resistance to the passage of air, and winds there frequently attain very high velocities. Conversely, a forested, mountainous surface generates a high degree of friction, and wind speeds are correspondingly much reduced.

Surface friction not only reduces wind speeds, it also produces wind **turbulence.** This refers to the eddying or swirling motions generated in near-surface air currents by the uneven blocking and redirecting influences of frictional barriers such as trees, hills, and buildings. Turbulence is responsible for the gustiness of a windy day. The generally smoother surfaces of water bodies cause average wind speeds to be greater over water than over land and also reduce turbulence. This often is quite evident on the seashore or the

shore of a large lake. The highest average wind speeds are encountered in the upper atmosphere, above the frictional influence of the surface. Here, friction is produced only by collisions between air molecules, and the wind normally flows in a smooth, steady stream that is almost free of turbulence. A pressure gradient sufficient to produce a gusty 18 m/sec (40 mph) wind over a land surface typically will produce a smoother wind of approximately 30 m/sec (65 mph) over water and a steady 45 m/sec (100 mph) wind above a height of 1,000 meters (3,300 ft).

Wind Direction Controls

The wind direction is always stated as the direction from which the wind is coming, for reasons described in the Focus Box on pages 86–87. The direction of the wind is determined by three factors: the orientation of the pressure gradient, the Coriolis effect produced by the Earth's rotation, and the frictional resistance to the flow of the wind.

The primary factor controlling wind direction is the orientation of the pressure gradient. Air flowing from high to low pressure tends to do so along the most direct route down the steepest pressure gradient. Again, an analogy can be drawn with the flow of a stream. Just as a stream tends to flow down the steepest slope from high to low elevation, wind tends to flow down the steepest pressure gradient from high to low pressure (Figure 5.10).

If the pressure gradient were the only factor controlling the wind direction, the wind would blow perpendicularly across the isobars at any location, just as a stream tends to flow perpendicularly across contour lines. In the time it takes the wind to flow from a high pressure system to a low, however, both pressure systems are moving in circular paths because of the rotation of the Earth. The Earth's rotation thus constantly reorients the direction of the steepest pressure gradient. If the wind is to continue to flow down the steepest pressure gradient, it must gradually turn in a similar arc. Its angular momentum, though, keeps it from fully doing so. As a result, to an observer on the Earth's surface, it appears that the wind starts by blowing directly down the pressure gradient, but, with time, gradually changes its direction to a path that increasingly parallels the pressure gradient (Figure 5.14 on page 88). In reality, however, it is the underlying Earth, not the wind, that is turning. This apparent deflection of moving fluids on the Earth's rotating surface, which affects ocean currents as well as winds, is termed the **Coriolis effect.**

The counterclockwise rotation of the northern hemisphere, as observed from above the North Pole, deflects winds in the northern hemisphere to the right of their original paths, regardless of their directions of flow. Conversely, the clockwise rotation of the southern hemisphere deflects all southern hemisphere winds to the left. The Coriolis effect at the equator is zero, so no wind deflection exists there. The Coriolis effect increases poleward because

WIND MEASUREMENT

The wind varies in both direction and speed. Therefore, two different systems of measurement and two types of meteorological instruments must be employed to fully describe this important weather element. A crucial and sometimes misunderstood aspect of wind description is that its direction is always stated as the *direction from which the wind comes*. Hence, a west wind is traveling eastward, and a southeasterly wind is blowing toward the northwest. This practice is sometimes confusing to people not familiar with wind measurement because we tend, when traveling, to give directions in terms of where we are going, rather than in terms of where we have been. A very good reason, however, exists for emphasizing the wind's direction of origin. The air's temperature and moisture are determined by its previous location, not its future destination. Thus, a north wind tends to bring colder than normal temperatures to the middle latitudes of the northern hemisphere, while a southerly breeze is typically associated with warm weather.

The Mariner's Compass (see page 15) is usually employed for stating wind directions in weather reports intended for the general public because of its widespread familiarity. Pilots and navigators, however, normally use the more precise azimuth system. The instrument commonly used to measure wind directions is the well-known **wind vane,** sometimes imprecisely termed a weather vane. This simple instrument consists of a flattened metal design mounted on a shaft about which it can rotate freely to point in the direction from which the wind is blowing (Figure 5.11).

Wind speeds in the United States normally are stated in miles per hour in public broadcasts, but are measured in nautical miles per hour, or **knots,** by the National Weather Service and other official agencies. A nautical mile is the mean length of one minute of latitude,

◆ **FIGURE 5.11** A wind vane is designed to point toward the direction from which the wind is blowing.
SOURCE: Courtesy of Qualimetrics, Inc., Sacramento, California

or 1,853 meters (6,080 ft). It is equal to 1.15 statute miles, so that a knot equals 1.15 statute miles per hour. In Canada and most other countries, metric system units of meters per second or kilometers per hour are employed.

The instrument most commonly used to measure wind speeds is the **anemometer.** It normally consists of three or four hemispherical cups

the speed of planetary rotation decreases progressively with increasing latitude, causing the rotational paths of high latitude locations to form "tighter" circles. Even in the high latitudes where the Coriolis effect is strongest, though, the deflection is gradual, and local winds are not significantly affected.

The Coriolis effect tends to continue turning the wind from its current direction, regardless of how much it already has been deflected. This tendency is resisted, however, by the inability of the wind to blow up the pressure gradient from lower to higher pressure. The ultimate tendency of the Coriolis effect therefore is to deflect the wind by 90°, so that it flows parallel to the isobars (Figure 5.15).

The third influence on wind direction is friction. Friction interferes with the Coriolis effect, causing it to deflect the wind less than a full 90°. With increasing friction, the angle of deflection is progressively reduced. The resulting wind therefore blows at some intermediate angle between that of the pressure gradient force (directly across the isobars) and the angle that would be produced by a full Coriolis effect (parallel to the isobars). Friction reduces the deflection of the wind because when wind speed is reduced, its momentum in that direction is also diminished, allowing it to turn more easily toward the new direction of the pressure gradient orientation. A similar reduction in momentum, for example, makes it much easier for an automobile to navigate a sharp turn when it is traveling slowly than when it is speeding.

Because land surfaces typically provide more frictional resistance to the wind than do water surfaces, the wind is mounted on a metal shaft so that they are free to rotate with the wind (Figure 5.12). While an observer can easily determine wind direction from a wind vane, the speed of the wind cannot be readily determined merely by looking at a whirling anemometer. For this reason, the typical anemometer powers a small electrical generator, which is wired to an instrument panel inside the building on which the anemometer is mounted. The generated electricity moves a pointer on a calibrated dial to indicate the correct wind speed.

Upper-level measurements also are of vital importance, not only to aircraft, but also to weather forecasters. Systematic observations are provided by releasing large hydrogen- or helium-filled balloons with a known rate of ascent. As they rise, these balloons also are carried downwind, so that the horizontal component of their motion indicates the wind directions and speeds for the various levels through which they ascend (Figure 5.13). The balloons may also carry instrument packages called **radiosondes** aloft. The radiosondes transmit temperature, humidity, and air pressure readings as they ascend, and their radio signals enable them to be tracked in order to determine wind direction and speed. ∎

◆ **FIGURE 5.12** An anemometer. As the wind blows, air is trapped within the cups, causing them to spin at a speed proportional to the wind speed.

<small>SOURCE: Courtesy of Qualimetrics, Inc., Sacramento, California</small>

◆ **FIGURE 5.13** The launching of a weather balloon carrying a radiosonde. As it ascends, the radiosonde automatically measures and transmits back information on temperature, air pressure, and humidity.

<small>SOURCE: NOAA</small>

generally deflected less over land than over water. Deflection angles average about 60° from the orientation of the steepest pressure gradient over land areas in general, but can be reduced to as little as 45° over extremely rough terrain. Over water, deflection angles range between 70° and 75°. In the upper atmosphere, where friction is minimal, the wind is deflected by almost 90° so that it flows nearly parallel to the isobars, as shown in Figure 5.15. The procedure for estimating wind directions for both land and water surfaces on a pressure map is illustrated in Figure 5.16.

General Circulation of the Atmosphere

Because differences in air pressure cause the wind to blow, the global pattern of wind currents is closely related to the world air pressure pattern. Therefore, keeping the air pressure pattern (Figure 5.7) in mind will make it much easier to understand the major wind currents.

Two factors are primarily responsible for maintaining the pressure imbalances that produce the global wind systems: the uneven latitudinal heating of the surface and the Coriolis effect resulting from the Earth's rotation. Before discussing the actual global wind pattern, it is desirable to examine the influence of each of these factors, which combine to produce what might be considered as a simplified theoretical global pattern of wind currents.

We might begin by imagining that the Earth is not rotating and that instead the Sun is circling the Earth. With no rotation there would be no Coriolis effect, and winds would blow directly across the isobars from high to low

◆ **FIGURE 5.17** The generalized pattern of global wind belts, both at the surface and aloft. Air is exchanged between the surface and the upper atmosphere in the vicinity of the pressure belts.

Global Surface Wind Belts

The global surface wind belts are the largest units of the Earth's surface wind system. They are closely associated with the world pressure belts, and have many features in common with them. Altogether, seven east/west-oriented wind belts are located in each hemisphere. They are zones of prevailing wind conditions; the wind pattern on any given day may differ considerably from the norm.

The two hemispheres, ideally speaking, have a mirror-image reversal of their wind belt patterns, just as they do with their pressure belt patterns. As is true of the pressure belts, the wind belts of the southern hemisphere are better defined than those of the northern hemisphere. A seasonal shift in latitude of the wind belts also occurs along with, and because of, the thermally induced seasonal shift of the pressure belts.

Our examination of the wind belts begins at the equator and progresses poleward. Figure 5.17 illustrates the positional relationships of the various belts.

Centered near the equator, and extending to about 5°N and S, is the **equatorial belt of variable winds and calms.** This is a weak wind belt associated with the low pressure center of the ITCZ. Winds here normally are light and variable in direction, with wind variations depending on small changes in air pressure patterns from day to day. The calm central portion of this wind belt is sometimes referred to as the **doldrums.** From a three-dimensional standpoint, this zone experiences a large-scale transfer of heat and moisture from the surface to the upper troposphere within the cloud columns of the numerous showers and thunder-

storms within this zone. It is here, more than anywhere else, that the Earth's surface releases the excess thermal energy it has gained from the Sun.

Just poleward of the equatorial belt of variable winds and calms, and surrounding it to both the north and south, are the broad expanses of the **trade winds,** which dominate the zones between 5° and 25° N and S. The trade winds received their name because they aided the European sailing ships in their trading voyages from Europe to the West Indies during the colonial period. They are generated by the pressure gradient between the subtropical highs of each hemisphere and the low pressure of the ITCZ. The trade winds are the major wind belts of the tropics and are noted for both their persistence and their steadiness of direction. In the northern hemisphere, the winds blow from a northeast or east-northeast direction, producing the **northeast trades.** In the southern hemisphere, the winds blow from the southeast or east-southeast to produce the **southeast trades.** The locations of the northeast and southeast trades are clearly depicted in Figure 5.18.

Within the centers of the subtropical highs, the winds again become light and variable due to the absence of a significant pressure gradient. The **subtropical belts of variable winds and calms,** as they are called, are, from a meteorological standpoint, the most tranquil latitudinal zones on Earth. They are characterized by gently subsiding dry air, clear skies, calm or light surface winds, and warm temperatures. They cover at most five degrees of latitude and are usually located between approximately 30° and 35° N and S. Their seasonal shift in latitude, however, typically carries

◆ **FIGURE 5.18** Advanced image processing techniques convert satellite data into a global picture of winds. Arrows show wind direction and colors represent wind speed, with yellow being the highest wind speed and blue the lowest. Notice that our atmosphere consists of whirling eddies of varying sizes.

SOURCE: NASA. The image was created by Peter Wolceshyn and Andrew Pursh of Jet Propulsion Lab, Professor Morton Wurtele of UCLA, and Dr. Stephen Peteherych of the Canadian Atmospheric Environment Service.

them several degrees poleward in summer and equatorward in winter. Legend has it that, on occasion, the crews of ships bound for the New World with cargoes of horses exhausted their supplies of drinking water and were forced to throw the horses overboard when the ships were becalmed for extended periods in the subtropical highs. As a result, mariners have named this wind belt the "horse latitudes."

Occupying the zone between the subtropical highs and the subpolar lows of each hemisphere are the **westerlies,** the

major wind belts of the middle latitudes. They dominate the zones from 35° to 60° N and S, making them second only to the trade winds in total areal extent. Although the westerlies are well-developed wind systems, they are neither as steady nor as persistent as the trade winds because the middle latitudes are much more prone to the development of traveling high and low pressure systems than are the tropics. Each such pressure system, while being carried eastward by the flow of the westerlies, maintains its own wind circulation

◆ FIGURE 5.19 Mountain and valley breezes. During the day, heated air slides upslope, producing a valley breeze. At night, air chilled by contact with the surface drifts downslope, producing a mountain breeze.

in much the same fashion that an eddy of water in a river maintains a circular flow pattern while being borne downstream.

Within the latitudinal zones dominated by the subpolar lows, centered at about 60° to 65° N and S, wind directions again become variable. In this region, sometimes termed the **polar front zone,** prolonged periods of light winds are uncommon. In fact, stormy conditions often prevail because the subpolar low pressure zone tends to contain well-developed traveling storm centers. These storms develop because of the convergence of warm, moist air masses from the lower latitudes and cold, dry air masses from the higher latitudes. The daily positions and intensities of these storms largely determine both the speed and direction of the winds. The storm centers, and their associated winds, are especially intense over ocean areas in the winter because the strongest latitudinal temperature contrasts exist at that time of year.

Proceeding poleward, a poorly developed wind belt termed the **polar easterlies** is next encountered. Dominating the latitudinal zone between 65° and 80° N and S, it exists because of the pressure gradient between the polar highs and the subpolar lows. Winds of the polar easterlies theoretically come from a direction slightly poleward of due east, but the constantly changing positions and intensities of high and low pressure centers in these zones actually produce considerable short-term variation in wind direction and speed.

A final pair of wind systems exist within the centers of the two polar highs. These systems, which can be termed the **polar zones of variable winds and calms,** are not beltlike in shape but ideally form roughly circular caps over both polar regions in the vicinity of 80° to 90° N and S. In reality, they are poorly developed because the position and strength of the polar highs vary and also because the ice-covered, low-friction polar surfaces promote strong winds.

Traveling High and Low Pressure Systems

The wind-flow patterns of the middle and high latitudes of both hemispheres are modified by the development and general eastward movement of traveling high and low pressure systems that are independent of the global pressure belts. Low pressure systems are also termed **cyclones** because they are characterized by a cyclonic pattern of converging and rising air (see Figure 5.3). High pressure systems, conversely, are termed **anticyclones** because of their reverse pattern of diverging and descending air (see Figure 5.2). The Coriolis effect deflects the wind fields associated with these pressure systems, giving clockwise or counterclockwise twists to their circulations. The development and characteristics of these weather systems, which play a vital role in our day-to-day weather, are discussed further in Chapter 7.

Local and Regional Winds

Further contributing to the overall complexity of the Earth's total wind pattern are various local and regional winds. Most of these winds are produced directly or indirectly by air pressure differences resulting from the differential heating or cooling of the Earth's surface. Surface temperature differences, in turn, usually result from variations in surface materials, elevation, or slope orientation. In this section we will briefly examine some of the most important of these winds.

Mountain and Valley Breezes

Mountain and valley breezes occur in areas of rugged terrain under conditions of fair weather and weak pressure gradients. They are thermally generated local winds that undergo a daily reversal in direction. The temperature contrasts producing them occur between air near the surface and air farther aloft (Figure 5.19). During the day, the slopes, especially those inclined toward the Sun's rays, are heated strongly. This warms the overlying air, which becomes buoyant and begins to drift upslope, producing a **valley breeze** (remember that winds are named after the place they come from). At night the slopes cool quickly, chilling the adjacent overlying air. Late at night and into the early morning, this air becomes sufficiently cool and dense to flow downslope under the influence of gravity, producing a **mountain breeze.**

Chinook and Föhn Winds

Another group of mountain-related winds is known by such names as **chinook** in the Rocky Mountains and **föhn** in the Alps. These winds are not produced in the mountain areas they influence, but result from the regional pressure gradient. However, they are modified by their passage over the mountains and subsequent descent into the adjacent lowlands. Typically, mild, moist air on the windward side of the mountains is carried over the mountain crests, where uplift causes precipitation and warming by the released heat of condensation (see Chapter 6). Upon descending the leeward flanks of the mountains, the air is further heated and dried by compression so that it enters the lowlands as a relatively warm, dry wind. The **Santa Ana** of southern California is an example of a related wind type whose characteristics are also altered by descent. It is discussed in the Focus Box on pages 94–95.

Land and Sea Breezes

Land and sea breezes are familiar to residents of coastal areas in many parts of the world. They are produced by daily differences in heating and cooling over adjacent land and water areas (Figure 5.20). During the day, the land warms rapidly, heating the overlying air and causing it to expand and rise. The resulting reduction in surface air pressure causes somewhat cooler, denser air, called a **sea breeze,** to flow in from the adjacent water. (The same process also occurs along the shores of large lakes such as the Great Lakes, where the resulting breeze is known as a **lake breeze.**) During a sunny afternoon, a sea breeze may penetrate inland as far as 15 to 25 kilometers (10 to 15 mi) before losing its identity. At night, a reverse process occurs, with the more quickly cooling land fostering the formation of an overlying mass of relatively cool, dense air. By the late night hours, a **land breeze** may develop as the cooler, denser air from the land spreads over the coastal waters and displaces the lighter, milder oceanic air. Land and sea breezes develop best when skies are clear, midday Sun angles are high, and the regional pressure gradient is weak. Consequently, they occur most frequently in summer, especially in the lower latitudes, where they are a daily phenomenon in many coastal areas.

Monsoons

The most important of the world regional winds not associated with the semipermanent global wind belts are the monsoons. A **monsoon** is a seasonally reversing wind system

◆ **FIGURE 5.20** Land and sea breezes are produced by air pressure differences resulting from land and water temperature variations. During the day, air heated over the land expands and rises, producing an onshore sea breeze. At night, the cooler and denser air from the land spreads over the adjacent water, producing a land breeze.

CALIFORNIA'S SANTA ANA WINDS

The Santa Ana is a desert-derived wind that periodically visits the lowlands of coastal southern California, where it causes considerable discomfort and sometimes major destruction. The Santa Ana typically develops in the winter half of the year when a strong high pressure system forms over the Great Basin region northeast of California (Figure 5.21). The clockwise flow of air around this high produces a flow of mild, dry air from the east that can attain great velocities as it is funneled through passes in the San Gabriel and San Bernardino Mountains east of Los Angeles. The air is heated and further dried by compression during its descent of 1,000 meters (3,000 ft) or more to the Los Angeles Basin, producing a hot, dust-laden wind that can blow for days. Wind gusts sometimes reach speeds of 130 to 140 kilometers per hour (80 to 90 mph), causing property damage and desiccating crops. The unsettling impact of this wind on the populace of southern California is suggested by police statistics that typically show a marked increase in crime rates and suicides during Santa Ana events. Satellite photos taken during a Santa Ana sometimes show plumes of smoke and dust extending well out over the adjacent Pacific Ocean (Figure 5.22).

Perhaps the most dangerous aspect of a major Santa Ana event is the accompanying high fire danger resulting from the combination of hot temperatures, strong winds, and extremely low relative humidities. The danger is particularly great when the Santa Ana develops during the fall, before the onset of the annual winter rains. One of the worst such fires was the infamous Bel Air fire of November 1961. It burned for three days, destroying 484 homes. In October 1991, a fire-storm driven by Santa Ana winds razed much of the Oakland Hills area just east of San Francisco, damaging or destroying over 3,000 homes, taking 25 lives, and causing almost $5 billion in property damage. In late October and early November 1993, wildfires driven by Santa Ana winds swept through a number of communities near Los Angeles. These fires blackened over 200,000 acres and caused an estimated $1 billion in damage. Hillsides denuded of their protective vegetation by such fires are particularly vulnerable to serious erosion and potentially deadly mudflows during ensuing rainstorms. Thus, although the Santa Ana is a short-term phenomenon, its effects on the landscape may remain for years. ■

produced by the differing thermal characteristics of continents and oceans. In terms of formation, it is the large-scale equivalent of a land and sea breeze system, where summer and winter are substituted for day and night. During the summer, land areas that experience a monsoon are strongly heated by the Sun. This causes the overlying air to expand, producing a thermal low pressure area. The existence of higher pressure over the adjacent ocean induces an onshore flow of cooler, denser air to produce the **summer monsoon.** In winter, the reverse process occurs as the land and the air overlying it become cooler than the adjacent ocean. This results in the formation of a thermal high over the land, producing the seaward flow of dry air that characterizes the **winter monsoon.**

The monsoon circulation tends to be most strongly developed over large landmasses in the lower latitudes, which seasonally heat and cool to greater extremes than do small landmasses. In addition, large landmasses develop larger and stronger thermal lows and highs as a result of their seasonal heating and cooling. It is the size and strength of these thermal pressure centers that produce the pressure gradients that set the monsoonal winds in motion. Asia, by far the largest continent, has the best developed monsoon system (Figure 5.23). Monsoonal circulations also are well developed in northern Australia and in sub-Saharan North Africa. Most other large landmasses, including North America, have weak monsoonal tendencies that alter somewhat their predominant wind directions between summer and winter.

The importance of the monsoon results primarily from its influence on the moisture content of the air. In areas where it is well developed, the summer monsoon causes the inland transport of vast quantities of moisture-laden oceanic air. The lifting of this air due to heating over land and the

◆ **FIGURE 5.22** Satellite photo showing the smoke from Santa Ana–induced forest fires in the hills east of Los Angeles being blown far out over the Pacific.

SOURCE: NOAA

◆ **FIGURE 5.21** The development of a high pressure system over the Great Basin produces a pressure gradient favoring the development of Santa Ana winds. On this weather map showing air pressure patterns for January 17, 1976, maximum temperatures at reporting stations are given in degrees Fahrenheit. Observe that downslope winds blowing into southern California resulted in temperatures in the upper 80s, while temperatures elsewhere were much lower.

possible influence of mountain barriers generates clouds and often copious precipitation. (The resulting association of the summer monsoon with the rainy season in Southeast Asia and other areas has led many people to believe erroneously that the monsoon is a rain system rather than a wind system.) In winter, the reversed offshore monsoonal flow produces conditions opposite to those that prevail in summer. Skies tend to be clear, humidities low, and daily temperature ranges large. This generally produces the dry season for areas dominated by the monsoon. The Case Study at the end of this chapter further explores the characteristics and human impact of the Asiatic monsoon.

Upper-Level Winds

Before the development of high-altitude aircraft during the World War II period, relatively little was known about the circulation patterns of the upper troposphere. In the succeeding decades, though, we have gained a great deal of knowledge about the upper-level winds, and it has become apparent that they exert a controlling influence on the development and movement of surface weather systems.

General Circulation

The air pressure and wind patterns of the upper troposphere are considerably simpler than those of the lower troposphere because of the absence of an adjacent surface. Because no land and water heating differences are present, and because local variations in wind direction and speed caused by frictional and blocking effects are absent, the winds in the upper troposphere normally flow as smooth, steady currents. In the absence of significant friction, moreover, the Coriolis effect is able to exert its full influence in deflecting the winds

◆ **FIGURE 5.23** The summer and winter wind-flow pattern of Southeast Asia. The blue arrows represent continental air from the interior of Asia, and the pink arrows indicate maritime air from the Indian and Pacific Oceans.

so that they blow at right angles to the pressure gradient (Figure 5.24). This causes the upper-level winds to be much more zonal than are the near-surface winds. It should be stressed, however, that the upper-level winds are closely associated with the low-level wind systems and that they also are powered largely by the absorption of surface insolation. Air is exchanged constantly between the upper and

lower levels by the rising air columns within low pressure systems and the descending air columns within highs.

Two wind systems dominate the upper tropospheric flow patterns of each hemisphere. The **tropical easterlies** form a broad belt of relatively light easterly winds that occupy the zone from about 15°N to 15°S. They actually are the upper-level continuation of the surface trade winds. The

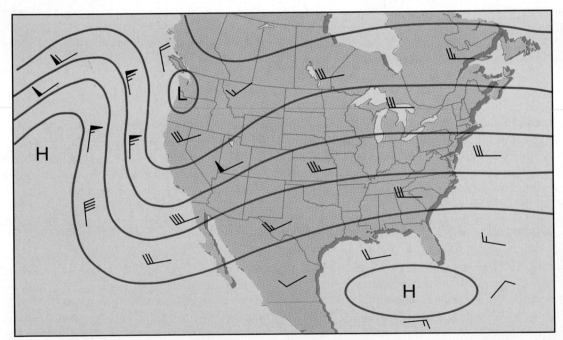

◆ **FIGURE 5.24** Upper troposphere pressure map showing wind direction and speed at selected stations. Shafts point into the wind, and indicate a Coriolis effect wind deflection of nearly 90° exists in the absence of surface friction. Wind speeds are indicated by the number of "feathers" on the wind shafts, as shown to the right of the map.

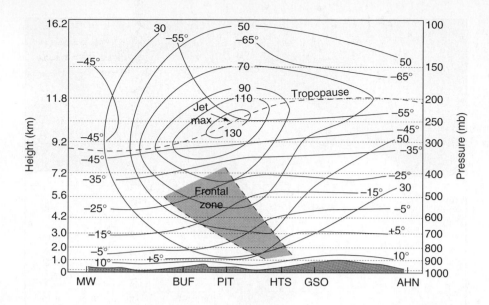

◆ **FIGURE 5.25** Vertical cross section through the atmosphere along a transect from Ottawa, Ontario (MW), to Athens, Georgia (AHN), at 7 A.M. on October 16, 1973. The red lines are lines of equal wind speed, and the blue lines are lines of equal temperature (isotherms). Wind speeds are given in knots, and temperatures in degrees Celsius. The jet stream is centered over Pittsburgh at a height of about 10.5 kilometers.

latitudinal extent of the tropical easterlies, however, is not nearly as great as that of the trade winds, which extend to approximately 25° N and S.

Poleward of approximately 20° latitude in each hemisphere, and extending all the way to the polar regions, is a vast belt of westerly winds termed the **upper-level westerlies.** This wind belt not only covers the majority of each hemisphere, it also contains by far the strongest winds in the upper troposphere. Although the upper-level westerlies have a mean direction of flow that is almost due west to east, they typically display a wavy pattern because they tend to develop a series of eastward-migrating ridges and troughs. These irregularities frequently cause winds to blow from the northwest or from the southwest. In the vicinity of the North and South Poles, but commonly offset from them by several hundred kilometers, are giant low pressure centers around which these westerly winds rotate.

The Jet Streams

The fastest-flowing upper-level winds, and those with the greatest impact on the surface weather, are the **jet streams.** They consist of relatively narrow bands of very strong winds centered in the upper troposphere at altitudes of approximately 9 to 13 kilometers (6 to 9 mi). When viewed three-dimensionally, a jet stream is tubular in shape, with wind speeds increasing toward the center from all directions (Figure 5.25). Maximum wind speeds vary considerably along the path of a jet stream but usually range between 80 and 240 kilometers per hour (50 to 150 mph). The winds of the jet streams generally flow from west to east in both hemispheres but, like the upper-level westerlies as a whole, tend to develop large-scale north/south-oriented waves termed **Rossby waves** (Figure 5.26). The entire undulating pattern of a jet stream typically shifts slowly eastward at about 25 to 40 kilometers per hour (15 to 25 mph). As this

shift occurs, existing waves change their amplitude, new waves may form, and old ones disappear. A jet stream thus exhibits a complex, dynamic flow pattern.

At most times there are two separate jet streams in each hemisphere. These are termed the subpolar jet and the subtropical jet. The **subpolar jet** is produced by the temperature contrasts between the high and low latitudes, and its position tends to coincide closely with the surface location of the polar front. It is strongest during the winter because the temperature contrast between high and low latitude air masses is greatest then. The jet stream also shifts equatorward during the winter because the frontal boundary between warm and cold air masses is located at a lower latitude. In North America, for example, the subpolar jet in midwinter typically extends over the southern United States; by midsummer it usually is positioned over southern or central Canada.

The **subtropical jets** are typically centered above the subtropical highs at a somewhat higher altitude than the subpolar jets. Their origin is less certain than the subpolar jets. They are apparently formed by the thermal contrast existing along the contact zone between warmer air that was advected poleward by the Hadley Cell and colder air transported equatorward by the Ferrell Cell (see Figure 5.17 for the location of these cells).

The practical significance of the jet streams is associated primarily with their influence on weather conditions at the Earth's surface. They act as upper-level steering currents or "highways," largely controlling both the speed and the direction of movement of the migratory highs, lows, and fronts that influence our surface weather. These surface weather systems extend high enough into the atmosphere to be carried along by the jet streams. As a result, weather systems located beneath a jet stream tend to travel along its path, and their forward speed is determined by the jet stream's strength. The speed of movement of surface

◆ **FIGURE 5.26** A deep trough in the subpolar jet stream associated with a cold air outbreak over the central United States appears on this map of February 22, 1975, showing wind-flow patterns at the 300-millibar pressure level. Dashed lines indicate wind speeds in knots. The jet stream maximum (thick line) separates warm surface areas (shaded pink) from cold surface areas (shaded blue).

weather systems is normally between one-half and one-third the maximum speed of the jet stream because of the restraining influence of surface friction.

The subpolar jet is particularly important to the weather of the middle latitudes because it is located above the surface boundary between warm and cold air masses. Consequently, an equatorward loop in this jet will draw a mass of air from the higher latitudes unusually far equatorward, producing an outbreak of cold weather (Figure 5.26). A poleward loop, conversely, will result in the poleward flow of a mass of warm air. In addition, the loops of the subpolar jet influence the locations of storm development. Because the subpolar jet is critically important to surface weather conditions, information on the jet streams is regularly gathered and mapped, and a great deal of research has been conducted to better understand the jet streams and predict their behavior.

❘ SUMMARY

This chapter has examined the causes and global distributions of two weather elements—air pressure and wind—that are perhaps most important because of their influence on other aspects of the weather. Air pressure is the only weather element to undergo changes that can be neither seen nor felt, yet it is of great significance because of its effect on wind direction and speed. The wind, in turn, is important both because of its direct effects on the surface and because it transports energy and moisture, often over great distances. Daily changes in temperature and the potential for precipi-

tation are strongly influenced by short-term changes in wind patterns. On a global scale, regions of high and low precipitation are associated, respectively, with zones of rising air within lows and descending air within highs.

For a more complete summary, we return to the Focus Questions appearing at the beginning of the chapter:

1. *What causes variations in air pressure, and how do these variations affect the weather?*

Air pressure varies largely because of global heating differences and the Earth's rotation. Heating causes air to expand and the air pressure to fall, while cooling causes air to contract and the pressure to rise. The Earth's rotation produces the Coriolis effect. This causes the air to be deflected into some latitudinal zones, producing high pressure, and to be deflected away from other latitudinal zones, producing low pressure.

2. *What is the basic pattern of global air pressure belts?*

In general terms, a symmetrical pattern of global pressure belts exists. The Intertropical Convergence Zone (ITCZ), a zone of low pressure that is produced primarily by high temperatures, straddles the equator. Surface air pressures rise poleward until the centers of the subtropical highs are reached at about 30° to 35° N and S. Mean air pressure values then decline poleward to about 60° to 65° N and S, where the centers of the subpolar lows are found. Finally, polar highs are located in the vicinities of the North and South Poles. This beltlike pattern is somewhat idealized and is generally best developed in the southern hemisphere, where land/water temperature differences are less pronounced than in the northern hemisphere.

3. *What factors control wind speed and direction?*

The wind is produced by differences in air pressure. Air tends to flow from high to low pressure at speeds proportional to the steepness of the pressure gradient existing between the two pressure systems. The wind speed is greatly reduced, however, by the frictional resistance it encounters at the Earth's surface.

Wind direction is controlled by three factors: the orientation of the pressure gradient, the rotation of the Earth, and surface friction. The net result of these factors is that surface winds in the northern hemisphere are deflected clockwise as they flow from high to low pressure. The angles of deflection vary from about 45° to 80° from the orientation of the steepest pressure gradient. In the southern hemisphere, a similar counterclockwise deflection occurs.

4. *What is the basic pattern of global wind belts?*

The pattern of global wind belts closely resembles that of the global pressure belts. Each pressure belt is associated with a central zone of variable winds. This wind variability results from a changeable, and often weak, pressure gradient. The broad zones of relatively steep pressure gradients between the pressure belts are occupied by three belts of stronger and steadier winds in each hemisphere. The trade winds occupy the latitudinal zone between the ITCZ and subtropical highs; the westerlies occupy the zone between the subtropical highs and the subpolar lows; and the polar easterlies occupy the zone between the subpolar lows and the polar highs.

5. *What are the jet streams, and how do they influence surface weather conditions?*

The jet streams are narrow, meandering belts of very strong westerly winds centered in the upper troposphere. They result largely from the surface temperature contrasts between the low and high latitudes. The jet streams act as steering currents that influence the direction and speed of movement of surface weather systems.

▌QUESTIONS FOR THOUGHT AND REVIEW

1. Explain why air pressure exists. What characteristic weather conditions are associated with high pressure and low pressure systems, respectively? How do temperature changes affect air pressure?
2. What are the two major categories of instruments used to measure air pressure? Which is currently most popular, and why? What scales are commonly employed to measure air pressure? How are air pressure patterns depicted on weather maps?
3. Name the four global pressure belts and describe their latitudinal locations. In which hemisphere is the beltlike pattern best developed, and why?
4. How could you tell by looking at today's weather map whether or not it will be windy? Explain how winds are generated.
5. What is the Coriolis effect, and how does its influence on wind direction differ between the northern and southern hemispheres?
6. Why do winds in most of North America come from the west?
7. Draw a diagram of the Earth showing the idealized pattern of global surface air pressure belts. Label the wind belts that exist both within each pressure belt and between each pair of pressure belts. Indicate with arrows the wind-flow pattern for those wind belts that have a prevailing direction.
8. Why are the monsoon winds better developed in Asia than in North America? Why do they reverse direction between summer and winter? With what types of weather are the summer and winter monsoons, respectively, associated? Explain why.
9. Under what meteorological conditions does southern California experience Santa Ana winds? Why are these winds associated with a high danger of fires?
10. What are the jet streams? Where and why do they form? How do they significantly affect our daily activities?

▌PROBLEMS

1. If you were aboard a ship in the North Atlantic, with high pressure centered due north of you and low pressure centered due south, from what direction would you expect the wind to come? (Give your answer in both the Mariner's Compass and the azimuth system.)
2. If you were in a New Zealand forest, with high pressure centered due west of you and low pressure centered due east, what would be your most probable wind direction?
3. If the wind is coming from the south, in what compass direction from your location is a high pressure center most likely to be found?

▌KEY TERMS

air pressure	**trade winds**
pressure gradient	**westerlies**
Intertropical Convergence	**polar easterlies**
Zone (ITCZ)	**land and sea breezes**
subtropical highs	**summer and winter mon-**
subpolar lows	**soons**
polar highs	**jet streams**
wind	**subpolar jet**
Coriolis effect	

The Impact of the Asiatic Monsoon

The Asiatic monsoon is a thermally induced wind system. Climatically, however, it is most important as the mechanism controlling the annual precipitation pattern of southern and eastern Asia. The "monsoon lands" of Asia, where the effects of the monsoon are especially dominant, occupy a broad arc from Pakistan to extreme eastern Siberia. Approximately half the world's population lives within this area.

The Asiatic monsoon is caused by the seasonal temperature differences that develop between the air over the continent of Asia and that over the adjacent Pacific and Indian Oceans. During the late fall and early winter, the Asian landmass cools rapidly, causing surface air pressures to increase. The clockwise outflow of air from the Asian mainland produces cool, dry, north-to-northeasterly winds over most of monsoon Asia. The relative dryness of this continental air is increased by its descent from the land toward the sea, so that in many areas no rain may fall for several months. Along its outer margins, however, the outflowing dense, cold air encounters warmer oceanic air, generating storms that bring substantial winter rains and snows to Japan.

As the Sun climbs higher in the sky and the length of daylight increases during the late winter and early spring, the Asian landmass warms rapidly. This causes a reduction in air pressures over eastern Asia and weakens the northeast monsoon winds. By late spring, temperatures have become hot over southern Asia, and inland air pressures have begun to fall to values lower than those over the cooler adjacent oceans. This pressure gradient initiates an onshore flow of tropical maritime air over extreme southern and eastern Asia, which gradually strengthens and expands northward and westward as the summer progresses (Figure 5.23).

In India, rains begin on the Bay of Bengal coast in early June and spread northwestward to the Pakistani border by mid-July. The onset of the rains associated with the summer monsoon is often very sudden and is sometimes referred to as the "burst" of the monsoon. In Bombay, for example, mean monthly rainfall totals increase from 18 millimeters (0.7 in) in May to 485 millimeters (19.1 in) in June (Table 5.1). It is estimated that 85 percent of India's annual rainfall occurs during the three-month period from mid-June until mid-September. Rainfall in most areas is not steady, but instead takes the form of frequent, often torrential, showers and thunderstorms. By September the rains begin to retreat southeastward, ending where they began on the Bengali coast in November as high pressure again becomes dominant over Asia.

Several factors account for the intensity of the summer monsoon rains of Southeast Asia. One is the large water vapor content of the tropical oceanic air masses that are drawn into the continent by the summer monsoon airflow. A second factor is that the ITCZ, with its associated wind convergence and uplift, is drawn northward deep into Asia during the summer (Figure 5.9). Third, and probably most important, is the forced lifting of the air as it passes from the ocean at sea level to the Asian landmass, much of which is at a high elevation. Rainfalls can be extremely heavy where mountain ranges lie directly across the path of the monsoon winds. For example, the town of Cherrapunji, located on the southern slopes of the Khasi Hills in eastern India, has a mean annual rainfall of 1,087 centimeters (428 in). Cherrapunji holds the world's record for the most rainfall ever recorded in a single 12-month span. From August 1860 through July 1861, it received an amazing 2,645 centimeters (1,042 in) of rain! This is about 30 times the world average annual total.

The monsoon precipitation cycle is especially critical to the inhabitants of Southeast Asia because of the dominance of subsistence agriculture in that part of the world. In India, for ex-

◆ **TABLE 5.1** Monthly Mean Precipitation Totals for Bombay India

	JAN.	FEB.	MAR.	APR.	MAY	JUNE	JULY	AUG.	SEPT.	OCT.	NOV.	DEC.	YEAR
Inches	0.1	0.1	0.1	0.0	0.7	19.1	24.3	13.4	10.4	2.5	0.5	0.1	71.3
Centimeters	0.3	0.3	0.3	0.0	1.8	48.5	61.7	34.0	26.4	6.3	1.3	0.3	181

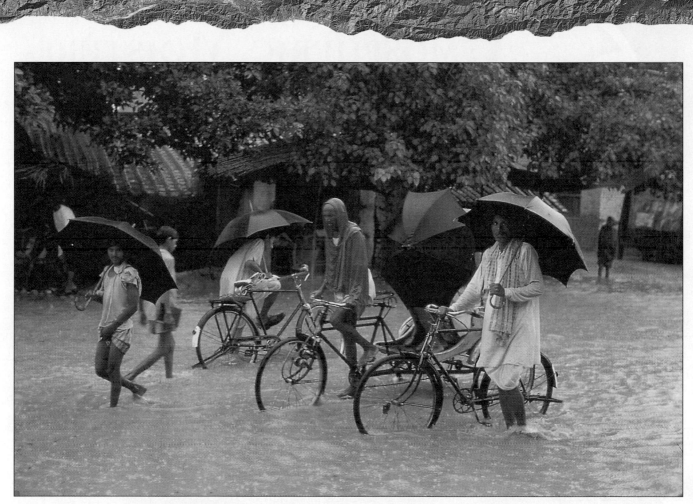

◆ **FIGURE 5.27** Residents of Benares, India, wade and pedal through torrential monsoon rains.
SOURCE: ©Paolo Koch/Photo Researchers, Inc.

ample, two-thirds of the population is engaged in agriculture; in Pakistan, three-fourths of the population is agricultural. Land holdings are small, and high population growth rates continue to increase the pressure on the land to provide ever-larger crop yields.

The agricultural calendar of events is dictated by the rainy season. Little can be done in the hot, dry spring months, when the land lies parched, cracked, and bare. The onset of the monsoon rains brings about a frenzy of planting and plowing, as the crops must have maximum exposure to the available moisture. Most harvesting takes place in the fall at the beginning of the winter monsoon dry season. As winter arrives, the land again lies fallow until the spring and the annual return of the rains.

6 ATMOSPHERIC MOISTURE

OUTLINE

FOCUS QUESTIONS

1. Under what conditions do water phase changes occur, and what role do they play
   in the global distribution of energy?
2. What is the hydrologic cycle?
3. How do clouds and precipitation form?
4. What is the geographic pattern of precipitation, and why does it exist?

Water is one of the most vitally important substances on Earth. It is also one of the most abundant; it covers 71 percent of the Earth's surface, and it comprises two-thirds of our own body weight. Although water is an essential component of all four terrestrial environmental spheres—the atmosphere, hydrosphere, lithosphere, and biosphere—the amount of water present at any given place changes through time. This variability exists because of the unique ability of water to undergo phase changes between the three physical states of matter in response to energy fluxes between the Earth's surface, the atmosphere, and outer space.

The resulting circulation of water between land, sea, and air makes life on land possible. It also distributes tremendous quantities of mass and energy that play key roles in the operation of most terrestrial surface processes. The speed at which this circulation system operates is ultimately linked to temperature. Temperatures at the Earth's surface and within the atmosphere not only determine the quantity of energy available to power phase changes of water, they also control the amount of water vapor the atmosphere is capable of holding at any given location.

PHASE CHANGES OF WATER

All matter exists in one of three physical states, or phases: solid, liquid, or gaseous (vapor). Although any substance, with sufficient gain or loss of energy, can theoretically change from any one of these states to any other, water is the only atmospheric component that undergoes phase changes within the existing range of atmospheric temperatures and pressures.

Energy is absorbed or released when water changes either temperature or phase. This energy is commonly measured in terms of calories. A **calorie** is the amount of heat needed to raise the temperature of one gram of liquid water by one Celsius degree (1.8 F°). Heat is absorbed by water when it is warmed and released when it cools. Large quantities of energy are involved in the six phase changes in which water can participate. Energy is absorbed as latent heat by the water when it changes from a lower to a higher phase and released as sensible heat to the environment when the water changes from a higher to a lower phase. Figure 6.1 illustrates these changes. They consist of three pairs of opposing processes, which we shall now examine:

◆ *Evaporation and condensation.* **Evaporation,** the change of liquid water to water vapor, occurs as individual molecules in a water body acquire sufficient kinetic energy through random molecular collisions to escape into the atmosphere. For every gram of water that evaporates at normal

◆ **FIGURE 6.1** Water phase change terminology and energy transfers per gram of H_2O.

Earth temperatures, approximately 585 calories of energy are transferred from the liquid water to the atmosphere.[1] Because most evaporating water is located on the Earth's surface, this process produces a major transfer of energy from the surface to the atmosphere. The addition of this water vapor does not directly heat the atmosphere, however, because the absorbed energy is used to change the physical state of the water molecules rather than to raise their temperature.

Condensation, the change of water vapor back to liquid water, is the exact opposite of evaporation. When condensation occurs, the energy absorbed by the water molecules during evaporation is released as the **heat of condensation.** Most condensation takes place well above the Earth's surface during the process of cloud formation; it therefore is a major heating process for the troposphere.

◆ *Melting and freezing.* A much smaller but still important quantity of energy is associated with the melting of ice and the freezing of water. Termed the **heat of fusion,** it involves the transfer of 80 calories per gram of water. (It is this transfer of energy that makes your hand cold when you hold an ice cube as it melts.) When water freezes, this energy is released as sensible heat to the atmosphere. At the surface freezing is often a slow process because, even at temperatures well below the freezing point, considerable time is required for a large water body to radiate enough energy into the atmosphere to freeze. However, freezing can occur rapidly in the atmosphere as water droplet clouds are lifted to higher and colder levels, where they are converted to ice-crystal clouds.

◆ *Sublimation.* Under certain conditions ice can change directly to water vapor, or vice versa, without passing through the liquid state. Both changes are known as **sublimation.** Sublimation involves the transfer of a

---

1. The lower the temperature of the evaporating water, the greater the quantity of energy required for evaporation. At the boiling point, 540 calories are required to evaporate a gram of water; at the freezing point, 600 calories are required.

greater amount of energy per unit of mass than does any other change of state: 665 calories per gram. (This is the sum of the heats of condensation and fusion.) In the atmosphere, sublimation occurs primarily when ice-crystal clouds form directly from water vapor at temperatures well below freezing. Although it liberates large amounts of energy per gram of water involved, sublimation is much less important as a source of atmospheric heating than is condensation because it occurs only at cold temperatures, which greatly reduce the capacity of the air to hold water vapor.

Taken together, the phase change processes perform three vital atmospheric functions. First, they transfer from the Earth's surface to the atmosphere much of the energy that powers the atmospheric processes, including storms and the global wind currents. Second, they add water vapor to the atmosphere. Third, they make possible the conversion of water vapor to the liquid and solid states, thereby enabling precipitation to occur.

THE HYDROLOGIC CYCLE

Earth's surface water supply is involved in an endless cycle of movement, called the **hydrologic cycle,** which is powered by solar energy and made possible by water's ability to change phase. As a result of this cycle, water is transported from the oceans to the continents, the land surface is sculptured by erosion, and the Earth's supply of solar energy is redistributed over the face of our planet. Most of the topics discussed in the remainder of this book are influenced at least indirectly by this water circulation system.

The hydrologic cycle is diagrammed in Figure 6.2. The details of the cycle are complex because water can follow a number of alternative routes depending on local conditions. Basically, though, the cycle consists of five key steps; when they are completed, the water has returned to its point of origin, and the stage is set for a repetition of the cycle.

The hydrologic cycle is essentially a closed material system because it involves the continuous recycling of the same water. It therefore is possible to begin the discussion with any of the cycle's five steps. Perhaps the best starting point, however, is with the ocean waters. The oceans contain an estimated 96.7 percent of the Earth's near-surface water supply, and they represent by far the longest stop that water normally takes in moving through the hydrologic cycle (Table 6.1).

In our illustration, the evaporation of water from the oceans and its subsequent upward diffusion into the atmosphere represent the first step of the hydrologic cycle.[2] Rates of evaporation from oceans vary greatly in different parts of the world for reasons that will be discussed shortly, but a mean annual rate of 100 cm (40 in) can be used as an approximation. At any one time, the atmosphere contains about 14,200 cubic kilometers (3,400 mi^3) of water, mostly in the vapor state.

The second step in the hydrologic cycle is the condensation (and/or sublimation) of atmospheric water vapor to form clouds. When condensation occurs, the water returns to its original liquid physical state, but is far from having its original appearance. Where before it was part of an extensive, interconnected body of water covering approximately 71 percent of the Earth's surface, it is now dispersed high in

2. As indicated in the diagram, only 14 percent of global evaporation takes place from land surfaces. We shall temporarily ignore evaporation from the land in order to concentrate on the basic steps of the hydrologic cycle.

◆ **TABLE 6.1** **Distribution of Global Near-Surface Water Supplies**

LOCATION	VOLUME		PERCENT
	Cubic Miles	*Cubic Kilometers*	
Oceans	286,230,000	1,192,625,000	96.7%
Fresh water	9,769,600	40,707,000	3.3%
DISTRIBUTION OF FRESH WATER			
	Cubic Miles	*Cubic Kilometers*	
Ice caps and glaciers	7,327,500	30,531,000	75.00%
Ground water below 800 meters (2,500 ft)	1,327,000	5,529,000	13.60%
Ground water above 800 meters (2,500 ft)	1,074,700	4,478,000	11.00%
Rivers and lakes	32,000	133,000	0.32%
Atmospheric moisture	3,400	14,000	0.03%
Soil moisture	5,000	20,800	0.05%
Total	9,769,600	40,707,000	100.00%

94,000 km³
(22,560 mi³)
20

106,000 km³
(25,440 mi³)   Precipitation   Atmospheric advection
22   of water vapor
                                          12
Land   57,000 km³
                                   (13,680 mi³)
                    Evaporation
                    Transpiration.   Evaporation
                  14   69,000 km³   86   419,000 km³
                       (16,560 mi³)   (100,560 mi³)
        Runoff   Precipitation
8   Runoff   78   382,000 km³
Groundwater   (91,680 mi³)
                          Lakes
                          37,000 km³
                          (8,800 mi³)
                                      Ocean

◆ **FIGURE 6.2** The hydrologic cycle. Values indicate mean annual volumes of water passing through each portion of the cycle. Figures in squares are percentages of total water flow.

the atmosphere in myriad tiny water droplets centered around dust particles that serve as condensation nuclei.

Most clouds do not gain enough moisture through condensation or sublimation to produce precipitation, but instead simply evaporate back into the air. Clouds that do generate precipitation, though, return water to the surface. In contrast to its long stay of perhaps thousands of years in the oceans, water remains in the atmosphere only a short time before precipitating out. Its estimated average residency time in the atmosphere is 10 days. If the clouds are still above the ocean, a three-step hydrologic cycle of evaporation, condensation, and precipitation results. Indeed, since 78 percent of the world's precipitation falls into the oceans, this is the most common pathway within the hydrologic cycle. However, we are usually much more concerned with the 22 percent of the precipitation that falls on land. The third step of the hydrologic cycle therefore involves the transport of the moisture from sea to land by wind currents either before or after condensation has occurred.

The fourth step is the precipitation of the water onto the land surface. Land areas in different parts of the world receive greatly differing amounts of precipitation, but the average is approximately 76 centimeters (30 in) of liquid water per year. Much of this moisture has been transported hundreds or even thousands of kilometers from the places where it was evaporated.

The fifth and final step of the hydrologic cycle is runoff, which eventually returns the water to the oceans. Once again, complications can arise, with the likelihood that much of the water will evaporate before it reaches the ocean. Flow to the sea may be on or beneath the surface. The water can also be stored for varying periods in rivers, lakes, and glaciers, or as groundwater. Eventually, though, the water completes its journey and is available to begin the cycle anew.

The hydrologic cycle, as described here, provides an outline for the material covered in the remainder of this chapter, as well as that treated in several subsequent chapters of this book. Each of the final three weather elements to be discussed—humidity, clouds, and precipitation—first appears at a particular point in the hydrologic cycle. Their analyses in a sense constitute studies of specific stages of the

Annual effective evapotranspiration in millimeters

0 – 499 mm (0 – 19.9 inches)	1000 – 1499 mm (40 – 59.9 inches)	2000 – 2500 mm (80 – 100 inches)
500 – 999 mm (20 – 39.9 inches)	1500 – 1999 mm (60 – 79.9 inches)	Data unavailable

◆ **FIGURE 6.3** World distribution of mean actual annual evapotranspiration rates. Globally, an estimated 14 percent of the atmosphere's water vapor is derived from the continents (including inland water bodies), while 86 percent evaporates from the ocean.

cycle. Later chapters on surface and subsurface waters (Chapter 9), weathering and mass wasting (Chapter 15), fluvial processes and landforms (Chapter 16), glacial processes and landforms (Chapter 17), and coastal processes and landforms (Chapter 19) cover the surface stages of the hydrologic cycle as well as its direct effects on the Earth's surface.

Evaporation

Evaporation obviously is a critical step in the hydrologic cycle, since it is the means by which the atmosphere obtains most of its water vapor. For the Earth as a whole, the amount of moisture entering the atmosphere by evaporation must, in the long run, approximate that leaving it as precipitation. However, although evaporation and precipitation may be considered equal in quantity on a global basis, their geographic distributions are very different.

A number of factors control the rate of evaporation from any point on the Earth's surface. One is the availability of water. In many land locations, the lack of surface moisture results in low evaporation rates. Water that evaporates from

land areas comes from the surface (including the soil), from water bodies such as rivers and lakes, and from water evaporated from plants—a process known as **transpiration.** In well-vegetated regions, water losses from transpiration can considerably exceed those from direct surface evaporation. The combined water loss from both evaporation and transpiration is termed **evapotranspiration.** Globally, an estimated 14 percent of the total atmospheric water vapor supply evaporates from the continents, including inland water bodies such as lakes and rivers, while 86 percent evaporates from the oceans. This means that, as a world average, the evaporation rates over land areas are about half those over the oceans per unit of area (Figure 6.3).

A second factor affecting evaporation rates is the temperature of surface water bodies because warm water evaporates more quickly than cold water. The amount and intensity of solar energy receipts are important in determining the supply of available heat.

Additional factors include the temperature and water vapor content of the air overlying the surface. Warm air can retain much more water vapor than cold air can (Fig. 6.5). Water therefore evaporates much more rapidly into warm air than into cold air, if other factors are equal. Evaporation rates at any given temperature also tend to be greater when the air is dry.

A final influence on evaporation rates is wind speed. Under calm conditions, evaporation produces a shallow layer of moistened air overlying the surface. The wind either removes this layer or mixes it with drier air from farther aloft, allowing evaporation to proceed at an increased rate. Well-known mechanical devices that use a combination of heat and airflow to promote rapid evaporation include electric hair driers and clothes driers.

Taking all these factors and relating them to what we know about the world distribution of land and water, temperature, pressure, and winds, we can make some generalizations concerning the global pattern of evaporation rates. One is that most evaporation occurs from the warm oceans of the lower latitudes (Figure 6.4). In fact, the leading source of the moisture received as precipitation for much of the world is the ocean surface in areas dominated by the subtropical highs. With their combination of clear skies, warm temperatures, and relatively dry air, the subtropics lose vast quantities of water to evaporation, and many subtropical areas receive little back as precipitation. On a global basis, evaporation greatly exceeds precipitation at latitudes between 10° and 40° in both hemispheres, while precipitation exceeds evaporation in all other latitudinal zones largely as a result of the transport of subtropical moisture into those zones, where it is released by storms. Because of temperature and vegetation factors, evaporation rates outside the tropics also are generally higher in summer than in winter and during the day than at night.

Humidity

When water evaporates, it contributes to the air's **humidity,** the amount of water vapor contained by the air. The atmosphere's water vapor content can be stated in a variety of ways, so several different types of humidity measurements exist. The two most commonly used in the atmospheric sciences are specific humidity and relative humidity.

Specific Humidity

Specific humidity is the ratio of the mass of water vapor in a parcel of air to the total mass of the air, including the water vapor. It usually is given in grams per kilogram, so that the numerical value is in easily calculable parts per thousand (Figure 6.5). Specific humidity values give the actual water vapor content of the air. They are unaffected by changes in temperature or air pressure—factors that influence the values of most other humidity measurements, including relative humidity. The geographic distribution of specific humidity by latitude is shown in Figure 6.6 on page 110. Because warm air can hold more water vapor than can cold air, mean specific humidity values are highest in the tropics and decline poleward.

◆ **FIGURE 6.4** Continental and oceanic pattern of annual evaporation and precipitation totals by 10-degree increments of latitude. For the world as a whole, evaporation and precipitation totals are equal. Note the influence of the global air pressure belts; evaporation exceeds precipitation only in the subtropics. Continents at all latitudes experience excess precipitation over evaporation.

Relative Humidity

Relative humidity is the most commonly used measure of atmospheric water vapor; it is also the one most familiar to the general public. Its popularity probably results from the fact that relative humidity readings correspond rather well to the sensible moisture content of the air; that is, to the apparent moisture content felt by people. It is also strongly correlated with evapotranspiration rates and with vegetative water requirements. It is, however, a poor indicator of the air's total water vapor content.

By definition, the **relative humidity** is the percent ratio of the amount of water vapor actually in the air to the maximum amount of water vapor the air could hold at that temperature and pressure. Put another way, the relative humidity indicates what percent of its capacity of water vapor the air is currently holding. The mean relative humidity of the air near the Earth's surface is about 78 percent. The atmosphere therefore is holding an average of 78 percent of its capacity of water vapor at its existing temperature and pressure.

Unlike the specific humidity, the relative humidity is directly affected by changes in air temperature. Because the air's capacity for holding water vapor increases with a rise in temperature, as shown in Figure 6.5, heating the air causes the relative humidity to fall. This occurs because the amount of water vapor contained by the air, though unchanged, now constitutes a smaller percentage of the air's increased capacity to hold this moisture. Similarly, if the air becomes cooler, its water vapor holding capacity is reduced, and the relative

CALCULATING HUMIDITY VALUES

As we have seen, the water vapor content of the air can be expressed in a variety of ways. The two most important for physical geographers are the specific humidity and the relative humidity. Fortunately, with the help of the accompanying graph (Figure 6.5), it is easy to calculate either of these values or to convert from one measure to the other.

It must be remembered that the relative humidity (RH) value does not represent a fixed amount of atmospheric water vapor; rather, it indicates, as a percentage value, the air's present water vapor content compared to its maximum capacity to hold water vapor at that temperature. We can state this relationship by means of a formula, as follows:

$$RH = \frac{\text{water vapor content}}{\text{water vapor capacity}} \times 100\%$$

◆ **FIGURE 6.5** Maximum specific humidity values for air at different temperatures.

humidity rises, even though no new water vapor is added to the air. As a consequence, relative humidity readings usually fall during the day as the air is heated by insolation, and rise during the night as the atmosphere cools (Figure 6.7).

Figure 6.8 depicts the mean distribution of relative humidity values on a latitudinal basis. While the global distribution of specific humidity values is closely correlated with temperature, the distribution of relative humidity values is strongly correlated with the global air pressure belts. The highest relative humidity values are associated with the ITCZ and the subpolar lows, where cloudiness and precipitation are prevalent. The lowest values are associated with the subsiding air of the subtropical highs. Because maritime air passes over large surface water bodies, it typically has higher relative humidity values than does continental air. By the same token, relative humidities in the ocean-dominated southern hemisphere average slightly higher than those in the northern hemisphere.

Specific humidity (SH) values, when known, can also be used to determine the relative humidity by means of the following formula:

$$RH = \frac{\text{current SH}}{\text{maximum SH at that temperature}} \times 100\%$$

Likewise, the specific humidity can be easily calculated if the current temperature and relative humidity are known, as follows:

$$SH = \frac{\text{maximum SH at current temperature}}{} \times \text{current RH}$$

Let's use these relationships to solve a few simple problems:

1. What is the current relative humidity reading if the temperature is 10°C (50°F), and the specific humidity is 2g/kg?

Solution: Our graph indicates that a kilogram of air at 10°C has a maximum water vapor holding capacity of 7.5 g/kg. Therefore, the current elative humidity is ($^2/_{7.5}$) × 100%, or *26.7 percent.*

2. You may recall that the mean global surface air temperature is approximately 15°C (59°F), and the mean relative humidity is 78 percent. Under these conditions, what is the specific humidity reading?

Solution: According to the graph, the maximum specific humidity at 15°C is approximately 10 g/kg. Seventy-eight percent of this value is *7.8 g/kg,* which is the approximate global mean surface specific humidity value.

3. If the current specific humidity is 15 g/kg and the relative humidity is 65 percent, what are the present air temperature and the dew point?

Solution: We first need to determine the air's water vapor capacity at the present temperature. If we divide 15 g/kg by 65 percent (or .65), we get 23.1 g/kg as the air's maximum capacity. Referring to the graph, we see that the temperature with that value as the maximum water vapor capacity is 27°C (80°F). We consult the graph again to see at what temperature the air would be saturated with a specific humidity of 15 g/kg. This value is approximately *20°C (68°F),* the current dew point. ∎

Condensation and Sublimation: Processes and Products

Water that evaporates from the Earth's surface eventually returns as precipitation. In order for precipitation to occur, water vapor, one of the atmosphere's lightest gases, must first be converted to the much denser liquid or solid states. The phase change processes involved in these conversions are condensation and sublimation.

Under what conditions do condensation and sublimation occur? Obviously, they do not continually remove water vapor from the atmosphere, because the weather often is fair and dry. At other times, though, the occurrence of precipitation or the formation of dew or frost makes it apparent that water vapor is indeed being removed from the air. The key to these processes is cooling. It was noted earlier that cooling the air reduces its water vapor holding capacity, causing an increase in the relative humidity. If cooled sufficiently, the air will

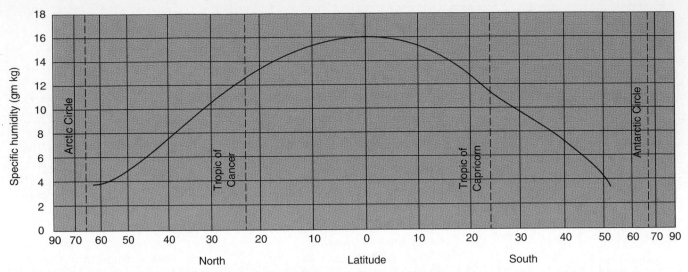

◆ **FIGURE 6.6** Mean specific humidity values by latitude. The dominant influence on this pattern is the effect of temperature on the water vapor holding capacity of the atmosphere.

reach a temperature at which the relative humidity is 100 percent and the air is saturated with water vapor. Any further cooling reduces the ability of the air to hold water vapor even more, producing an excess of moisture that must be removed by condensation or sublimation.

The temperature at which a given parcel of air will be saturated is its **dew point.** By indicating how much the air must be cooled in order to initiate condensation or sublimation, the dew point is important in predicting the likelihood of precipitation, dew, frost, or fog, and the amount and height of certain types of clouds. When the relative humidity is low, the dew point is much lower than the air temperature, and a large amount of cooling must occur before the air becomes saturated. When the air is saturated, the dew point and air temperature values are identical.

The forms taken by the condensed or sublimed moisture vary with temperature and location. Condensation occurs when the air is cooled below a dew point that is near or above freezing. The products of condensation are dew, fog, water droplet clouds, drizzle, and rain. Sublimation normally occurs when the dew point is well below freezing. Its three major products are frost, ice-crystal clouds, and snow. The remainder of this chapter is devoted to an examination of the causes and global distributions of these forms of atmospheric moisture.

Adiabatic Processes and Atmospheric Stability

Atmospheric cooling that leads to the formation of clouds and precipitation nearly always results from the rising of air. As a parcel of air rises, it enters areas of reduced atmospheric pressure and consequently expands. The expansion of the rising air requires energy, which is taken from the internal kinetic (or thermal) energy of the air molecules, making the air parcel cooler. When this cooling has reduced the air temperature to the dew point, the air has become saturated,

and any further rising will cause condensation and cloud formation to begin. Clouds thus are the visible tops of rising air currents. The cooling of air resulting from its expansion is termed **adiabatic cooling.** This cooling process involves no actual loss of energy, because the energy has not left the air but has merely changed form from sensible heat to mechanical energy. As long as the air keeps rising, it continues to cool, causing the condensation or sublimation of additional water vapor. This allows the cloud to grow larger and its constituent water droplets and ice crystals to increase in size, so that eventually they may become large enough to fall as precipitation.

When condensation and sublimation take place within a cloud, heat energy is released as the water vapor changes phase. This energy slows, but does not stop, the atmospheric cooling that occurs because of the expansion of the rising air. The released heat of condensation and sublimation is derived from solar energy that was originally used to evaporate water at the Earth's surface. It is a major source of energy

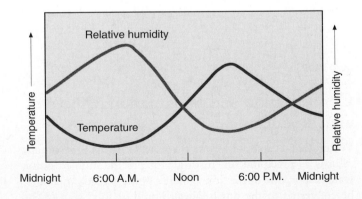

◆ **FIGURE 6.7** Typical temperature and relative humidity variations during a 24-hour period. Note the inverse relationship between these two weather elements.

◆ FIGURE 6.8 Mean relative humidity values by latitude. The dominant influence on this pattern is the distribution of the global air pressure belts.

that is available to power winds, storms, and other atmospheric phenomena.

Air that rises must eventually descend. When the air descends, it undergoes **adiabatic heating** because the surrounding air forces the descending parcel to contract, returning the energy it used in expansion.[3] As the descending air warms, its water vapor holding capacity increases, so that a cloud within a descending airflow will soon evaporate. This is why the descending air associated with high pressure systems produces fair, dry weather.

The stability of the atmosphere also plays a critical role in the formation of clouds and precipitation. Atmospheric stability is determined by the vertical distribution of temperatures within an air column. If temperatures near the surface are much warmer than those farther aloft, the relatively warm and buoyant near-surface air has a tendency to rise to the top of the air column, making this air **unstable.** If temperatures decline only slowly with height, however, or actually increase with height because of a temperature inversion, the near-surface air resists rising, making it **stable.** If forced to rise by any lifting mechanism, stable air will return to its original altitude when this becomes possible. Unstable air, however, tends to rise, and when given an initial lift, it will continue to rise on its own. The factors affecting the stability of the atmosphere are discussed in the Focus Box on page 119.

Dew and Frost

Dew and frost both form on the Earth's surface. Because they do not fall from the atmosphere, they are not techni-

cally considered forms of precipitation. **Dew** consists of water droplets that have condensed on surface objects because the overlying air has cooled below the dew point. It nearly always occurs at night and is most likely when the sky is clear, the wind light or calm, and the air moist. Under these conditions, terrestrial radiation causes the land surface to cool rapidly. Conduction with the cold surface then chills the overlying layer of air, forcing the excess atmospheric moisture to condense on available surfaces such as leaves and blades of grass. Dew normally evaporates in the morning as the air is warmed and its water vapor holding capacity increases. Dew is a relatively minor source of surface moisture, but typically provides land areas in the middle latitudes with 12 to 50 millimeters (0.5 to 2.0 in) of water per year.

Frost forms under nearly identical conditions, but when the dew point is below freezing. Frost is not frozen dew, but instead forms by sublimation directly from water vapor. All products of sublimation have a crystalline structure, and the feathery patterns formed by frost on windowpanes and other surfaces are well known (Figure 6.9).

Fog

Dew and frost, as we have seen, form when a layer of air just above the surface is cooled below its dew point, and the resulting moisture is deposited on objects at the surface. When a deeper layer of air is cooled, however, condensation or sublimation occurs on atmospheric dust particles. The result is fog. **Fog** is a visible accumulation of minute water droplets or ice crystals suspended in air immediately overlying the surface. It is actually a cloud at ground level and can reduce visibilities to less than 1 km (3,300 ft). In addition to the problems it causes by reducing visibilities, fog is generally associated with surface temperature inversions that may

---

3. An analogy might be made to a compressed spring, which uses energy when it is released and expands, but regains this energy when it is again compressed.

◆ FIGURE 6.9 The crystalline patterns formed on a window pane by a heavy frost indicate that it is a product of sublimation.
SOURCE: Edward F. D'Arms, Jr./Photo Researchers, Inc.

trap unhealthful quantities of air pollutants over urban and industrial areas.

The most common fog type over land areas is **radiation fog** (Figure 6.10). It forms under the same clear, calm, moist nighttime conditions that result in the formation of dew and frost. Its presence indicates that conditions for terrestrial radiation favor the cooling of an unusually deep layer of near-surface air, and it is associated with heavy dew or frost. Local topographic features often control the distribution of radiation fog. Over flat surfaces, it can form an extensive blanket covering many square kilometers. In more rugged areas, though, the downhill drainage of air chilled by contact with the cold surface often causes radiation fog to accumulate in valley bottoms, while the surrounding uplands remain clear.

Another important fog type, **advection fog**, forms when moist, mild air flows over a colder surface. Although it occurs less frequently worldwide, advection fog typically develops much more extensively than radiation fog and sometimes covers hundreds or even thousands of square kilometers. It also is generally deeper than a radiation fog and is likely to persist for an extended period.

The most favorable locations for advection fogs are ocean areas where warm and cold waters lie adjacent to one another. One notorious site is the Grand Banks area southeast of Newfoundland, where air that has traveled over the warm Gulf Stream passes over the adjacent, much colder water of the Labrador Current. Another well-known site is the California coast, where fog forms as mild Pacific air is chilled by its passage over cold water that has upwelled near the coast (Figure 6.11). Extensive advection fogs can also be produced when mild maritime air is transported over a cold, perhaps snow-covered, land surface in winter or a glacier surface in summer. A southerly flow of air from the Gulf of Mexico often produces dense winter fogs over the central United States in this manner.

The geographical pattern of fog frequency in North America is depicted in Figure 6.12. Fog is particularly prevalent in two types of settings: areas near large bodies of cold water, and mountainous regions. The cold-water coasts notably include the entire West Coast from southern Alaska to southern California, the northern Great Lakes states, coastal New England, and the Arctic Coast of Canada and Alaska. In each of these areas, fog typically is produced by damp air that is chilled below the dew point as it passes over the cold water. The chief mountain region frequented by fog is the Appalachians, where the air is cooled by being lifted by the elevated surface. It is interesting to note that even

◆ FIGURE 6.10 Radiation fog in the Appalachians is typically concentrated in valleys.
SOURCE: J. Irwin/H. Armstrong Roberts

◆ **FIGURE 6.11**   Advection fog drifting inland from the chilly Pacific partially shrouds San Francisco's famed Golden Gate Bridge.
SOURCE: NOAA

though the mountains of the western United States are much higher than the Appalachians, the western mountains typically experience much less fog because of their continental climates and dry air.

Clouds

Clouds, with their endless variety and constantly changing patterns of form and color, add substantially to the beauty of the natural environment. They also provide valuable information about the operation of often complex atmospheric processes, and most important of all, they produce virtually all of the Earth's precipitation.

Like fog, **clouds** are accumulations of minute water droplets or ice crystals, suspended in the air, that have condensed or sublimed around dust particles because of atmospheric cooling. Clouds may be differentiated from fog by the fact that they are not in contact with the surface. The diameter of a typical cloud droplet is on the order of 0.01 millimeters (0.0004 in), and its mass is about one-millionth of a gram. Particles of this size and mass can remain suspended in the air almost indefinitely.

Most clouds consist largely or entirely of water droplets. This is true both because warmer air typically contains more water vapor for cloud formation and because tiny cloud droplets, with their high surface tension values, can remain in the liquid state at temperatures well below freezing. Clouds are composed mostly of water droplets down to a temperature of about −7°C (20°F), a mixture of supercooled water droplets and ice crystals between −7°C and −39°C (20°F and −38°F), and entirely of ice crystals at temperatures

below −39°C (−38°F). Because very cold temperatures are needed for their formation, ice-crystal clouds generally are high clouds, located in the upper troposphere.

CLOUD CLASSIFICATION. Although an unlimited variety of clouds exists, 10 basic types are commonly recognized on the basis of their altitude and appearance. Their names are derived from five Latin root words: *cumulus, stratus, cirrus, nimbus,* and *altus.*

Cumulus means "pile" or "heap." All cumuliform clouds have an uneven, puffy appearance produced by individual parcels of unstable air that rise to form discrete cloud masses. *Stratus* means "layer." Stratiform clouds have a smooth, sheetlike appearance produced by the forced gentle uplift of stable air. They often are relatively thin, but may cover vast areas. *Cirrus,* which literally means "curl of hair," refers to high, ice-crystal clouds, often with a wispy appearance. *Nimbus,* which means "rain cloud," is used in conjunction with the two cloud types that produce nearly all of the Earth's precipitation. Finally, *altus,* which means "high," is used as the prefix "alto" in the names of two middle-level relatives of common low cloud types.

The 10 cloud types often are separated into four categories on the basis of height, as indicated in Figure 6.13. It is difficult to provide adequate verbal descriptions of clouds, so you should carefully examine the accompanying photographs of each cloud type in Figure 6.14 on pages 116−117 as you read the following paragraphs.

Three high cloud types are recognized. All are relatively thin, whitish in color in sunlight, and composed entirely of ice crystals. **Cirrus** clouds are the highest and most common

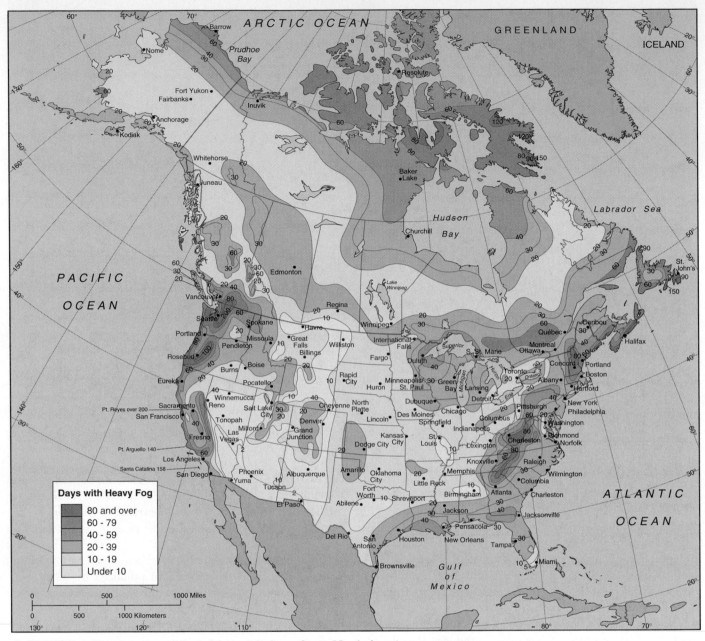

◆ **FIGURE 6.12**   Average number of days with dense fog in North America.

of the three. They have a distinctive wispy, hairlike, or "broom-swept" appearance. **Cirrostratus** clouds cover most or all of the sky in a smooth, thin white veil. The Sun and Moon shine through these clouds fairly brightly, but often are surrounded by a halo. **Cirrocumulus** clouds consist of a large number of very small, rounded, and sometimes iridescent cloud masses arranged in groups or lines. Cirriform clouds are often forerunners of storms in the middle latitudes, especially during the colder half of the year.

Two types of middle clouds exist, both with the prefix "alto." **Altocumulus** clouds are rather similar to cirrocumulus in that they consist of small, circular cloud masses arranged

in a linear pattern, but the individual masses appear considerably larger, darker in their centers, and have more distinct margins. **Altostratus** typically forms a smooth, textureless layer of gray cloud over the entire sky. The Sun may be visible as a diffuse bright spot, and a light, steady rain or snow may occur.

There are three types of low clouds, all stratiform in nature. **Stratus** is a dense, uniformly gray cloud, usually covering the entire sky and capable of producing a light drizzle. Common on dull, damp mornings, it is the lowest cloud type and often is formed by the lifting of a fog layer. **Stratocumulus** clouds form a whitish or grayish layer with distinct cloud masses, usually in the form of rolls, between which patches of clear sky may show. These clouds sometimes produce light rain showers or snow flurries. **Nimbostratus** clouds are the important winter precipitation producers of the middle and high latitudes; they are associated with the large-scale forced uplift of stable air. They form a dark, generally textureless cloud layer often partially obscured by precipitation. From the surface, nimbostratus clouds often appear identical to stratus, but may be distinguished by their continuously falling precipitation.

The final two cloud types differ from the others in that they display a considerable degree of vertical development. Both are cumuliform and are associated with unstable air. **Cumulus** clouds are the world's most familiar cloud type. A common sight on summer afternoons, they consist of distinct individual cloud masses with flat bases and uneven sides and tops. The flat bases of cumulus clouds indicate the level at which rising and expanding masses of warm, near-surface air have cooled below the dew point, thereby initiating condensation. Cumulus clouds typically have sharply defined outlines and can assume a great variety of shapes. If especially large, they can generate showers. The **cumulonimbus** is the most visually impressive cloud type. It is the chief precipitation producer of the tropics and, probably, of

the world as a whole. It is associated with heavy rain, lightning and thunder, and occasionally hail, destructive winds, and even tornadoes. Cumulonimbus clouds form in moist air that is unstable to great altitudes. They look like overgrown cumulus clouds, which is exactly what they are. A large cumulus cloud (cumulus congestus) becomes a cumulonimbus when its top grows high enough for the supercooled water droplets to freeze, or **glaciate.** Cumulonimbus clouds are by far the deepest cloud types. They are usually at least 6,000 meters (20,000 ft) thick and can occasionally attain depths of 15,000 or even 18,000 meters (50,000 to 60,000 ft), as they extend from perhaps 1,000 meters (3,000 ft) above the surface up to the tropopause. Because their great thickness prevents most light from penetrating to the surface, their bases typically appear ominously black.

CLOUD DISTRIBUTION. Clouds have a complex and constantly changing geographic pattern that is influenced by the interplay of a variety of factors. These include surface land and water features, pressure and wind systems, fronts, and daily and seasonal variations in temperature. Satellite observations have indicated that an average of about 52 percent of the world is covered by clouds at any one time. Distinctive cloud bands typically demarcate the locations of the ITCZ

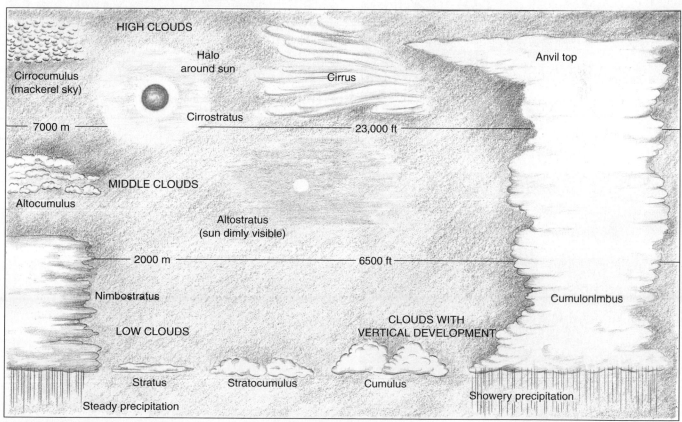

◆ **FIGURE 6.13** Generalized depiction of the 10 major cloud types based on height above the surface and extent of vertical development.

◆ **FIGURE 6.14A** Cirrus
SOURCE: Camerique/H. Armstrong Roberts

◆ **FIGURE 6.14B** Cirrostratus

◆ **FIGURE 6.14C** Cirrocumulus
SOURCE: Richard Jacobs/© JLM Visuals

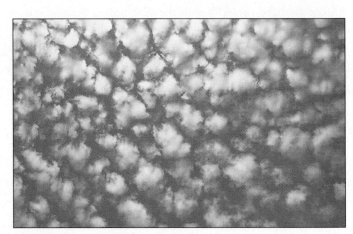

◆ **FIGURE 6.14D** Altocumulus
SOURCE: Graham Macfarlane/Tony Stone Images

◆ **FIGURE 6.14E** Altostratus

◆ **FIGURE 6.14F** Stratus
SOURCE: Carl Wilmington

◆ **FIGURE 6.14**G Stratocumulus

◆ **FIGURE 6.14**H Nimbostratus

◆ **FIGURE 6.14**I Cumulus

◆ **FIGURE 6.14**J Cumulonimbus
SOURCE: Jerry Scott

and the subpolar lows. Areas of cold ocean waters in the subtropics and the tops of high mountain ranges also are frequently cloud covered. In addition, the middle and higher latitudes experience a constant progression of distinctive cloud patterns attending traveling cyclonic storms and their associated frontal systems. These are examined in more detail in the next chapter. As a general rule, atmospheric heating causes cumuliform clouds to be more common in the lower latitudes, during the day, and in the summer. The existence of cooler, more stable conditions causes stratiform clouds to predominate in the higher latitudes, at night, and during the winter.

The tropopause forms a "lid" on vertical cloud development. Tropospheric temperature conditions favor cloud formation because temperatures decline with altitude, thereby encouraging the uplift of relatively warm and buoyant near-surface air. Above the tropopause, however, the temperature trend is reversed (see Chapter 3). The characteristic flattened "anvil" shape associated with thunderheads often develops as the rapidly rising tops of cumulonimbus clouds encounter the stable tropopause and spread out (Figure 6.14j).

Precipitation

Precipitation is water, in either liquid or solid form, falling through the atmosphere toward the surface of the Earth. If it did not occur, life on land likely would never have evolved. Without the physical and chemical influences of water, Earth's landforms would be strikingly different in appearance, and a true soil cover would not have formed. Even the oceans, without the receipt of nutrients eroded from the land, would be far less capable of supporting life than they are today. The condensation and sublimation from which precipitation is derived are also associated with the release of huge amounts of heat energy to the atmosphere. For example, a rainfall of one centimeter (0.4 in) on one square kilometer (0.4 sq mi) of surface is associated with the release of a quantity of heat equivalent to that produced by burning approximately 1,400,000 kilograms (3,000 tons) of coal.

ATMOSPHERIC STABILITY

Convectional lifting is closely associated with the vertical distribution of temperatures within the troposphere. This factor controls the stability of the atmosphere. We saw earlier that the average environmental lapse rate within the troposphere is 6.4C° per 1,000 meters (3.5F° per 1,000 ft). We also noted that rising air cools adiabatically as it enters areas of decreased air pressure aloft. If the rising air is unsaturated, it cools at a constant rate of approximately 10C° per 1,000 meters (5.5F° per 1,000 ft). This value is termed the **dry adiabatic rate.**

Let us imagine that a parcel of unsaturated air at sea level with a temperature of 70°F is forced to rise to an altitude of 4,000 feet on a day when the average environmental lapse rate exists, and is then released to test its stability (Figure 6.15). What will happen? Recall that if it is stable, the air will descend to its original level, but if it is unstable, it will continue to rise and perhaps produce clouds and precipitation. While being lifted, the air expands and cools adiabatically by a total of 5.5F° × 4, or 22F°, so that its temperature at 4,000 feet is 48°F. The surrounding air, however, is only becoming cooler at a rate of 3.5°/1,000 feet, so that at a height of 4,000 feet, the outside air temperature is 56°F. The parcel of air that was lifted is thus substantially cooler and denser than the air into which it was lifted. It therefore is stable. When released, it

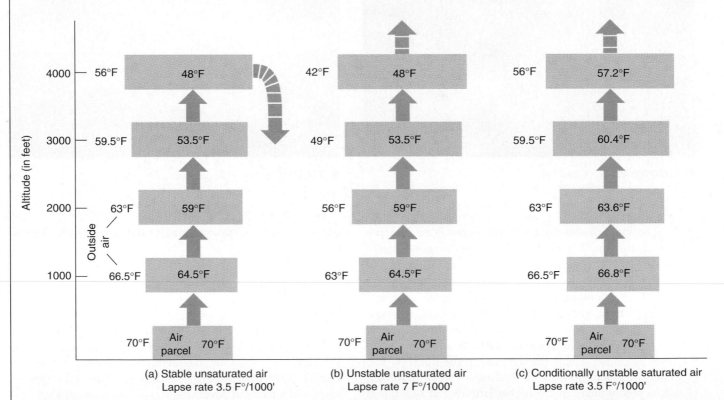

◆ **FIGURE 6.15** Temperatures of three rising air parcels as compared to the temperatures of the surrounding air. The air parcels are unstable if they are warmer than the surrounding air, and will continue to rise on their own. If they are cooler than the surrounding air, they are stable and will have a tendency to descend.

will return to the surface, warming at the dry adiabatic rate, so that when it reaches its original level, its temperature is once again 70°F.

The experiment can be repeated with a similar air parcel on a day when the environmental lapse rate is 7F°/1,000 feet. Once again the air parcel is lifted a total of 4,000 feet and cools adiabatically to 48°F. This time, however, the surrounding air at the 4,000-foot level is only 42°F. As a result, the released air parcel is warmer and lighter than the surrounding air. It therefore is unstable and will continue to rise on its own.

A complication occurs if the air parcel is saturated or becomes saturated while rising. In this case, adiabatic cooling causes the condensation or sublimation of some of the water vapor, releasing heat to the air. The released heat slows the cooling process to an average rate for air in the lower troposphere of 6C°/1,000 meters (3.2F°/1,000 ft). This reduced rate of cooling for rising, saturated air is termed the **wet adiabatic rate.** Unlike the dry adiabatic rate, the value of the wet adiabatic rate varies with the rate of condensation or sublimation. More condensation occurs per unit of cooling for warm air than for cold air (Figure 6.5); therefore, the wet adiabatic rate becomes progressively higher

as the air becomes colder. To avoid undue complication, however, we will assume a constant wet adiabatic rate of 6C°/1,000 meters (3.2F°/1,000 ft).

If we now repeat our first rising air experiment with saturated air, we find that the outcome has changed. At a level of 4,000 feet, the surrounding air temperature is still 56°F, but because of the released heat of condensation, our rising air parcel has cooled only to 57.2°F. It therefore is unstable and will continue to rise. This illustrates the fact that saturated air is considerably less stable than unsaturated air.

The conclusion that can be drawn from these examples is that *air is unstable if the environmental lapse rate exceeds the operative adiabatic rate.*

More specifically, unsaturated air is unstable if the lapse rate exceeds 10C°/1,000 meters (5.5F°/1,000 ft), and saturated air is unstable if the lapse rate exceeds the operative wet adiabatic rate. If the air in either case has a lapse rate lower than these values, it is stable (Figure 6.16).

A point to note is that the lapse rate is likely to vary considerably at different altitudes, and that a parcel of air will continue to remain unstable and to rise only as long as it remains warmer than the surrounding air. For this reason, cumulonimbus clouds and heavy convectional precipitation, caused by air that has risen for many kilometers, can occur only if the air is unstable to great heights. ■

◆ **FIGURE 6.16** Graphic depiction of the three categories of relative atmospheric stability. If the existing lapse rate is greater than 5.5F°/1000 ft (as illustrated by line A), the air is absolutely unstable. If the existing lapse rate is less than 3.2F°/1000 ft (as illustrated by line C), it is absolutely stable. If the existing lapse rate is between 3.2F and 5.5F°/1000 ft, the air is conditionally unstable—that is, it is unstable if saturated, but stable if unsaturated. (This situation is illustrated by line B.)

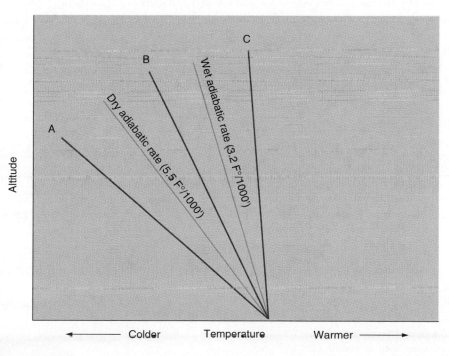

Line A—Absolutely unstable air (unstable either saturated or unsaturated)

Line B—Conditionally unstable air (unstable if saturated; stable if unsaturated)

Line C—Absolutely stable air (stable either saturated or unsaturated)

◆ **FIGURE 6.17** The pattern of convectional cumulus and cumulonimbus clouds over the Florida peninsula on a summer afternoon closely outlines the boundary between the warmer land and slightly cooler ocean.

SOURCE: NASA

Much of this energy escapes into space, but the rest is widely distributed to the Earth's atmosphere and surface. The great importance of predicting precipitation intensities and accurately measuring precipitation amounts has led to the development of a variety of weather instruments, as discussed in the Geographic Techniques Box on page 125.

Causes of Precipitation: Atmospheric Lifting Processes

We have noted that clouds and precipitation are produced by the rising and adiabatic cooling of air, but we have not yet examined the factors causing air to rise. Four basic atmospheric mechanisms can cause sufficient uplift to produce precipitation. They are convectional lifting, orographic lifting, frontal lifting, and cyclonic lifting.

Convectional lifting refers to the rising of unstable air because it is warmer, and therefore lighter, than the surrounding air. The chief way by which the air is warmed is by absorbing energy from a warm surface. Over land, this frequently occurs during the midday or afternoon hours because of surface heating by the Sun (Figure 6.17). Over water, convectional lifting may occur at almost any hour of the day or night because of the constancy of surface temperatures, but it is restricted largely to areas of relatively warm water. Convectional lifting is associated with cumuliform clouds and localized, showery precipitation. The summertime afternoon showers and thunderstorms commonly experienced in many areas in the middle latitudes are mostly products of this atmospheric lifting process. The brevity of a convectional shower results largely from the fact that the shower cools and stabilizes the air, thereby reducing its ability to continue to rise and prolong the precipitation (see the Focus Box on page 118). The other three lifting processes are produced by the forced uplift of air; they are typically characterized by lighter and more prolonged precipitation.

Because it is generally associated with warm, unstable air, convectional lifting occurs most frequently in the tropics, where it is the leading cause of precipitation in most areas. In the middle and especially the higher latitudes, it is less common and is restricted largely to the warm season. Convectional precipitation, however, is not always associated with warm weather. This is illustrated by the heavy convectional snowfalls that may be triggered by the passage of extremely cold arctic air over an unfrozen water body. (As an example, see the Case Study at the end of the chapter.)

Orographic lifting is the lifting of air over a mountain range or other elevated surface (Figure 6.18). In contrast to convectional lifting, orographic lifting involves the forced ascent of air due to the physical barrier imposed by the topography. As a result, the stability of the air is not a crucial

◆ **FIGURE 6.18** The orographic effect. The windward sides of mountains are associated with rising air, clouds, and precipitation; the leeward sides with descending air and dry conditions.

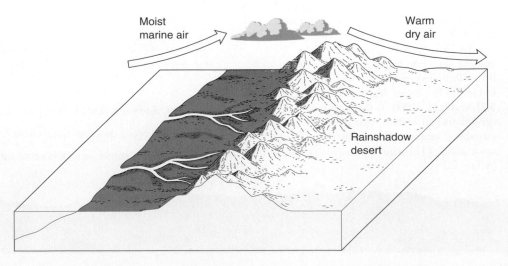

Moist marine air

Warm dry air

Rainshadow desert

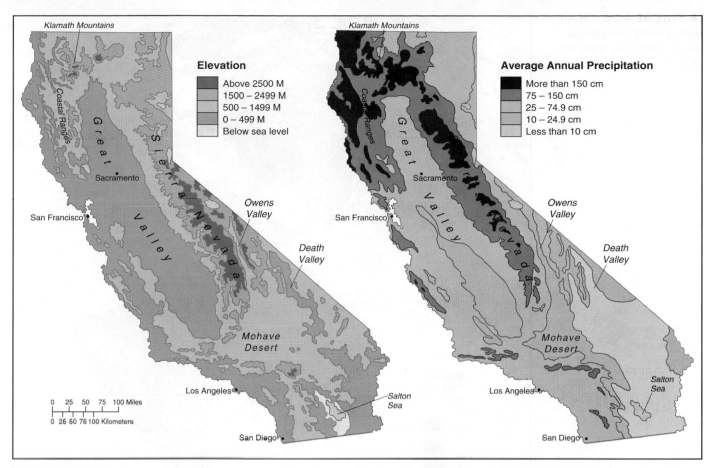

◆ **FIGURE 6.19** A comparison of these topographic and mean annual precipitation maps of California clearly illustrates the effect of surface topography on precipitation. The highest precipitation totals are associated with the Coast Ranges and Sierras.

factor, although unstable air is more effectively lifted by any lifting mechanism than is stable air.

Most clouds and precipitation produced by the orographic effect are concentrated on the side of the mountain range facing the wind (the **windward** side), because the forced uplift occurs here. As the air passes over the crest and descends the downwind, or **leeward,** side of the mountains, it is adiabatically heated, and progressively drier conditions prevail. An area that has a dry climate because it is on the leeward side of an orographic barrier is said to experience a **rainshadow effect.** The orographic effect can therefore either increase or decrease precipitation amounts.

Figure 6.19 illustrates the importance of the orographic effect in the mountainous state of California. A comparison of the two maps indicates that California's precipitation pattern is strongly controlled by the state's topography. Each of California's major mountain ranges has higher precipitation means than the surrounding regions, while lowland areas, especially those in the eastern part of the state, are relatively dry. On a larger scale, orographic lifting is responsible for the narrow band of heavy precipitation that occurs along the mountainous west coast of North America from southern Alaska to northern California, while the rain-

shadow effect to the east of these same mountains produces the aridity of the Great Basin region of the interior western United States and the interior valleys of western Canada (Figure 6.20).

Although orographic lifting can be an extremely effective precipitation generator, it is the most geographically restricted of the four atmospheric lifting mechanisms. It is normally strongly developed only in the vicinity of relatively high mountain ranges or plateaus, which comprise at most 10 percent of the Earth's surface. Because orographic lifting is tied to the Earth's surface features, it is incapable of movement. The other three lifting mechanisms are readily transported by the wind currents, greatly increasing the total surface area they influence, but decreasing the amounts of precipitation likely to be produced in any one location.

Frontal lifting is associated with the lifting of relatively light warm air by denser cold air along a weather front (Figure 6.21 on page 124). It is the primary cause of precipitation in most of the middle and higher latitudes. In the middle latitudes, frontal lifting is most prevalent in winter and early spring, when fronts are much stronger and more numerous than they are during the summer. In the high latitudes, fronts are fairly numerous throughout the

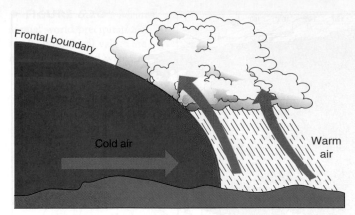

◆ **FIGURE 6.21** Frontal lifting occurs as a warm air mass is forced to rise over a denser cold air mass.

the water droplets by a mechanism known as the **Bergeron process** (Figure 6.22). The vapor pressure, or pressure to evaporate, is slightly higher for water than for ice at the same temperature. This is because water requires less energy to evaporate than does ice. As a result, the water droplets in a mixed cloud tend to evaporate; this water vapor is then sublimed onto the growing ice crystals, which eventually fall as snow. Because of the Bergeron process, a cloud composed almost entirely of water droplets, and with only a small amount of ice crystals, will produce snow rather than rain. In fact, nearly all precipitation outside the tropics begins falling as snow and simply melts to become rain as it passes through warmer air on its way to the surface.

Forms of Precipitation

Precipitation can assume a variety of forms, depending on atmospheric conditions both within the clouds and between the clouds and the Earth's surface. The two types of liquid precipitation are rain and drizzle. **Rain** is by far the most common and important form of precipitation; it comprises about 90 percent of the Earth's precipitation. By definition, it consists of liquid drops with diameters of at least 0.5 millimeters (0.02 in). **Drizzle** consists of droplets with diameters less than 0.5 millimeters (0.02 in). It falls only from stratus clouds developed from weak updrafts.

There are four major forms of frozen precipitation. Rain falling onto a subfreezing surface will freeze quickly on contact with whatever object it strikes to form a solid coating of ice. This makes **freezing rain** one of the most potentially dangerous forms of precipitation. Severe ice storms, although they may transform the landscape into a crystalline wonderland, are capable of causing widespread damage to power lines and trees because of the sheer weight of the accumulated ice (Figure 6.25).

When a considerable depth of the lower atmosphere has subfreezing temperatures, raindrops will freeze while falling to form sleet. **Sleet** consists of small, hard pellets of ice that

collect to form a crunchy, granular surface.[4] Both freezing rain and sleet indicate the presence of a temperature inversion and form when above-freezing air overlies subfreezing air (Figure 6.26).

The most common form of solid precipitation, comprising nearly a tenth of the Earth's precipitation total, is snow. **Snow** consists of crystalline ice formed directly from water vapor by sublimation. Its occurrence indicates that subfreezing (or near-freezing) conditions exist from the upper cloud levels all the way to the surface.

The most spectacular of the major precipitation forms is hail. **Hail** consists of relatively large, rounded particles of ice, which typically have a layered structure and are associated with severe thunderstorms. Hailstones form as strong updrafts in a thunderstorm, which may exceed 45 meters per second (100 mph) in velocity, carry falling ice pellets repeatedly above the freezing level in a cumulonimbus cloud. This allows multiple layers of ice to freeze onto the growing

---

4. In English-speaking countries outside North America, sleet is commonly considered to be a mixture of rain and snow.

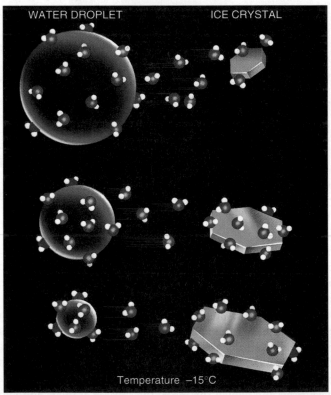

◆ **FIGURE 6.22** The Bergeron process. Water evaporates more quickly from super-cooled water droplets than from ice crystals because the phase change from liquid to vapor requires less energy than does the change from solid to vapor. Many of these water vapor molecules then sublime onto the ice crystals, causing them to grow larger, while the water droplets, which are losing water molecules faster than they are gaining new ones, grow smaller.

PRECIPITATION MEASUREMENT

The accurate measurement of precipitation is vital because of the influence of precipitation on soil moisture levels and crop growth, water supplies, and flood potentials. As a result, precipitation is measured at thousands of meteorological field stations around the world.

The instrument most commonly used for collecting and measuring precipitation in North America is the standard 20-centimeter (8 in) **rain gage** (Figure 6.23). It is a simple apparatus consisting of a funnel-shaped collector with a diameter of 20 centimeters (8 in), that feeds into a measuring tube 50 centimeters (20 in) long.

A much more sophisticated instrument that is increasingly used to estimate the intensity of falling precipitation is **radar.** Radars transmit pulses of microwave energy that are reflected from falling precipitation. The returning radar "echo" is plotted on a screen to depict the size, shape, and intensity of the precipitation area (Figure 6.24). Observed over a period of time, changes in the radar echoes indicate the direction and speed of movement of storms. ■

◆ **FIGURE 6.23** A standard 20-centimeter (8 in) rain gage consists of a narrow collecting tank with a mouth 20 centimeters in diameter. The barrel-shaped outer container reduces evaporation and provides basal support.
SOURCE: NOAA

◆ **FIGURE 6.24** Color radar depiction of a rain area fronted by a line of thunderstorms extending from southwestern Georgia into the Gulf of Mexico.
SOURCE: Courtesy of Alden Electronics, Inc., Westboro, Mass.

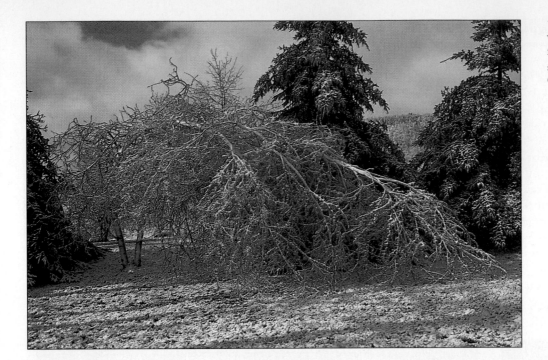

◆ **FIGURE 6.25** Heavily weighted tree branches following an ice storm.
SOURCE: Visuals Unlimited/ © W. Banaszewski

hailstone. Eventually, the hailstone either grows too heavy or misses a strong updraft and falls to the ground.

Hailstorms are most common in continental climate areas of the middle latitudes. In North America, they are particularly numerous in the western Great Plains and the Great Basin of the United States (Figure 6.27). Within the middle latitudes, they occur most frequently during the spring and early summer, when the atmosphere may be highly unstable and the freezing level is lower than it is later in the summer. Within the tropics, they are most common in uplands, where the hail has less time to melt before reaching the surface. Kericho, a town in the Nandi Hills of Kenya, may be the world's most hail-prone site, with an average of 132 hail days a year.

Most hailstones have diameters of a centimeter (0.4 in) or less, but some that were larger than baseballs have been recorded. Severe hailstorms can do great damage to crops, livestock, and buildings and can occasionally take human lives. The deadliest hailstorm on record killed 246 people in India in April 1888. Hail fatalities are rare in North America, but the United States probably leads the world in property damage caused by hailstorms. Damage in 1991, for example, was estimated at $412 million, mostly to crops. Most hailstorms last for only a few minutes, but this is not always the case. The longest-lasting storm on record in the United States struck Seldon, Kansas, on June 3, 1959. It lasted 85 minutes and covered the surface to a depth of 46 centimeters (18 in).

◆ **FIGURE 6.26** Vertical temperature profiles associated with various forms of precipitation.

Less than 1 day
1–3 days
3–5 days
More than 5 days

◆ **FIGURE 6.27** Average annual number of days with hail.

World Precipitation Distribution

The total amount of precipitation received at any given location depends upon the moisture content of the atmosphere and the duration and intensity of atmospheric lifting. The estimated average annual precipitation for the Earth as a whole is 90 centimeters (35 in). This value provides a useful point of comparison for assessing the annual total at any given site.

In the next several paragraphs, the basic characteristics and causes of the global pattern of precipitation are examined. As you read this material, it is a good idea to make frequent reference to Figure 6.20. More detailed information on regional precipitation patterns and their causes is provided in Chapter 8.

The equatorial zone, located between approximately 15°N and 15°S, is the world's rainiest zone of latitude. Most locations receive 125 to 300 centimeters (50 to 120 in) of rainfall annually. The air in the tropics is very warm and often unstable; typically, it has a large water vapor content, giving it a high rainfall potential. Rainfalls are further augmented by the convergence of air from the trade wind zones of both hemispheres into the ITCZ. The heaviest rainfall averages occur in areas where high mountains produce substantial orographic uplift.

The subtropics, comprising those areas between 15° and 35°N and S, are substantially drier as a whole than the equatorial zone, although great regional variations in precipitation means exist. The predominant precipitation influences here are the subtropical belts of high pressure with their subsiding air. In places where the highs are especially well developed, precipitation totals are so greatly reduced that arid or semiarid conditions prevail. Along continental

WORLD PRECIPITATION RECORDS

Precipitation records of various types have been set at a large number of meteorological stations around the world. Despite their widespread distribution, a certain geographic similarity is evident in the physical settings of these stations. For example, most rainfall and snowfall records were established at sites with a strong orographic effect. Most rainfall records were set in the tropics and subtropics. Most rainfall records for periods extending from several hours to several days have been associated with tropical storms and hurricanes. Most records for dryness have been established in subtropical deserts under the dominance of the subtropical highs.

A sampling of especially significant precipitation records includes the following:

◆ Greatest amount of rain received in one minute was 3.8 centimeters (1.50 in) at Barst, Guadeloupe, in the Lesser Antilles (November 26, 1970).

◆ Greatest 24-hour rainfall total was 187.0 centimeters (73.62 in) at Cilaos, Réunion, in the Indian Ocean (March 15–16, 1952).

◆ Greatest rainfall total received in a one-month period was 930.0 centimeters (366.14 in) at Cherrapunji, India (July 1861).

◆ Greatest rainfall total received in a period of one year was 2,646.1 centimeters (1,041.78 in) at Cherrapunji, India (August 1860–July 1861).

◆ Heaviest mean annual precipitation is 1,168 centimeters (460 in) at Mt. Waialeale, Kauai, Hawaii (1920–1972).

◆ Greatest 24-hour snowfall was 193 centimeters (76 in), at Silver Lake, Colorado (April 14–15, 1921).

◆ Greatest one-year snowfall was 3,110 centimeters (1,224.5 in), at Mt. Rainier, Washington (February 19, 1971–February 18, 1972). ■

west coasts paralleled by cold ocean currents, atmospheric lifting is further discouraged by the chilling of the air at the surface, and the world's driest deserts occur here. In contrast, the warm ocean currents that flow along the east coasts of continents in the subtropics serve to heat and destabilize the air as well as to add large quantities of water vapor. Precipitation amounts in these areas generally are substantial, especially if orographic or monsoonal influences are present.

The middle latitudes, which include areas between 35° and 65° N and S, are characterized by moderate precipitation amounts. Many areas receive annual totals near the global mean of 90 centimeters (35 in) per year. This is due largely to the influence of two factors. First, temperatures in the middle latitudes are mild because of the intermediate quantities of insolation received. This causes the air's water vapor holding ability, and therefore its potential for precipitation, to be moderate. In addition, the middle latitudes are located between high and low pressure belts—specifically, the sub-

tropical highs and subpolar lows. Because the effect of highs is to decrease precipitation amounts and that of lows is to increase them, the tendency exists for moderate totals. As is true of all latitudinal zones, however, orographic factors and proximity to water can have a profound influence on precipitation receipts. The heaviest precipitation totals are typically experienced along windward (west-facing) coasts backed by high mountains.

In the high latitudes, extending from 65°N and S to the poles, precipitation means are lower than in any other latitudinal zone. This is primarily a function of temperature. The cold air overlying these areas simply does not contain enough water vapor to generate large precipitation totals.

Other Important Precipitation Variables

Up to this point, the discussion of precipitation patterns has centered on the precipitation amounts received in different parts of the world. This is undoubtedly the most important

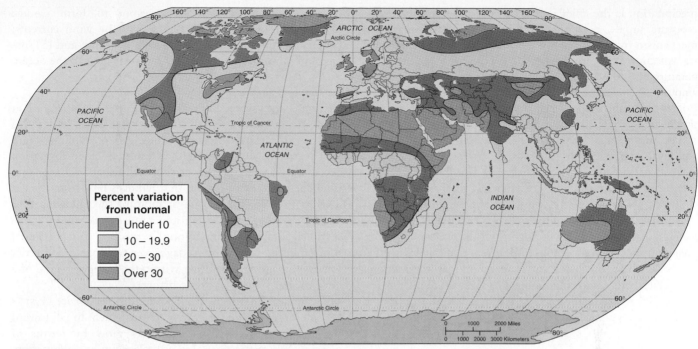

◆ **FIGURE 6.28** Mean annual precipitation variability. Values represent the average departure of a given year's precipitation from the long-term mean.

of the geographic precipitation variables, but several others also are noteworthy.

The *form* of the precipitation is of great practical significance. Most of the world's precipitation is rain, but large quantities fall in solid forms, especially as snow. Because of the greater availability of atmospheric moisture, annual snowfall totals typically are heavier in parts of the upper middle latitudes and subpolar regions than near the poles.

The *intensity* and *duration* of precipitation play major roles in determining how much moisture can be absorbed by the soil and how much will become runoff. High-intensity rainfalls may occur at rates exceeding the absorption capacity of the soil, especially in arid areas or in areas disturbed by human activities. This will lead to erosion and possible flooding. On the other hand, steady, gentle rainfalls allow the soil to absorb more moisture, causing runoff to be reduced and spread over a greater time span. Intense rainfalls occur only when substantial amounts of water vapor are available, and are usually produced by the convergence of unstable air. They are most common in the low latitudes and during the summer months in the middle latitudes.

Another factor that has a great effect on water supplies and agricultural activities is the **seasonal distribution** of precipitation. In many areas precipitation is relatively evenly distributed throughout the year, but other areas have distinctive wetter and drier seasons. Where precipitation is highly seasonal, the maintenance of a continuous water supply is a problem, and cycles of human activities, especially those relating to agriculture, are strongly affected. (The Case Study at the end of Chapter 5 provides an illustration.)

The **annual variability** of precipitation refers to the average percentage deviation of an area's precipitation from the mean. For example, if a location has a mean annual precipitation variability of 30 percent, a 50 percent chance exists that the total precipitation it receives in any given year will be at least 30 percent higher or lower than its long-term annual mean. A low annual variability of precipitation means that the area can be more confidently expected to receive a quantity of precipitation close to its mean value. A high variability, on the other hand, means that much less certainty exists as to the amount of precipitation to be received, and floods or droughts are more likely to occur. Figure 6.28 depicts the pattern of the mean annual variability of precipitation for the world.

A final factor that is not actually a precipitation factor, but nonetheless has a tremendous effect on surface water supplies, is the **potential evapotranspiration rate.** It may be defined as the rate at which water would be lost through the combined processes of evaporation and transpiration if the surface were constantly supplied with an unrestricted quantity of water. The balance between precipitation and evapotranspiration largely determines the "wetness" of an area. Of the total precipitation falling on land areas of the world, approximately 75 percent is evaporated or transpired; only 25 percent reaches the oceans as runoff. Therefore, while precipitation is the chief means by which the land gains its moisture supply, evapotranspiration is the most important mechanism by which this moisture is lost. Potential evapotranspiration rates are influenced primarily by the air temperature and to a lesser extent by the relative humidity and amount of sunshine. In general, they are highest in the tropics and subtropics, especially in dry areas, and become progressively lower through the middle and high latitudes.

The necessity of examining both precipitation amounts and evapotranspiration rates to determine the moisture balance of an area is well illustrated by the case of the interior of Antarctica. This area annually receives the liquid equivalent of only a few centimeters of precipitation, but is currently buried under approximately 3 kilometers (2 mi) of ice. The Antarctic ice has formed from the accumulation of precipitation over many thousands of years, coupled with the virtual absence of losses by evaporation or sublimation. Conversely, areas with high annual or seasonal temperatures require substantial amounts of precipitation to support a continuous vegetation cover.

▌SUMMARY

This chapter has examined the role of water in the atmosphere. Water is a unique atmospheric component because of its ability to undergo phase changes between solid, liquid, and gaseous states. In so doing, it produces a variety of atmospheric phenomena and, more importantly, makes possible the transfer of great quantities of water from the oceans to the land and of energy from the Earth's surface to the atmosphere. In order to review more systematically some of the most important information presented in the chapter, we now return to the chapter-opening Focus Questions:

1. *Under what conditions do water phase changes occur, and what role do they play in the global distribution of energy?*

Water undergoes phase changes when environmental conditions favor a gain or loss of energy. Energy is gained when ice melts or sublimes to water vapor, or when liquid water evaporates. Energy is released as heat to the environment when water vapor condenses or sublimes to ice, or when liquid water freezes. Because most surface water exists as liquid water and most atmospheric water exists as water vapor, the net result of the phase changes is the transfer of great amounts of energy from the surface to the atmosphere. This energy provides heat to the atmosphere, powers atmospheric processes, and is widely distributed around the Earth by wind currents.

2. *What is the hydrologic cycle?*

The hydrologic cycle is the continuous closed cycle of water transport between the surface and atmosphere. Although a number of alternate routes are taken, the following five steps comprise the basic elements of the hydrologic cycle as it involves land areas: (1) evaporation of ocean water;

(2) condensation of this water vapor to form clouds; (3) transport of the clouds to the land by wind currents; (4) precipitation of the moisture onto the land; and (5) surface and subsurface runoff of the moisture back to the ocean.

3. *How do clouds and precipitation form?*

Clouds form when air is cooled below the temperature at which it can hold all its moisture as water vapor. When this happens, the excess moisture condenses or sublimes onto dust particles. Most cooling that results in cloud formation occurs adiabatically as the air rises and expands. Expansion is an energy-using operation that removes heat from the air. The air rises because of one or more of the following four factors: atmospheric heating and resulting instability (convectional lifting); lifting along a weather front (frontal lifting); lifting caused by passage over a topographic barrier (orographic lifting); or lifting within the rising air spiral of a low pressure system (cyclonic lifting).

Precipitation forms as cloud water droplets or ice crystals become large enough for gravity to cause them to fall toward the Earth's surface. Cloud droplets grow by means of collision and coalescence. Ice crystals grow by sublimation, which is encouraged by the lower vapor pressure over ice crystals as compared to that over water droplets within a mixed cloud.

4. *What is the geographic pattern of precipitation, and why does it exist?*

The global precipitation pattern is complex because it is affected by many factors. In general, the largest totals occur in the tropics, where the air is very moist and low pressure prevails. Most portions of the subtropics experience varying degrees of dryness because of the influence of the subtropical highs. The middle latitudes typically receive moderate precipitation amounts caused largely by traveling frontal cyclones and summertime convectional showers. The high latitudes are the driest latitudinal zone because their cold air has limited water vapor holding capacity. In general, precipitation is highest near warm ocean currents and on the windward sides of mountain ranges.

▌QUESTIONS FOR THOUGHT AND REVIEW

1. Explain how water phase changes result in the transfer of energy from the Earth's surface to the atmosphere. How is this energy released within the atmosphere? What atmospheric processes does this energy subsequently power?
2. Draw a diagram of the hydrologic cycle and label the cycle's five basic steps. In which portion of the cycle does water remain longest? In which portion does it remain for the briefest period, on the average?
3. What is evapotranspiration? In what parts of the world are evapotranspiration rates highest and lowest? Explain why in each case.

4. Why is the relative humidity lower during the day than at night? Is the specific humidity also lower during the day? Explain why or why not.

5. What happens to the temperature of air as it rises? Why? What happens to the temperature of air as it descends? Why?

6. What environmental conditions favor the formation of fog? List three diverse geographical areas where fogs frequently form, and explain why.

7. What are clouds? How and why do they form? How are clouds classified? What are the two primary precipitation-producing cloud types, and how do their precipitation characteristics differ?

8. Why do some clouds produce precipitation, while others do not? Does most precipitation originate from near the bottom or the top of a cloud? Explain why.

9. Describe the six basic forms of precipitation and briefly explain the atmospheric and/or surface conditions under which each forms.

10. What factor determines whether the atmosphere is stable or unstable? Why is air generally less stable during the day than at night? Is air typically less stable when it is saturated or unsaturated? Why?

11. Explain the reasons for each of these aspects of North America's precipitation pattern:
    A. Most summer thunderstorms occur during the afternoon hours.
    B. The west sides of mountain ranges typically are wetter than the east sides.
    C. Winter storms are often followed by major temperature changes.
    D. The United States gets more precipitation, on average, than does Canada.
    E. The southeastern United States gets more precipitation, on average, than does the southwestern United States.

12. What precipitation factors other than annual mean amounts received have important implications for the inhabitants of a region? Why is more precipitation generally needed for agricultural purposes in the tropics than in the high latitudes?

13. Explain why the Great Lakes snow belts develop. Would there be a tendency for heavier snow belt snowfalls in early winter or in late winter under equally favorable atmospheric conditions? Why?

PROBLEMS

1. How many calories of heat are required to evaporate 25 grams of water from the Earth's surface?

2. How many calories of energy are released to the atmosphere by the formation of 40 grams of frost?

3. If a kilogram of air near the surface has the capacity to hold 23 grams of water vapor but is currently holding only 14 grams, what is its relative humidity? What are its current temperature and its dewpoint?

4. If the current air temperature is 20°C and the specific humidity reading is 10 grams of water vapor per kilogram of air, what is the current relative humidity? (Hint: Refer to Figure 6.5.)

5. Is unsaturated air with a lapse rate of 7.5C°/1,000 meters (4F°/1,000 ft) stable or unstable? Is saturated air with this same lapse rate stable or unstable?

KEY TERMS

evaporation	dew
condensation	frost
sublimation	radiation fog
hydrologic cycle	advection fog
evapotranspiration	cloud types (ten major
specific humidity	types)
relative humidity	convectional lifting
dew point	orographic lifting
adiabatic temperature	frontal lifting
changes	cyclonic lifting
stable and unstable air	

Snow Belts of the Great Lakes

The Great Lakes moderate the climate of the areas that surround them, and because the prevailing winds blow from the west, the influence of the lakes is especially strong along their eastern shores. During the spring and summer, the lake waters are colder than the land, and the shorelines receive cooling lake breezes. During the fall and winter, the lakes are warmer than the surrounding land and warm the areas to their east. This has the beneficial effect of delaying autumn frosts and has permitted the planting of extensive orchards of cold-sensitive fruit trees. The same warming influence, however, is responsible for the formation of the heavy lake-effect snowstorms that plague the lee shores of the Great Lakes each winter. Each of the five lakes has a distinctive snow belt along its southern and eastern shores.

The lake-effect snowstorms are highly localized storms of a predominantly convectional nature. They form following the passage of strong cold fronts, when frigid arctic air is carried on gusty west or northwest winds over the unfrozen waters of the Great Lakes. The air picks up large quantities of water vapor in its lower levels during its passage over the lakes, providing the moisture for heavy snow squalls.

The key factor responsible for lifting this moistened air is the air's great temperature contrast with the water of the Great Lakes. Arctic air, as it reaches the lakes, may have temperatures well below –18°C (0°F). As it passes over the water, which usually is a few degrees above freezing, it suddenly encounters a surface 15 to 30C° (30 to 60F°) or more warmer than it is. Large quantities of this heat are conducted to the overlying air, making it much lighter and more buoyant than the unmodified air farther aloft. This produces conditions of extreme instability. Convectional uplift of the air results, producing cumuliform clouds and heavy snow squalls. Latent heat energy, released as the water vapor condenses and sublimes, enhances conditions for uplift.

When the air reaches the hilly terrain along the eastern and southern shores of the lakes, orographic and frictional influences come into play to produce further lifting. The low hills, trees, and buildings act as barriers to reduce the speed of the strong winds that have developed over the open waters of the lakes, causing the air to pile up and rise. Higher hills are encountered farther inland, especially leeward of Lakes Erie and Ontario, and produce a further significant orographic uplift.

The configuration of the topography and the wind direction seem to be the two factors that exert the greatest control on the distributional pattern of snow squalls at any given time. If the wind blows across the narrow axis of a lake, the amounts of heat and water vapor it picks up are limited, and the resulting snow showers tend to be widely scattered and of light-to-moderate intensity. When the wind blows down the long axis of a lake, however, large quantities of heat and moisture are absorbed, and narrow bands of heavy snowfall typically result. Along the eastern shores of Lakes Erie and Ontario in upstate New York, these bands usually are 15 to 50 kilometers (10 to 30 mi) wide and extend inland from 80 to 160 kilometers (50 to 100 mi). Their edges are often very sharply defined, so that one site may experience a heavy snowfall, while another site a short distance away receives little or nothing. These localized snowstorms are sometimes accompanied by lightning and thunder, and on occasion funnel clouds may form.

One of the most remarkable aspects of the lake-effect snowstorms is their persistence in any given area, despite their localized nature and the rapid movement of the clouds that produce them. In most parts of the world, a convectional storm will influence a given location for only a short period, perhaps an hour or less. This occurs largely because these storms are produced by the surface heating of the air, and the heat is soon dissipated by the cooling effect of the storm. The waters of the Great Lakes, however, generate a continuous supply of heat, and as long as the wind direction is constant, new snow clouds tend to form along the same band to replace those that are streaming inland and dissipating. Under extreme conditions, this can produce continuous heavy snowfall for a period of several days, resulting in amazingly large snow depths.

The winter of 1976–77 was a record-setting year for lake-effect snowstorms in many portions of the eastern Great Lakes region. It was caused by a jet stream pattern that produced a nearly continuous flow of arctic air over the Great Lakes for most of the winter. The season was capped by the great snowstorm of January 28 through February 1, which produced snowfall totals exceeding 250 centimeters (100 in) in some locations southeast of Watertown, New York (Figure 6.29). This storm, which paralyzed the city of Buffalo, produced snowdrifts more than 7.5 meters (25 ft) high and resulted in 29 deaths. Winds gusted to 137 kilometers per hour (85 mph), creating zero visibilities in blowing snow for much of the five-day period, and producing wind-chill readings as low as –51°C (–60°F). For the entire 1976–77 winter season, Buffalo received a record 506.5 centimeters (199.4 in) of snowfall, but Hooker, New York—a small town east of Lake Ontario—was buried under 1,185.9 centimeters (466.9 in) of snow! Figure 6.30 depicts the 1976–77 snowfall totals for the snow belt areas of Lakes Erie and Ontario. The much higher totals to the east of Lake Ontario resulted from the fact that relatively shallow Lake Erie froze over in midwinter, cutting off its moisture supply, while Lake Ontario, which is much deeper, remained unfrozen.

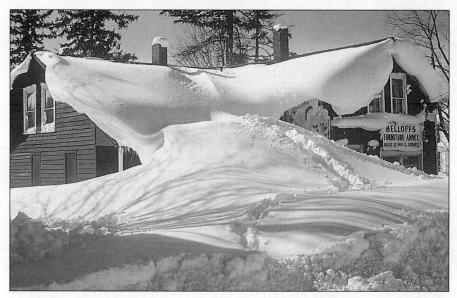

◆ **FIGURE 6.29** A view of snow depths in Adams, New York, on February 2, 1977.
SOURCE: NOAA

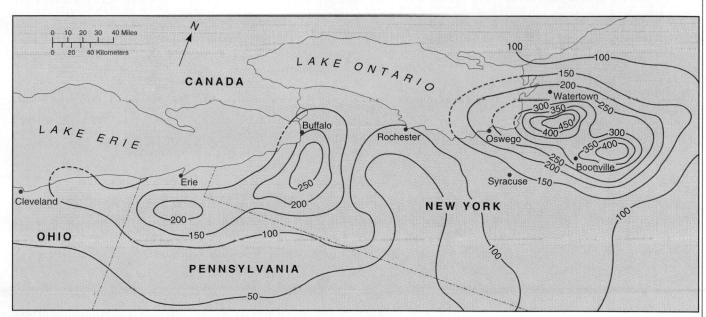

◆ **FIGURE 6.30** Snowfall totals in inches for the 1976–77 winter season in the snow belt region of Lakes Erie and Ontario.
SOURCE: *Weatherwise*, "Lake-effect Snowstorms. . ." by Kenneth F. Dewey, Dec. 1977. Reprinted with permission of the Helen Dwight Reid Educational Foundation. Published by Heldref Publications, 4000 Albemarle St., N.W., Washington, D.C. © 1977.

7 WEATHER SYSTEMS

OUTLINE

FOCUS QUESTIONS

1. What are air masses? How and why do they form, and what influence do they have on our weather?
2. What are the most important weather systems of the middle and higher latitudes, and what types of weather are associated with them?
3. How do thunderstorms and tornadoes form, and what areas are most affected by them?
4. How and why do weather conditions of the tropics differ from those of the higher latitudes?
5. What are hurricanes? How and where do they form, and what land areas do they most affect?

In the predawn hours of August 24, 1992, Hurricane Andrew, one of the most powerful Atlantic hurricanes of the twentieth century, swept ashore just south of Miami, Florida. Andrew's peak winds, gusting to speeds of well over 78 meters per second (175 mph), devastated South Florida, producing the costliest natural disaster in U.S. history. (The story of this storm is recounted in greater detail in the Case Study at the end of the chapter.) How do major storms such as Hurricane Andrew develop? As terrestrial supplies of moisture and thermal energy are transported over the face of our planet, they tend to become organized into discrete patterns that comprise **weather systems** such as air masses, weather fronts, and storms. Now that the components of the weather have been introduced, it is time to put them together to discover how and why weather systems develop and what areas of the Earth they influence.

Some weather systems, notably air masses, occur in all parts of the world. Because they influence weather conditions everywhere, they will be examined first. Most other weather systems are restricted primarily to certain zones of latitude. For example, fronts, frontal cyclonic storms, and tornadoes are predominantly nontropical phenomena; easterly waves and hurricanes originate within the tropics. Following the discussion of air masses, the weather systems of the middle and high latitudes will be treated. This will be followed by a discussion of the systems of the tropics.

AIR MASSES

As television weather forecasters make clear, air masses are important controllers of present or future weather conditions. Statements such as "A frigid Canadian air mass should reach our area by tonight . . ." or "A continued onshore flow of cool, damp Pacific air can be expected . . ." are familiar to all of us. But what exactly are air masses, and why do they form?

An **air mass** is a large body of air with relatively uniform temperature and moisture characteristics. A typical air mass has a diameter of 800 to 2,400 kilometers (500 to 1,500 mi). Its properties develop gradually over several days as the air lies relatively stationary over a large land or water body. During this period, the temperature and moisture characteristics of this surface area, termed the **source region** of the air mass, are gradually imparted to the overlying air.

Air Mass Formation and Types

Air masses form with much greater frequency over some areas than others. An important source region has several attributes. Like the air masses that form over it, it must be large in size and homogeneous in its temperature and moisture characteristics. For this reason, it must be either a land or a water body, not a combination of both. The source region should frequently experience light upper-level winds, because strong winds will transport air near the surface out of the region too rapidly for the air mass to attain its characteristics. Finally, a source region should be an area that favors air subsidence and the development of high pressure. The centers of most air masses, in fact, appear on weather maps as high pressure systems. The diverging airflow associated with a high pressure system helps an air mass maintain its homogeneity because this flow pattern constantly pumps air from the center of the air mass over an expanding area. Because the air originates from one central location, it has similar temperature and moisture values throughout. Conversely, the converging winds associated with a low pressure system draw air of dissimilar characteristics inward toward a common central area. For this reason, lows are frequently associated with two or more air masses.

The principal source regions for the world's air masses are large land and water bodies in the high and low latitudes. This fact led to the development of a simple four-type classification system for air masses based on the nature of their source regions; the classifications also implicitly describe the temperature and moisture characteristics of the air masses themselves. Each classification consists of two words. The first indicates the surface nature of the source region. An air mass that forms over land is called *continental* and can be expected to contain dry air. Conversely, a *maritime* air mass forms over water and can be expected to have a high water vapor content, at least near the surface. The second word indicates the general latitudinal zone in which the air mass forms. A *polar* air mass forms over a polar or subpolar source region and is therefore relatively cold. A *tropical* air mass forms over a tropical or subtropical source region and is characterized by warm temperatures.

Combining these terms, we have an air mass classification system that consists of four types (see Figures 7.1 and 7.2). On weather maps, these air masses often are abbreviated by using the lowercase letters *c* and *m* for continental and maritime and uppercase *P* and *T* for polar and tropical as follows: continental polar (cP), maritime polar (mP), maritime tropical (mT), and continental tropical (cT). The source regions of these air masses are depicted in Figure 7.1 and the summer and winter characteristics of each is summarized in Table 7.1.

Meteorologists employ two other temperature terms for additional accuracy in classifying air masses. They are *arctic* (abbreviated *A*) and *equatorial* (abbreviated *E*). Use of these terms gives rise to two additional air masses. One is continental arctic (cA) air, which is extremely cold and dry. In the northern hemisphere, it forms over northern Canada, Alaska, and Siberia only in the colder half of the year. It is responsible for the occasional winter cold waves that affect most of the United States and Europe. In the southern hemisphere, it forms over the continent of Antarctica, where

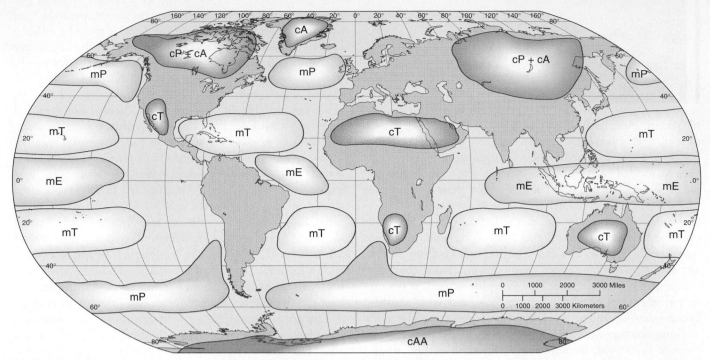

◆ **FIGURE 7.1** Major air mass source regions of the world.

(a)

(b)

(d)

(c)

◆ **FIGURE 7.2** Typical views of the source regions of the four major air mass types, clockwise from upper left: cP— (a) black spruce forest in interior Alaska; cT—(b) sandy desert near Parachilna, South Australia; mP—(c) the coast of Scotland near Glenborrodale; mT—(d) beach scene at Seychelles, Aride Island.

SOURCE: a: © Glenn M. Oliver/Visuals Unlimited; b: © Bill Bachmann/Photo Researchers, Inc.; c: K. Scholz/H. Armstrong Roberts; d: Colin Prior/Tony Stone Images

	CONTINENTAL POLAR (cP)	MARITIME POLAR (mP)	MARITIME TROPICAL (mT)	CONTINENTAL TROPICAL (cT)
Source regions	Continental interiors in high latitudes	Oceans in subpolar regions	Oceans in tropics and subtropics	Continental interiors in dry portions of tropics and subtropics
Basic attributes	Cold and dry	Cool and moist	Warm and moist	Warm to hot and very dry
Associated summer weather in middle latitudes	Pleasantly cool, low humidities, excellent visibilities, scattered cumulus	Cool to mild, moderately high humidities, mostly sunny in lowlands, showers likely in mountains	Uncomfortably warm and humid, hazy conditions inland, considerable cumulus cloudiness, scattered showers and thunderstorms	Very hot and dry, good visibilities, scattered daytime cumulus, very rare showers or thunderstorms
Associated winter weather in middle latitudes	Cold, dry, and often windy, excellent visibilities; clear to scattered or broken cumuliform cloudiness, snow showers likely in mountains or leeward shores of warm water bodies	Chilly and damp, frequently windy, considerable cloudiness, rain or snow showers especially in mountains	Mild and moist, fog or low stratiform cloudiness common especially night and morning hours; substantial precipitation along frontal boundaries	Normally does not reach middle latitudes in winter

it is labeled *continental Antarctic* (cAA) air. The second air mass type is maritime equatorial (mE) air, which forms near the equator in the vicinity of the Intertropical Convergence Zone (ITCZ) and does not affect the middle latitudes. It is similar to maritime tropical air, but typically is somewhat warmer, moister, and less stable.

Five air masses influence the weather of North America. *Maritime polar (mP)* air is the chief air mass affecting the West Coast from California through southern Alaska, particularly in winter. Although it is normally relatively stable, this cool, moist air mass can release abundant precipitation when forced to rise over mountainous terrain. Atlantic mP air masses are also sometimes drawn onshore along the east coast of Canada and the northeastern United States. *Continental polar (cP)* air masses dominate central and eastern North America in the winter and are common even in summer in Canada and the northeastern United States. They are typically associated with cool, dry weather. During winter, frigid *continental arctic (cA)* air masses form over northern Canada and Alaska. They periodically flow southward into the eastern United States and occasionally reach the West Coast north of California. *Maritime tropical (mT)* air dominates the southeastern United States in summer and is often drawn northward into the eastern United States in winter by the circulation systems of cyclonic storms. The lifting of this air mass provides the majority of precipitation received in the United States east of the Rockies. Finally, hot

and dry *continental tropical (cT)* air masses develop over the deserts of the Southwest during the summer season.

Air Mass Movement and Modification

Air masses are frequently transported out of their source regions by the upper-level winds. As the air masses move, they bring their weather characteristics to new areas. Successive passages of air masses from diverse source regions are largely responsible for day-to-day changes in our weather conditions. A basic task of weather forecasters is to predict accurately the direction and speed of these air mass movements. Although air masses rarely form in the middle latitudes, those latitudes are constantly invaded by air masses from both high and low latitudes. Indeed, the conflict between dissimilar air masses makes the weather of the middle latitudes more variable on a day-to-day basis than that of either the high or low latitudes. The boundaries between air masses are indicated on maps by weather fronts, and the differences in temperature and humidity conditions attending a frontal passage result largely from the change in the controlling air mass.

As an air mass leaves its source region, it undergoes gradual modification as it passes over surfaces of differing temperature and moisture characteristics. Topographic barriers also may remove moisture through the orographic lifting process, which may generate vertical wind currents that mix air in the lower levels. Just as the air mass took several days to gain its characteristics, however, it normally takes several days to lose them. As a result, when it arrives over a new location, an air mass has temperature and moisture values that reflect the conditions of both the source region and the areas over which it has passed during its journey. The degree of modification that takes place depends on the speed at which the air mass moves, the distance it travels, and the extent of the dissimilarity between the areas over which it passes and its source region. As an illustration of the modification of a southward-moving cP air mass, note the progressively warmer temperatures that exist from Canada southward to Texas in Figure 7.15 on page 151. (The temperature in degrees Fahrenheit for each reporting station is to the upper left of the circle that indicates the station.) An example of very rapid air mass modification that sometimes leads to extreme weather conditions is the sudden injection of heat and moisture into the lower portions of arctic air masses as they pass over the Great Lakes. This phenomenon was discussed in the Case Study at the end of Chapter 6.

Eventually, an air mass is so extensively modified that it becomes in effect a different air mass. The formation, movement, modification, and dissolution of air masses is a continuous process that produces a constantly changing and never exactly repeated pattern of weather conditions over the Earth's surface.

SECONDARY CIRCULATION SYSTEMS OF THE MIDDLE AND HIGH LATITUDES

The global air pressure and wind belts, which play a crucial role in the distribution of the Earth's long-term climatic characteristics, are sometimes referred to as primary circulation systems of the atmosphere. While the primary systems exert their control over the climates of the world, a traveling and constantly changing array of smaller secondary circulation systems is responsible for the variations in daily weather conditions at a given site.

The atmosphere at any one time contains a large number of high and low pressure systems that vary in size, shape, and intensity. On occasion, pressure systems remain nearly stationary over one area for an extended period. More commonly, they actively move, or "build," across the Earth's surface at speeds ranging up to approximately 55 kilometers per hour (35 mph). In general, the pressure systems of the middle and high latitudes are better defined and move more rapidly than those of the low latitudes.

The direction and speed of motion of the pressure systems are largely controlled by wind currents in the upper atmosphere. With the passage of time, pressure systems form and disappear, strengthen and weaken, change size and shape, and vary in their speed and direction of motion. The Earth, and in particular the middle latitudes, are therefore covered by an ever-changing kaleidoscope of pressure patterns—one that never exactly repeats itself. The dynamic nature of our atmosphere, as well as the number and complexity of the forces acting upon it, make accurate weather forecasting a highly challenging task.

The changing patterns of secondary circulation systems are especially important in controlling the weather of the middle and high latitudes, which tends to be much more variable than that of the tropics. There are three major types of systems: anticyclones (high pressure systems), weather fronts, and frontal cyclones (low pressure systems). Outside the tropics, all three are transported generally eastward by the upper-level steering currents, and each is associated with its own sequence of weather conditions.

Anticyclones

A high pressure system, or **anticyclone,** can be described as a mound of air under higher pressure than the surrounding air. Highs develop where atmospheric conditions favor the upper-level convergence and subsidence of air. Because highs are generally located in the centers of air masses, they normally contain air that is relatively homogeneous in temperature and water vapor content.

High pressure systems have distinctive wind-flow patterns that differ between the northern and southern hemispheres because of the influence of the Coriolis effect. Highs in both hemispheres contain air that converges in the upper levels, subsides, and then diverges in the lower levels. In the northern hemisphere, the Coriolis effect imparts a clockwise spiral to these diverging winds, while in the southern hemisphere, it gives them a counterclockwise spiral (Figure 7.3). The descent of the air results in the compression and adiabatic heating of the atmosphere. This, in turn, is

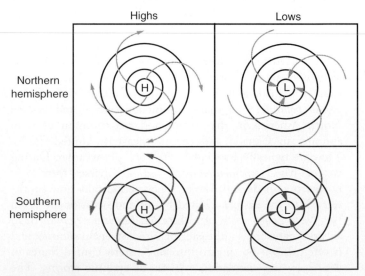

◆ **FIGURE 7.3** Wind-flow patterns for high and low pressure systems in the northern and southern hemispheres.

◆ **FIGURE 7.4** A portion of the surface weather map for March 28, 1982, showing a large Canadian anticyclone consisting of cP air centered over the central Appalachians.
SOURCE: NOAA

responsible for both the high barometric pressure and the fair weather normally associated with highs.

Anticyclones affecting North America generally come from one of three source regions. During the colder portion of the year, when weather systems typically are in active movement across the continent, high pressure systems containing Pacific mP air masses frequently enter western North America. As they cross the western mountains, they dry out and are reclassified as cP air masses, but their temperatures remain relatively mild for the time of year.

Troughs in the jet stream frequently cause cold cP or cA anticyclones from central or northern Canada to invade the United States during the winter season (Figure 7.4). They may bring subfreezing conditions deep into the South and East. An orderly sequence of weather conditions typically is associated with the approach and passage of a cold Canadian anticyclone. As the cold front that precedes the high pressure

system passes through the area, the weather turns cold, with blustery northwest winds and considerable cumuliform cloudiness. A day or two later, as the high pressure center approaches and the pressure gradient relaxes, the winds decrease in speed, the skies clear, temperatures remain cold, and visibilities are excellent. If the center of the high passes nearby, the winds become light and variable for a time and then gradually turn to the south or southwest as the center moves to the east of the area. The change in wind direction, coupled with the gradual modification of the cP or cA air within the system, causes a warming trend. As the high continues moving eastward, an increase in cloudiness is likely as the next low pressure system, with its associated fronts, approaches.

The final type of anticyclone develops chiefly during the summer. At this time, the southeastern and south-central portions of the United States normally come under the

◆ **FIGURE 7.5** A pair of low pressure
systems with their associated fronts over
the North Pacific as depicted both by
satellite and by surface weather map. The
clouds in the upper photo are white,
while the ocean surface appears black.
SOURCE: NOAA

domination of a quasi-stationary poleward extension of the
northern subtropical high. This Bermuda High, as it is
called, contains warm, humid mT air derived largely from
the Gulf of Mexico. Daytime surface heating may reduce the
stability of this air mass sufficiently to produce scattered
afternoon showers and thunderstorms.

Weather Fronts

A **weather front** is the interface between adjacent air masses.
It is a narrow zone along which air masses of different
temperature and moisture characteristics come into contact.
Because it is situated between the high pressure centers of
two air masses, a front is a trough of low pressure. This low
pressure causes air to flow from both air masses into the
frontal zone. As the air converges, the warmer and lighter air
mass is forced to rise over the cooler and denser air mass;
this frequently leads to the formation of clouds and frontal
precipitation (Figure 7.5).

The surface position of a front appears as a line on a
weather map, but a front is actually a three-dimensional
surface with height and width as well as linear extent. From
its surface location, the frontal boundary extends aloft at a
very gentle angle from the horizontal, so that it resembles an
inclined ramp (Figure 7.7). As a result, a few kilometers
above the surface the front may be located several hundred

◆ **FIGURE 7.6** Profile (side) view of a cold front, showing the relative positions of the warm air, cold air, clouds, and precipitation. The direction of frontal motion is from left to right.

◆ **FIGURE 7.7** Profile view of a warm front. Note that a warm front, unlike a cold front, is tilted forward from its surface position. Most precipitation occurs on the poleward (cold) side of the front.

horizontal kilometers from its surface position. This frontal inclination is caused by the differing densities of the two air masses, with the warmer air, which is lighter, always lying above the colder air. Fronts also have width because some mixing of the adjacent air masses takes place, and several hours may be required for the full transition of air masses to occur at a specific site following a frontal passage.

Fronts are especially numerous and important in the middle latitudes because this portion of the world is within reach of both the cold air masses of the polar and subpolar regions and the warm air masses of the subtropics. The rapid weather changes for which the middle latitudes are noted usually are associated with frontal passages. These weather changes are most dramatic from late autumn through early spring, when the latitudinal variation in insolation, and therefore in temperature, is especially large. The stronger upper-level wind currents of the winter also transport fronts faster than they travel in the summer, allowing less time for modification and mixing of the opposing air masses.

The primary frontal boundary separating air masses of subpolar and subtropical origin (usually cP and mT air) is called the *polar front*. Although its position is subject to large daily and seasonal fluctuations, it is generally located within the middle latitudes. This front is not straight, but develops large-scale waves that travel eastward along it, causing it to shift alternately northward and southward. The constant latitudinal shifting of the polar front allows for the redistribution of excess moisture and thermal energy from the tropics to the higher latitudes (Figure 4.5).

The passage of a front is usually attended by a number of changes in weather conditions. The extent and speed of the changes are subject to considerable variation, depending on factors such as the temperature contrast on opposite sides of the front and its speed of motion. These weather changes can be grouped into the following five categories:

1. Temperature
2. Atmospheric water vapor content
3. Wind direction and speed
4. Air pressure tendency
5. Cloud cover and precipitation

By observing changes in these weather elements, a knowledgeable weather observer, even without sophisticated meteorological instruments, usually can tell when a front has passed (Table 7.2).

Types of Fronts

Weather fronts, which separate warmer and colder air masses, are named according to which air mass, if either, is advancing. Four types of fronts are recognized.

The **cold front** is the most common and typically the most sharply defined front. It is a surface along which advancing cold air is displacing warm air. The cold, denser air wedges beneath the warm air, lifting it well above the surface (Figure 7.6). If the warm air has sufficient water vapor, this uplift will produce frontal clouds and precipitation. Surface friction tends to slow the advance of the cold air near the surface, steepening the slope of the front to an angle that is usually between 1:40 (that is, 1 vertical unit to 40 horizontal units) and 1:80. This relatively steep frontal slope causes the warm air to rise rapidly, typically producing a brief period of heavy precipitation followed by rapid clearing. Fair, cooler, and drier weather may be expected over the next few days.

Along a **warm front,** advancing warm air is displacing cold air. Warm fronts tend to move considerably more slowly than cold fronts, because in this case the advancing air is lighter than the air it is displacing. The warm air rides up over the retreating wedge of cold air, causing the front aloft to be tilted in the direction of motion. (Figure 7.7. Note that a cold front aloft is tilted in the opposite direction.) As the warm air rises, it is cooled adiabatically, and with sufficient moisture, clouds and precipitation again develop. By retarding the movement of the lower air, surface friction stretches and flattens the front, rather than steepening it. This produces a frontal slope of only 1:100 to 1:200. Because the warm air gradually slides up this gentle incline, warm frontal precipitation usually falls from nimbostratus clouds and tends to be lighter in intensity and longer in duration than cold frontal precipitation. If the rising warm air is unstable, however, heavier showers and possibly thunderstorms may

◆ TABLE 7.2 Idealized Frontal Weather Conditions*

COLD FRONT

Temperature: Warm before passage; progressively colder after passage.

Moisture: High humidity and dew point before passage; progressive lowering after passage.

Wind direction: Southwest winds before passage; northwest winds after passage.

Air pressure: Slow fall before passage; rapid rise after passage.

Clouds and precipitation: Brief period of heavy showers or a thunderstorm; rapid clearing after passage.

WARM FRONT

Temperature: Cold before passage; warmer after passage.

Moisture: High relative humidity but low dew point before passage; high relative humidity and high dew point after passage.

Wind direction: Easterly winds before passage; southwest winds after passage.

Air pressure: Rapid fall before passage; steady readings after passage.

Clouds and precipitation: Gradual increase and lowering of cloudiness followed by extended period of steady precipitation; slow clearing after passage.

STATIONARY FRONT

Weather: Extended period of cloudiness and light precipitation, particularly poleward of the front; temperature, moisture, wind, and air pressure remaining relatively steady.

OCCLUDED FRONT

Temperature: Cold before and after passage.

Moisture: High relative humidity but moderate dew point before passage; lowering relative humidity and low dew point after passage.

Wind direction: Southerly before passage; west to northwest after passage.

Air pressure: Falling before passage; rising after passage.

Clouds and precipitation: Gradual increase and lowering of clouds, followed by extended period of precipitation; short period of heavy precipitation at end, followed by clearing.

*Conditions described are typical, but do not occur in all cases. For example, weak fronts or fronts separating dry air masses may pass with no precipitation and little cloudiness. Wind directions are those typical of middle latitude areas in the northern hemisphere. The north-south directional component should be reversed for southern hemisphere fronts. Terms such as warm and cold are *relative* to typical conditions.

be interspersed within the area of precipitation. In winter, rain falling from the warm air above the frontal boundary may freeze as it descends through the underlying cold air, producing sleet. If freezing does not occur until the rain comes into contact with subfreezing objects on the surface, a glaze of freezing rain will result.

The mean forward speed of cold fronts is about 40 kilometers per hour (25 mph). That of warm fronts is about

25 kilometers per hour (15 mph). Average speeds are somewhat faster in winter than in summer because of the greater strength of the upper-level steering currents.

A **stationary front,** as the name implies, shows little movement. Although the front itself may be stationary, warm air continues to be lifted up the frontal surface, generally resulting in cloudiness and precipitation. In the vicinity of a stationary front, moderate to heavy rains may persist for an extended period, sometimes leading to flooding. The Midwest floods of 1993 were triggered by a jet stream pattern that caused a succession of weather fronts to become stationary over the Upper Mississippi Valley, producing a protracted period of heavy rainfalls over a large area. This subject is explored in the Focus Box on page 143.

Weather fronts that remain stationary for several days tend to weaken and eventually lose their identity as the opposing air masses gradually mix and moderate. Frequently, however, a front that has been stationary will begin to move as an imbalance between the opposing forces develops. As it moves, it is reclassified as a cold front or a warm front.

Any of these three types of fronts can, by a change in its motion, become any of the other types. In fact, different portions of the same front can be classified as different types of fronts because of their motions. This situation is analogous to that of a battlefront between two opposing armies. On one day, the battlefront may move in one direction, and on the next day, a counterattack may cause it to move in the opposite direction. In addition, an army may be making gains by advancing the front in one location, while at the same time it is losing ground in another. Weather "fronts" received their name because these air mass boundaries, like battlefronts, are linear zones of conflict between opposing forces.

The fourth type of front differs from the others in that it is formed by the meeting of three, rather than two, air masses. It was noted earlier that cold fronts tend to move somewhat faster than warm fronts. On occasion, a cold front will overtake a warm front, lifting the intervening mass of warm air above the surface (Figure 7.8). The merging of the warm and cold fronts forms an **occluded front.** Although a newly formed occluded front may produce substantial precipitation, the warm air that supplies most or all of this

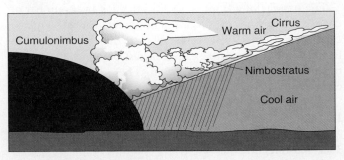

◆ **FIGURE 7.8** Profile view of an occluded front. These fronts are formed by the meeting of three air masses.

THE SUMMER OF 1993—A SEASON OF FLOOD AND DROUGHT

During the summer of 1993, the Upper Mississippi River Valley experienced some of the worst flooding in its history. Beginning in mid-June, wave after wave of moisture-laden cyclonic storms battered the region over a two-month span, chasing tens of thousands of families from their homes and causing well over $10 billion of destruction. The Mississippi River, swollen by eight times its normal volume of flow, resembled an inland sea (Figure 7.9). In late July, the combined flow from the Mississippi and Missouri Rivers resulted in three record flood crests at St. Louis within a few days. A peak crest of 5.9 meters (19.4 ft) above flood stage was reached on August 1, and the river never fell below flood stage at St. Louis during the entire month of August. Eventually, half of the nearly 800 counties in Iowa, Missouri, Wisconsin, North Dakota, South Dakota, Nebraska, Minnesota, Illinois, and Kansas were declared federal disaster areas.

Flood crests along the river were made higher by flood control structures that confined the water within its channel, rather than allowing it to spill out onto the surrounding floodplain. Eventually, however, the river

continued on next page

◆ **FIGURE 7.9** The points of juncture of the Mississippi, Illinois, and Missouri Rivers during the drought of 1988 (top photo) and the flood of 1993 (bottom photo). St Louis appears at the lower right of both photos.

SOURCE: Landsat imagery courtesy of the Earth Observation Satellite Company, Lanham, Maryland

July 18, 1993

breached most of these structures. Although almost the entire Upper Mississippi Basin was affected, the major flooding was largely limited to the portion of the river located north of its confluence with the Ohio River. The Ohio, although considered a tributary of the Mississippi, normally carries a much greater volume of water. The water-flow capacity of the Mississippi River channel therefore increases greatly south of the juncture of the two rivers at Cairo, Illinois. Fortunately, in the summer of 1993, the Ohio River's volume of flow was actually somewhat below normal.

At the same time that the Upper Midwest was being deluged by rain, record hot and dry weather plagued the Southeast. Columbia, South Carolina, reached a daily maximum of 38°C (100°F) or higher on 17 days in July. Richmond, Washington, D.C., and Philadelphia endured their hottest month on record. Newark, New Jersey, registered a temperature of 41°C (105°F) on two days. The heat and an almost total lack of rainfall in some areas withered crops in the South Atlantic states, leading President Clinton to declare South Carolina and most of North Carolina and Georgia federal disaster areas.

CAUSE OF THE FLOODS AND DROUGHT

The midwestern flooding and the southeastern drought were interconnected phenomena associated with an abnormal pattern of the jet stream. In midsummer the jet stream is usually weak and is typically centered over southern or central Canada. The United States, to the south of the jet stream flow, is usually dominated by warm, moist, maritime tropical air. During the summer of 1993, however, the jet stream remained unusually far to the south, sandwiched between a high pressure ridge over the eastern Pacific and a pronounced low pressure trough dipping southward into the western states (Figure 7.10). At the same time, the Bermuda High over the tropical North Atlantic strengthened, and its center shifted further west than normal, blanketing the southeastern states with hot, humid air. A strong flow around the western margin of this ridge injected a continuous supply of warm, moisture-laden maritime tropical air into the central states, where it collided with the unusually cool Canadian and Pacific air associated with the western jet stream trough. Frontal cyclonic storms, riding along the boundary between the warm and cold air, periodically lifted the tropical air over the polar air, producing torrential rainfalls. The southern states, meanwhile, baked beneath the stable center of the Bermuda High, temporarily immune from the daily afternoon thunderstorms that are typical midsummer phenomena in that part of the country.

The pattern the jet stream took in the summer of 1993 is common during spring and is not particularly rare in summer—in some years it develops on several occasions. What was unusual in 1993 was that the pattern persisted for much of the summer, bringing heavy rains or excessive heat to the same area for a period of more than two months. Some atmospheric scientists believe that the springlike pattern may have been a lingering legacy of the June 1991 eruption of the Philippine volcano Mt. Pinatubo. This eruption injected 18 million metric tons of sulfur dioxide into the stratosphere and veiled the Earth with a global haze. This haze was still partially intact in the summer of 1993 and may have kept the jet stream from shifting as far north as usual.

Another possible culprit was the continuation of a weak El Niño event in the tropical Pacific (see the Case Study starting on page 238). The presence of an El Niño encourages the formation of a high pressure center off the west coast of the United States, often leading to the formation of an upper-level trough to its east over the western and central states. ∎

◆ **FIGURE 7.10** Meteorological conditions associated with the Midwest floods and Southeastern drought during the summer of 1993.

moisture has been cut off from its surface source of supply. As a consequence, precipitation soon diminishes as the remaining moisture is exhausted. Occluded fronts are weak fronts, since at the surface they separate two cold air masses. Weak fronts tend to become weaker as the opposing air masses mix and modify and usually lose their identities within a few days.

Frontal Cyclones

The importance of weather fronts is not solely the result of their associated precipitation and the changes in temperature and humidity that accompany their passage. They also produce conditions favoring the development of **frontal cyclones.** These storms are among the most important secondary circulation features of the atmosphere and are the dominant type of weather disturbance in both size and significance for areas outside the tropics. Frontal cyclones produce most of the large storms of rain and snow in the middle and high latitudes. Their low air pressures cause diverse air masses to converge, often setting the stage for severe weather. In addition, they transport thermal energy from the surface to the upper atmosphere and help to carry oceanic moisture to the continents.

Formation and Developmental Cycle

A weather front provides a favorable setting for the development of cyclonic disturbances because it is a trough of low pressure and therefore a zone of air convergence. Low pressure centers can form only in areas where surface convergence occurs. In addition, the presence of adjacent air masses of differing densities on opposite sides of a front provides a mechanism for releasing the heat of condensation of atmospheric water vapor through the process of frontal lifting.

The critical factor in determining whether a frontal cyclone will develop is the pressure and wind-flow pattern of the upper troposphere. Specifically, an area of wind divergence must exist above the area where the cyclone is forming. This is necessary because near the surface, any low consists of converging and rising air. As the air rises, it must have a place to go. If the air in the upper troposphere is not diverging, the rising air cannot spread out to make room for the continued upward flow of air. The updraft therefore will be blocked, and the surface low will fill and disappear. This situation is analogous to the flow of air up a chimney. This flow cannot occur if the chimney vent is closed. If, on the other hand, a diverging flow exists in the upper levels, air rising into the low will be able to flow outward, making room for more air from the lower levels. In order for a low to deepen, upper-level divergence must exceed surface convergence.

A zone of air convergence that forms a favorable site for the formation and development of frontal cyclones is the eastern flank of a trough in the subpolar jet stream (Figure 7.11). The flow of the jet stream above the middle latitudes tends to develop equatorward-projecting troughs that move eastward. The pressure and wind patterns are generally most compressed at the axis of such a trough, creating a zone of

◆ **FIGURE 7.11** The typical position of a surface frontal cyclone on the forward flank of an upper-level low pressure trough. Red lines are surface isobars, and black lines are upper-level isobars.

upper-level air convergence on the trough's western flank and a zone of upper-level divergence on its eastern flank. Surface lows therefore tend to form on the eastern flanks of these troughs, ahead of the main trough axis. The converging airflow on the western flank of the trough causes the air to accumulate there. It can then subside to the surface and diverge to produce a high pressure system.

The cycle of formation and dissolution of a frontal cyclone can be compared to that of an ocean wave, because both tend to build, break, and dissipate. In Figure 7.12, various stages of the cycle are diagrammed as they would appear on a weather map. The entire process normally takes about four or five days to complete, so the time interval between each diagram is approximately 20 to 24 hours. It should be stressed that both the diagrams and the description of the cycle in the next several paragraphs pertain to "model" cyclones, for no two frontal cyclones are exactly alike. Nevertheless, the differences are more in degree than in kind, and most frontal cyclones go through the basic stages described here.

In panel (a), a portion of the polar front extends in an east-west direction through an area in the middle latitudes of the northern hemisphere. This front delineates the boundary between cold cP air to the north and warm mT air to the south. The wind in the opposing air masses flows parallel to the front, with only a slight tendency toward convergence. This slight convergence is responsible for the presence of a zone of cloudiness and light precipitation along the front. A balance of forces exists between the two air masses, so the front is stationary.

In panel (b), a bend or **wave** has formed on the front, perhaps in response to the approach of a trough in the jet stream. The position of the wave is indicated by both the actual bend or wave on the front and by the formation of a center of slightly lower barometric pressure. The reduced pressure begins to cause the winds on opposite sides of the front to blow inward, and the influence of the Coriolis effect

◆ **FIGURE 7.12** Six stages in the "life cycle" of a frontal cyclone. (See text for a discussion of each panel.) Air pressures are indicated in millibars, and zones of cloud cover (blue) and precipitation (tan) are also shown.

on this converging air initiates a counterclockwise spiral. Because warm air to the east of the center of the wave flows northward into the front, this portion of the front begins to move northward as a warm front. At the same time, cold air to the west of the center is being directed southward, causing that portion of the front to advance as a cold front. The area of cloudiness and precipitation associated with the front undergoes a marked expansion as the wind-flow patterns of the two opposing air masses converge more directly on the front, increasing frontal lifting of the warm air mass. This, in turn, increases the release of heat of condensation available to power the low.

Panels (c) and (d) illustrate stages in the development of a mature open-wave cyclone (one that has not yet occluded). The developing wave is carried northeastward by the upper-level winds and continues to intensify as more and more heat energy is released by condensation and sublimation. This causes air pressures within the cyclone to fall and wind speeds to increase. The warm front, extending southeastward from the center of the low, advances northeastward, preceded by a broad band of precipitation. The trailing cold front, which extends southwestward from the low's center, advances southeastward at a more rapid rate, producing a narrow band of heavy showers and thunderstorms. Because the cold front advances more rapidly than the warm front, the mT air that supplies moisture and energy to the storm is gradually reduced in size to a narrowing wedge, and the storm center is increasingly surrounded by cold air.

In panel (e), the cold front has overtaken the warm front in the vicinity of the low pressure center to form an occluded front, and the process of occlusion is continuing at the point of intersection between the warm and cold fronts. As this occurs, the warm air is lifted from the surface, and its supply of water vapor is cut off. The low pressure center begins to weaken as its source of energy is slowly depleted.

Panel (f) depicts the dissipating stage of the cyclone. The occluded front has weakened and lost its identity, and the low pressure center, now entirely surrounded by cold air, continues to fill and weaken. The polar front once again has stabilized, and as the last remnants of the cyclonic circula-tion die away, the situation returns to that depicted in panel (a). The stage is now set for the formation of another frontal cyclone that will repeat the pattern.

Frontal Cyclone Distribution and Movement

Frontal cyclones develop primarily within the middle lati-tudes. Their specific areas of formation vary during the course of the year because they are closely associated with the position of the polar front. The most vigorous storms develop during the winter and early spring, when the temperature contrast between air masses of subpolar and subtropical origins is greatest.

Frontal cyclones affecting the United States in winter frequently develop in the North Pacific, in the Gulf of Mexico, along the Atlantic coast, and in the eastern Rockies and adjacent sections of the Great Plains (Figure 7.13). One well-known site for cyclonic development is in the Alberta prairie just to the east of the Rocky Mountains. These storms often travel rapidly southeastward into the northern and central United States, earning the name "Alberta Clip-pers." Although they have access to limited moisture and usually produce only modest snowfalls, Alberta Clippers are usually followed by outbreaks of frigid arctic air.

During the summer, the major storm track lies well to the north, near the border of the United States and Canada, and the storms typically are much weaker because of the reduced temperature contrast between the opposing air masses. The frontal cyclones that develop in winter over the central and southern United States or along the Atlantic coast typically deepen and strengthen as they move northeastward. They eventually reach their peak development over the subpolar

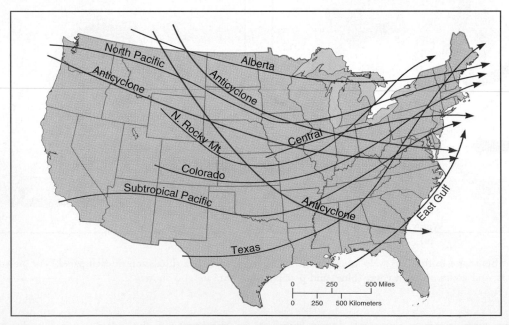

◆ **FIGURE 7.13** Typical winter tracks of cyclones (in blue) and of anticyclones (in red) across the United States. Note that while all systems track generally eastward, most cyclones have a poleward directional component, while anticyclones have an equatorward component of motion.

◆ **FIGURE 7.14** Time-lapse view of multiple lightning strokes illuminating the interior of an intense thunderstorm at night.
SOURCE: Dave Baumhefner/NCAR

North Atlantic south of Greenland. In similar fashion, storms developing along the east coast of Asia move northeastward, often attaining massive proportions over the waters south of Alaska. In the southern hemisphere, the great temperature contrast between the frigid continent of Antarctica and the relatively mild adjacent ocean waters often causes that landmass to be surrounded by a ring of well-developed frontal cyclones.

Localized Storm Types of the Middle and High Latitudes

Frontal cyclones have a profound impact on the middle and high latitudes because of their large size, widespread geographical distribution, and frequent occurrence. For most areas outside the tropics, they and their attendant fronts produce more precipitation than do orographic and convectional sources combined. In terms of their effects on people and property, though, two much smaller and generally more intense storm types, the thunderstorm and the tornado, are especially noteworthy. In actuality, all three storms are interrelated because thunderstorms outside the tropics often are associated with frontal cyclones, and tornadoes are produced only by severe thunderstorms.

Thunderstorms

A thunderstorm is a convectional storm accompanied by lightning and thunder and associated with cumulonimbus clouds and unstable air. Most thunderstorms produce heavy precipitation, usually in the form of rain, but are quite localized, with diameters of only a few kilometers. Severe thunderstorms also are capable of producing large hail and tornadoes. These storms have undoubtedly impressed more people with the power of the forces of nature and have created more general interest in the weather than has any other single meteorological phenomenon (Figure 7.14).

The lightning that accompanies thunderstorms is the most deadly of all storm-related natural hazards in North America, but its cause is still not completely understood. It is known to be initiated when a charge separation occurs within a cumulonimbus cloud. Scientists believe that hailstones exist within most of these storms, and that the smaller stones have a tendency to develop positive electrical charges, while the larger hailstones develop negative charges. Strong updrafts within the storm are able to carry the smaller hailstones to the upper portion of the storm, producing a buildup of positive charge in that area, while a negative charge develops in the cloud's lower portion. This negative charge, in turn, induces the formation of a positive

THE WEATHER MAP AS A FORECASTING TOOL

The modern science of forecasting meteorology is highly complex, and meteorologists employ a variety of sophisticated instruments to gather and analyze the tremendous quantity of weather data needed for accurate weather forecasting. Currently, most weather analysis and prediction is performed by high-speed computers that can simultaneously digest and synthesize data from a large number of sources.

Nevertheless, a relatively old forecasting tool—the surface "synoptic" weather map—is still widely used by forecasters. Although it appears complex, a map such as that shown in Figure 7.15 chiefly indicates the current weather conditions at a large number of reporting stations in the United States and Canada. From the combined station data, large-scale weather patterns can be discerned and drawn on the map. These include the surface air pressure pattern (indicated by isobars with pressure units in millibars); the positions of highs, lows, and weather fronts; areas of current precipitation; and the locations of the freezing and 0°F lines. With a basic knowledge of weather systems, a person who can translate the weather information shown on such a map

often can make reasonably accurate short-term forecasts using only this source of data. Appendix B provides an explanation of the various map symbols used.

Before we try to forecast weather using Figure 7.15, note the following points:

1. Each reporting station is indicated by a circle. The amount of cloudiness is indicated by the proportion of the circle that is blackened in. The temperature in degrees Fahrenheit is given to the upper left of the station circle. The direction the wind is *coming from* is indicated by the "shaft" protruding from each station circle, and the wind speed in knots is indicated by the number of "feathers" at the end of the shaft.
2. Areas of current precipitation on the map are shaded gray, and precipitation at any station is indicated to the left of each station circle. Dots indicate rain and stars indicate snow.
3. Weather fronts are shown as thick lines. Triangles on a front indicate a cold front, half-circles indicate a warm front, and an alternation of these symbols indicates a stationary front.

4. Weather systems in the middle latitudes typically travel eastward at an average speed of 30 to 45 kilometers per hour (20 to 30 mph) or roughly 800 to 950 kilometers (500 to 600 mi) per day.

Let's now try to forecast the weather for the next 24 hours at a couple of representative cities:

- *Oklahoma City.* We see that Oklahoma City lies to the west of the cold front cutting through the center of the country. The sky is cloudy, the temperature is 46°F, and the wind is from the northwest at about 15 knots. We would expect the cold front to continue to move eastward, away from the city, and the large high pressure area containing cold cP air to approach from the northwest. Accordingly, we would forecast that the sky will gradually clear and the temperature turn colder.
- *Pittsburgh.* Pittsburgh lies in the warm mT air mass to the east of the cold front. The sky is mostly cloudy, the temperature is 58°F, and the wind is light and southerly. We would expect the cold front to pass through the Pittsburgh area within the next 24 hours, placing the city

electrical charge on the ground below. The intervening atmosphere acts as an insulator, but when the attraction between these opposite charges grows strong enough, its insulating abilities are overcome, and a brilliant flash of charged particles ensues. Approximately 90 percent of all lightning flashes occur within the cloud, but the remaining 10 percent are between the cloud and the Earth's surface. The passage of the lightning bolt almost instantaneously

heats the atmosphere to temperatures as high as 27,000°C (50,000°F), causing the air to expand explosively; this expansion results in the crack of thunder.

The thunderstorm is by far the world's most common storm type. The key factor required for thunderstorm development is the vigorous convectional lifting of a deep layer of moist, unstable air. Conditions favoring thunderstorm formation occur frequently, and it is estimated that, at any

◆ **FIGURE 7.15** Surface weather map for November 12, 1982. See Appendix B for an explanation of the map symbols.
SOURCE: NOAA

in the cP air mass. The forecast, then, should go something like this: "Cloudy and mild with increasing southerly winds over the next few hours. A period of rain or showers will begin shortly and last for several hours, followed by gradual clearing and turning much colder. Winds will shift to the northwest and become strong and gusty."

You may wish to try your own forecasts for Atlanta, New York City, and your own city. ■

given moment, 1,800 thunderstorms on average are in progress over the Earth's surface.

Meteorologists commonly recognize two general categories of thunderstorms. **Air mass thunderstorms** are strictly convectional in origin and may form anywhere when the atmosphere is sufficiently warm, moist, and unstable. Over land areas, they are most numerous in the afternoon and early evening, when insolation has increased the temperature and lapse rate of the lower troposphere to its daily maximum values. Air mass thunderstorms are usually distributed within an area in a poorly organized fashion; currently, it is impossible to predict accurately the precise localities they will affect on any given day. **Frontal thunderstorms** are triggered by the frontal lifting of unstable air. Unlike the air mass type, they tend to form in a line. This line may be located either along the front itself or at some distance ahead of it as a prefrontal

(a) Cumulus stage

(b) Mature stage

(c) Dissipating stage

◆ **FIGURE 7.16** The three stages of a single-cell thunderstorm "life cycle." During the *cumulus stage,* the developing cell consists of a single strong updraft of air that has been heated by both the ground and by the released heat of condensation. No rain has yet reached the surface. During the *mature stage,* lightning and thunder begin, and heavy rain begins to reach the ground. The frictional drag of the heavy rainfall columns produces strong downdrafts that gradually choke off the initial updraft feeding the cloud. In the *dissipating stage,* the downdraft has spread throughout the base of the cloud. This causes the cloud's lower portion to evaporate as its upper ice-crystal portion gradually spreads out. Rainfall tapers off and eventually ends.

squall line. Some of the world's most violent and destructive thunderstorms develop along squall lines that form about 240 kilometers (150 mi) in advance of strong cold fronts in the middle latitudes. The causes of squall line formation are complex, but are basically associated with the compression or "crumpling" of the warm air by the approaching cold air.

Air mass and frontal thunderstorms have similar structures. Each is composed of one or more **cells** that originally form from a roughly circular updraft of air several kilometers in diameter. Larger thunderstorms or those that have formed into a line are composed of multiple cells located adjacent to one another. The life cycle of a thunderstorm cell, which typically lasts less than an hour, is illustrated in Figure 7.16.

Thunderstorms occur in nearly all parts of the world, but vary greatly in frequency in different areas. Figure 7.17 illustrates two important aspects of their global distribution. The first is that the frequency of thunderstorms is correlated with latitude. They are most numerous near the equator and become progressively less common poleward. This distribution is related to that of both global mean temperatures and specific humidity values—an indication of the great importance of heat and moisture to thunderstorm development. Second, thunderstorms are far more common over land than over water. This is primarily a thermal factor. Land areas attain maximum temperatures considerably higher than those experienced over water, making the air correspondingly lighter and more unstable. Orographic lifting can also trigger thunderstorms when the air is unstable. This lifting mechanism is, of course, restricted to land areas.

The global pattern of thunderstorms is shown in more detail in Figure 7.18. As can be seen, the continent of Africa leads the world in the frequency of thunderstorms, followed closely by the world's wettest continent—South America. Thunderstorms occur almost daily in many parts of the trop-

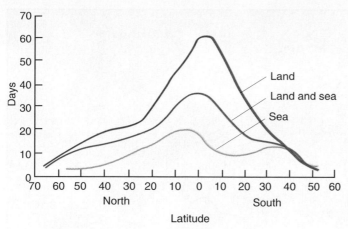

◆ FIGURE 7.17 Latitudinal distribution of mean annual thunderstorm days for land areas, water areas, and the Earth's surface as a whole. Note that thunderstorms are most common near the equator and are also far more frequent over land than over water.

ics and provide much of the rainfall received in that part of the world. In the middle and higher latitudes, thunderstorms occur primarily during the late spring and summer, when insolation is sufficiently strong to make the air unstable. Most of the middle latitudes have an intermediate frequency of thunderstorms, while the high latitudes, some desert regions, and areas dominated by cold ocean currents have the fewest.

In the United States, thunderstorms are most common in the Gulf Coast area, especially in central Florida, where strong solar heating in summer induces the convergence of unstable mT air from both the Atlantic and the Gulf of Mexico (Figure 7.19). They are least frequent along the Pacific coast, where the cold California Current keeps the air cool and stable throughout most of the year.

In Canada, a largely latitudinal thunderstorm distribution pattern exists. Thunderstorms are most common in southern Canada, particularly in Ontario's Niagara Peninsula and in the southern Canadian prairie. They decline in frequency poleward and are quite rare in the Canadian Arctic.

Tornadoes

The **tornado** is by far the smallest of the world's major storm types, but it is also the most violent, producing the strongest winds of any storm on Earth. It takes the form of a tubular vortex of whirling air surrounding a central core of extremely low pressure that extends downward from the base of a cumulonimbus cloud (Figure 7.20). In a sense, the tornado should not be considered a separate storm type. Along with lightning, thunder, and hail, it is a phenomenon uniquely associated with thunderstorms. Since the specific causes, characteristics, and world distributions of tornadoes differ from those of thunderstorms, however, they are treated here as a distinctive storm type.

Most tornadoes form in thunderstorms that have developed in advance of strong cold fronts in the middle latitudes. (Figure 7.21 provides further information on optimal conditions for tornado development.) The atmosphere must be highly unstable, giving the air a tendency to rise rapidly. For

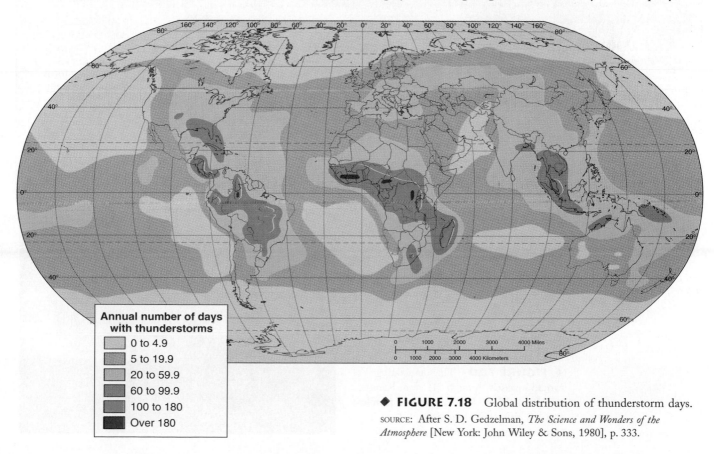

Annual number of days with thunderstorms
- 0 to 4.9
- 5 to 19.9
- 20 to 59.9
- 60 to 99.9
- 100 to 180
- Over 180

◆ FIGURE 7.18 Global distribution of thunderstorm days.
SOURCE: After S. D. Gedzelman, *The Science and Wonders of the Atmosphere* [New York: John Wiley & Sons, 1980], p. 333.

◆ **FIGURE 7.19** Average annual number of days with thunderstorms.

◆ **FIGURE 7.20** This devastating tornado passed through Hesston, Kansas, on March 13, 1990. It injured 13 people and left about 300 homeless.

SOURCE: © Wade Balzer/Weatherstock

◆ **FIGURE 7.21** An outbreak of severe weather, including possible tornadic activity, is most likely to occur in the area shown here in orange. It precedes the "dry line" (the boundary between mT and cT air masses) ahead of a cold front in the southern Great Plains.

this reason, tornadoes are most frequent during the spring, when a large temperature contrast exists between opposing mT and cP air masses. In addition, air in spring heats much more rapidly near the surface than it does in the upper troposphere, producing steeper lapse rates than are typical during the summer.

A tornado apparently is initiated in the middle levels of a mature cumulonimbus cloud when a vigorous updraft begins to rotate, producing a **mesocyclonic** circulation that may be several kilometers in diameter. This occurs as updrafts feeding into the cloud are given a cyclonic spiral by wind-shear conditions (opposing directions of wind flow) at different altitudes. The mesocyclone gradually extends downward through the cloud and narrows, causing its rotational speed to increase. Eventually, the cyclonic vortex protrudes from the base of the cloud as a **funnel cloud,** which becomes a tornado if it reaches the ground.

A tornado funnel generally tapers from cloud to ground somewhat like an elephant's trunk. It usually moves along the surface at speeds of 30 to 70 kilometers per hour (20 to 45 mph). Typical diameters of tornado funnels at the surface are 100 to 600 meters (300 to 2,000 ft), although diameters of more than 1,500 meters (1 mi) have been observed. Wind speeds associated with tornadoes are difficult to determine with any accuracy, but currently are believed to range from about 22 to 135 meters per second (50 to 300 mph).

The paths of tornadoes frequently are highly erratic and vary greatly in length. On the average, tornadoes are on the ground for a distance of only about 7 kilometers (4 mi) over a time span of perhaps 10 minutes, but some remain on the surface for a much greater period. The record in this respect is held by a tornado that carved a path 470 kilometers (292 mi) long through Illinois and Indiana on May 26, 1917. It was on the ground for more than seven hours.

The great destructiveness of tornadoes is attributable to several factors. One is the tremendous winds of the tornado funnel. These winds are capable of blowing down buildings, leveling swaths of trees, and even lifting railroad locomotives from their tracks. Another tornado attribute that can have a deadly destructive effect on large structures is the wind shear within the funnel, which can literally rip buildings apart (see Figure 7.24 on page 159). A third factor, which is largely responsible for the substantial death toll annually associated with these storms, is their unpredictability. The extremely localized nature of tornadoes, coupled with their erratic movement, rapid formation, and generally short spans of existence, make accurate tornado forecasts extremely difficult. Although the capability of reliably pinpointing specific areas that will experience tornadoes lies well in the future, the recent development of Doppler radar holds considerable promise (see the Geographic Techniques Box on pages 162-163).

Tornadoes are restricted largely to land areas within the middle latitudes, where air masses of subtropical and subpolar origin are most likely to collide. Most tornadoes occur in the United States, which has the dubious distinction of experiencing more of these storms than the rest of the world combined. The concentration of tornadoes within the United States results largely from its geographic location between major source regions of both cP and mT air, coupled with an absence of east/west-oriented mountain barriers. These two contrasting air masses are often drawn into contact by the circulation systems of traveling frontal cyclones, sometimes producing major outbreaks of dozens of tornadoes within a one- or two-day period.

The distribution of tornadoes within the United States is shown in Figure 7.23 on page 158. Although tornadoes can occur in all 50 states, the area of greatest frequency is the southern Great Plains, especially the states of Texas, Oklahoma, and Kansas. This area is sometimes referred to as "Tornado Alley." Tornadoes also are quite common throughout the southeastern states. Within any given region, tornadoes tend to be more common in flat, treeless areas than in hilly or forested topography that offers higher surface friction. The tornado season begins in March in the Gulf Coast area and advances slowly northward with the retreat of the polar front, reaching the vicinity of the Canadian border in June or July. Most tornadoes occur between 2 and 6 P.M., when daytime heating and atmospheric instability are at their peak.

THE TORNADO SUPEROUTBREAK OF APRIL 3–4, 1974

When large-scale clashes between maritime tropical and continental polar air masses take place over the central United States, the stage may be set for major tornado outbreaks. One such outbreak of at least 60 tornadoes occurred on February 19, 1884, causing destruction from Mississippi to Virginia and taking as many as 800 lives. Another, the so-called Palm Sunday outbreak of 31 tornadoes on April 11, 1965, produced 256 fatalities. In terms of sheer magnitude, however, the tornado "superoutbreak" of April 3–4, 1974, is the largest on record.

During the 16-hour period from 2 P.M., April 3, to 6 A.M., April 4, a total of 148 tornadoes formed in advance of a strong cold front moving through the Ohio and Tennessee River valleys. Most of the storms developed along three squall lines well in advance of the front and, under the influence of strong upper-level winds, raced toward the northeast at 65 to 100 kilometers per hour (40 to 60 mph). Not only did this outbreak contain more tornadoes than any other on record, it affected a larger geographical area than any had before. The total path length of all tornadoes combined was estimated at 3,200 kilometers (2,000 mi), about half the average annual total path length for the entire country in a year.

Hardest hit was the town of Xenia, Ohio (population 27,000). At approximately 4:40 P.M. a tornado struck Xenia and, in a five-minute period, carved a path a kilometer (0.6 mi) wide and 5 kilometers (3 mi) long. The storm damaged or destroyed 2,400 homes as well as most of the business district, killed 32 people, and injured nearly 600 others.

The paths of the 148 tornadoes that are known to have occurred during the superoutbreak were mapped by University of Chicago meteorologist T. Theodore Fujita (Figure 7.22). The Xenia tornado has path number 37. The very small numbers appearing along each path are the estimated **Fujita Scale** intensities of the tornado at those locations. The Fujita scale is a numerical scale that rates the severity of a tornado from a minimum of 0 (weak) to a maximum of 5 (devastating). Tornado 37 had a Fujita rating of 4 as it passed through Xenia. ∎

TROPICAL WEATHER

The subtropical belts of high pressure serve to some extent as a meteorological barrier, dividing the Earth into two zones that are similar in size, but display marked differences in weather and climate. On the equatorward side of the subtropical highs, the weather is much less susceptible to dramatic change, especially with respect to temperatures. The tropics also experience their own storm types, which are triggered more by the instability of the intensely heated atmosphere than by the conflict between opposing air masses.

General Weather Characteristics of the Tropics

When examined over an extended period, several important differences in weather and climate between tropical and nontropical regions become apparent. Nearly all the differences in the following list result from the year-round receipt of large quantities of insolation within the low latitudes:

1. Temperatures are warmer on average for the year as a whole in the tropics than in the higher latitudes. No cold season occurs, so annual temperature ranges tend to be small and often are exceeded by daily temperature ranges.
2. Day-to-day temperature fluctuations are small and usually are related to variations in cloud cover. The constancy of temperatures exists largely because cold fronts normally do not penetrate into the low latitudes. In fact, genuine weather fronts of any type are rare within the tropics, and mT, mE, or cT air masses permanently control most areas.
3. Summer and winter differ chiefly in cloudiness and precipitation, rather than in temperature. Thus, while temperature is the leading seasonal climatic variable in the higher latitudes, precipitation is the leading seasonal climatic variable in the tropics. In most locations, summer is the wettest season and winter is the driest.

◆ **FIGURE 7.22** Paths of the 148 tornadoes that formed during the "superoutbreak" of April 3–4, 1974.

SOURCE: Reprinted by permission of T. Theodore Fujita

SURVIVING THE TORNADO

Sunday morning, March 27, 1994, dawned summerlike over northern Alabama and Georgia. A warm front had passed northward across the region overnight ushering in a moist unstable maritime tropical (mT) air mass from the Gulf of Mexico. Temperatures at dawn were already in the low 70's Fahrenheit, some 30° warmer than the previous day. Dew points near 70° indicated very humid air. A strong cold front was approaching the region from the west as afternoon temperatures climbed to 86° in Atlanta and 87° at Birmingham.

This ominous mix of moist, unstable, tropical air ahead of a cold front, combined with 80 mph winds blowing from the southwest at the 500-millibar level, had the attention of

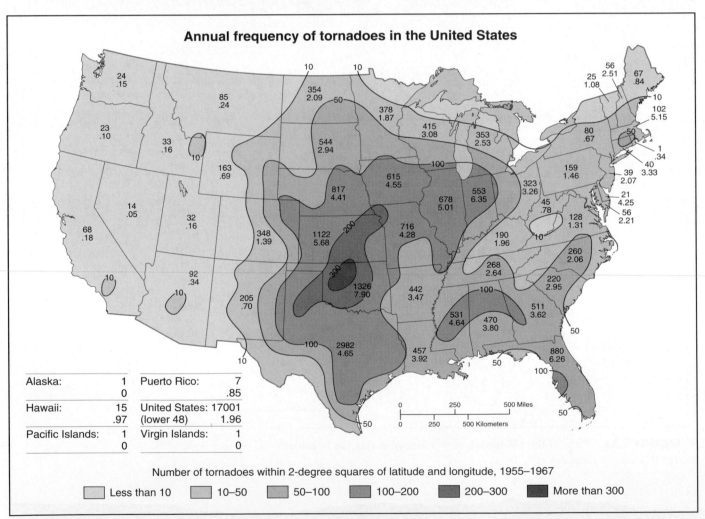

Annual frequency of tornadoes in the United States

Alaska:	1	Puerto Rico:	7
	0		.85
Hawaii:	15	United States: 17001	
	.97	(lower 48)	1.96
Pacific Islands:	1	Virgin Islands:	1
	0		0

Number of tornadoes within 2-degree squares of latitude and longitude, 1955–1967

Less than 10 | 10–50 | 50–100 | 100–200 | 200–300 | More than 300

◆ **FIGURE 7.23** Distribution of tornadoes within the United States. The upper number in each pair indicates the total number of tornadoes in each state between 1955 and 1967. The lower number represents the average annual number of tornadoes per 10,000 square miles. On the average, more than 700 tornadoes occcur in the U.S. each year.

◆ **FIGURE 7.24** Remains of the Goshen United Methodist Church near Piedmont, Alabama, which was struck by a tornado on March 27, 1994. Twenty persons died here as the roof and a concrete block wall collapsed onto the congregation during Palm Sunday services. Large rooms are often the first to fail during strong winds.

SOURCE: Thomas Schmidlin

meteorologists at the National Severe Storms Forecast Center in Kansas City. Five tornado watches were issued through the day for northern Alabama and Georgia, the first beginning late Sunday morning.

Their concern was warranted. At least 13 tornadoes in two or more "tornado families" touched down in northern Alabama and Georgia between 11 A.M. and 8 P.M. Five of these tornadoes were strong or violent (F3 or F4 on the Fujita scale) and were on the ground for 40 to 50 miles as they moved toward the northeast at 50 mph. These five long-track tornadoes killed 40 persons and injured over 300.

We traveled to the disaster sites a week after the storms to conduct research on the tornadoes, the 40 persons killed by the tornadoes, and the actions of survivors and victims prior to and during the tornadoes. Our purpose was to identify characteristics of individuals in the tornado path, focus-

ing on their responses to the tornado threat. Our hypothesis, based on our previous research and the work of other scientists, was that certain personal characteristics and responses to the tornado threat had caused some to die in the tornadoes and allowed others to survive. Our goal was to identify high-risk groups and behaviors and to suggest improvements in tornado preparedness and safety procedures. The research was funded by a "Quick Response" grant from the National Hazards Research and Applications Information Center in Boulder, Colorado.

Our research confirmed that many of the established tornado safety rules are appropriate, but also cast doubt on one. Half of the 40 tornado deaths in Alabama and Georgia on March 27, 1994, occurred in the collapsed Goshen United Methodist Church near Piedmont, Alabama (Figure 7.24). Sixteen others died in destroyed mobile homes, two were killed outdoors,

one died in a collapsed frame house, and one was killed as he was ejected from a van hit by a tornado.

The large sanctuary at the Goshen United Methodist Church apparently offered little resistance to strong winds, and the roof and a concrete block wall collapsed onto the congregation during the Palm Sunday service. Smaller rooms of the church remained intact after the storm, as did nearby brick homes. This strengthens the argument that large rooms, such as those found in churches, school auditoriums, factories, or shopping centers, are the first to fail in strong winds and should be evacuated during severe storms.

Tornado safety rules have evolved through experience, research, and tradition. When a tornado is approaching, entering a basement or underground

continued on next page

shelter is recommended for protection from flying debris. In the absence of a basement, an interior room without windows, such as a closet, hallway, or bathroom, provides some protection. The key factor is to surround yourself with a sturdy structure.

In addition to the deaths at the church, all of the other persons killed in buildings across northern Alabama and Georgia were in first-floor rooms with windows where walls and ceilings collapsed or were blown away by the tornadoes. In contrast, 63 percent of the survivors we surveyed were in first-floor rooms *without* windows or in rooms below ground. Survivors were generally in rooms where walls did not collapse and where they were protected from flying debris.

Some tornado safety rules have changed as new information has become available. For example, experts once recommended opening windows prior to a tornado's arrival to relieve pressure in a house and to take shelter in the southwest corner of a basement. Research, however, has shown that houses do not explode, so opening windows is not useful, and that the center of a basement is safer than the southwest corner.

New information is now casting doubt on the recommendation that motorists and mobile home residents should leave their vehicles or homes and lie in a ditch. In the Alabama and Georgia tornado disaster we studied, there were several instances in which a vehicle was left standing with little damage next to the spot where a mobile home was swept away and its occupants killed. At least 10 of the deaths in mobile homes could have been prevented if the residents had walked to their vehicles and buckled up to ride out the storm.

Based on these observations, perhaps the best advice for a resident of a mobile home is to get into a vehicle, buckle safety belts, and drive to a sturdy building. Even if struck by the tornado before reaching shelter, persons buckled into a vehicle are safer than they would be in a mobile home and may be safer than if they were outside where a ditch may not be available and debris can be lethal.

Our results also showed that persons who were watching television prior to the storm were more likely to survive. This is probably because they were more likely to have received advanced warning of the storm's approach.

Field work following natural disasters can help geographers learn more about the behaviors and responses that increase the chance that a person will survive. That knowledge can help improve disaster preparedness and save lives in the future. ▪

SOURCE: This focus box was written by Thomas W. Schmidlin, Department of Geography at Kent State University, and Paul King, Boyce Thompson Institute.

4. While most precipitation in the higher latitudes is generated by frontal lifting, most tropical precipitation is convectional. As a result, it tends to be showery, localized, and often intense.

5. Frozen moisture forms, such as frost, snow, and sleet, do not occur in most tropical areas because of a lack of subfreezing temperatures. The only natural form of ice ever seen by many residents of the tropics is hail.

6. Although the prevailing direction of movement of wind and weather systems is from west to east in most nontropical areas, it is from east to west within the tropics. The weather systems of the tropics tend to move relatively slowly as well, because the upper tropospheric steering currents are weaker, on the average, than those of the higher latitudes.

All in all, weather conditions of the tropics are much more constant than those of the higher latitudes. Many areas tend to experience a well-defined, repetitious daily cycle of morning sunshine and heating, afternoon clouds and scattered showers, and nighttime clearing and cooling.

Weak Tropical Weather Disturbances

The predominant tropical weather disturbances are localized convectional showers and thunderstorms. Convectional activity in general and thunderstorms in particular are more common in the tropics than anywhere else on Earth (Figure 7.18). Large-scale weather disturbances also develop within the lower latitudes, but the relative lack of conflicting air masses causes them to be less numerous and generally weaker than those of the higher latitudes. Because of the instability and high moisture content of the air in much of the tropics, however, these disturbances are substantial rainfall producers and contribute significantly to the precipitation totals of many areas. Three of the more important larger tropical disturbances are easterly waves, equatorial lows, and hurricanes.

Easterly waves are weak weather disturbances that are most likely to develop during the summer and early autumn over ocean areas. They consist of north/south-oriented troughs in the trade wind belts of each hemisphere that drift slowly westward (Figure 7.25). Their western sides contain diverging air and generally fair weather. To the east of the trough line, the inflowing air converges as it approaches the axis of the wave, reducing its stability and producing numerous showers and thunderstorms. Easterly waves are important producers of precipitation for the islands and east coasts of continents that lie between approximately 10° and 30° N and S. Those over ocean areas are watched closely during the summer and early autumn because they occasionally develop into hurricanes.

Weak centers of low pressure, termed **equatorial lows,** frequently develop along the ITCZ and drift westward. Although they normally are situated too near the equator to develop into hurricanes, they usually are associated with extensive areas of cloudiness and heavy convectional precipitation.

Hurricanes

The hurricane, the world's most destructive storm, is the only large, strong storm type that develops in the low latitudes. By definition, a **hurricane** is a tropical cyclonic storm with maximum sustained winds of at least 33.5 meters per second (75 mph or 65 knots). In the western North Pacific, where they are most numerous and intense, they are termed **typhoons.** Other names for this storm type include "cyclone" in the Indian Ocean and "chubasco" off the west coast of Mexico.

Hurricanes gain energy and moisture through a combination of convectional and cyclonic lifting. Tremendous quantities of warm, moisture-laden air that spiral into these storms because of their very low central barometric pressures are lifted into the upper troposphere and are cooled adiabatically to release torrential rainfalls.[1] The heat of condensation released during this process powers the storms and helps to maintain the global energy balance by transferring excess thermal energy from the surface to the upper atmosphere and, as the storms progress poleward, from the lower to the higher latitudes.

Development and Characteristics

Hurricanes form only over tropical or subtropical oceans with surface water temperatures of 26°C (79°F) or more. Unlike the frontal cyclones of the middle and higher latitudes, a hurricane does not develop on the frontal boundary

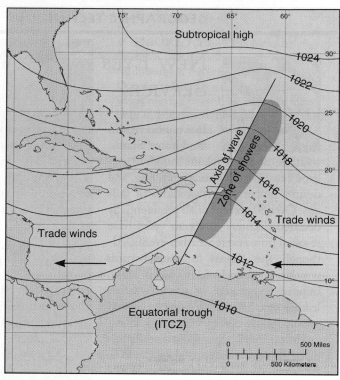

◆ **FIGURE 7.25** Weather map depiction of an easterly wave in the western Atlantic. Isobars are drawn at 2-millibar intervals.

between opposing air masses, but instead forms within a single mass of maritime tropical air. In its earliest stages of development, it is little more than a large, unorganized mass of convectional showers and thunderstorms. The rising of the air within the showers reduces surface pressures and initiates a weak inflow of air, which is given a cyclonic rotation by the Coriolis effect. The development of a cyclonic wind flow raises the disturbance to the category of a **tropical depression.** If maximum sustained winds reach 18 meters per second (40 mph), the developing storm is given a name and is classified as a **tropical storm.**[2] Upon reaching sustained winds of 34 meters per second (75 mph), the storm officially becomes a hurricane (Figures 7.27 and 7.28 on page 164). If conditions remain favorable for intensification, the hurricane eventually may attain wind speeds exceeding 67 meters per second (150 mph); the most powerful storms on record have had sustained winds in the vicinity of 90 meters per second (200 mph).[3] The Saffir-Simpson scale

---

1. World records for both low barometric pressures and high rainfall amounts have been set in hurricanes. The world's lowest sea level air pressure of 870 millibars (25.69 inches of mercury) was recorded during a 1979 typhoon in the Philippine Sea. A record 24-hour rainfall total of 187 centimeters (73.62 in) was established during a 1952 storm on the island of Réunion in the Indian Ocean.

2. The practice of naming tropical storms and hurricanes according to alphabetical lists was begun in 1953 by the U.S. National Weather Service. Separate lists are maintained for the western Atlantic area, the eastern Pacific, and the western Pacific. Originally, only female names were used, but since the late 1970s, male and female names have been alternately employed.

3. Only one storm this powerful is known to have occurred during the twentieth century in the Atlantic Ocean. This was the Labor Day Storm of 1935 that hit the Florida Keys. Two relatively recent Gulf of Mexico hurricanes—Camille in 1969 and Allen in 1980—had peak winds of at least 83 meters per second (185 mph).

◆ FIGURE 7.27 Satellite photograph of Hurricane Gilbert in the Gulf of Mexico on September 15, 1988. The spiral banding and central eye are clearly visible.

SOURCE: NOAA

move poleward, they approach the subtropical highs. The steering currents in the subtropics are weak and often erratic, but most hurricanes gradually turn more poleward and reduce their forward speed to 10 or 15 kilometers per hour (5–10 mph). Eventually, the storms enter the westerly wind belt and recurve toward the northeast in the northern hemisphere or toward the southeast in the southern hemisphere. Their forward speed gradually accelerates to perhaps 50 or 60 kilometers per hour (30–40 mph), carrying them rapidly poleward. Eventually, they pass over colder waters, which causes them to lose strength and to slowly dissipate because of the loss of the water vapor that supplies their energy.

Although the majority of hurricanes reach colder waters and dissipate without ever encountering land, others strike coastal areas, often with disastrous consequences. The great destruction and loss of life associated with these storms

◆ FIGURE 7.28 Cross-sectional view through the center of a hurricane. Letter A indicates the location of the eye, with its descending air currents and clear skies. B is in the eyewall clouds—the most intense part of the storm. C is on one of the updrafts producing rings of squalls that surround the storm's center.

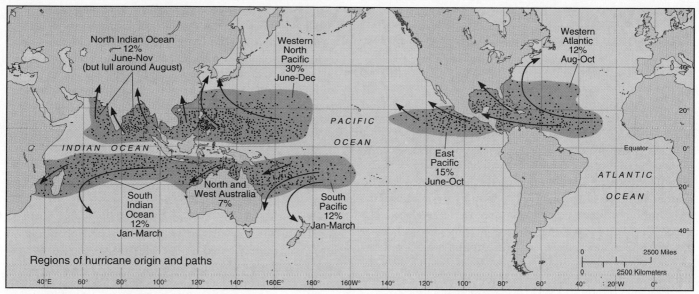

◆ **FIGURE 7.29** Areas of hurricane formation and typical storm tracks. Each dot indicates the point where a tropical storm initially reached hurricane intensity. The average percentage of the global hurricane total occurring within each region is indicated.

results from three factors. Best known are the hurricane-force winds, which are capable of uprooting trees, collapsing weaker buildings, and turning loose objects such as boards and strips of sheet metal into lethal missiles. Most of the destruction caused by 1992's Hurricane Andrew, the most costly hurricane in U.S. history, was wind related. The story of this storm is presented in the Case Study at the end of this chapter. A second factor is the extremely heavy rainfall associated with hurricanes. Rainfall totals, which may exceed 25 centimeters (10 in) with slow-moving storms, frequently produce extensive flooding, landslides, and loss of life. By far the most deadly feature of hurricanes, however, is the **storm surge** that accompanies their landfall. The very low air pressure within a hurricane, coupled with strong air convergence, causes the ocean surface to bulge upward as much as 10 meters (30 ft) above normal. This water may then be blown onshore by winds well in excess of 45 meters per second (100 mph), inundating the coast. Flooding is most extensive on low-lying coastlines and on those occasions when a hurricane crosses the coast during the time of normal high tide. The greatest loss of life caused by a hurricane in the United States occurred in Galveston, Texas, in 1900, when 6,000 people were drowned by a storm surge. Table 7.4 lists the 10 deadliest hurricanes to strike the United States in the twentieth century. In recent decades, improved warnings and evacuation procedures have resulted in greatly decreased losses of life in the United States. Monetary losses from these storms have escalated, however, due to the continuing development of beachfront property and increased values resulting from inflation.

From a global perspective, the largest numbers of human casualties on record are associated with hurricanes (locally labeled "cyclones") that have struck the Bay of Bengal coast

in what is now Bangladesh in 1737 and again in November 1970. Approximately 300,000 people died in the 1737 storm, and the death toll for the 1970 storm was estimated at a staggering 550,000! Recently, in spring 1991, another hurricane in the same area was responsible for 180,000 deaths. Most of the lives lost in all three storms resulted from the storm surge, which in the 1737 storm was said to have raised water levels by 13 meters (37 ft), causing the ocean to advance inland for many kilometers.

◆ **TABLE 7.4** Deadliest Hurricanes of the Twentieth Century in the United States

LOCATION	YEAR	NUMBER OF FATALITIES
Galveston, Texas	1900	6,000
South Florida	1928	1,836
Florida Keys and Corpus Christi, Texas	1919	600–900
New England	1938	600
Florida Keys	1935	408
Louisiana and Texas	1957	390
East Coast—Virginia to Massachusetts	1944	390
Grand Isle, Louisiana	1909	350
Galveston, Texas	1915	275
New Orleans, Louisiana	1915	275

▌Summary

This chapter has examined the Earth's major weather systems. Weather systems are in a continuous state of flux—forming; changing size, shape, and intensity; and dissipating—as they drift across the face of our planet largely in response to the upper-level wind currents. As these systems pass over any given location, they bring their attendant weather conditions to that location for a period of a few hours or days.

Each weather system is an organized assemblage of the weather elements discussed in the previous chapters. Each exists because of the differential receipt of insolation in various portions of the Earth system and plays a role in the distribution of energy and moisture over the globe. The seven weather systems emphasized in this chapter are air masses, anticyclones, weather fronts, frontal cyclones, thunderstorms, tornadoes, and hurricanes. In order to review the formation, characteristics, and distribution of these systems, we return to the Focus Questions posed at the beginning of the chapter:

1. *What are air masses? How and why do they form, and what influence do they have on our weather?*

Air masses are large bodies of air characterized by internal homogeneity in temperature and water vapor content. They form over large land or water bodies that gradually impart their surface temperature and moisture conditions to the overlying air. After they have formed, air masses may be transported to other parts of the world by the upper-level winds. Most air masses form in either the high or low latitudes. The middle latitudes experience frequent incursions of both polar and tropical air masses.

2. *What are the most important weather systems of the middle and higher latitudes, and what types of weather are associated with them?*

The major large-scale weather systems of the middle and higher latitudes are anticyclones (high pressure systems), weather fronts, and frontal cyclones. Anticyclones generally are associated with the centers of air masses. They are areas of descending and diverging air and generally produce fair weather. Weather fronts are linear boundaries where air masses from different source regions meet. They typically are associated with cloudiness and precipitation as air from the warmer and more buoyant air mass rises over the adjacent colder, denser air mass. Frontal cyclones are low pressure systems that form on weather fronts. Their convergent wind-flow patterns typically draw in warm air on their equatorward and eastern sides and cold air on their poleward and western sides, forming warm and cold fronts, with their associated precipitation.

3. *How do thunderstorms and tornadoes form, and what areas are most affected by them?*

Both thunderstorms and tornadoes are associated with unstable air and are most likely to form when the air has been strongly heated at the surface, especially during the afternoon. This produces vigorously rising convectional air currents that, in the case of tornadoes, are apparently given a cyclonic spiral by wind-shear conditions (opposing directions of wind flow) at differing altitudes. Thunderstorms are most abundant over land areas in humid tropical regions and generally decrease in frequency poleward. Tornadoes, however, are most frequent in the middle latitudes, especially in the central United States, where they typically form in advance of strong cold fronts.

4. *How and why do weather conditions of the tropics differ from those of the higher latitudes?*

Tropical weather differs from that of the higher latitudes largely because the tropics experience large quantities of insolation throughout the year. This makes the air warmer and more buoyant and increases its water vapor holding capacity. As a consequence, temperatures are warm throughout the year, fronts generally are absent, and precipitation is often heavy, coming largely in the form of afternoon convectional showers.

5. *What are hurricanes? How and where do they form, and what land areas do they most affect?*

Hurricanes are powerful cyclonic storms that form over ocean areas at latitudes between approximately 5° and 30°. They are initiated when a region of moist, unstable air containing numerous showers and thunderstorms develops a cyclonic wind-flow pattern that gradually intensifies. Like most tropical weather systems, hurricanes tend to move in a westward direction, so that the east coasts of land masses are most likely to be affected. If they enter the middle latitudes, however, they usually recurve to a poleward and eastward direction. They typically dissipate after striking land or passing over colder water.

▌Questions for Thought and Review

1. What is an air mass? Where and how does it attain its characteristics?
2. Why are most air mass source regions in the high and low latitudes rather than in the middle latitudes? How do air masses reach the middle latitudes?
3. What type of air mass dominates in your area in the summer? In the winter? What are the source regions of these air masses?
4. What is a weather front? Explain how weather fronts are classified, and briefly describe each of the four types of fronts.
5. Indicate several ways by which you could tell whether a weather front that passed through your area was a cold front or a warm front.
6. During what time of year are thunderstorms most frequent in your area? Explain why. What essential charac-

teristics must the atmosphere exhibit in order for thunderstorms to develop?

7. What are tornadoes? Under what meteorological conditions are they likely to occur? Why do they occur more frequently in the United States than in other parts of the world?

8. Discuss the basic ways in which you would expect weather conditions to differ from those typical of your own area if you lived near the equator.

9. What is a hurricane? Where do these storms form, and why? Why are they experienced frequently on the East Coast of the United States, but not on the West Coast?

KEY TERMS

air mass

source region

anticyclone

cold front

warm front

stationary front

occluded front

frontal cyclone

thunderstorm

tornado

easterly wave

hurricane

Hurricane Andrew of 1992

The entire Atlantic and Gulf Coast of the United States, from New England to the Mexican border, is vulnerable to the devastation produced by hurricanes. Despite continuing advances in hurricane-forecasting technology, the unprecedented pace of human development of this coastal region continues to increase the risk to lives and property. During the 1970s and most of the 1980s, hurricanes were especially numerous and intense along the Gulf Coast, while the Atlantic Coast received relatively few major storms. This situation, however,

changed dramatically in September 1989, when powerful Hurricane Hugo slammed ashore near Charleston, South Carolina. Hugo was followed two years later by Hurricane Bob, which wreaked havoc in southern New England in August 1991.

The portion of the Atlantic Coast most renowned for hurricanes is undoubtedly the Gold Coast of Southeast Florida, dominated by the city of Miami. In the six-year time span from 1945 to 1950, five major hurricanes struck this area. After 1950, however, no severe storms passed over the South Florida mainland for more than four decades. During this period, the Gold Coast experienced a tremendous building boom as its population quadrupled to three million. The result is a continuous strip of urban and subur-

ban development from West Palm Beach southward 160 kilometers (100 mi) to Homestead.

The storm that would finally end the Gold Coast's period of immunity from major hurricanes began as an area of thunderstorms over West Africa on August 13, 1992. Transported to the west by the trade winds, this area of disturbed weather entered the eastern Atlantic on August 14 as an active easterly wave. These weak troughs of low pressure are common summertime phenomena over the tropical North Atlantic; some 60 to 70 are tracked each hurricane season by the National Hurricane Center in Coral Gables, Florida.

The wave, steered rapidly westward by the trade winds, passed south of the Cape Verde Islands on August 15.

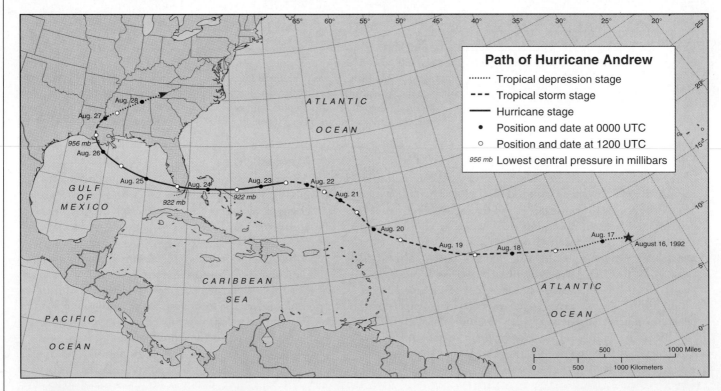

◆ **FIGURE 7.30** Path of Hurricane Andrew.

◆ **FIGURE 7.31** Ruins of Florida City, a coastal community hit hard by Hurricane Andrew.
SOURCE: AP/Wide World Photos

Rising air currents within the thunderstorms of the easterly wave lowered sea level air pressures, causing an inflow of air at the surface. This inflow, given a counterclockwise twist by the Earth's rotation, organized the disturbance into a closed low pressure center, or tropical depression, on August 16. By August 17, the strengthening circulation became the first tropical storm of the 1992 hurricane season, as its maximum sustained winds topped 18 meters per second (40 mph), and it was named Andrew.

The newly developed storm continued moving rapidly over the open Atlantic toward the west-northwest (Figure 7.30). Between August 19 and 21, Andrew passed south of a strong upper-level trough centered southeast of Bermuda. Interaction with the trough caused the storm to slow, while its path shifted to the northwest, thus sparing the islands of the Lesser Antilles. At the same time, upper-level southwesterly winds produced wind shear that ripped the tops off Andrew's westward moving thunderstorms, keeping the storm from strengthening.

By August 21, however, environmental conditions changed. The upper-level low weakened and shifted north, and at the same time the Bermuda High strengthened to the north of the storm. These events redirected Andrew to an almost due westward course at an increased forward speed and allowed the storm to strengthen rapidly. In a 48-hour period, Andrew's peak winds increased from 23 meters per second (52 mph) to 63 meters per second (140 mph), as the storm quickly developed an eye and became a major hurricane. Peak intensity during this period was reached just east of the Bahamas, as the storm attained maximum sustained winds of 69 meters per second (155 mph) and a minimum central air pressure of 922 millibars.

Drawing a bead on South Florida, Andrew passed over Eleuthera Island in the northern Bahamas on August 23, leaving four people dead and over 1,700 homeless. The storm's passage through the Bahamas caused it to weaken slightly, but it rapidly regained its strength over the Florida Straits in the last few hours before its eye made landfall in Florida at approximately 5 A.M. about 40 kilometers (25 mi) south of downtown Miami (Figure 7.31).

At the time it struck South Florida, Andrew had maximum sustained winds of 65 meters per second (145 mph) with gusts of at least 78 meters per second (175 mph). The storm's central barometric pressure had once again fallen to 922 millibars, the third lowest in a hurricane to hit the United States in the twentieth century.* The storm's approach was well publicized, and a massive evacuation of over 1,200,000 residents was ordered.

The hurricane devastated Homestead and other communities in the southern Miami metropolitan area. In contrast to most hurricanes striking low-lying coasts, Andrew's destruction resulted primarily from wind and flying debris, not from its storm surge, which reached a height of 5.2 meters (16.9 ft). Entire neighborhoods were reduced to rubble by winds that may have locally gusted to speeds approaching 90 meters per second (200 mph). Mobile homes proved particularly vulnerable, and many disintegrated completely. In all, over 80,000 South Florida homes were destroyed or severely damaged, more than 200,000 people were left homeless, and 43 lost their lives.

continued on next page

*The Labor Day Hurricane that passed over the Florida Keys in 1935 had a central air pressure of 892 millibars, and Hurricane Camille, which struck the Louisiana coast in August 1969 had a minimum air pressure of 909 millibars.

◆ FIGURE 7.32 Color-enhanced satellite photograph of Hurricane Andrew making landfall south of Miami, Florida, on the morning of August 24, 1992.
SOURCE: NOAA

As the storm proceeded westward through the Everglades, high winds destroyed some 70,000 acres of mangroves. Vegetation in the Everglades and elsewhere in South Florida proved particularly susceptible to hurricane damage because it had already been placed under stress by declining water tables resulting from drainage and flood control diversion canals as well as groundwater withdrawals. Animals also suffered greatly; thousands of dogs and cats were left behind as their owners hurriedly evacuated their homes. Numbers of larger animals and birds also escaped from zoos and research facilities, and for weeks after the storm the threat of unexpected encounters with animals such as snakes, apes, lions, and panthers made life especially interesting for South Florida residents.

Its passage over land had weakened Andrew slightly by the time the storm crossed the southwest Florida coast and emerged into the Gulf of Mexico during the daylight hours of August 24. By this time, the storm's peak winds had diminished to about 54 meters per second (120 mph), and its central pressure had risen to 950 millibars. Fortunately, Andrew did not reintensify significantly as it passed over the Gulf. As high pressure to the north weakened and a frontal trough to the northwest approached, the storm slowed somewhat and turned gradually toward the northwest. Hurricane warnings were issued along portions of the Louisiana and Texas coasts, and another massive evacuation was ordered.

At approximately 11 P.M. on August 25, Andrew's eye passed inland on a sparsely populated portion of the Louisiana coast about 40 kilometers (25 mi) west of Morgan City. Once again, tremendous destruction to homes and property ensued; the state's agricultural and fishing industries sustained major damage, and 15 people were killed.

Weakening rapidly after its final landfall, Andrew was downgraded to a tropical storm after 10 hours and to a tropical depression after another 12 hours. As it came under the influence of the westerlies, the storm continued to curve to the north, and then to the northeast, while accelerating in forward speed. Heavy rains of up to 25 centimeters (10 in) continued along its track. By midday of August 28, Andrew's remnants merged with a frontal system over the mid-Atlantic states, and the storm gradually lost its identity.

Hurricane Andrew has been the costliest natural disaster in U.S. history, with total dollar losses estimated at $19.6 billion. Destruction would undoubtedly have been even greater if not for Andrew's tight organization and high forward speed. The total of 62 human lives lost is surprisingly small for a storm of this magnitude and is a tribute to both the accuracy of the warnings and the efficiency of the evacuations that preceded the landfalls of this great hurricane.

8 CLIMATES OF THE WORLD

OUTLINE

FOCUS QUESTIONS

1. What are the chief advantages and shortcomings of climate classification systems?
2. What are the major climatic characteristics of each of the global climate regions, and why do these characteristics occur?

Throughout history, climate has been closely associated with patterns of human activities. With few exceptions, the densely populated regions of the world have warm summers and sufficient precipitation to ensure the success of agriculture. In contrast, most of the world's cold and dry regions remain lightly populated and little developed. Within the past several decades, the population of the United States has undergone a dramatic climatically induced shift toward the "Sun Belt" states extending from the Carolinas southward to Florida and westward to California. This shift, based largely on the rapid expansion of tourist-related activities and the continued increase in the number of retirees, appears likely to continue for the foreseeable future.

In this chapter, the global pattern of climates is examined on a regional basis. A region's climate consists of its characteristic long-term weather conditions. Separating weather and climate often is difficult, however, because only the time factor differentiates the two subjects. Much of the material presented in the previous four chapters, especially the discussions of the global distributions of the various weather elements, was actually climatic information. In contrast to the approach in those chapters, however, where weather and climate phenomena were discussed in a topical fashion, we now examine the long-term condition of the atmosphere as a whole in a spatial context by dividing the Earth into a number of climatic regions.

A basic goal in physical geography, as noted in Chapter 1, is to develop an understanding of the world as a single functioning system. Such an understanding is developed largely through a knowledge of the relationships that exist among the Earth's natural phenomena. The role of climate in these relationships is vital because climate is probably the leading independent environmental variable operating on our planet's surface. It therefore exerts a powerful influence on the other components of the environment. For example, climate is the primary factor controlling the world distribution of natural vegetation. Climate, vegetation, and rock type together largely determine the type of soil that will develop within a given area. Climate also exerts a major influence on landform characteristics.

The distribution of Earth's human population also is strongly influenced by climate. Especially important in this respect are the climatic characteristics (such as extremes of temperature and precipitation) that restrict the growth of the cultivated plants that form the basis of our food supply. Continuing rapid advances in science and technology are reducing the degree to which climate influences the distributional pattern of the human population, but its continuing importance cannot be denied.

CLIMATE CLASSIFICATION

Examined in detail, the climate of any locality is unique because its climatic characteristics will not be precisely duplicated anywhere else. The key to understanding regional climate patterns and to developing climatic classification systems lies in generalization. Of necessity, climate types are based not on specific values, but on ranges of values. All areas having conditions falling within the stated ranges are, perhaps somewhat arbitrarily, considered to have the same type of climate. Climatic systems are devised to simplify the study of climate by reducing the number of climate types to a manageable total.

Climate descriptions normally stress mean values because they can be readily calculated and easily described and generally are indicative of the most common atmospheric conditions. Regional climatic analyses are thus based largely on the past weather data compiled by weather reporting stations within the region. It is generally assumed that no significant changes in climate have occurred and therefore that the weather conditions of the past will occur with statistically similar frequencies in the future. Recent concerns about global warming and other global- and regional-scale climate changes make this assumption increasingly questionable, especially if climates are examined in detail. For the broad global overview of climate provided in this chapter, however, we may assume that the climates of most areas have not changed significantly since the period of record keeping began. In general, the longer climatic data have been gathered, and the more comprehensive these data are, the more accurately a region's climatic characteristics may be determined.

The division of the Earth into climatic zones allows regional associations to emerge. It becomes apparent that the global distribution of climates is not random; rather, climates are spatially organized, with similar climatic conditions occurring in widely separated parts of the Earth. This, in turn, points to the existence of unifying controls that operate in similar fashions in areas that may be distant from one another.

The use of classification systems to regionalize climates has limitations that must be understood by those employing them. One of the basic shortcomings is loss of detail. The more climatic characteristics are generalized, the larger the individual climate regions tend to become, and the more detail is sacrificed. Nevertheless, generalization is a necessary part of geographic analysis, especially at the global scale. A second shortcoming is the danger of taking climatic regions too literally. It should be understood that the regions and boundaries shown are correct only for the system being used, and that other, equally valid systems may divide the area under consideration into quite different regions. It must also be realized that climates do not change suddenly as bound-

THE COLLECTION AND DISPLAY OF WEATHER DATA

Weather data such as temperature, precipitation, air pressure, and humidity are gathered daily at hundreds of locations throughout North America. The more sophisticated weather collection centers have automated sensing and recording instruments and often are located at airports. Most weather data, however, are collected by cooperative weather observers located on farms, in suburban neighborhoods, on college campuses, and in a variety of other settings.

Outdoor weather instruments typically are housed in shelters like that shown in Figure 8.1. The standard weather instrument shelter is constructed of white painted wooden slats that are louvered to allow for free airflow. It has a double-layered roof and is set on a long-legged stand about 1.5 meters (5 ft) above the ground. The shelter generally is located in an open grassy area well away from trees, buildings, or other obstacles to the wind. The objective is to protect the instruments and at the same time allow air temperatures inside the shelter to remain the same as those on the outside. A weather instrument shelter normally contains a variety of thermometers, including maximum and minimum thermometers, and a sling psychrometer. A rain gage is usually located nearby. Readings are taken

◆ **FIGURE 8.1** A standard weather instrument shelter.
SOURCE: Ross DePaola

once or twice a day at established times.

To graphically depict a station's climatic characteristics, a **climagraph** such as the one in Figure 8.2 is often employed. Most climagraphs depict monthly temperature and precipitation means. For clarity, the temperature data usually are indicated by points that are connected to form a line graph, while the precipitation values are indicated by the heights of 12 bars.

Additional information that may be provided on climagraphs includes station latitude, longitude, and elevation; a location map; annual temperature means and ranges; and annual precipitation means. Climagraphs providing this information are used throughout this chapter to illustrate representative conditions in each of the climatic regions discussed. ■

Lansing, Michigan
42°45'N 84°35'W
Elev: 852 feet (260 m)

Humid Continental
Köppen: Dfb

Mean temp: 46.9°F (8.3°C)
Mean range: 49F° (27C°)
Mean precip: 31.2" (79 cm)

◆ **FIGURE 8.2** Climagraph for Lansing, Michigan. Mean temperatures throughout the year are depicted by a line graph, and mean precipitation amounts for each month are indicated by bar graphs.

ary lines are crossed, but tend to vary gradually over large distances.

The data employed to develop climate classification systems also have limitations. Climatic data are inadequate for large portions of the Earth. This includes the 71 percent of our planet covered by oceans as well as extensive areas in the polar regions, subtropical deserts, and tropical forests. As data from these areas become increasingly available, climatic

Distribution of world climate types.

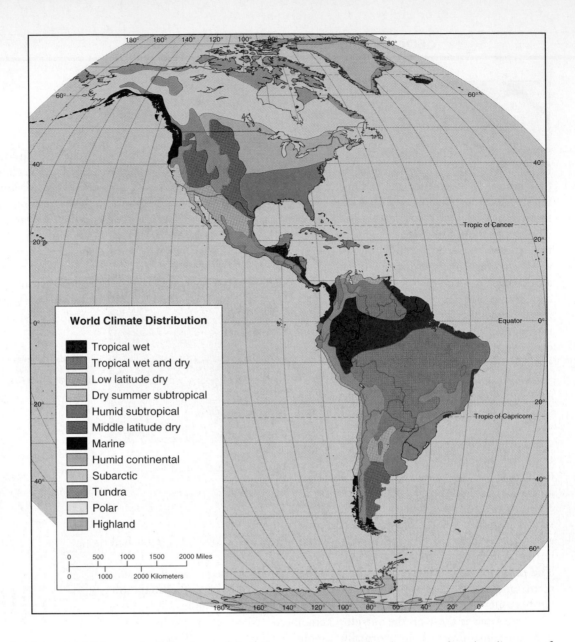

World Climate Distribution

- Tropical wet
- Tropical wet and dry
- Low latitude dry
- Dry summer subtropical
- Humid subtropical
- Middle latitude dry
- Marine
- Humid continental
- Subarctic
- Tundra
- Polar
- Highland

0   500   1000   1500   2000 Miles

0   1000   2000 Kilometers

"boundaries" must on occasion be revised. The methods used to collect and display climatic data are discussed in the box on page 173.

A final important shortcoming is our lack of detailed understanding of the causes of global climate patterns. All of the many climate influences, such as the winds and ocean currents, the vast fluxes of energy and moisture between the Earth's surface and atmosphere, and even the effects of human activities, are almost infinitely complex in themselves. The modification of climatic patterns in urban areas, for example, influences the lives of hundreds of millions of people; this topic is examined in the Case Study at the end of the chapter.

Climate results from the global linkage of all these factors, producing a multiplication of complexity. With the aid of modern supercomputers, climatologists are currently developing and refining general circulation models (GCMs) of the atmosphere in an attempt to better understand current global climate patterns and to better predict the climates of the future.

Because of the complexity of world climate patterns and the diversity of human interests, no single classification system can satisfy everyone. Various systems have been developed through the years, and the choice of the system to be utilized in any specific instance depends largely on its intended purpose.

The most widely known climatic classification system was first published in 1918 by the German botanist and climatologist Wladimir Köppen. Köppen, who was interested in gaining a better understanding of the causes of the world pattern of natural vegetation, realized that vegetative distribution was more closely tied to climate than to any other environmental control. His climate system therefore stressed annual means and seasonal variations of temperature and precipitation, the climatic factors that have the greatest influence on global vegetative characteristics. The system has

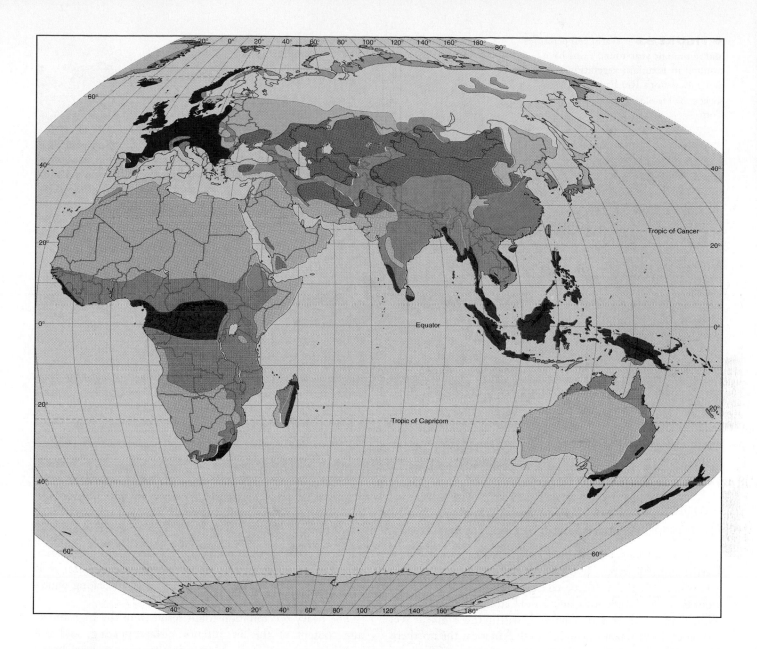

been revised and refined on several occasions. Details of the Köppen climatic classification system, as well as a brief discussion of the system's chief advantages and shortcomings, are provided in the box beginning on page 177.

Our need in this chapter is for a global climate classification system that is relatively simple, stresses geographic patterns of climates, and corresponds to the world patterns of natural vegetation and soils examined in subsequent chapters. The system used is primarily descriptive and is based closely on the Köppen classification system. It divides the world into 11 climate regions defined largely on the basis of their temperature and moisture characteristics. In each region, the following factors are stressed:

1. Geographic locations
2. Basic atmospheric temperature and moisture characteristics
3. The reasons for these characteristics

Following the precedent established in the discussion of air pressure and wind belts, our examination of world climate regions begins at the equator and progresses poleward. The locations of the various climate types are mapped in Figure 8.3.

THE LOW LATITUDE CLIMATES

Approximately half the Earth's surface is located within the low latitudes. The basic climatic characteristics of this part of the world were discussed at the end of the preceding chapter; in essence they consist of continuously warm temperatures, along with showery precipitation that is predominantly convectional in nature. The placement of the subtropical highs and the ITCZ are largely responsible for the differences between the three major regional climates of the

Warm temperatures and abundant year-round rains have resulted in luxuriant vegetation growth along Ecuador's Rio Santiago.
SOURCE: © Haroldo Castro/FPG International Corp.

low latitudes. Near the equator, the ITCZ produces year-round cloudy and humid conditions, with generally abundant rainfalls. Within the subtropical highs, clear skies, low relative humidities, and scanty rainfalls generally prevail. In much of the low latitudes, however, climatic conditions are controlled by the seasonal migration of these two pressure belts. This produces a pattern of alternating rainy and dry seasons.

Tropical Wet Climate (Köppen Af and Am)

The Tropical Wet Climate is the most equatorward of the Earth's climatic types. It occurs in extensive areas between 10° N and S, and in some coastal areas it extends to the vicinity of the Tropics of Cancer and Capricorn (Figure 8.3). Three large land areas are contained within the Tropical Wet Climate: the Amazon Basin in South America, the northern Congo Basin in western Africa, and the majority of the large islands of the western Pacific, including most of Malaysia and Indonesia and the Philippines. Additional smaller areas include coastal strips in eastern Central America and Brazil, the Guinea coast of West Africa, southwestern India, the eastern coastal margin of the Bay of Bengal, central Vietnam, and northeastern Australia.

The Tropical Wet Climate may be characterized as warm and rainy. Most locations near sea level have monthly mean temperatures near 27°C (80°F) throughout the year. Constantly high midday Sun angles and daylight periods of nearly equal length throughout the year are responsible for the warm temperatures as well as for the very small annual temperature ranges. Ranges are 3C° (5F°) or less for most locations, and even those areas situated near the poleward margins of the climate almost never experience annual ranges of more than 6C° (10F°).

Not only are seasonal differences in temperature very small, but day-to-day variations are minor and result primarily from differences in cloud cover. Air masses of nontropical origin rarely if ever penetrate the inner tropics to produce frontal temperature contrasts, and maritime equatorial (mE) air dominates throughout the year. The result is that one day is very much like another with respect to temperature. Only the constant rhythm of daily heating and nightly cooling provides thermal variety, leading some writers to describe day and night as the "seasons" of the tropics. In all portions of the Tropical Wet Climate, daily temperature ranges substantially exceed annual ranges.

Yearly rainfall totals usually average 125 to 500 centimeters (50 to 200 in) a year, making the Tropical Wet Climate the world's rainiest climate type (Figure 8.4). Large variations in amount frequently occur over short distances, with the heaviest totals in mountainous regions and along windward coasts.

The heavy precipitation totals result from the high moisture content of the air, intense surface heating, and the proximity of the ITCZ. Most rainfalls occur as brief heavy showers and thunderstorms interspersed with periods of sunshine. Along coastal areas, showers are likely to occur at any hour of the day or night. Inland areas, on the other hand, generally experience distinct afternoon precipitation maxima caused by surface heating, which results in the convective uplift of the unstable air. Especially heavy rainfalls occur where mountains provide an additional lifting mechanism. In fact, it is the orographic lifting of the trade winds along mountainous windward coasts such as eastern Central America and Brazil that results in the poleward extensions of the Tropical Wet Climate in those areas.

Most locations within the Tropical Wet Climate experience distinct periods of increased and decreased rainfall during the year. Indeed, many of the more poleward and coastal locations have a short dry season, with little or no rainfall over a two- or three-month period; in the Köppen system, they are classified as Tropical Monsoon (Am) climates. The annual precipitation distribution pattern of

THE KÖPPEN CLIMATIC CLASSIFICATION SYSTEM

The Köppen climatic classification system was first published by the German biologist and climatologist Wladimir Köppen in 1918. It has been revised and refined on several occasions, most notably by Köppen's students R. Geiger and W. Pohl. The version currently in most widespread use, which is presented in simplified form here, was first published in 1953 and is frequently referred to as the Köppen-Geiger system. For convenience, we refer to it simply as the Köppen system.

ADVANTAGES AND SHORTCOMINGS

The popularity of the Köppen system is based on several advantages that it offers the user. Foremost is its relative simplicity. Each climate type is defined on the basis of mean monthly or annual temperature and precipitation data, which are readily available from thousands of locations around the world. The definitions are precise and the formulas relatively simple, allowing for easy determinations of climate types. Because the climates are defined precisely, no decisions on the part of the user are necessary, nor is any previous knowledge of climate required. A major attraction for geographers is that the system relates well to global distributions of natural vegetation and soils. This factor is especially important when these two subjects are to be studied along with climate in a single academic course. Finally, the system

has the advantage of widespread familiarity and use.

Among the several shortcomings often cited for the Köppen system is its relative lack of detail. The system divides the world into 16 major climate types. For most of these climates, secondary and tertiary divisions are available, but compared to several alternative classification systems, the Köppen system is rather generalized, and its climatic regions contain distinctive unaddressed regional variations. Another criticism is that many important climatic factors are not taken into account. These include the radiation balance, evapotranspiration rates, precipitation type and intensity, winds, and temperature extremes. Associated with this factor is the Köppen system's lack of direct applicability to most practical uses in areas such as agriculture, forestry, and hydrology. Finally, the correlation with natural vegetation—the reason for which the system was initially devised—has always been imperfect.

Despite its shortcomings, the Köppen system remains an effective tool for organizing the study of general world climate patterns. Because such a study is precisely the goal of this textbook, the system is included here for those wishing to employ it.

CLASSIFICATION PROCEDURE

The Köppen system uses various uppercase and lowercase letters to represent mean monthly and annual tem-

perature and precipitation data. Five major climatic categories, represented by the uppercase letters A–E, are recognized. They broadly differentiate world climatic zones in an ascending latitudinal sequence. Descriptions and definitions are as follows:

◆ A: Tropical climates with abundant precipitation during at least a portion of the year. Annual precipitation exceeds evapotranspiration, so that an overall moisture surplus exists. By definition, the mean temperature of each month must exceed 18°C (64.4°F).

◆ B: Arid and semiarid climates of the low and middle latitudes in which potential evapotranspiration exceeds precipitation. The boundaries between the B climates and adjacent humid climates take into account annual mean precipitation (P), annual mean temperature (T), and seasonal distribution of precipitation. This last factor is calculated by the following formulas, with the first formula in each pair employing metric values (centimeters of precipitation and degrees Celsius). The second formula (in parentheses) uses inches for precipitation and degrees Fahrenheit.

continued on next page

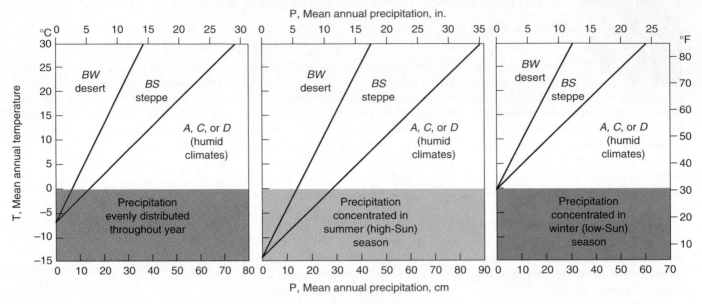

◆ **FIGURE 8.5** Boundaries between the B climate subtypes for stations with differing annual temperature and precipitation means.

$$P \leq C + 14 \qquad (P \leq 0.44T - 3)$$

when at least ten times as much rain falls in the wettest summer month as in the driest winter month.[1]

$$P \leq C \qquad (P \leq 0.44T - 14)$$

when at least three times as much rain falls in the wettest winter month as in the driest summer month.

$$P \leq C + 7 \qquad (P \leq 0.44T - 8.5)$$

when seasonal precipitation distribution is too even to meet either of the first two formula criteria.

◆ *Note:* If the value of P exceeds the formula values, the climate is not a B climate. The B climate bound-

---

[1]"Winter" refers to the colder half of the year. This includes the months October through March in the northern hemisphere and April through September in the southern hemisphere. The "summer" months are the opposite six months for each hemisphere.

aries may also be determined by using Figure 8.5. The B climate classification takes precedence over all the others. Therefore, when classifying a climate, it should first be determined if a B classification is warranted before proceeding further.

◆ C: Mild and humid (mesothermal) climates primarily in the lower middle latitudes. The mean temperature of the coldest month must be between 18°C (64.4°F) and –3°C (26.6°F), and at least one month must average 10°C (50°F) or higher.

◆ D: Humid continental climates with cold winters (microthermal), located within the upper middle latitudes and subpolar regions of the northern hemisphere. The coldest month must average below –3°C (26.6°F) and the warmest must average 10°C (50°F) or higher.

◆ E: Cold climates of the high latitudes. All months must average below 10°C (50°F).

Note: Approximately 20 percent of the Earth's land surface has an A climate, 26 percent a B climate, 16 percent a C climate, 21 percent a D climate, and 17 percent an E climate.

Secondary and sometimes tertiary letters are employed in association with each of the five primary letters to allow for further differentiation into individual climate types. The most important of these secondary and tertiary letters are as follows:

A Climates

◆ f: Mean precipitation of each month is at least 6 centimeters (2.4 in).

- m: Short dry season compensated for by surplus precipitation in wetter months.
- w: Well-defined dry season producing a period of water shortage.

The boundary between Am and Aw climates may be determined from Figure 8.6.

B Climates

- W: Arid. Annual precipitation totals less than half the amounts calculated from the B climate formulas.
- S: Semiarid. Annual precipitation totals more than half the amounts calculated from the B climate formulas.

The boundaries between the BW and BS climates may be determined from Figure 8.6.

- h: Annual mean temperature is above 18°C (64.4°F).
- k: Annual mean temperature is below 18°C (64.4°F).

C Climates

- s: Summer dry season. Precipitation during the wettest winter month is at least three times that of the driest summer month; precipitation in the driest summer month is less than 3 centimeters (1.2 in).
- w: Winter dry season. Wettest summer month receives at least ten times the precipitation of the driest winter month.
- f: No pronounced dry season. Criteria for neither s nor w are met.
- a: Hot summer season. Temperature of warmest month must average at least 22°C (71.6°F).
- b: Long mild summer. Warmest monthly mean is below 22°C (71.6°F) and four or more months are above 10°C (50°F).
- c: Short cool summer. Warmest monthly mean is below 22°C (71.6°F) and one to three months are above 10°C (50°F).

D Climates

- s, w, f: Defined as in the C climates.
- a, b, c: Defined as in the C climates.
- d: Extremely cold winter. Mean temperature of the coldest winter month is –38°C (–36.4°F) or lower.

E Climates

- T: Tundra or Periglacial climate. Warmest monthly mean is between 0°C (32°F) and 10°C (50°F).
- F: Polar or Icecap climate. Mean temperature of every month is below 0°C (32°F).

continued on next page

◆ **FIGURE 8.6** Boundaries between the A climate subtypes.

MAJOR KÖPPEN CLIMATE TYPES

The names of the major Köppen system climate types are listed here. The global distribution of these climates on land is depicted in Figure 8.7.

- ◆ Af: Tropical Rainforest climate
- ◆ Am: Tropical Monsoon climate
- ◆ Aw: Tropical Savanna climate
- ◆ BWh: Tropical Desert climate
- ◆ BSh: Tropical Steppe climate
- ◆ BWk: Mid Latitude Desert climate
- ◆ BSk: Mid Latitude Steppe climate

- ◆ Csa, Csb: Mediterranean climate
- ◆ Cwa, Cwb: Subtropical Monsoon climate
- ◆ Cfa: Humid Subtropical climate
- ◆ Cfb, Cfc: Marine climate
- ◆ Dfa, Dwa: Humid Continental–Warm Summer climate

◆ **FIGURE 8.7**
Global distribution of major Köppen climate types.

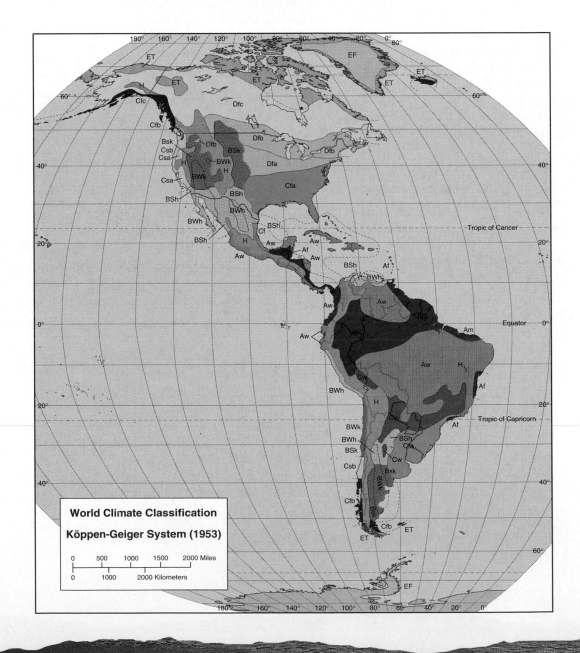

World Climate Classification
Köppen-Geiger System (1953)

- Dfb, Dwb: Humid Continental–Cool Summer climate
- Dfc, Dwc, Dfd, Dwd: Taiga (or Subarctic) climate
- ET: Tundra climate
- EF: Polar (or Icecap) climate ■

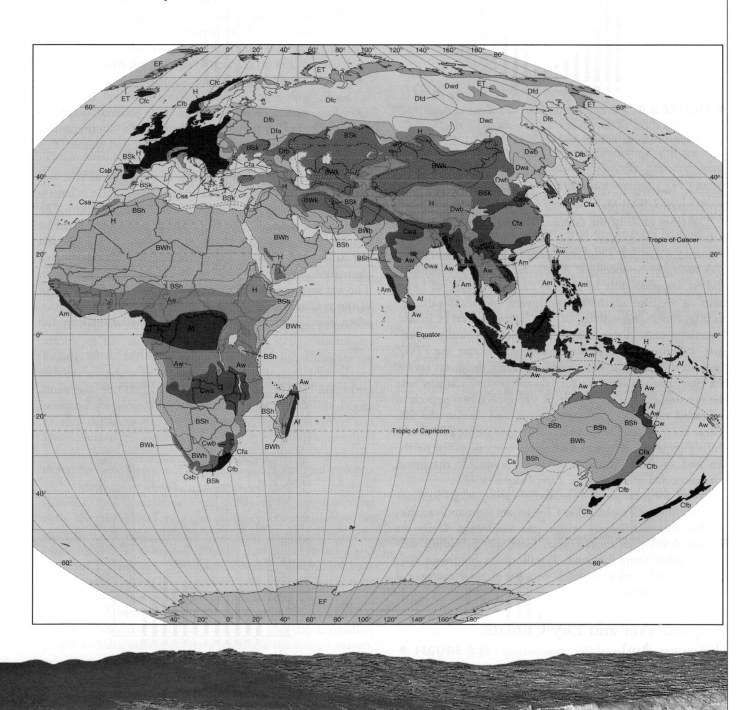

are only moderately dry are termed **semiarid.** They commonly have well-developed and often fertile soils, support a natural grass cover, and offer considerable potential for certain types of agricultural development. The very dry areas are termed **arid** and generally can be described as deserts. Arid regions typically have poorly developed soils, contain at most a scattering of drought-resistant natural vegetation, and are normally capable of supporting only limited human populations. The Low Latitude Dry Climate is centered between 20° and 30° N and S, but in places extends equatorward to 15° and poleward to 35°. It is the chief climate of the subtropics, although it should be noted that large portions of the subtropics have humid climates. In North America, the areas with a Low Latitude Dry Climate include most of the southwestern United States and northern Mexico. In South America, this climate is restricted primarily to the narrow Atacama Desert of coastal Peru and northern Chile. By far the most extensive land areas with this climate are to be found in Africa, Asia, and Australia. In

(a)

(a)

(b)

(b)

◆ **FIGURE 8.13** (a) Arid climatic conditions in Mexico's Baja peninsula have allowed the growth of only a sparse cover of xerophytic shrubs. (b) Semiarid conditions, including a brief rainy season, have permitted this portion of Kenya to support a cover of grasses as well as a few shrubs and small trees. The photo was taken during the dry season.

SOURCE: a: © Mark Richards/PhotoEdit; b: © Susan Van Etten/ PhotoEdit

◆ **FIGURE 8.12** Wet and dry season contrasts within Mikumi National Park, Tanzania, located within the Tropical Wet and Dry Climate.

SOURCE: Wayne McKim

Africa, it includes the vast expanse of the Sahara and its borderlands, the Kalahari region of Southwest Africa, and the easternmost African "horn," located mostly within Somalia. The low latitude dry lands of Asia are in the southwestern part of the continent and include the Arabian Peninsula, most of Iraq, Iran, Pakistan, and western India. In Australia, sometimes called the "desert continent" because a greater proportion of this continent is desert than any other, all of the interior and much of the western and southern margins are dry. Finally, it should be noted that vast ocean areas within the tropics and subtropics experience a dry climate.

The relative lack of precipitation in Low Latitude Dry Climate areas results primarily from the year-round dominance of the subtropical highs. The climate is especially extensive on the west sides of the continents and adjacent east sides of the oceans where the subtropical highs are most

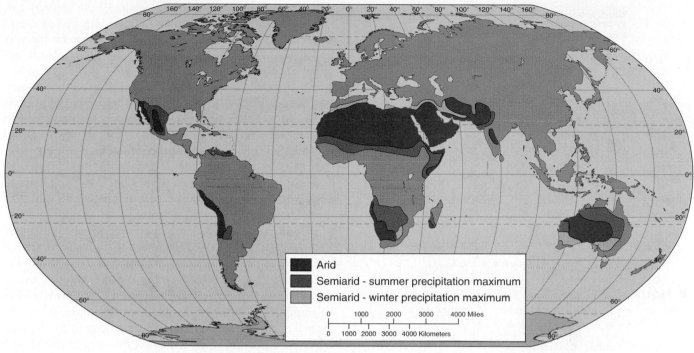

◆ **FIGURE 8.14** Distribution of arid and semiarid subtypes of the Low Latitude Dry Climate.

strongly developed and cold ocean currents help stabilize the atmosphere. Dryness is further accentuated in most areas by high temperatures, high potential evapotranspiration rates, continentality, and, in some cases, mountain barriers. Arid areas, which typically have annual precipitation means averaging less than 25 centimeters (10 in), form the central or core areas within the regions having this climate. They are surrounded by broad peripheral rims of semiarid conditions that generally have annual precipitation means of 25 to 75 centimeters (10 to 30 in).

Three distinct precipitation patterns exist within the Low Latitude Dry Climate. These correspond with the three bands of differing dryness shown in Figure 8.14. On the climate's equatorward margins, semiarid conditions with a pronounced midsummer rainfall maximum and almost totally dry winter predominate. Areas with this rainfall pattern are adjacent to the Tropical Wet and Dry Climate, and their precipitation regime represents a poleward extension of the rainfall pattern of this climate. During most of the year in this zone, the subtropical high dominates and suppresses precipitation, but for two or three months during the summer, the poleward shift of the ITCZ brings substantial convectional rainfall. Rainfall is insufficient to offset the dryness of the other seasons, however, so a semiarid climate prevails. This pattern is illustrated by the climagraph of N'Djamena, Chad (Figure 8.15).

The arid core of this climate occupies the largest total area. Rainfalls here are very scanty and no distinctive pattern may be evident. The passage of time commonly produces an endless succession of days with deep blue skies, perhaps with scattered cirrus or afternoon cumulus clouds, and excellent

visibilities. On occasion, however, visibilities may be reduced by airborne dust particles or even full-fledged dust storms. During the summer, intense heating may be sufficient to produce scattered convectional showers and thunderstorms, especially over upland regions. These storms have high bases, and the air between the clouds and ground is often so dry that the rainfall evaporates as it descends. By contrast, if

◆ **FIGURE 8.15** Climagraph for N'Djamena, Chad. This city has a double maximum temperature pattern caused by the advent of the rainy season. Rains are associated with the summertime approach of the ITCZ.

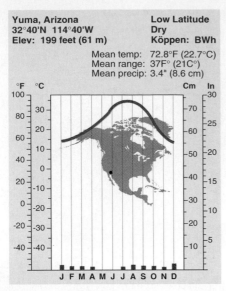

Yuma, Arizona
32°40'N 114°40'W
Elev: 199 feet (61 m)

Low Latitude
Dry

Köppen: BWh

Mean temp: 72.8°F (22.7°C)
Mean range: 37F° (21C°)
Mean precip: 3.4" (8.6 cm)

◆ **FIGURE 8.16** Climagraph for Yuma, Arizona, a station dominated by the northern hemisphere subtropical high.

rain falls in winter, it generally is produced by the unusually deep equatorward penetration of frontal cyclonic storms from the middle latitudes. The winter rainfalls typically are lighter and more widespread than those produced by summer season convectional storms. The climagraph for Yuma, Arizona (Figure 8.16), is typical of a low latitude desert station. Although its scanty precipitation pattern is rather difficult to discern on the climagraph, Yuma actually has a bimodal precipitation distribution. The city has a midwinter maximum caused by the passage of Pacific frontal cyclones as well as a second late summer maximum resulting from convectional thunderstorms.

Semiarid regions on the poleward margins of the Low Latitude Dry Climate have a precipitation pattern that is the reverse of the pattern of the climate's equatorward margins. In these areas, which are transitional between the low and middle latitudes, summers generally are dry because the subtropical high shifts poleward at this time of year to blanket the area. During the winter, however, the high shifts equatorward, allowing the westerlies with their eastward-traveling cyclonic storms to penetrate. Occasional one- or two-day periods of light to moderate precipitation can result, and in some of the cooler locations significant snowfalls are sometimes experienced. The climagraph for Tripoli, Libya (Figure 8.17), illustrates the pattern.

The Low Latitude Dry Climate is as well known for heat as for dryness. The subtropical deserts are the source regions for continental tropical air masses, and their summer temperatures are the hottest in the world. Many Low Latitude Dry Climate stations have means in the mid 30s Celsius (the 90s Fahrenheit) for several consecutive summer months, and a few reach a mean of 38°C (100°F) in their hottest month. Daytime highs of 46° to 50°C (115° to 120°F) are not uncommon in some inland desert areas at low elevations, and these daytime maximum readings often are remarkably consistent for weeks at a time.

Associated with the extreme summer heat are very low relative humidity values. Low humidities are caused not so much by a lack of atmospheric water vapor as by the very high water vapor holding capacity of the air at such hot temperatures. Daytime relative humidity readings in the desert often fall below 10 percent in the afternoon, and a low of 2 percent has been recorded. The relatively dry air may make the heat more bearable, but it results in extremely high evaporation rates from moist surfaces such as reservoirs and irrigated fields and causes inhabitants to need large quantities of liquids.

Semiarid areas, with their somewhat larger amounts of evaporable surface moisture, experience slightly less extreme heat. This is especially true for areas on the equatorward side of the subtropical highs that receive substantial convectional summer rainfalls (such as N'Djamena, Chad) and for coastal regions with a maritime influence (such as Tripoli, Libya).

In contrast to the excessive heat of the summer, winter temperatures within the Low Latitude Dry Climate tend to be surprisingly cool. The combination of clear skies and dry air makes the atmosphere relatively transparent to outgoing terrestrial radiation, while the lower Sun angles and shorter periods of daylight considerably reduce insolation receipts. As a result, midwinter temperatures average 10° to 18°C (the 50s to low 60s Fahrenheit) in all but the equatorward portions of this climate. (Compare, for example, the winter temperatures of Yuma and Tripoli with those of N'Djamena, which lies 20° closer to the equator.) Winter days generally are pleasantly warm, especially in the sunlight, but nighttime lows can be quite cold, and frost and freezing temperatures are occasionally experienced in many of the more poleward locations. Daily temperature ranges within subtropical deserts during the winter half of the year are among the highest in the world. These ranges usually are between 18

Tripoli, Libya
32°50'N 13°13'E
Elev: 56 feet (12 m)

Low Latitude
Dry (Semiarid)

Köppen: BShs

Mean temp: 67.1°F (19.5°C)
Mean range: 31F° (17C°)
Mean precip: 14.7" (37.3 cm)

◆ **FIGURE 8.17** Tripoli, Libya, has a semiarid climate because of its location on the poleward margin of the northern hemisphere subtropical high.

and 35C° (30 and 60F°), and temperatures have been known to rise from below freezing to highs exceeding 38°C (100°F) in a single day. Little wonder, then, that the desert is sometimes described as "broiling during the day and freezing at night"!

Strong surface heating during the summer half of the year produces very steep near-surface lapse rates, resulting in strong, gusty daytime winds and the occasional formation of dust devils. These winds are capable of picking up clouds of dust and stinging fine sand, adding one more discomfort for the desert traveler. Hot, dust-laden winds from the Sahara, for example, are periodically experienced all along that desert's margins. The problems of atmospheric dust have been compounded in recent decades by the large-scale human removal of protective vegetation from low latitude semiarid areas for agriculture and for fuel (see the Chapter 18 Case Study).

THE MIDDLE LATITUDE CLIMATES

Between the zones dominated by the subtropical highs and the subpolar lows lie the broad expanses of the middle latitudes. This area often is subject to fast-changing weather conditions as both high pressure systems and cyclonic storms with their associated fronts pass by in endless succession in the westerly wind flow. Climatic conditions are determined largely by continentality and latitude and in some places are strongly modified by the topography. In general, more poleward locations are farther from the subtropical highs and experience increased cyclonic and frontal activity.

Insolation receipts in the middle latitudes are sufficiently low in winter to produce a cold season that substantially lowers annual temperature means. The net result of the lowered temperatures and increased storminess is that annual precipitation totals exceed potential evapotranspiration rates within most of the middle latitudes, bringing a return to humid or subhumid climatic conditions.

Dry Summer Subtropical Climate (Köppen Csa and Csb)

The Dry Summer Subtropical Climate (also called the Mediterranean Climate) is found in the western portions of continents in the lower middle latitudes, generally between 30° and 45° N and S. This climate is transitional in location between the subtropical deserts and the humid middle latitudes; in effect, the climatic characteristics of each of these neighboring regions dominate for a portion of the year. The mountainous nature of most land areas in which this climate occurs typically restricts it to rather narrow coastal zones. As a result, the Dry Summer Subtropical Climate occupies the smallest land area of any of our climate types. Its human significance, however, is great since most areas are densely populated and economically prosperous.

Portions of five continents have areas with a Dry Summer Subtropical Climate. Although separated by vast distances, these areas have similar relative geographical positions and remarkably similar climatic conditions (Figure 8.18). In North America, the climate extends from the vicinity of Los Angeles north to near the Oregon border. In general, it is restricted to the coastal zone west of the Sierra Nevada Mountains. In South America, it is located in central Chile west of the high Andes. It occurs throughout the Mediterranean Basin region of southern Europe and northwestern Africa, east into parts of the Middle East. This is the only area in the world with both enough land and a sufficient lack of restricting mountain barriers to allow this climate to cover a large area. Finally, the Dry Summer Subtropical Climate occupies the coastal portions of southern and southwestern

(a)

(b)

◆ **FIGURE 8.18** (a) The town of Capileira, in Andalusia, southern Spain, is located in the Dry Summer Subtropical Climate; (b) California's rugged Big Sur coast has cooler and more humid summer conditions than do sites located farther inland.

SOURCE: a: Stephen Johnson/Tony Stone Images; b: ©Trevor Wood/Tony Stone Images

◆ FIGURE 8.19 Climagraph for San Francisco, California. San Francisco's exposed coastal location causes it to have unusually cool summers and mild winters for its latitude.

◆ FIGURE 8.2O Climagraph for Red Bluff, California. Red Bluff's interior location causes it to have three times the annual temperature range of San Francisco.

Australia and brushes the southwestern tip of Africa in the vicinity of Cape Town.

Temperatures within the Dry Summer Subtropics generally are held within moderate limits by the presence of nearby water bodies. The maritime influence lowers both daily and annual ranges. During the summer, large differences in temperature frequently exist between locations near the coast and those farther inland. Coastal areas often are kept quite cool by the adjacent water, but readings rise rapidly inland, and interior locations usually have hot temperatures with warmest monthly means in the upper 20s Celsius (upper 70s or lower 80s Fahrenheit). Temperature differences are especially striking along coasts paralleled by cold ocean currents and backed by hills or low mountains that block the inland penetration of cool ocean breezes. Such is the case in California and to a lesser extent in Chile and Portugal. (Compare, for example, the climagraphs for San Francisco and Red Bluff in Figures 8.19 and 8.20.)

The Mediterranean coast has much warmer summer means than do locations such as San Francisco that are situated on an oceanic coast. This is because the Mediterranean Sea is a restricted interior body of warm water that experiences intense summer insolation and receives only a very limited inflow of cooler Atlantic water. As a result, most locations experience summer means in the mid 20s Celsius (the high 70s Fahrenheit). Split, a city on the Adriatic coast of Croatia, is a typical example (Figure 8.21).

Although daytime highs in the summer commonly rise into the mid 30s Celsius (the 90s Fahrenheit) and sometimes exceed 38°C (100°F) in interior locations, nighttime low temperatures in most areas are pleasantly cool, with readings in the low to mid teens Celsius (the 50s to lower 60s Fahrenheit). These cool temperatures result from the

generally clear skies of summer and reasonably low atmospheric water vapor conent even along the coasts produced by the combination of subsiding air in the subtropical highs and relatively cool water. These factors promote radiational cooling.

Winter temperatures within the Dry Summer Subtropics average mostly between 5° and 12°C (the 40s and lower 50s Fahrenheit). Temperature variations at this time of year between coastal and inland locations are much smaller than they are in summer, and generally the interiors are now cooler. (Compare winter values for San Francisco and Red Bluff.) Winter daily temperature ranges are also considerably higher inland than near the water. As a result, frosts and freezes, though rare in coastal areas, are common inland. This is especially true for locations situated well above sea level. Maritime polar air masses dominate all areas during the winter, but on the west coast of North America and in the northern Mediterranean Basin area, occasional invasions of continental polar air from the northeast can bring killing frosts even to the coasts.

The most distinguishing characteristic of the Dry Summer Subtropical Climate is its annual precipitation pattern of dry summers and humid winters. During the summer season, the subtropical high in each hemisphere shifts poleward with the Sun to blanket the Dry Summer Subtropics. The summer months are almost rain-free along cold water coastal regions, such as California and Chile, where cool temperatures further reduce the already low lapse rates. The climagraph for San Francisco illustrates this situation. Summer advection fogs prevail along these coasts because of the cold water. The dry season also is exceptionally long and dry along the equatorward margins of the climate, because these areas lie closest to the centers of the subtropical highs.

In more poleward locations, the influence of the subtropical highs is not as strong, and occasional rainfalls from convectional showers or weak frontal passages may occur. Monthly precipitation means, however, are generally less than 2.5 centimeters (1 in) during midsummer. The climagraph for Split illustrates this situation.

During the winter, the subtropical highs retreat equatorward, and a moderate onshore flow of less stable maritime polar air develops. Fronts and frontal cyclones also extend their influence into the subtropics at this time of year, providing the lifting needed to trigger periods of precipitation. Storms typically occur every three or four days, producing moderate to occasionally heavy rainfall. The periods between storms usually consist of sunny, mild weather, and most areas receive 50 percent or more of their maximum possible winter sunshine. Monthly precipitation means during the winter are commonly close to 7 to 10 centimeters (3 to 4 in), with amounts increasing poleward (away from the subtropical highs) and with increasing elevation. Most areas receive at least 75 percent of their total precipitation during the winter half of the year.

Annual precipitation totals in most cases average 40 to 90 centimeters (15 to 35 in), with the lower totals most common in the more equatorward locations. These amounts are quite low, and some of the drier areas would be classified as semiarid were it not for the fact that most of the precipitation occurs in the winter, thus maximizing its effectiveness in supplying moisture to the surface by reducing evapotranspirational losses. Yearly variations in the position of the subtropical high and the storm tracks result in a large annual variability in amounts received in all areas. As a result, droughts and floods are not uncommon (see the Focus Box on page 190).

Split, Croatia
43°30'N 16°28'E
Elev: 420 feet (128 m)

Dry Summer
Subtropical
Köppen: Csa

Mean temp: 60.4°F (15.8°C)
Mean range: 33F° (18C°)
Mean precip: 27.8" (70.6 cm)

◆ **FIGURE 8.21** Climagraph for Split, Croatia.

Humid Subtropical Climate (Köppen Cfa)

The Humid Subtropical Climate is the east coast counterpart of the Dry Summer Subtropics. It occupies the eastern sides of continents in the lower middle latitudes between approximately 25° and 43° N and S. The leeward position of these areas results in a moderate continental influence, but this is tempered by large adjacent bodies of warm water (usually oceans with warm currents) that supply sufficient water vapor to produce humid conditions.

The most extensive land area within the Humid Subtropics is the southeastern one-third of the United States, extending from extreme southeastern New England south through most of Florida and west to Texas and eastern Kansas. Another large area occupies southeastern South America from southern Brazil through Uruguay and northeastern Argentina. In Asia, much of eastern China, southern Japan, and extreme southern Korea experience a Humid Subtropical Climate. A fourth relatively extensive area is found in eastern Australia (Figure 8.23 on page 192). Finally, small patches of Humid Subtropics occur in southern and southeastern Europe, especially around the northern portions of the Adriatic Sea and the Black Sea, and in Africa, along a portion of the southeast coast of the Republic of South Africa.

Temperatures within this climate are relatively similar to those experienced in inland portions of the Dry Summer Subtropics, with warm to hot summers and cool to mild winters, depending upon latitude and elevation. Higher humidities, however, make the Humid Subtropics less comfortable, especially in summer. Poleward areas also generally have colder winters and larger annual temperature ranges than do comparable west coastal locations.

The long period of summer heat and humidity is one of the most notable characteristics of the Humid Subtropics. Oceanic centers of the subtropical high far to the east pump in a nearly continuous flow of maritime tropical air. (In the case of southeastern Europe, moisture is derived from the Atlantic Ocean and the Mediterranean and Black Seas.) Adding to the discomfort produced by the humidity is heat resulting from high Sun angles and long daylight periods. Although weak incursions of continental polar air bring occasional interludes of pleasant weather to the more poleward areas, the heat and humidity are almost unbroken equatorward of approximately 35° N and S for a two- or three-month period. Coastal locations are slightly cooler, and daily sea breezes frequently develop, but humidities are typically higher near the coast than inland.

Summer monthly means depend largely upon latitude and range through the 20s Celsius (the lower 70s to mid 80s Fahrenheit). Daily highs usually average from the upper 20s to middle 30s Celsius (middle 80s to lower 90s Fahrenheit), but in some areas the hottest summer days have readings in excess of 38°C (100°F). Nights are warm and sultry, with overnight lows frequently remaining in the mid or high 20s

Water Shortages in California

Despite serious flooding in parts of California in 1993 and 1995, the southern portion of the state has experienced general drought conditions for much of the past decade. California's precipitation pattern is notoriously fickle because the state is situated in the area of transition between the westerlies, with their traveling cyclonic storms, and the arid climates associated with the eastern Pacific center of the subtropical high. In order for California's winter rains to arrive on schedule, the subtropical high must shift southward over Mexico, and the jet stream also must shift southward so that it enters North America in California, rather than in Canada or the Pacific Northwest.

Major precipitation events are related to the pattern of Rossby waves that develop within the flow of the jet stream. In some years, deep wintertime troughs propagate inland at regular intervals over California from the Pacific, bringing copious rains and mountain snows. In other years the jet stream remains largely north of the state; this has been the case for most of the late 1980s and early 1990s.

The human impact of the California drought has been greatly augmented by the state's continued rapid economic growth. California currently has a population of about 31 million—by far the largest of any state—and is expected to reach 39 million by the year 2010. This population is distributed very differently than is water availability; three-fourths of the people live in southern California, while three-fourths of the precipitation falls in the northern half of the state. California has developed a sophisticated water collection and distribution system to make use of its limited water resources; it has 1,200 reservoirs, a system of deep wells, and long-distance aqueducts to carry water from sources such as the Colorado River and the Sierra Nevada to the parched south. Unfortunately, however, the state is running out of expansion potential for its water supply system because there are few rivers left to dam and groundwater levels in some areas are declining as rapidly as 0.6 meters (2 ft) a year.

Another basic problem affecting future water availability in California is the fact that 80 to 85 percent of the state's water is used for irrigation agriculture (Figure 8.22). Although California is the leading agricultural state in the nation and its farmers grow half the fruits and vegetables consumed in the United States, agriculture generates only 3 percent of the state's revenues. Water allocation laws passed decades ago allow farmers to purchase vast quantities of water at bargain rates. The laws also make it difficult or impossible in many cases to transfer agricultural water to urban supply systems.

Californians will continue to face water supply problems for many years to come because no end is in sight for the state's rapid economic and population growth. Any significant global warming would likely compound the situation by producing a northward shift of both the subtropical high and

Celsius (middle 70s to about 80°F). The climagraphs for Baltimore, Maryland, and Jacksonville, Florida (Figures 8.24 and 8.25 on pages 192–193), illustrate the influence of latitude on the length and intensity of the warm season.

During the last few weeks of summer and extending into early autumn, invasions of cool continental polar air masses become increasingly frequent, and by late autumn and winter, cP air dominates in most areas. Autumn is a delightful time of year in the Humid Subtropics, with extended periods of fair, pleasantly mild weather. Temperatures fall rapidly in the more poleward areas and more slowly farther equatorward to midwinter means ranging from about –3° to 13°C (the upper 20s to the mid 50s Fahrenheit). Weather fronts are active during the winter half of the year, resulting in substantial daily temperature variations. Northern hemisphere areas are especially prone to sharp temperature fluctuations because they are within reach of arctic (cA) air masses, and most locations experience winter days in which temperature readings remain near or below freezing. Temperatures normally begin to become milder a few weeks

◆ **FIGURE 8.22** Vast quantities of irrigation water are used to support agriculture in California. Most of this water comes from snowmelt in the Sierra Nevada.

SOURCE: ©Andy Sacks/Tony Stone Images

the westerlies. This would reduce winter precipitation while at the same time increasing temperatures and potential evapotranspiration rates.

A variety of solutions to the water supply problem are being explored. One is the large-scale construction of plants to desalinate seawater; unfortunately, this option is very expensive, costing an estimated $1,000 to supply an average household with water for a year. An even more expensive option that has been discussed is importing water from British Columbia in supertankers. More realistic in the short term is a policy of water conservation and reallocation. It appears that the large-scale use of vast quantities of water for low-value agricultural purposes such as keeping livestock pastures green and growing alfalfa will have to be phased back. California, like several other western states, must come to grips with the fact that its water supplies are finite and are now approaching the limits of their potential for regional development. ■

before the vernal equinox, and the spring brings another period of pleasant temperatures before the return of summer heat and humidity.

The most pronounced climatic difference between the Dry Summer Subtropics and the Humid Subtropics is in their annual precipitation patterns. While the former climate experiences a well-defined dry season in summer, the latter receives substantial amounts of precipitation throughout the year, with summer most commonly the season with the heaviest totals. Annual means within the Humid Subtropics are about twice those of the Dry Summer Subtropics, with most areas receiving 75 to 165 centimeters (30 to 65 in). Also in contrast to the Dry Summer Subtropics, precipitation amounts are greatest near the climate's equatorward margins. They are lightest poleward and inland, and interior portions of the Humid Subtropics frequently adjoin semiarid areas.

During the summer, the subtropical highs dominate, but their centers are situated well to the east, producing a weak onshore flow of unstable mT air. The monsoonal tendency

(a)

(b)

◆ **FIGURE 8.23** (a) A view of the countryside in the rolling piedmont of Virginia, located within the humid subtropical climate that dominates the southeastern United States. (b) Paddy fields in Guilin, China, also within the Humid Subtropics.

SOURCE: a: © Kenneth Garrett/FPG International Corp. b: C. Barr/H. Armstrong Roberts

serves to reinforce this wind-flow pattern, especially in eastern Asia. The resulting weather is warm, humid, and mostly sunny, but daytime heating frequently produces scattered afternoon and evening convectional showers and thunderstorms. Cumulonimbus clouds are observable almost every summer afternoon in more equatorward portions of the Humid Subtropics, such as the U.S. Gulf coastal states, and are responsible for heavy downpours. Rainfalls are highly variable in amount because these storms are localized and are slow moving due to the very weak upper-level winds that prevail at this time of year. During the winter, most precipitation is frontal and cyclonic in nature, causing it to be lighter, steadier, and much more widespread than the convectional rains of summer. Most winter precipitation falls as rain, but occasional snowstorms are experienced, especially in the more poleward areas.

Along the coasts of the United States, Asia, and Australia, summer and early autumn bring the threat of hurricanes. These tropical storms occasionally produce great destruction and loss of life within the Humid Subtropics, as illustrated by the Case Study at the end of Chapter 7.

Mid Latitude Dry Climate (Köppen BWk and BSk)

Nearly three-fourths of the world's dry climates—those areas where potential evapotranspiration rates exceed precipitation amounts on a yearly basis—are in the low latitudes. The remaining one-fourth of the dry climates occur in the middle latitudes, mostly within continental interiors in the northern hemisphere. Their dryness is not associated with the influence of the subtropical highs, but instead results chiefly from the presence of orographic barriers on their windward sides. A secondary factor in some areas is their remoteness from oceanic sources of maritime air.

Three widely separated areas are included within the Mid Latitude Dry Climate. The largest is the central portion of Eurasia extending from just north of the Black Sea eastward some 8,000 kilometers (5,000 mi) to north-central China. This vast region is cut off from moisture-bearing winds from the Indian Ocean and Mediterranean Sea by an imposing rim of mountains extending along its southern and southwestern margins. In addition, it is separated from Atlantic moisture by vast distances. Central Asia, in the heart of this area, is more remote from oceans than any other place in the world. In North America, the Great Plains and Great Basin are largely blocked from Pacific moisture by mountain ranges to their west and thus experience a dry climate (Figure 8.26). These areas adjoin the drylands dominated by

◆ **FIGURE 8.24** Climagraph for Baltimore, Maryland, near the northern border of the Humid Subtropical Climate in the United States.

Jacksonville, Florida
30°25'N 81°39'W
Elev: 38 feet (12 m)

Humid
Subtropical
Köppen: Cfa

Mean temp: 69.4°F (20.8°C)
Mean range: 27F° (15C°)
Mean precip: 53.3" (135.4 cm)

◆ **FIGURE 8.25** Climagraph for Jacksonville, Florida. The differences between Baltimore's and Jacksonville's temperature and precipitation means result largely from their differing latitudes.

◆ **FIGURE 8.26** Sagebrush grassland in the foreground gives way to snow-clad peaks in the distance near Muddy Gap, Wyoming, within the Mid Latitude Dry Climate.
SOURCE: © Tom M. Hugh/Photo Researchers, Inc.

the subtropical high in the southwestern United States. The third, and smallest, area with this climate type is the Patagonian region of southern and western Argentina. It is geographically distinct from the other two in that it is located on a relatively narrow body of land. Patagonia's dryness is associated primarily with the rainshadow effect of the Andes, but the potential for precipitation is further reduced by the stabilizing influence of the cold Falkland Island Current off the east coast.

Most portions of the Mid Latitude Dry Climate experience highly continental conditions because maritime air masses do not normally enter these areas. Annual mean temperatures vary considerably from place to place, as illustrated by the climagraphs for Saratov, Russia; Reno, Nevada; and Mendoza, Argentina (Figures 8.27–8.29). Readings depend most upon latitude, degree of continentality, and elevation.

Summer temperatures in the Mid Latitude Dry Climate generally are warm, with occasional hot spells. Warmest monthly means are as low as the low teens Celsius (the mid 50s Fahrenheit) in southern Argentina and as high as 30°C (86°F) in some of the desert basins of eastern Asia. Clear skies and dry air allow daytime readings to soar well above 38°C (100°F) on occasion in the hottest areas. Nighttime lows, however, generally fall to comfortable readings in the upper teens to low 20s Celsius (the 60s or lower 70s Fahrenheit) in the warmer areas and to between about 7° and 15°C (the upper 40s and 50s Fahrenheit) in cooler locations at higher latitudes and elevations. Even in the warmest locations, the summer heat is not constant since these areas, unlike the dry climates of the low latitudes, are within reach of occasional air masses of subpolar origin.

Winter temperatures are cold in most places, but the degree of coldness is highly dependent upon location. The mildest of the three regions is Argentina, where the surrounding ocean causes virtually all areas to have means that remain above freezing. In contrast, most stations in North America and nearly all those in Eurasia have subfreezing temperatures during the winter, with January means in portions of eastern Asia below −18°C (0°F). Both North America and Eurasia are within reach of wintertime arctic air masses and experience cold waves when this frigid air moves south from its source regions. Temperatures during these cold air outbreaks can get as low as −40°C (−40°F) in the northern Great Plains and −45°C (−50°F) in eastern Asia. Coupled with the bone-chilling cold are winds that

Saratov, Russia
51°30'N 45°30'E
Elev: 295 feet (90 m)

Mid Latitude
Dry
Köppen: BSk

Mean temp: 42.1°F (5.6°C)
Mean range: 61F° (33.9C°)
Mean precip: 14.8" (37.6 cm)

◆ **FIGURE 8.27** Climagraph for Saratov, Russia.

Reno, Nevada
39°32'N 119°49'W
Elev: 4,404 feet (1,342 m)

Mid Latitude
Dry
Köppen: BWks

Mean temp: 50.1°F (10.1°C)
Mean range: 38F° (21C°)
Mean precip: 7.7" (19.6 cm)

◆ **FIGURE 8.28** Climagraph for Reno, Nevada. Reno is located close enough to the Dry Summer Subtropical Climate of California to experience a modest winter precipitation maximum.

frequently reach very high velocities over these flat, treeless surfaces, producing potentially deadly wind-chill values for humans and unprotected livestock. Interspersed with the cold periods, however, are intervals of milder weather as air masses from the south or west penetrate into the regions. Sudden onsets of Chinook winds in the northern Great Plains, for example, are capable of producing dramatic rises of temperature as they displace arctic air masses. An extreme example of a Chinook occurred in Spearfish, South Dakota, on January 22, 1943. At 7:30 A.M., the temperature was –20°C (–4°F). Two minutes later, at 7:32 A.M., it had risen to 7.2°C (45°F)!

Precipitation amounts within different portions of the Mid Latitude Dry Climate depend largely on the strength of the rainshadow effect. Both arid and semiarid conditions occur. As an approximation, arid areas receive less than 20 centimeters (8 in) of precipitation annually, and semiarid areas, 20 to 50 centimeters (8 to 20 in). These amounts are somewhat lower than the equivalent values for the arid and semiarid climates of the low latitudes because the cooler temperatures of the middle latitudes reduce evapotranspiration rates and thus increase precipitation effectiveness. As is true in the subtropics, the Mid Latitude Dry Climate contains arid core areas surrounded by extensive semiarid rims. Each of the three continents has one or more arid cores (Figure 8.30).

The seasonal distribution of precipitation is highly variable and generally is similar to that of the nearest humid climate area. For example, Reno, located near the margin of the Dry Summer Subtropics, receives a winter precipitation maximum from Pacific storms that retain some moisture as they cross the Sierra Nevada. Saratov, near the border of the Humid Subtropical Climate, has a weak summer maximum. The majority of locations receive a warm season precipita-

tion maximum, largely because the air contains more water vapor at this time of year.

As is typical of the middle and higher latitudes, most precipitation is associated with traveling frontal cyclones. These storms usually are capable of generating only light amounts of rain or snow because they do not have access to major sources of moisture. Winter snowstorms, however, occasionally produce substantial accumulations, with accompanying strong winds creating blizzard conditions.[1] Spring sometimes brings turbulent weather resulting from the strong temperature contrasts that develop at this time of year between opposing warm and cold air masses. In the U.S. Great Plains, especially, this conflict can produce severe hailstorms and tornadoes. During the summer, weather fronts are weaker and less numerous, but the air may be sufficiently moist and unstable to trigger convectional showers and thunderstorms.

Marine Climate (Köppen Cfb and Cfc)

Moving poleward into the upper middle latitudes, we enter the heart of the westerly wind belt. Here, frontal cyclonic storms are both numerous and well developed, and weather systems tend to move rapidly eastward under the influence of strong upper-level winds. Situated in this zone on the west, or windward, sides of continents generally between 40° and 60° N and S, the aptly named Marine Climate has the strongest maritime influence of any of the climate types (Figure 8.31). An almost constant onshore flow of maritime

---

1. A **blizzard** is a heavy snowstorm accompanied by winds of more than 16 meters per second (35 mph) and visibilities that are reduced to below 150 meters (500 ft) by falling and blowing snow.

Mendoza, Argentina
32°48'S 68°45'W
Elev: 2,713 feet (827 m)

Mid Latitude
Dry
Köppen: BWk

Mean temp: 61.5°F (16.4°C)
Mean range: 29F° (16.1C°)
Mean precip: 7.6" (19.3 cm)

◆ **FIGURE 8.29** Climagraph for Mendoza, Argentina.

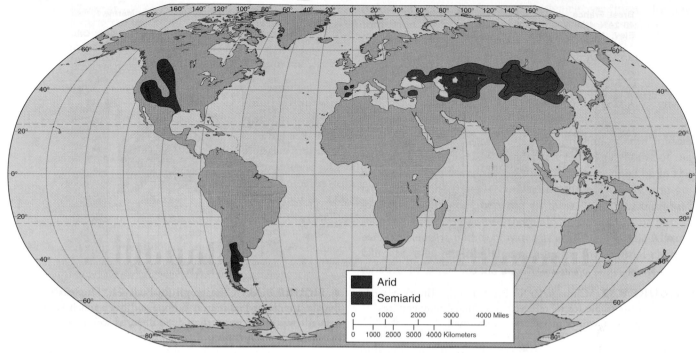

◆ **FIGURE 8.30** Distribution of arid and semiarid regions within the Mid Latitude Dry Climate.

polar air provides adequate moisture to allow precipitation totals to exceed potential evapotranspiration rates through most or all of the year so that no dry season is experienced. In addition, the onshore wind flow produces cool summers but unusually mild winters.

Geographically, the Marine Climate is generally located just poleward of the Dry Summer Subtropics. Like that climate, it is restricted in size, except in Europe, by the blocking influence of coastal mountain ranges and, in the southern hemisphere, by a lack of landmasses at the proper latitude. In North America, the Marine Climate occupies a narrow coastal strip beginning in extreme northern California and extending northward and westward along the west coast of Canada and the south coast of Alaska to the Aleutian Islands. In South America, it occurs in a similarly restricted zone in the portion of Chile west of the Andes and south of 40°S. The Marine Climate extends well inland in Europe, except in Scandinavia, where it is orographically restricted to western Norway and southern Sweden. Elsewhere in Europe it exists from southern Iceland southward through England and France to northern Spain and eastward as far as Poland, Romania, and Bulgaria. In both Africa and Australia, it occurs in portions of the southern coastal regions; it also occupies most of New Zealand. It should also be noted that vast ocean areas in the middle latitudes have a Marine Climate. Indeed, these areas are the source regions for the mild, moist mP air masses that are eventually transported onto the continental margins to the east.

Perhaps the most noteworthy characteristic of the temperatures in Marine Climate areas is their relative constancy.

The resistance of the nearby ocean waters to temperature changes causes both daily and annual ranges to be exceptionally small for the latitude. Summertime high temperatures for Marine Climate stations typically average between about 12° and 20°C (the upper 50s and 60s Fahrenheit); nightly lows are commonly in the lower teens Celsius (the 50s Fahrenheit). In the winter, daytime highs generally range between about 5° and 11°C (the 40s or lower 50s Fahrenheit), while nightly lows range from about −4° to

◆ **FIGURE 8.31** Chilkat Inlet, in southeast Alaska, has a Marine Climate with heavy orographic precipitation.
SOURCE: Charlie Ott/Photo Researchers.

FIGURE 8.36 Aerial view of fertile agricultural land in Illinois, near the southern margin of the Humid Continental Climate region in North America.

SOURCE: © Tony Stone Images/Gary Irving

(80°F) in the central United States to near 16°C (60°F) in south-central Canada. Even the most poleward areas have hot spells, with daytime temperatures reaching the middle 30s Celsius (the 90s Fahrenheit). Unlike the southern margins, however, these areas also are within reach of refreshingly cool summer cP air masses that frequently produce highs in the upper teens to low 20s Celsius (the 60s to lower 70s Fahrenheit).

Winters are cold, with the lowest means in poleward and inland locations. Temperatures average well below freezing everywhere, and the ground is frozen and snow-covered for one to five months. Fronts are very active during the winter, so that day-to-day temperatures are subject to large variation. The dominant wintertime air mass is cP, which generally is associated with clear, moderately cold weather. On occasion, though, outbreaks of frigid arctic (cA) air bring temperatures below –18°C (0°F) to all areas, and nighttime lows in the coldest regions can plummet to –40°C (–40°F) or lower. At the opposite extreme, midwinter thaws

FIGURE 8.37 Climagraph for Des Moines, Iowa. Des Moines has the large annual temperature range and spring precipitation maximum typical of the Corn Belt.

sometimes occur when traveling cyclones draw mT air northward. Variations in the path of the jet stream may cause extended periods, or even the entire winter season, to be unusually cold or mild.

Because available moisture is limited, precipitation totals within Humid Continental Climate areas generally are modest. The long, cold winters reduce evapotranspiration rates sufficiently to produce an annual moisture surplus, but summer water deficits occur in all areas except near-coastal regions. Precipitation means are greatest along the southern

FIGURE 8.38 Climagraph for International Falls, Minnesota. This city has the large annual temperature range and strong warm season precipitation maximum typical of a continental climate location in the upper middle latitudes.

◆ **FIGURE 8.39** Climagraph for Saint John, New Brunswick. Saint John is on the North Atlantic storm track and often receives heavy winter snowfalls.

◆ **FIGURE 8.40** Climagraph for Harbin, China. Harbin has a very large annual temperature range and a strong monsoonal summer precipitation maximum.

and coastal margins and gradually decline poleward and inland. The largest precipitation totals occur in Japan, one of the few areas within this climate with a strong orographic influence. Annual totals there average 180 centimeters (70 in) or more in some localities. The New England and Canadian Maritime region has access to Atlantic moisture and also is quite wet, with typical annual totals of 100 to 125 centimeters (40 to 50 in), as at Saint John (Figure 8.39). Amounts in both North America and East Asia gradually decline inland, and means near 43 centimeters (17 in) occur in the northern interiors of all three Humid Continental regions.

Seasonal precipitation distributions are most strongly influenced by the extent of continentality. Along east coasts and in the eastern Great Lakes area of the United States and Canada, winter storms are frequent and substantial cold season precipitation occurs. The climagraph for Saint John illustrates this situation. Progressing inland, the tendency for a summer precipitation maximum increases because the much higher temperatures of summer greatly increase the water vapor holding capacity of the air and reduce its stability. In addition, a weak monsoonal tendency, the poleward shift of the oceanic subtropical highs to the east, and the northward displacement of the polar front all combine to produce an influx of moist mT air. The tendency for a summer maximum is strongest in continental East Asia, as in Harbin (Figure 8.40) where the monsoonal tendency produces a winter dry season. Elsewhere, however, winter evapotranspiration rates are so low that a winter moisture surplus exists despite the low precipitation totals at this time of year.

Precipitation during the fall, winter, and spring is almost entirely cyclonic and frontal and normally occurs at intervals

of two to four days. Most winter precipitation is in the form of snow. Winter snowstorms sometimes are heavy and, if accompanied by strong winds, can produce blizzard conditions capable of bringing outdoor activities to a standstill.

In summer, cyclonic and frontal activity weakens and shifts northward, and most areas come under the dominance of mT air. The weather is much clearer at this time of year than during the cold season, but strong surface heating frequently triggers convectional showers and thunderstorms. In contrast to the general rains and snows of winter, the localized nature of these storms may produce greatly varying amounts and leave localized pockets of drought. More general droughts occur in some years when shifts in the pressure and wind patterns draw in cT, rather than mT, air masses. The cooler northern fringes of the Humid Continental regions receive less summertime convectional precipitation, but this is largely offset by the combination of lower evapotranspiration rates and increased frontal precipitation.

THE HIGH LATITUDE CLIMATES

The high latitude climates are dominated by cold weather during most or all of the year. Low temperatures produce difficult living conditions and severely restrict agricultural activities. As a consequence, most areas in the high latitudes are lightly populated or uninhabited. During the brief summer, many inland areas are surprisingly mild, but coastal regions as well as the ice plateaus of Antarctica and interior Greenland remain cold. During the winter, cold and nearly constant darkness reign everywhere. Low temperatures limit the water vapor content of the air, so that precipitation totals

◆ **FIGURE 8.41** The peaks of the Alaska Range tower above a forest of spruce in southern Alaska, within the Subarctic Climate.

SOURCE: © Charlie Ott/Photo Researchers

are generally low despite the extended periods of rain or snow that frequently accompany high latitude storms. Evapotranspiration rates are also greatly reduced by the cold air, however, so an annual moisture surplus exists nearly everywhere.

Subarctic Climate (Köppen Dfc, Dfd, Dwc, and Dwd)

The Subarctic Climate exists only in the northern interiors of North America and Eurasia, where it is found in a broad zone situated mostly between 50° and 70° N. As is true of the Humid Continental Climate, no Subarctic Climate regions exist within the southern hemisphere because of the absence of sufficient land in the proper latitudes. West winds transport in enough oceanic warmth to shift the climate poleward on the western sides of the continents, and latitudes gradually decline eastward. In North America, the climate extends from central Alaska east through most of interior Canada to Newfoundland (Figure 8.41). In Eurasia, it exists from northern and eastern Scandinavia east in a widening swath through the northern interior of Russia to the Kamchatka Peninsula.

One of the most important thermal characteristics of the Subarctic Climate is its extreme seasonal temperature variations. Portions of this climate experience the world's largest annual temperature ranges. In North America, differences between the summer and winter monthly means are generally 33 to 44C° (60 to 80F°). Ranges are much more diverse in Eurasia, varying from less than 17C° (30F°) in coastal northern Norway, where strongly maritime conditions prevail, to well over 56C° (100F°) in interior eastern Siberia. The climagraphs of three representative Subarctic Climate stations (Figures 8.42–8.44) illustrate the diversity of ranges.

Winter, the dominant season in the Subarctic Climate, is long, dark, and extremely cold. The cold winter temperatures, which are lower than those anywhere else in the world except for the ice plateaus of Antarctica and Greenland, result from a combination of several factors. One is the lack of insolation. Most areas are near or north of the Arctic Circle and receive at most only a few hours of low-intensity heating from a Sun barely above the horizon. A second factor is the continentality produced by the location of most areas in the interiors of large, snow-covered landmasses. Finally, the thermal compaction of the cold air results in the formation of high pressure centers over Canada and Siberia.

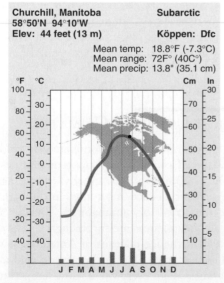

◆ **FIGURE 8.42** Climagraph for Churchill, Manitoba.

◆ **FIGURE 8.43** Climagraph for Kuopio, Finland.

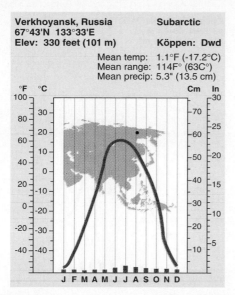

Verkhoyansk, Russia **Subarctic**
67°43'N 133°33'E
Elev: 330 feet (101 m)   **Köppen: Dwd**

Mean temp: 1.1°F (-17.2°C)
Mean range: 114F° (63C°)
Mean precip: 5.3" (13.5 cm)

◆ **FIGURE 8.44** Climagraph for Verkhoyansk, Russia. Verkhoyansk has one of the world's largest annual temperature ranges.

These high pressure systems generally are associated with clear skies and light winds that favor surface radiation.

Winter readings in most locations average below –18°C (0°F), and subfreezing mean temperatures last anywhere from five to eight months. The lowest temperatures of all occur in eastern Siberia; in the Russian city of Verkhoyansk, for example, the January mean temperature of –50°C (–58°F) makes it one of the coldest permanently inhabited locations in the world (Figure 8.44). In western Canada and Alaska, northern Europe, and the east coasts of Canada and Siberia, the proximity to water raises winter means (see the climagraph for Kuopio in Figure 8.43), but the reduction in cold is partially offset by an increased prevalence of clouds, winds, and heavy snows.

In contrast to the frigidity of the winters, summers are brief but very mild. At this time of year, the midday Sun is fairly high in the sky, and daylight is almost continuous. As a result, total daily insolation receipts are temporarily equivalent to those of the tropics, and only the thermal inertia of the atmosphere and surface keeps temperatures from rising to tropical levels. Means are in the teens Celsius (the 50s and 60s Fahrenheit), with the warmest readings in the southern interiors.

Fronts are active throughout the year so that day-to-day temperature variations are substantial. Under the influence of warm winds from the south, afternoon summertime highs sometimes reach the upper 20s to mid 30s Celsius (the 80s to lower 90s Fahrenheit). Fairbanks, in interior Alaska, recorded that state's all-time high temperature of 38°C (100°F) under such conditions. In contrast, cold air from the ice-covered Arctic Ocean, the subpolar North Atlantic, or Canada's Hudson Bay can keep highs in the single digits Celsius (the 30s or low 40s Fahrenheit), and frosts and light freezes can occur in most places throughout the summer.

The combination of cold temperatures and continentality results in generally light precipitation. Heaviest totals occur along the coasts, where annual mean temperatures are warmest, water sources are present, and winter storms are frequent. Annual means of up to 100 centimeters (40 in) occur in eastern Newfoundland and northern Norway. Precipitation declines rapidly inland, and in the northern interiors, amounts less than 25 centimeters (10 in) are the rule.

Most precipitation is frontal and cyclonic. Despite the low totals, storms with limited moisture supplies frequently pass through the subpolar regions, producing a large number of days with cloudy skies and light rain or snow. Except for coastal areas, a summer precipitation maximum occurs almost everywhere. Winter snows generally are light in the interiors, but are frequently associated with deep low pressure systems that produce strong winds and blizzard conditions. Coastal areas sometimes receive heavy snowfalls from storms with access to oceanic moisture. The coasts in general are cloudier, foggier, and damper year-round than the interiors.

Tundra Climate (Köppen ET)

The last two climates in our world survey are differentiated only by degrees of coldness. Located in the subpolar and polar regions, they receive smaller quantities of insolation than any other places on Earth. As noted earlier, the high latitudes have a net radiation deficit. A substantial proportion of their thermal energy is supplied by the poleward advection of heat from the lower latitudes by winds and ocean currents.

Within the Tundra Climate, the mean temperature of the warmest month is between 0° and 10°C (32° and 50°F). As in the case of the previous two climates, the great majority of land areas experiencing a Tundra Climate are in the northern hemisphere. Most form a coastal rim around the Arctic Ocean. Included are northern Canada and Alaska, coastal Greenland, northern Iceland, and the north coast of Eurasia from Norway eastward to the Bering Strait (Figure 8.45). Land areas within the southern hemisphere are limited to the southern tip of South America, the Antarctic Peninsula, and a number of small islands. Most high latitude ocean areas also have a Tundra Climate. Finally, this climate occurs in high mountains and plateaus in various parts of the world, notably on the Tibetan Plateau. Its characteristics within these areas are discussed more fully in the Highland Climates section of this chapter.

The Tundra Climate can be described as a high latitude maritime climate. Few areas are more than 160 kilometers (100 miles) from the ocean. The proximity to water reduces annual temperature ranges from the extremes encountered in the Subarctic Climate, causing summers to be colder and winters to be less frigid than in the former climate, despite the higher latitude. The major seasonal temperature control (and one of the most important aspects of the climate) is the

◆ **FIGURE 8.45** A field of rhododendrons colors the tundra in the Richardson Mountains, within the Canadian Yukon.
SOURCE: © Steve McCutcheon/Visuals Unlimited

◆ **FIGURE 8.46** Climagraph for Chesterfield Inlet, Northwest Territories, Canada. This town's annual precipitation total is limited by cold temperatures and a continental location.

duration of sunlight. Most areas are poleward of the Arctic Circle; as a result, sunlight varies from 24 hours a day near the time of the summer solstice to 0 near the winter solstice. The Sun's motion is much more horizontal (or circular) than vertical, so little diurnal variation occurs. Temperatures trend irregularly upward in spring and downward in fall, with the irregularities caused largely by frontal passages.

During the summer most places average between 4° and 10°C (the 40s Fahrenheit). The Sun, though low in the sky, is out all or nearly all day. As a result, insolation receipts in late spring and early summer can reach daily values equivalent to those of the tropics. A south wind in mainland North America and in Eurasia will bring in warm continental air and occasionally can cause temperatures to reach the mid 20s to low 30s Celsius (the 70s and 80s Fahrenheit). The ice-covered Arctic Ocean is always nearby, though, and a shift to northerly winds can quickly drop readings to near freezing, with the likelihood of frost. Coastal fogs frequently form in summer when warm air from the land encounters cold water or the Arctic ice pack.

Winter temperatures are regionally much more variable than those of summer. For the considerable majority of the tundra lands of North America, Eurasia, and northern Greenland, winter means are well below –20°C (0°F). The climagraph for Chesterfield Inlet, on the northwest shore of Hudson Bay, serves as an example (Figure 8.46). In these areas, mean temperatures are below freezing eight or nine months of the year, and readings can fall as low as –46° to –50°C (–50° to –60°F). Some areas, however, are near water bodies that remain partially or entirely unfrozen throughout the winter. Temperatures here are considerably less cold. One large area that experiences such conditions is located where the North Atlantic merges with the Arctic Ocean. Here the last remnants of the Gulf Stream keep water temperatures above freezing and air temperatures near or

above –18°C (0°F) on the adjacent land. The climagraph for Svalbard, an island in the Arctic Ocean north of Norway, provides an example (Figure 8.47). Extreme examples of the maritime influence occur in the southern hemisphere. Macquarie Island, located southwest of New Zealand (Figure 8.48), has a warmest monthly mean of 6.4°C (43.5°F), but its coldest month averages 2.8°C (37.0°F), producing an annual range of only 3.6C° (6.5F°).

Cold temperatures help reduce annual precipitation means in most places to below 38 centimeters (15 in). The exceptions are mountainous windward coasts such as southern Greenland, northern Norway, parts of western Alaska, and the southern hemisphere islands. Macquarie Island, for

◆ **FIGURE 8.47** Climagraph for Svalbard, Norway.

Macquarie Island, Australia
54°30'S 108°57'E
Elev: 20 feet (6 m)

Tundra
Köppen: ET

Mean temp: 40.0°F (4.4°C)
Mean range: 6.5F° (3.6C°)
Mean precip: 41.8" (106.2 cm)

◆ **FIGURE 8.48** Climagraph for Macquarie Island, Australia. Macquarie Island's extreme maritime location causes it to have a very small annual temperature range despite its high latitude.

example, receives a mean of 106.2 centimeters (41.8 in). These same coastal areas generally have a winter maximum caused by storms associated with the subpolar low and receive heavy winter snowfalls. Most areas, though, have a strong summer precipitation maximum (as in the case of Chesterfield Inlet) because the winter cold causes the Arctic Ocean to freeze over, removing the primary moisture supply. Virtually all precipitation is frontal and cyclonic. During the fall, winter, and spring, cyclonic storms can be very strongly developed and generate blizzard conditions because the flat, treeless tundra offers little frictional resistance.

Polar Climate (Köppen EF)

The Polar Climate is the most poleward and the coldest of the world's climate types. It also is the only one with no permanent human inhabitants. Monthly mean temperatures are below freezing all year, and the landscape is covered almost everywhere by a deep layer of ice. Two widely separated areas—Antarctica and interior Greenland—have a polar climate; together they comprise about 10 percent of the Earth's land surface.

The extreme coldness of this climate is only partially caused by low Sun altitudes. It also is produced by the high reflectivity of the ice and snow cover and by the high elevation of the surface. Most areas are situated between 1.5 and 2 kilometers (1 and 2 mi) above sea level.[2]

Although the unrelenting cold is by far the dominant overall characteristic of this climate, the extreme variation in the duration of daylight is its major seasonal attribute. Except for southern Greenland, all areas lie poleward of the Arctic or Antarctic Circles and experience the full 24-hour range in the duration of sunlight between the times of the solstices.

Summer temperatures average below freezing on a monthly basis, but in some parts of coastal Antarctica and the margins of the Greenland ice sheet, mild spells occasionally occur, with above-freezing conditions lasting for hours or even several days. The high-elevation interiors of both areas remain well below freezing. In interior Antarctica, temperatures remain below –18°C (0°F) throughout the year. Antarctica as a whole is colder than Greenland because of its higher latitude, somewhat higher average elevation, and greater degree of continentality.

Winter temperatures, especially in Antarctica, are the lowest on Earth. The coldest readings of all occur not at the South Pole but at the so-called "Pole of Cold" located in the center of the continent near 78° S 107° E. A research station named Vostok established at this location in the 1950s by the Soviets has recorded the lowest temperatures in the world. The current record is –89.2°C (–128.6°F), set on July 21, 1983.

Although the Sun remains below the horizon for several months in the late fall and winter in interior Antarctica, it is near enough to the horizon during much of this period to produce some twilight. Additional light often is provided by the southern lights or **aurora australis** (the southern hemisphere counterpart of the **aurora borealis** of the northern hemisphere).[3] Because of these light sources, the polar regions, though receiving less insolation than any other part of the world, also have less total darkness than any other place.

The Polar Climate is not only the coldest, but also the driest of the world climate types in terms of mean precipitation amounts received. Evapotranspiration rates are so low, however, that a major surface moisture surplus exists. The answer to the poetic question "Where are the snows of yesteryear?" is obvious in Greenland and Antarctica (Figure 8.50). They bury the surface to a depth of up to 4 kilometers (2.5 mi)! These areas provide an excellent example of the inadequacy of measuring aridity solely in terms of total precipitation, without taking temperatures and evapotranspiration rates into account.

Precipitation is cyclonic and frontal in nature and, of course, virtually all is frozen. The combination of cold, continentality, and thermal high pressure blocks most storms from the interior areas, especially in Antarctica. As a result,

---

2. Because of their elevation, these areas could well be included within the highland climates classification, discussed in the next section, as are the glaciated areas of the lower latitudes. By convention, however, because of their location and large size, they are treated as a distinct climate type.

3. Auroral displays are produced when electrons streaming outward from the Sun as part of the solar wind collide with and ionize atoms in the Earth's atmosphere in the vicinity of the magnetic poles. Light is emitted when these atoms reconstitute and release the energy they absorbed during the collisions.

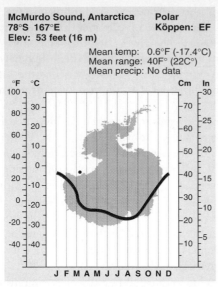

McMurdo Sound, Antarctica
78°S 167°E
Elev: 53 feet (16 m)

Polar
Köppen: EF

Mean temp: 0.6°F (-17.4°C)
Mean range: 40F° (22C°)
Mean precip: No data

◆ **FIGURE 8.49** Climagraph for McMurdo Sound, on the coast of Antarctica.

precipitation is heaviest near the coasts and decreases rapidly inland. Amounts received are uncertain due to a lack of reporting stations and the difficulty involved in making accurate measurements of wind-blown snowfalls. It is believed that precipitation totals in the interiors are mostly 5 to 13 centimeters (2 to 5 in) per year in terms of water equivalent. Precipitation totals increase to means of perhaps 50 to 65 centimeters (20 to 25 in) near the Antarctic coast. Heavy amounts in excess of 125 centimeters (50 in) occur near the southern edge of the Greenland ice cap because it offers an orographic barrier to storms moving along the North Atlantic winter storm track.

Winter storms frequently produce gale- and even hurricane-force winds, largely because snow and ice offer

◆ **FIGURE 8.50** Nearly all Polar Climate areas have a surface cover of ice and snow. Here scientists in Antarctica conduct research in front of a wall of glacial ice, while volcanic Mt. Erebus steams in the distance.

SOURCE: © Jeanette Thomas/Visuals Unlimited

less frictional resistance than any other surface on Earth, including water. Coastal storms also are numerous and often very strongly developed; some coastal stations in Antarctica report gale-force winds nine days out of ten. Under such conditions, the air becomes filled with fine, horizontally driven snow and ice crystals, which sometimes produce whiteouts that reduce visibilities to zero.

HIGHLAND CLIMATES

It is widely known that mountain regions have distinctly different climates than adjacent lowlands (Figure 8.51). Mountains are popular summer vacation destinations for people seeking to escape the heat of lowland areas, and many people also visit mountainous areas in winter to take advantage of winter sports opportunities. In this section the climatic differences between highland and lowland regions are examined, and the reasons for these differences are analyzed.

The highland climates as a group differ in one basic respect from the world climate types already discussed. Instead of consisting of a single large, relatively homogeneous region, they form a complex mosaic of diverse climate types located in close proximity to one another. In fact, it is their diversity, rather than their homogeneity, that causes them to be grouped together here. Climatic conditions in the world's highland areas vary so greatly and change so rapidly with distance that their total pattern is too complicated to depict on a world map or to analyze fully in a section of a chapter.

Distribution

Describing the global distribution of highland climate types is complicated by a lack of agreement on how greatly the climatic conditions of an area must be topographically modified to warrant its inclusion as a highland climate region. In general, the greater the topographic relief of an area, the greater the diversity of its climatic conditions.

All of North America's high western mountains, as well as the mountains and interior plateaus of Mexico, are climatically distinct from the surrounding lowlands. In South America, the Andes produce a continuous wide band of highland climates almost 8,000 kilometers (5,000 mi) long. The world's greatest concentration of highland climates occurs in southern Eurasia, extending from western China to northern Spain, and includes the Himalayas and its associated ranges, the Tibetan Plateau, the Zagros and Elburz Mountains of Iran, the Caucasus of Georgia and Azerbaijan, and the European Carpathians, Alps, and Pyrenees. In Africa, important areas include the Atlas Mountains and Ethiopian Highlands.

Lower mountain and upland areas that modify their climates to a lesser degree also are extensive. Some of the most notable are the Appalachians, the highlands of south-

◆ **FIGURE 8.51** A variety of vegetation zones controlled by elevation are evident in this view of Arizona's San Francisco Peaks.

SOURCE: © D. Newman/Visuals Unlimited

◆ **FIGURE 8.52** Climagraph for Bogotá, Colombia. Bogotá's location near the equator accounts for its double precipitation maximum (caused by passages of the ITCZ) as well as its very small annual temperature range.

east Brazil and southern Africa, the Russian Urals, the Italian Apennines, the mountains of Scandinavia, and the Dividing Range of eastern Australia. Smaller highland areas, often located on mountainous islands, are too numerous to list fully, but include portions of Japan, Madagascar, the East and West Indies, and New Zealand. It therefore is evident that regions with topographically modified climates are extensive and widely distributed.

Climatic Controls

The climate of any specific highland location is controlled by the combined influences of elevation, latitude, orographic effect, and local topography. The most important of these is elevation. Its significance stems from the fact that, in the troposphere, the temperature declines at an average rate of 6.4C° per 1,000 meters (3.5F° per 1,000 ft). As a result, temperature means in highland areas are substantially lower than those of adjacent lowlands. An example is provided by the climagraph for Bogotá, Colombia (Figure 8.52). Most sea level stations located at Bogotá's latitude experience annual means near 27°C (80°F). Bogotá's elevation of more than 2,500 meters (8,000 ft), however, causes its mean temperatures to remain in the low to mid-teens Celsius (mid- to high 50s Fahrenheit) throughout the year. High mountains can cause temperatures to be reduced sufficiently to produce polar climatic conditions, along with permanent ice and snow, even at the equator, as is the case on the highest peaks of the Andes.

Latitude, a second important climatic control, affects highland areas in much the same way as it does lowlands: as latitude increases, decreased insolation produces lower temperatures. As a result, the higher the latitude of a place, the lower the elevation at which a given climate may be expected to occur (Figure 8.53). An increase in latitude also is generally associated with larger annual temperature ranges. For example, Bogotá, despite its cool temperatures, has the

very small annual temperature range that one would expect of a city at its low latitude. Säntis, in the Swiss Alps (Figure 8.54), is at nearly the same elevation as Bogotá, but its latitudinal position causes it to have a much lower annual mean and a much higher annual range of temperature.

The orographic effect is the dominant control on precipitation patterns within a highland area. In general, the windward sides of mountains tend to be wet, the leeward sides dry, and large differences in amounts typically occur over small distances. The combination of orographic uplift and reduced evapotranspiration rates resulting from cooler temperatures generally causes highlands to be much wetter than nearby lowlands. Most of the world's major river systems, in fact, have their headwaters in highland regions. The Ethiopian Highlands in which Addis Ababa (Figure 8.55) is located, for instance, are largely surrounded by arid lands, yet they provide much of the water for the Nile. Säntis receives about four times as much precipitation as do lowland European stations at the same latitude. On the

◆ **FIGURE 8.53** Idealized relationship of climate type to latitude and elevation.

Säntis, Switzerland   **Tundra**
47°15'N 9°21'E
Elev: 8,208 feet (2,502 m)   Köppen: ET

Mean temp: 27.5°F (-2.5°C)
Mean range: 25F° (13.9C°)
Mean precip: 95.7" (243.1 cm)

◆ **FIGURE 8.54** Climagraph for Säntis, Switzerland. Säntis receives heavy orographic precipitation, most of it in the form of snow.

Addis Ababa, Ethiopia   **Tropical Wet**
9°02'N 38°45'E   **and Dry (modified)**
Elev: 7,899 feet (2,408 m)   Köppen: Cwb

Mean temp: 62.9°F (17.2°C)
Mean range: 7.4F° (4.1C°)
Mean precip: 43.5" (110.5 cm)

◆ **FIGURE 8.55** Climagraph for Addis Ababa, Ethiopia.

other hand, high interior valleys and plateaus may be relatively dry if surrounded by higher mountains that block moisture-bearing winds. This is the reason for Bogotá's modest precipitation totals.

Two other climatic characteristics of highland areas also are associated with orographic uplift. Fog is common in mountain areas, especially in humid regions. On many windward slopes, the continuous generation of fog can obscure mountain peaks for days at a time. The combination of orographic barriers and unstable air can also lead to dangerously changeable weather. Sudden violent thunderstorms and, at high elevations, local blizzards frequent many mountainous areas.

The last of the major climatic controls is the local topography. Topographic variation obviously is much greater in mountainous areas than in flat areas, and the variety and complexity of localized climates are correspondingly increased. Two of the most important topographic factors were discussed in Chapter 4. These are (1) slope aspect, which causes slopes facing the Sun to be warmer and thus generally drier than more shaded slopes; and (2) the nature of the surface, which can vary greatly in material constituency and vegetative cover. For example, a bare rocky slope will likely be warmer, drier, and windier during the day than a forested slope. A third topographically related factor is a site's degree of exposure. This especially influences wind speeds and directions. A sheltered valley nestled beneath high mountains, for instance, is protected from strong winds, but an isolated peak is exposed to the full force of the winds, unreduced by surface friction. A good example of the effects of exposure has been provided by a weather station located atop Mt. Washington, New England's highest peak. It recorded the world's highest wind speed of 90 meters per second (231 mph).

❚ SUMMARY

This chapter has explored the global pattern of climates. World climatic characteristics and resulting climate regions are distributed in a systematic geographic pattern. This pattern in turn results largely from the controlling influences of spatial variations in insolation, continentality, air pressure, wind patterns, and, in some areas, orographic influences and ocean currents.

In order to summarize the most important information presented in the chapter, we return to the Focus Questions:

1. *What are the chief advantages and shortcomings of climate classification systems?*

Climate classification systems are devised largely as an organizational tool for the systematic analysis of world climates. Their basic purpose is to simplify the study of climate by reducing the number of climatic types to a manageable total. By so doing, geographical patterns of climatic characteristics become evident, and a greater understanding of global climate controls can be gained. A basic shortcoming of all climatic classification systems is their artificiality. The world's climates, which in reality display tremendous variety, are grouped into a limited number of more or less broadly defined types, with each type occupying a certain geographical region. This setup may give a misleading impression of regional homogeneity, with sudden climatic changes occurring as boundary lines are crossed. This is not the case. In reality, climatic conditions vary gradually with distance within climatic regions and do not generally change in any dramatic respect when boundary lines are crossed.

2. *What are the major climatic characteristics of each of the global climate regions, and why do these characteristics occur?*

In this chapter, world climates were grouped into low latitude, middle latitude, and high latitude types, with highland climates treated as a separate category. Three low latitude climate types exist. The Tropical Wet Climate, under the domination of the ITCZ, is characterized by continuous warmth and by abundant precipitation of a predominantly convectional nature. The Tropical Wet and Dry Climate also is warm throughout the year, but displays a distinctive pattern of rainy summers and dry winters. The Low Latitude Dry Climate, dominated by the subtropical highs, is characterized by mild winters, very hot summers, and year-round dryness.

The middle latitudes contain five climate types. The Dry Summer Subtropical Climate has a pattern of dry, warm summers dominated by the subtropical high and mild, humid winters under the influence of the westerlies, with their traveling cyclones and anticyclones. The Humid Subtropical Climate has hot, humid summers with scattered convectional precipitation, and cool to mild winters with alternating periods of fair weather and frontal and cyclonic precipitation. The Mid Latitude Dry Climate is cut off from major moisture sources by orographic barriers and therefore is dominated by dry weather with large annual temperature ranges. The Marine Climate, in contrast, is characterized by adequate to abundant precipitation from cyclonic storms and by small annual variations in temperature. The most poleward of the middle latitude climates is the Humid Continental Climate. It has cold winters, mild to warm summers, and modest but generally adequate precipitation that is predominantly frontal and cyclonic in origin.

Three high latitude climates were recognized. All are characterized by varying degrees of coldness and by rather low precipitation amounts from cyclonic storms and their associated fronts. The Subarctic Climate has severely cold winters and cool to mild summers of rather short duration. The Tundra Climate experiences above-freezing mean temperatures for only a few months in summer. Finally, the Polar Climate is characterized by subfreezing temperatures and a surface ice and snow cover throughout the year.

The climates of the highland regions of the world display great variability, often over small distances. This variability exists largely because of differences in elevation, orographic influences on precipitation, and local topographic conditions such as slope aspect and exposure.

QUESTIONS FOR THOUGHT AND REVIEW

1. What are the chief advantages of using a regional climate classification system to study world climate characteristics? What are the shortcomings of such systems?

2. How is the Köppen system of climate classification organized? Why has this system become popular among geographers? What are the major strengths and weaknesses of the Köppen system?

3. Explain how the location of the Tropical Wet and Dry Climate controls its annual precipitation distribution pattern.

4. What factors determine the dryness of an area? Why are there no arid and semiarid climates in the high latitudes?

5. The Low Latitude Dry Climate is located between approximately 18° and 33° N and S. Describe the changes in both amounts and seasonal distributions of precipitation you would encounter if you traveled from the climate's equatorward margin to its poleward margin, making a one-year stop at every 7° of latitude.

6. Describe the annual precipitation distribution pattern of the Dry Summer Subtropical Climate. What causes this pattern?

7. How and why do annual temperature characteristics differ between coastal and interior California?

8. How and why does the annual precipitation distribution pattern of the Humid Subtropics differ from that of the Dry Summer Subtropics?

9. How do the causes of the dryness of the Low Latitude Dry Climate and the Mid Latitude Dry Climate differ? In which climate are the annual temperature ranges larger? Why?

10. Why is the only extensive land area with a Marine Climate located in Europe? What changes in climatic conditions occur as one travels eastward from the west coast of France to the western border of Russia? Why do these changes occur?

11. What are the reasons for the great seasonal temperature contrasts within the Humid Continental Climate? What are the causes of the large day-to-day temperature contrasts that sometimes occur?

12. Why do Tundra Climate areas remain cold during the summer, despite receiving large amounts of insolation? Why do the Polar Climates remain so cold in summer?

13. In what basic way do the highland climates as a group differ from the other world regional climate types?

14. In what respects would your own climate be different from what it is now if your elevation were increased by 1,500 meters (5,000 ft)?

15. List several reasons why climatic characteristics often vary substantially over small distances in mountainous areas.

16. In what important respects do the climates of cities differ from those of the adjacent countryside?

KEY TERMS

climagraph   semiarid
arid

The Human Impact on the Climates of Cities

The Case Study at the end of Chapter 4 dealt with possible global temperature changes resulting from human activities, notably the addition of carbon dioxide and other greenhouse gases to the atmosphere. These concerns, although serious, are still speculative because thus far human actions have produced no large-scale worldwide changes in temperature. However, human activities already have resulted in the formation of distinctive climatic conditions in urban areas, where a large and rapidly growing segment of the world's population lives.

The most significant impact that an urban area has on its climate is to increase local temperatures by producing what has been termed an **urban heat island.** It has been found that temperatures in the centers of large cities are frequently at least 6C° (11F°) warmer than those of the surrounding countryside. A combination of several factors is responsible for this occurrence. First, and probably most important, human activities produce large quantities of energy that are eventually released to the atmosphere. Most of these activities, such as heating homes and businesses, running factories and power plants, and operating vehicles, involve the burning of fossil fuels. Heat-producing operations have an especially large impact in the winter, when heating needs are greater and proportionally less solar energy is present. Studies have shown that the artificial

heat produced in cities in northern Europe and North America during the midwinter months may rival or exceed that provided by insolation.

A second factor contributing to the formation of a heat island is the abundance of building materials such as brick, concrete, and asphalt. These materials readily absorb heat and reradiate it for long periods. A third factor is the buildup of heat-absorbing pollutants over the city. The major product of the combustion of fossil fuels is carbon dioxide—the chief gas responsible for the greenhouse effect. A final important factor is the surface dryness of the city. The materials from which a modern city is constructed are relatively impervious to moisture, and most rain water quickly flows into storm sewers. As a result the surface soon dries, and subsequent insolation that otherwise would be used to evaporate water instead is converted to sensible heat. In contrast, the evapotranspiration of water from soil and vegetation in rural areas uses a large proportion of the Sun's energy—often more than is converted directly to heat.

The urban heat island typically is best developed at night, largely because energy absorbed by city materials during the day continues to be released at this time. The difference in nocturnal temperatures between adjacent urban and rural areas sometimes is large enough that forecasters give separate predicted lows for each.

Cool temperatures, clear skies, and light winds all aid in heat island development. Cool temperatures result in a greater consumption of fuel for heat production, clear skies allow the day's heat to radiate quickly into space in rural areas, and light winds keep the warmer city air from being blown away.

The heat island patterns of a number of cities have been mapped (Figure 8.56). Most patterns consist of

roughly concentric circles. The highest readings occur in the center of the city, where the heat-producing factors are most concentrated, and progressively lower values are found toward the suburbs. In detail, the patterns are complicated by the presence of especially warm temperatures over industrial or other areas with high concentrations of activity, and by cool pockets, often associated with open areas such as parks and cemeteries.

Urban climates also differ from those of rural areas in a number of aspects involving atmospheric moisture. Relative humidities are somewhat reduced because the warmer city temperatures increase the air's water vapor holding capacity. On the other hand, specific humidity values are at least as high as those of surrounding rural areas because combustion releases large amounts of water vapor.

Mean amounts of cloudiness, fog, and precipitation all are higher in urban areas than in rural areas. One of the factors encouraging the formation of these condensation products is the large quantity of dust released by human activities. Dust particles sometimes are present within large cities in sufficient quantities to restrict visibilities greatly and form what has been called an "urban dust dome." The dust particles reduce the intensity of sunlight, and many are also **hygroscopic** (water attracting) so that they serve as condensation nuclei. City heat also encourages convectional lifting, especially in summer, resulting in the development of cumuliform clouds and sometimes of convectional precipitation. A number of studies have determined that urban heat production is responsible for a 5 to 10 percent increase in mean annual precipitation totals in and immediately downwind of large North American cities.

Enhanced urban precipitation amounts coupled with the presence of

Built-up area

| 0 | 5 | 10 Miles |
| 0 | 5 | 10 Kilometers |

London

52 (11.1)

50 (10.0)

48 (8.9)

46 (7.8)

44 (6.7)

42 (5.6)

40 (4.4)

Thames

◆ **FIGURE 8.56** Map of the London urban heat island as it developed on the night of May 14, 1959. Temperatures are given in Fahrenheit degrees, with Celsius equivalents in parentheses.

SOURCE: T. J. Chandler, *The Climate of London,* 1965, Hutchinson Press, London.

impervious surfaces and efficient drainage systems have also considerably increased the urban flood potential in many areas. This is because storm water is now conducted quickly as surface flow to urban streams that are usually hemmed in by city development. As a result, streams that formerly overtopped their banks only rarely now do so frequently.

Wind patterns in a city also are considerably altered from those of the surrounding countryside. A large city has an extremely irregular surface profile, with the vertical walls of buildings acting to block and redirect the wind. The large amount of surface friction reduces average wind speeds, often by a factor of 20 or 30 percent. On the other hand, the wind currents are redirected down streets, which literally form "avenues of least resistance" to the wind. Most people who are outdoors are on the streets, so that urban wind speeds in areas frequented by pedestrians actually may be higher than in rural areas.

The combination of explosive world population growth, major urban immigration, and a continued rise in per-capita energy consumption will all further increase the extent and significance of urban climatic modification.

9 WATER ON AND BENEATH THE EARTH'S SURFACE

OUTLINE

FOCUS QUESTIONS

1. What factors control the movement and storage of subsurface water?
2. How do lakes form and disappear?
3. What factors determine geographic and temporal variations in stream discharge?
4. What are the chief physical and chemical characteristics of ocean water?
5. What are the causes of the three oceanic motions of currents, waves, and tides?

As we have seen, water vapor is a major atmospheric gas and is the direct source of the Earth's precipitation. At any given time, however, only a comparatively minuscule 0.001 percent of our planet's near-surface water is located in the atmosphere: the overwhelming majority is to be found on and beneath the Earth's surface (see Table 6.1). In terms of surface characteristics, Earth is much more a planet of water than of land. Water covers over seven-tenths of the surface, to a mean depth of more than 3,800 meters (2.5 mi). Life evolved in the oceans, and the transfer of water from the oceans to the continents by means of the hydrologic cycle makes life on land possible.

In this chapter, we begin the study of the terrestrial portion of the hydrologic cycle by examining the location, characteristics, and movement of the Earth's surface and subsurface water supplies. Succeeding chapters examine the crucial roles this water plays in influencing the planetary distributions of vegetation, soils, and landforms.

Most precipitation falling on land does not flow immediately into rivers; instead, it enters the soil and sometimes the underlying rock layers, where it may stay for an extended period before eventually returning to the surface. The first portion of the chapter is therefore devoted to the Earth's subsurface waters. We next examine lakes and rivers, which derive their water from a combination of surface and subsurface sources. The final topic is the ocean, which is the ultimate destination of runoff from the land and the feature that dominates the surface geography of our planet.

SUBSURFACE WATER

The hydrologic cycle is a vast distillation mechanism for water. Salts and other dissolved substances are removed through the process of evaporation, and the water subsequently is returned to the surface as precipitation. Fresh water, both on and beneath the Earth's surface, therefore has not had the opportunity to gain large quantities of dissolved substances since being precipitated as rain or snow.

Although large quantities of fresh surface water exist in lakes and rivers, the great majority of the Earth's liquid fresh water lies beneath the surface. The total quantity of this water has not been accurately determined, but its volume is estimated at some 10^7 cubic kilometers (2,400,000 mi^3). This is approximately 77 times the water in all of the Earth's surface rivers and lakes and is about one-third of the quantity locked up in the Antarctic and Greenland ice caps. If evenly spread, it is enough water to cover the Earth's surface, continents and oceans alike, to a depth of about 10.5 meters (35 ft). Nearly half this water is within 0.8 kilometers (0.5 mi) of the surface and is considered to be accessible for extraction.

It is convenient to separate the Earth's subsurface water into two categories based on depth, movement, and origin. Nearest the surface is **soil water,** which occupies the pore spaces between individual soil particles. Some soil water adheres in a molecular film to the soil particle surfaces, while the rest moves slowly down through the soil under the influence of gravity. Soil water is located in the **zone of aeration,** where both water and air occupy the pore spaces between the individual soil particles in constantly varying proportions. Below this is the **zone of saturation.** Here, all pore spaces or other openings in the soil or rock are completely filled with **groundwater** (Figure 9.1). The zone of saturation, which extends to about 3 kilometers (2 mi) beneath the surface, contains most of the Earth's fresh subsurface water. Both soil water and groundwater are derived primarily from precipitation and are in constant slow movement. Most of this water eventually returns to the surface to continue its participation in the hydrologic cycle.

Soil Water

Although nearly all water in the soil is derived from precipitation, much precipitation on land does not fall directly onto the soil. Instead, it is **intercepted** by the

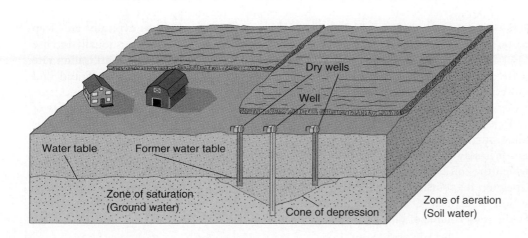

◆ **FIGURE 9.1** The water table separates the zone of aeration from the zone of saturation. Excessive water extraction from wells temporarily lowers the water table in their vicinity.

◆ **FIGURE 9.2** Water may be intercepted or may be stored on the surface before infiltrating the soil. During rainy periods, considerable subsurface water travels laterally near the surface as throughflow, rather than descending to the water table to become groundwater. Most throughflow soon returns to the surface.

surface cover of vegetation and vegetative litter (Figure 9.2). These overlying materials must be thoroughly wetted before an excess is produced that drips or flows down to the soil level. The interception process operates most effectively at the beginning of a rainfall, so a light shower falling in a heavily vegetated area may be almost entirely intercepted and the moisture subsequently lost to evaporation.

When the vegetation cover is fully wetted, additional moisture drips or flows to the soil surface. At this point, the water may collect on the surface and eventually evaporate, it may flow directly into rivers or lakes, or it may be absorbed by the soil. The maximum rate at which the soil can absorb water at any given time depends largely on its porosity and permeability. The **porosity** of the soil is defined as the volume of soil pore spaces divided by the total volume of the

soil itself. The more porous a soil, the more void space it contains that is capable of retaining water. Undisturbed soils in humid, well-vegetated environments typically are deep and relatively porous. As a result, only exceptionally intense storms have rainfall rates that exceed the soil infiltration rate and produce surface flow. The thinner soils of arid and semiarid regions, however, more commonly experience surface runoff as a result of heavy rainfalls. The **permeability** of a soil refers to its ability to transmit flow. Permeability is determined by the number and size of the pore spaces, and especially by the extent to which they are interconnected.

Both the porosity and permeability of the soil can be greatly reduced by surface disturbances such as compaction by machinery or livestock. Even the removal of vegetation allows the splash of raindrops on the bare soil to dislodge fine particles and wash them into surface pore spaces, clogging them and considerably reducing the soil's permeability. Soil disturbances related to agricultural activities have significantly lowered overall soil infiltration rates, producing increased surface runoff and greatly increased erosion rates.

Water that has infiltrated the soil is drawn slowly downward by gravity. The water moves in very circuitous routes around the soil particles, with some moisture adhering to the surfaces of the soil grains. With depth, soils become increasingly compacted by the weight of the overlying material, resulting in a progressive decline in porosity and permeability.

During a prolonged rainstorm, the continued downward percolation of soil water may temporarily cause the lower soil to be saturated. Since this water can no longer move downward rapidly enough to offset the addition from above, it is diverted laterally so that it flows parallel to the surface as **throughflow.** Throughflow is probably the major type of hillside flow in humid climates. This flow may intersect the surface along a stream bank or other steep slope and seep out to continue downhill as surface flow (Figure 9.2).

Soil water that does not seep out at the surface or adhere to the soil particles continues to sink slowly down until it reaches the **water table,** which is the upper boundary of the zone of permanently saturated soil or rock. In most cases, water reaches the water table within 48 hours of the time it initially entered the soil. At that point, the soil is left containing only water that adheres to the soil particles and can therefore be held indefinitely against the pull of gravity. This "held" water is considered to be an integral constituent of the soil. Although the amount of soil water varies because of evaporative losses and extraction by plant roots, it always is present to some extent, even in the driest areas. Its characteristics and behavior are further considered in Chapter 12.

Groundwater

Groundwater saturates the soil and rock layers below the water table. In general, it is most abundant in areas underlain by extensive layers of porous rock such as sandstone. The quality and quantity of water present as well as the depth and thickness of the zone of saturation, however, vary greatly in different localities. In some places, the upper surface of the zone of saturation intersects the Earth's surface, producing an outflow of groundwater. In other places, the zone of saturation exists hundreds of meters beneath the surface. Any hole, such as a well, that is dug into the zone of saturation will fill with water up to the level of the water table, which forms the upper boundary of the zone of saturation.

In arid areas, little or no water is currently being added to existing groundwater supplies. Groundwater in such areas is generally a legacy of more humid climates in the past and is sometimes referred to as **fossil water.** Like oil, this water is a nonrenewable resource because it cannot be replaced under existing climatic conditions. Its extraction permanently lowers the water table and reduces groundwater supplies.

Even in semiarid and subhumid regions, such as the southern Great Plains of the United States, the large-scale extraction of water for irrigation is producing rapidly declining water tables and the threat of eventual groundwater exhaustion. Shallow wells in such regions create surrounding **cones of depression** in the water table as the water is withdrawn more rapidly than it can be replaced (Figure 9.1). The lowered water tables around wells are normally temporary and disappear after pumping stops. If wells are located too close to one another, however, adjacent cones of depression can coalesce to produce a regional lowering of the water table.

In humid areas, the water table is subject to continual fluctuations in level. During wet periods, the addition of water from precipitation causes the water table to rise until it reaches the surface in valleys or other low-lying areas, sometimes producing an outflow of groundwater known as a **spring.** If the outflow occurs in a valley that continues downhill, it will form a spring or add water to a river or stream. If the outflow occurs in an enclosed depression, it will form a lake or swamp. In this fashion, groundwater returns to the surface to continue its movement through the hydrologic cycle.

Most streams in humid areas are at the water table and have a substantial portion of their total volumes of flow provided by the outflow of groundwater. The dependable supply of water provided by this outflow enables larger streams to be **perennial streams,** which continue to flow even during extended droughts. The much higher flow volumes during wet periods are derived primarily from the temporary additions of surface runoff and throughflow. In contrast, in arid regions the water table frequently is far below the level of the river valleys, causing the streams to be intermittent or **ephemeral streams,** which flow only during wet periods when they are supplied by surface or soil water sources. For the world's land areas as a whole, it is estimated that groundwater outflow supplies 30 percent of total stream flow. Total land runoff is believed to average approximately 26.7 vertical centimeters (10.5 in) of water per year, so that the annual groundwater contribution to this flow averages about 8.0 centimeters (3.2 in).

Groundwater flow rates usually are imperceptible, ranging from a meter or two per day to a few meters a year. The speed of flow depends on both the pressure head that develops as a result of local differences in the water table level and on the frictional resistance the water encounters to its movement. The greater the friction (i.e., the lower the permeability of the strata through which the water is flowing), the larger and steeper the pressure head that can be maintained. The direction of water movement is determined by the slope of the water table, which generally reflects the contours of the land surface (Figure 9.3).

As already noted, different Earth materials vary tremendously in their ability to store and transmit groundwater. A

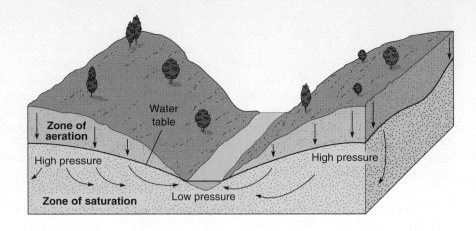

◆ FIGURE 9.3 Relationship between the surface topography and the movement of groundwater. The water table is higher beneath uplands, causing groundwater to flow slowly toward valleys or depressions. The water table typically intercepts the Earth's surface at the surface of a lake or stream, resulting in the outflow of groundwater.

porous and permeable rock layer that contains large quantities of groundwater and permits its ready movement is called an **aquifer.** In many parts of the world, large aquifers, often consisting of deep layers of sandstone or limestone, act as subsurface reservoirs that hold tremendous volumes of fresh water. Wells dug or drilled into these strata may be expected to yield large and dependable water supplies.

The world's largest known aquifer, the Nubian Sandstone, underlies much of the western Sahara. It contains an estimated 600,000 cubic kilometers (144,000 mi³) of water. Although its present recharge rate is almost zero, this vast underground reservoir is still virtually untapped and constitutes a natural resource of enormous potential for the countries of northwestern Africa.

The Ogallala aquifer, which underlies much of the southern and central Great Plains, is the most economically important aquifer in the United States. In contrast to the Nubian Sandstone, the Ogallala has been greatly lowered in many areas in recent decades by large-scale withdrawals of water for irrigation agriculture. This problem is explored in the Focus Box beginning on page 216.

Aquifers may be underlain or surrounded by relatively impermeable rock layers known as **aquicludes.** These strata prevent the aquifer from losing its water by acting as seals to its movement. Aquifers often serve as storage tanks, holding their water supplies in one area for extended periods of time. In some cases, however, aquifers can serve as conduits that transport water great distances. As was noted in Chapter 8, highlands often have substantial precipitation surpluses because of orographic lifting factors and cool temperatures. Aquifers extending from humid highlands to adjacent dry lowland areas therefore may serve as important water suppliers to the lowlands.

Human Impact on Groundwater

Since the beginning of civilization, humans have extracted groundwater, primarily for the irrigation of crops. Until less than a century ago, only near-surface water supplies were tapped through the digging of shallow wells. In recent decades, however, water extraction technology has advanced

rapidly. The development of deep drilling machinery and powerful pumps has made large quantities of groundwater available.

Groundwater supplies often are most extensively developed in arid and semiarid areas where surface water is inadequate or nonexistent. Maintaining reserves can be a problem, however, because the groundwater in these areas generally is fossil water. The overpumping of water in numerous drier areas of the world is producing rapidly declining water tables, increasing pumping costs, and decreasing yields. Three areas within the United States where these problems are especially acute are southern Arizona, the Central Valley of California, and the southern Great Plains.

Other problems resulting from the overuse of groundwater are saline water intrusion, land subsidence, and waterlogging. **Saline water intrusion** occurs in coastal regions where so much fresh water has been removed that the normal seaward flow of fresh water is reversed, and saline water from the ocean is drawn in instead (Figure 9.4).

When groundwater saturating certain fine-grained sediments is withdrawn, an irreversible dewatering and compaction of the sediments occurs, resulting in the sinking, or **subsidence,** of the surface. Surface subsidence has produced serious problems in places such as Mexico City, Bangkok, Venice, southeastern Texas, and the Central Valley of California. In parts of Mexico City, for example, subsidence has lowered the surface by more than 8 meters (25 ft), and the process continues at a rate of about 2.5 centimeters (1 in) a year. Subsidence in Venice, Italy, is especially critical because much of the city is situated at or near sea level; as a result, the famed city of canals has been partly inundated by the Adriatic Sea.

Waterlogging of the soil typically occurs in low areas in dry climates where the application of irrigation water over extended periods has raised the water table to the surface. This process often results in the dissolution and surface deposition of salts, rendering the land unfit for cultivation. Waterlogging and salt deposition are becoming an increasingly severe problem in parts of the western United States and have long existed in the Middle East. Large portions of the once fertile Tigris-Euphrates lowland of Iraq, for ex-

(a)

(b)

◆ **FIGURE 9.4** The process of saline water intrusion. Fresh water is less dense than salt water, and tends to overlie it, creating a fresh water "buffer" extending to the coast. Excessive extraction, however, can deplete the fresh water and pull up the underlying salt water.

ample, have been converted to a semidesert by this process. For a more detailed discussion of this situation, see the Focus Box on pages 300–301.

Another groundwater problem resulting from human activities is pollution. Liquid contaminants in an ever increasing variety are finding their way into groundwater supplies, especially in urban and industrialized areas with humid climates such as the eastern United States and western Europe. The chief sources of pollutants are waste disposal sites such as landfills, lagoons, and disposal pits. Nonpoint sources such as urban runoff and water from agricultural land treated with fertilizers and pesticides also contribute to pollution problems for both groundwater and surface water (see the Focus Box on pp. 226–227).

LAKES

A **lake** is a landlocked body of water with a horizontal surface water level. Lakes are short-lived landscape features, usually lasting only a few centuries or millennia. Although this may seem long in human terms, it is an extremely short time span from a geological standpoint. Despite the relative brevity of their existence, however, lakes currently contain more than 95 percent of our planet's surface supply of fresh liquid water.

Geographic Significance

Freshwater lakes cover an estimated area of 825,000 square kilometers (320,000 mi^2), or about 1.5 percent of the Earth's land surface. They contain more than 125,000 cubic kilometers (30,000 mi^3) of water. Although probably more than a million lakes are in existence, roughly four-fifths of the total volume of the Earth's fresh surface water is found in the 40 largest lakes.

The world distribution of lakes, like that of groundwater and the ocean, but unlike that of rivers, is related more closely to the presence of basins suitable for water accumulation than it is to climate. Most of the world's freshwater lakes exist on three continents—North America, Africa, and Asia (see Table 9.1 and Figure 9.7 on p. 218). The large lakes of North America contain approximately 32,000 cubic kilometers (7,700 mi^3) of water, of which over three-fourths is in the five Great Lakes. Large lakes in East Africa contain some 36,000 cubic kilometers (8,640 mi^3) of fresh water. Lake Baikal in southeastern Siberia is the world's deepest and most voluminous body of fresh water; it contains 23,000 cubic kilometers (5,600 mi^3) of water—18 percent of the world total. In contrast, lakes in South America, Europe, and Australia collectively contain only about 3,000 cubic kilometers (720 mi^3) of water, or about 2 percent of the total.

Although freshwater lakes are much more numerous and important for human activities than are lakes containing salt water, some of the world's largest lakes are saline. Saline lakes cover an estimated total area of 700,000 square kilometers (270,000 sq mi) and contain approximately 105,000 cubic kilometers (25,000 mi^3) of water. Both the area and volume of all saline lakes combined therefore are about 85 percent of the respective totals for freshwater lakes. Saline lakes are located mostly in Asia and are dominated by the Caspian Sea, which, despite its name, is considered the world's largest lake. By itself, the Caspian Sea contains 76 percent of the total volume of all the world's saline lakes.

Lakes with outlets nearly always contain fresh water, while those with no outlets usually are saline. In most cases, the salt has been supplied gradually by inflowing rivers and has accumulated because the evaporation of pure water has left the salt behind. Some lakes are only slightly saline, while others, like the Dead Sea, are saturated with salt. The Dead Sea is the world's most saline large lake. It contains about 24 percent dissolved solids, making it six times as salty as the ocean.

Origins of Lakes

Only two criteria are necessary for the formation of a lake. One is the presence of an enclosed surface depression or basin; the other is the availability of sufficient water from either surface or subsurface sources to fill it at least partially. Although some temporary lakes are formed by the surface accumulation of runoff from rainfall or melting ice, most

THE DEPLETION OF THE OGALLALA AQUIFER

The vast short grasslands of the American High Plains were first settled by farmers and ranchers in the 1880s. This region, lying in the rainshadow of the Rocky Mountains, has a semiarid climate, and for 50 years after its initial settlement, it supported a low-intensity agricultural economy of cattle ranching and wheat farming. In the early twentieth century, however, it was discovered that much of the High Plains were underlain by a huge aquifer. This aquifer was named the Ogallala aquifer after the Ogallala Sioux Indians who once inhabited the region.

The Ogallala aquifer is a calcareous sandstone formation that underlies some 583,000 square kilometers (225,000 mi²) of land extending from northwestern Texas to southern South Dakota (Figure 9.5). Although the aquifer is thin near its southern terminus, it gradually thickens northward, attaining a maximum vertical extent of over 300 meters (1,000 ft) in Nebraska. Water from precipitation and melting glaciers has been accumulating in the Ogallala for the past 30,000 years, and it is estimated that the aquifer contains enough water to fill Lake Huron. Unfortunately, under the region's present semiarid climatic conditions, current rates of addition to the aquifer are minimal, amounting to about half a centimeter (0.2 in) a year.

The first wells were drilled into the Ogallala during the drought years of the early 1930s. The ensuing rapid expansion of irrigation agriculture, especially since the 1950s, has trans-

◆ **FIGURE 9.5** Areal extent of the Ogallala aquifer, and changes in the aquifer's water level between the time of discovery and 1980.

formed the economy of the region. Some 150,000 wells now tap the Ogallala. Modern center-pivot irrigation devices, each capable of spraying 4.5 million liters (1.2 million gallons) of water a day, have produced a landscape dominated in many places by geometric patterns of circular green islands of crops (Figure 9.6). Ogallala water has enabled the High Plains region to supply 25 percent of the cotton, 38 percent of the grain sorghum, 16 percent of the wheat, and 13 percent of the corn grown in the

◆ **FIGURE 9.6** Tremendous quantities of water from the Ogallala aquifer are used by center-pivot irrigation systems. The resulting circular crop patterns form a distinctive landscape when viewed from the air.

SOURCE: Science VU/Visuals Unlimited

United States. In addition, 40 percent of American grain-fed beef cattle are fattened here.

This unprecedented development of a finite groundwater resource with an almost negligible natural recharge rate unfortunately has caused water tables in the region to fall drastically. In the 1930s, wells encountered plentiful water at a depth of about 15 meters (50 ft); currently, they must be dug to depths of 45 to 60 meters (150 to 200 ft) or more. In places, the water table is declining at a rate of a meter (3 ft) a year, necessitating the periodic deepening of wells and the use of ever more powerful pumps. It is estimated that, at current withdrawal rates, much of the aquifer will run dry within 40 years. The situation is most critical in Texas, where the climate is driest, the greatest amount of water is being pumped, and the aquifer contains the least water. More than 25 percent of the Texas portion of the Ogallala has now run dry, and pumping costs for the remainder increased by 400 percent between the early 1970s and the mid-1980s. According to projections, by the year 2030, the remaining Ogallala water will support only 35 to 40 percent of the irrigated acreage in Texas that it supported in 1980.

Farmers are reacting to the inevitable depletion of the Ogallala in various ways. Many have been attempting to conserve water by irrigating less frequently or by switching to crops that require less water. Others, however, have adopted the philosophy that it is best to use the water while it is still economically profitable to do so, and to concentrate on high-value crops such as cotton. Farmers who wish to conserve water have less incentive to do so as they realize that many of their neighbors are profiting by using great amounts of water and drawing down the entire region's water supplies in the process.

In the face of the upcoming water supply crisis, a number of grandiose schemes have been developed to transport vast quantities of water by canal or pipeline from the Mississippi, the Missouri, or the Arkansas Rivers. Unfortunately, obtaining water through any of these schemes would increase pumping costs at least tenfold, making irrigated agricultural products from the region uncompetitive on national and international markets. Somewhat more promising have been recent experiments for releasing capillary water above the water table by injecting compressed air into the ground. Even if this process proves successful, however, it would almost triple water costs. Genetic engineering also may provide a partial solution, as new strains of drought-resistant crops continue to be developed. Whatever the final answer to the water crisis may be, irrigation water will never again be the abundant, inexpensive resource on the High Plains that it was during the agricultural boom years of the mid-twentieth century. ■

◆ **TABLE 9.1** The World's Largest Lakes

NAME	BASIN ORIGIN	SURFACE AREA		VOLUME		MEAN DEPTH	
		Km^2	Mi^2	Km^3	Mi^3	M	Ft
1 *Caspian Sea	Uplift (cut off from sea)	371,000	148,000	79,340	19,035	182	597
2 Superior	Glacial erosion	82,100	31,700	12,150	2,916	149	489
3 Victoria	Uplift of margins	68,780	26,560	2,660	637	40	131
4 Huron	Glacial erosion	59,570	23,000	3,450	827	59	195
5 Michigan	Glacial erosion	57,750	22,300	4,840	1,161	85	279
6 *Aral Sea†	Uplift (cut off from sea)	41,000	15,830	374	90	9	30
7 Tanganyika	Faulting (graben)	33,990	13,120	19,420	4,659	572	1,876
8 Baikal	Faulting (graben)	31,490	12,160	23,260	5,581	740	2,427
9 Malawi	Faulting (graben)	30,790	11,890	8,370	2,009	273	895
10 Great Bear	Glacial erosion	30,400	11,740	2,200	529	72	238
11 Erie	Glacial erosion	25,670	9,910	480	116	19	62
12 Great Slave	Glacial erosion	25,390	9,800	1,550	373	62	204
13 Winnipeg	Glacial erosion	24,520	9,470	320	76	13	43
14 Ontario	Glacial erosion	19,010	7,340	1,640	393	86	283
15 Ladoga	Glacial erosion and uplift	18,731	7,230	917	220	91	300

SOURCE: Academic American Encyclopedia
*Saline lakes.
†Because the two major rivers entering the Aral Sea have been diverted to supply water for irrigation agriculture, this water body has been shrinking rapidly. Between 1957 and 1990, the Aral Sea decreased 35 percent in area and 67 percent in volume. In 1990, its area was estimated at 40,000 km^2 (15,440 mi^2), and its volume at 350 km^3 (84 mi^3).

permanent lakes, particularly those in humid climates, have their surface levels at the level of the water table.

Surface depressions capable of trapping water are formed in a variety of ways, but the two most important by far are movements of the Earth's crust and glacial action. Crustal movements, the origins and characteristics of which are discussed in Chapters 13 and 14, are notable for forming large lakes, sometimes of great depth. In some instances, the motion involves the gradual broad uplift of a portion of the Earth's surface that cuts off an arm of the sea to form a lake.

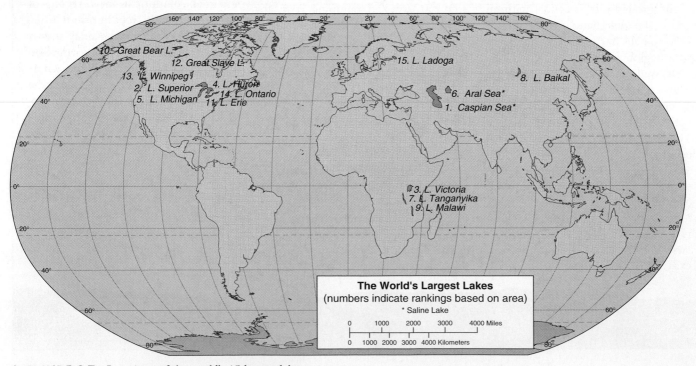

◆ **FIGURE 9.7** Locations of the world's 15 largest lakes.

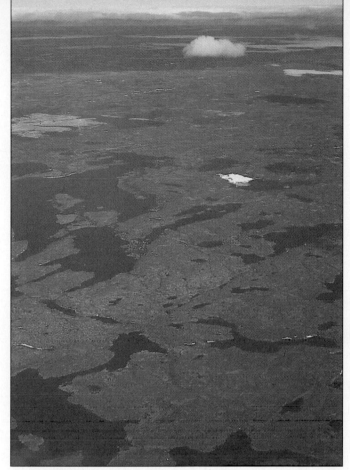

◆ **FIGURE 9.8** Repeated glaciation has left a mosaic of rocky land surfaces and water-filled depressions northeast of Yellowknife, in Canada's Northwest Territories.

SOURCE: Walt Anderson/Visuals Unlimited

The Caspian and Aral Seas formed in this fashion when crustal uplift gradually severed their connection with the Mediterranean. Tectonic activity also can result in a narrow wedge of land (a **graben**) being dropped sharply downward between a pair of faults. Lakes resulting from this type of movement tend to be long, narrow, and exceptionally deep. Lake Baikal was formed this way. The world's most extensive group of fault-produced lakes is located along the Great Rift Valley, which extends through Asia and Africa from Syria to Mozambique. Major lakes within this valley include the Dead Sea, Lake Albert, Lake Tanganyika, and Lake Malawi.

Lakes resulting from glacial erosion and deposition are more numerous than those from all other origins combined (Figure 9.8). They vary greatly in size, and most are located in northern portions of North America and Eurasia that were covered by continental glaciers between 10,000 and 20,000 years ago (Chapter 17). Most glacial lake basins were produced by the scooping action of ice either in areas of soft bedrock or in loose ice-deposited sediments. Others resulted from the blocking of surface drainage by glacial deposits. Literally hundreds of thousands of glacial lakes, large and small, are found within North America north of the Ohio and Missouri Rivers. They include some of the world's largest lakes, such as the five Great Lakes, Great Slave Lake, Great Bear Lake, and Lake Winnipeg. Large numbers of glacial lakes also exist in northern Europe, especially in Scandinavia, although they are less numerous and generally smaller than those of North America.

Lakes also can form in depressions produced by numerous other natural processes. Foremost among these are solution depressions in regions underlain by highly soluble rock types such as limestone and dolomite. Their characteristics and distributions are discussed in Chapter 15. Other lakes result from volcanic eruptions, landslides, or other mass movements that may block drainage outlets; changes in river flow patterns; wind erosion or deposition; and meteorite impacts.

Artificial lakes have become important features on the landscape in many areas. Some are produced inadvertently as a result of earth-moving activities such as mining or quarrying. Others, often of large size, are purposefully created. Large reservoirs such as Lake Powell and Lake Mead in the western United States serve a variety of purposes, including flood control, water supply, hydroelectricity generation, and recreation. In agricultural areas, great numbers of small farm ponds produced by the damming of small streams often dot the landscape. They provide water and recreational opportunities as well as helping to control runoff and erosion.

Disappearance of Lakes

The basic reason lakes have such a brief existence is that physical and biological factors work against the maintenance of enclosed basins containing standing water. In many cases, the inflow of water exceeds a basin's capacity to hold it and the ability of evaporation to remove it. If this occurs, a surface outflow of water results that gradually erodes the lip of the basin. Continued erosion will eventually destroy the basin's ability to hold standing water.

The relative lack of motion of lake water also makes the basin a sediment trap. Sediments carried by inflowing streams are deposited on the floor of the lake, gradually filling it. This process has shortened the expected useful life spans of even large reservoirs in various parts of the world to only a century or two. The process of infilling often is hastened by the growth of aquatic plants and the consequent deposition of organic matter. Most shallow lakes gradually become **swamps** when their water depths decline sufficiently to allow water-tolerant vegetation to cover the entire lake surface, and with continued infilling, they eventually become dry land.

A change in climatic conditions may also cause lakes to disappear. As discussed in Chapter 17, the gradual advent of warmer and drier conditions in the interior western United States over the past 10,000 years has caused many formerly extensive lakes to evaporate completely, leaving in their place enclosed saline basins.

STREAMS

Rivers and **streams** are channeled bodies of flowing water. They engage in the gravitational transport of excess surface moisture derived from precipitation. The water is ultimately transported either to interior basins on land or, more commonly, to the sea. Although the terms *streams* and *rivers* are often used interchangeably, a stream has no particular size connotation, whereas a river generally is considered to be large—often the master stream in a drainage system. Stream flow constitutes the second most dynamic, or rapid and short-term, phase of the hydrologic cycle, after water vapor transport by wind. It forms a vital final link in the cycle by returning vast quantities of water to the oceans.

In addition to carrying water, streams also transport sediments derived from the land. In doing so, they serve as the major agent responsible for the gradual erosion of the Earth's land surface. Stream-produced landscapes are so important that an entire chapter (Chapter 16) is devoted to them. In this section we concentrate on the hydrologic aspects of streams.

Geographic Distribution and Significance

Streams vary greatly both in number and in volume of flow in different parts of the world. They are generally absent only from the 10 percent of the Earth's land surface that is perennially covered by ice. Even those areas, located mostly within Antarctica and interior Greenland, have excess precipitation removed by rivers; the only difference is that they are rivers of frozen water, or glaciers. In addition, meltwater streams drain many glaciated areas during the summer. Elsewhere, the extent of river development is determined largely by the balance between precipitation amounts and evapotranspiration rates. Even the driest deserts, however, have stream channels that occasionally carry water.

The volume of water contained in the world's streams at any one time is very much smaller than the amount held in lakes and as groundwater. It amounts to only about 0.025 percent, or 1/4,000th, of the Earth's total near-surface supply. The practical significance of this water, though, is much greater than its volume implies, both because of its ready accessibility and because of its speed and power of flow. While water in lakes and beneath the surface can be considered to be in at least temporary storage, the water in streams is in active movement so that, in time, large quantities are involved. In one year, streams discharge a total volume of water estimated at 33,000 cubic kilometers (8,000 mi^3) into the ocean. This is approximately one-third of the total volume of the Earth's freshwater lakes.

Although virtually all stream flow ultimately is derived from precipitation, the water in a direct sense comes from several different sources. The most obvious of these is surface runoff from storms, which in some areas supplies the majority of the water entering stream systems. It currently is believed, however, that in most humid tropical and temperate regions, the largest volume of water in streams is supplied by the outflow of groundwater and by throughflow. This is especially true in areas of porous and permeable rock and of deep, relatively undisturbed soils. Some streams also

TABLE 9.2 The World's Major Rivers Ranked by Drainage Basin Area

NAME	DRAINAGE AREA (× 1,000)		PERCENTAGE OF WORLD'S LAND AREA DRAINED	LENGTH		MEAN DISCHARGE (× 1,000)		PERCENTAGE OF WORLD'S TOTAL DISCHARGE
	Mi^2	Km^2		Mi	Km	F^3/sec	M^3/sec	
1 Amazon	2,722	7,052	4.8	4,000	6,440	6,350	180	19.2
2 Rio de la Plata/Paraná	1,600	4,145	2.8	2,485	4,000	777	22	2.3
3 Congo	1,314	3,404	2.3	2,914	4,692	1,458	41	4.4
4 Nile	1,293	3,350	2.3	4,132	6,653	110	3	0.3
5 Mississippi/Missouri	1,244	3,223	2.2	3,741	6,020	650	18	2.0
6 Ob/Irtysh	1,149	2,977	2.0	3,362	5,413	558	16	1.7
7 Yenisey	996	2,580	1.7	3,332	5,365	671	19	2.0
8 Lena	961	2,490	1.7	2,734	4,402	575	16	1.7
9 Yangtze	756	1,959	1.3	3,434	5,526	1,200	34	3.6
10 Niger	730	1,891	1.3	2,600	4,186	215	6	0.7
11 Amur	716	1,855	1.3	1,755	2,826	438	12	1.3
12 Mackenzie	711	1,842	1.2	2,635	4,242	400	11	1.2
13 Ganges/Brahmaputra	626	1,622	1.1	1,800	2,898	1,360	39	4.1
14 St. Lawrence/Great Lakes	565	1,464	1.0	2,500	4,025	360	10	1.1
15 Volga	525	1,360	0.9	2,293	3,692	282	8	0.9

SOURCE: Encyclopaedia Britannica.

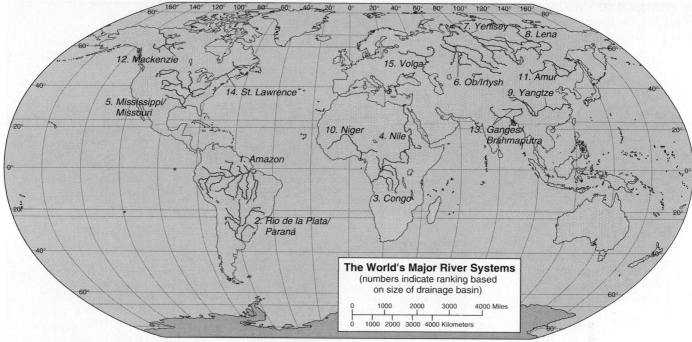

◆ **FIGURE 9.9** Locations of the world's 15 largest rivers.

are fed by glacial meltwater, and a small quantity of additional water is added directly by precipitation falling into the streams.

A stream's **discharge,** the volume of flow past a given point in a specified time interval, normally increases downstream because of the increase in the total area drained. Losses occur to seepage and especially to evaporation, however, and streams flowing through dry climate regions actually decrease in discharge in a downstream direction, sometimes even drying up completely. Two well-known examples of major rivers that decrease naturally in flow because their courses traverse arid regions are the Nile and the Colorado. (In recent times, the volumes of flow of both the Nile and Colorado have further diminished because of water withdrawals for human use. At present, in fact, no water from the Colorado River reaches its mouth at the head of the Gulf of California except during times of exceptionally high flow!)

The total global discharge of streams into the ocean is believed to be slightly less than 30 percent of all the precipitation that falls on the land areas of the Earth. Most of the remainder is evaporated, making evaporation the leading process by which moisture leaves the land. A relatively small amount of water also enters the ocean as subsurface flow.

Although many thousands of streams of all sizes flow into the ocean, total global discharge is dominated by a small number of large rivers (Table 9.2 and Figure 9.9). The Earth's major river system in all respects except length is the Amazon (Figure 9.10). By itself, it drains 4.8 percent of the Earth's land area and contributes nearly one-fifth of the total global discharge, a figure more than four times that of the

second largest river, the Congo (or Zaire). The Mississippi, North America's dominant river, ranks only seventh in total discharge among the rivers of the world. The Nile, with a length of 6,650 kilometers (4,132 mi), has the distinction of being the world's longest river, and it also has the fourth largest **drainage basin,** or area drained. Because of the arid nature of much of its course, however, it is not among the top 20 rivers in discharge.

The global distribution of stream discharge volumes is strongly correlated with climate, especially with precipitation. As might be expected, regions with high precipitation generally have streams with large discharge volumes. Temperature and humidity also are important criteria, since they largely determine evapotranspiration rates. The geographic influence of these two factors is to decrease discharge at any given level of precipitation in areas at low latitudes and elevations, as these areas tend to have warmer climates.

Several nonclimatic variables also significantly influence discharge rates. One of the most important is the amount and type of vegetation growing in the drainage basin. In general, a high vegetation density decreases the amount of water entering nearby stream systems by intercepting precipitation, by increasing friction for runoff (giving it more time to be absorbed or evaporated), and especially by transpiring large quantities of moisture. For these reasons, lumbering or the removal of vegetation for agricultural purposes typically increases runoff and erosion.

The nature of the surface also strongly influences local discharge volumes. Irregular surfaces, and particularly the presence of surface depressions, reduce runoff by promoting the storage of water in puddles, ponds, or lakes. Areas of steep slopes tend to experience greater stream discharge

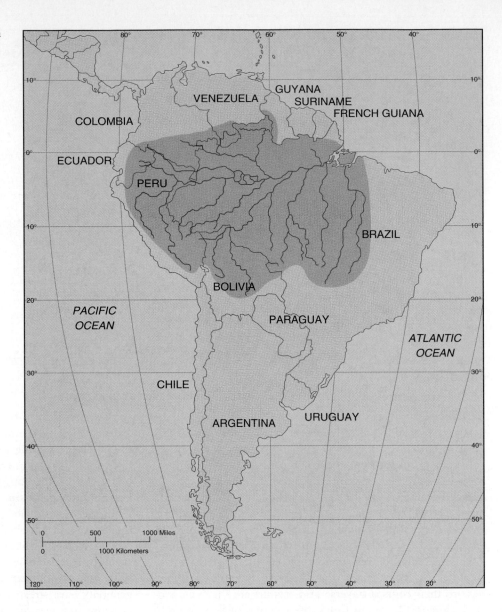

because rainwater reaches stream systems rapidly, giving it less time to be absorbed or to evaporate. The porosity and permeability of the soil and rock also play important roles in controlling rates of water absorption and movement because the pores provide conduits for subsurface flow. A crucial determinant of stream discharge is the size of its drainage basin. Within a given region, where other controlling factors are relatively constant, the volume of discharge of each stream is strongly correlated with the drainage basin area.

Temporal Variations in Stream Flow

Although the size of a stream can be described in a general way by citing its mean volume of flow, the discharge of most streams varies considerably on both a seasonal and a short-term basis. The causes of seasonal variations are mostly climatic, because the great majority of the Earth's land surface has distinctive seasonal fluctuations in precipitation, evapotranspiration, or both.

Near the equator, large rainfall totals produce high runoff rates throughout the year. Somewhat farther poleward, in the seasonally wet and dry tropics, streams tend to have a single well-developed summer discharge maximum. Many smaller streams here are intermittent, with no flow at all near the end of the dry season. Within the middle and higher latitudes, flow patterns vary considerably, depending on precipitation distribution, but streams most commonly exhibit winter and spring discharge maxima and late summer and autumn minima. High discharges in spring are associated with the combination of spring rains, snowmelt, the possible presence of impermeable frozen subsoils, and relatively low evapotranspiration rates. Peak flows tend to occur progressively later in spring with increasing latitude as warm moist air and melting conditions advance poleward with the Sun's rays.

In dry climates, especially within the subtropics, only larger rivers exhibit year-round flow. These rivers are **exotic;** their water is derived from runoff from more humid areas at

MEASURING STREAM DISCHARGE

Streams vary greatly in volume of flow over time as a result of variations in the amount, duration, and intensity of precipitation events as well as in the preexisting soil water and groundwater conditions. Each stream has a unique hydrologic "personality," which should be familiar to those involved in the design of dams and other river-modification projects, water-supply planning, and flood forecasting.

The U.S. Geological Survey (USGS) is the federal agency chiefly responsible for monitoring the discharge of the more important rivers within the United States. It maintains over 11,000 stream-gauging stations that regularly measure flow volumes. In addition, many individual states have hydrologic agencies that operate stream-gauging networks. Stream-flow variations over days or weeks often are displayed pictorially on **hydrographs** (Figure 9.11). Analyses of these graphs help hydrologists to better understand the hydrologic characteristics of individual streams and to determine the contributions of surface flow, throughflow, and groundwater flow to their total discharge.

Stream discharge is measured in cubic feet per second (cfs) or cubic meters per second (cms). The formula employed is $Q = wdv$, where "Q" is stream discharge, "w" is stream width, "d" is mean stream depth, and "v" is

◆ FIGURE 9.11 Hydrographs plot stream discharge over periods of time. These three hydrographs show the response of streams of differing size to the heavy rains associated with Hurricane Agnes in June 1972. Bald Eagle Creek, the smallest of the three, had the "flashiest" response, rising and then falling relatively quickly after the rainfall event. The largest of the three streams, the Susquehanna River, rose and fell much more slowly.

mean current velocity. Because a given stream cross section typically has an irregular shape and a different speed of flow at different points, it is normally divided into a number of evenly spaced segments for purposes of measurement. The discharge is determined separately for each segment, and the total discharge value is then derived by summing the segmental values.

Numerous stream-flow velocity readings must be taken at different points along a stream to obtain an accurate overall velocity figure. A variety of techniques have been developed

for this purpose. The width, depth, and velocity of a stream change almost continually with time, so that frequent measurements are required. Fortunately, all three of these parameters for a given stream tend to vary systematically with the stream's water level, or **stage.** Stream stage is easily determined by observing the height of the water on a measuring staff at the gauging station. Once the relationships between stage and discharge have been established, the discharge of a stream usually can be accurately estimated from its stage. ■

much higher elevations or in differing latitudes (the Nile and Colorado Rivers again serve as examples). Smaller local streams in dry climate areas usually are situated well above the water table and are ephemeral, carrying surface runoff

only for short periods during and immediately following storms. In most cases, their water either evaporates, is absorbed by the ground, or flows into saline lakes occupying basins of interior drainage (see Chapter 18).

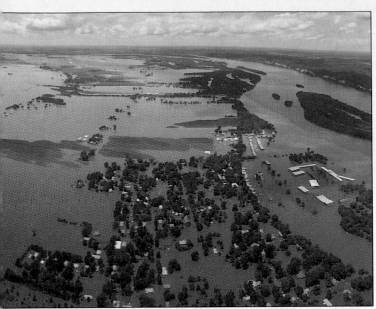

◆ **FIGURE 9.12** Portage des Sioux, St. Charles County, Missouri, during the Midwest floods of July, 1993. The channel of the Mississippi River is at the far right.

SOURCE: Photo courtesy of Michael Lawton

Short-term variability in flow is characteristic of virtually all streams and is especially significant for those fed principally by surface runoff. Changing stream-flow volumes result from variations in the quantity of water supplied to the stream systems primarily by storms and sometimes by snowmelt. The irregular timing and quantity of these water influxes make it much more difficult to predict short-term stream flow than the longer-term seasonal variations in flow. Because accurate assessments and forecasts of stream-flow volumes are crucial to people residing along the streams, they are carefully monitored by a variety of government agencies. This topic is examined in the box on the previous page.

Surface runoff from an intense storm is relatively rapid and usually does not travel far before it enters a tributary of the local drainage system. Small streams respond quickly and generally **crest,** or reach peak flow, shortly after the period of most intense rainfall. Crests normally are reached after progressively longer time intervals in streams draining larger basins, since most water must travel greater distances, and the rise in water levels is more gradual. Figure 9.11 illustrates this point. Stream levels within drainage basins of all sizes tend to fall more gradually than they rose following a precipitation event.

A **flood** occurs when the volume of flow of a stream exceeds the carrying capacity of its channel, and excess water spills over the banks to cover adjacent areas. Nearly all rivers flood occasionally, and for some it is an annual event. River flooding is so frequent largely because many rivers flow within **floodplains.** These are broad belts of low, flat land adjacent to rivers that are covered by river sediments and are periodically inundated by water during times of flooding. Floodplains, with their level surfaces, fertile soils, and proximity to river water for irrigation and transportation, are favorite settlement sites, but floodplain residents must cope with the hazard of occasional floods.

Major river floods rank with earthquakes and hurricanes among the world's foremost natural hazards (Figure 9.12). Between 1947 and 1967, for example, floods claimed approximately 173,000 human lives. In the United States, an estimated 50 to 100 people on average die in floods each year, and dollar losses exceed $1 billion. The great Midwest floods of 1993, discussed in the Focus Box beginning on page 143, were the most expensive in the nation's history, costing well over $100 billion. Currently, 20 million people, or 12 percent of the U.S. population, live in areas susceptible to periodic river flooding.

Human Impact on Stream Flow

Rivers and streams are our major source of fresh water. Unfortunately, however, the geographic correlation between areas of supply and demand for this resource has always been low. Large surpluses of stream water exist near the equator and in the high latitudes, but in many parts of the subtropics and middle latitudes, the demand exceeds the supply. The pressure on existing water supplies is greatest in subhumid, semiarid, and seasonally dry portions of the less-developed countries where soils are often fertile, and the population density is high and expanding rapidly. As a consequence, regional and national conflicts over existing water supplies have been increasing, and many scientists and politicians believe that this trend will continue.

During the twentieth century, the availability of water from most of the world's major river systems has been augmented by the development of water control systems that feature the construction of huge dams. Unfortunately, in addition to their high costs and the environmental disruption they produce, the reservoirs impounded by these dams typically lose tremendous amounts of water by evaporation and seepage. Future expansion of water supplies by water control structures is also limited by a lack of suitable undeveloped sites.

In the contiguous United States (the lower 48 states), streams transport an estimated 29 percent of the total precipitation to the sea. (The remainder is returned by evapotranspiration to the atmosphere.) At present, approximately half the volume of water added to these streams each day is withdrawn for human use. The chief categories of use are for industry (57 percent), irrigation (34 percent), and municipalities (9 percent). Most industrial use is nonconsumptive; that is, the water is returned to the stream after use. Much of the water used for irrigation and by municipalities, however, is consumed, resulting in a lowering of stream-flow volumes. Canada's stream-flow volumes are less affected by human usage because a much smaller volume of water is withdrawn, and 84 percent is used primarily for nonconsumptive purposes by industry. Unfortunately, a large proportion of the water returned to streams after use has been thermally or chemically polluted and poses a variety of problems and hazards to water users and residents farther downstream. Some aspects of stream pollution are discussed in the Focus Box beginning on page 226.

The net effect of humans on river volumes, at least over the short term, is not always one of flow reduction; the likelihood of flooding is often increased as well, usually by activities associated with agriculture and urbanization. The removal of vegetation and the compaction of the surface by the passage of humans, livestock, and vehicles reduce the soil's infiltration capacity and increase surface runoff. In suburban and urban areas, extensive areas are surfaced with impervious materials such as asphalt and concrete. This greatly increases surface runoff, which then enters storm drainage systems designed to conduct the water as quickly as possible to stream systems. As a result, most urban streams have "flashy" discharges, with very rapid rises and falls following precipitation events.

The Ocean

The **ocean** is the interconnected body of salt water that covers 71 percent of the surface of our planet. Even the 29 percent of the Earth's surface that consists of dry land owes some of its most basic characteristics to the ocean, largely because of its influence on global climates. The ocean moderates temperatures around the world, and most of the precipitation that falls on land is derived from water vapor that evaporated from the sea. In subsequent chapters, we will see that the quantity of precipitation received in any given locality plays a crucial role in determining the amount and type of vegetation clothing the surface, as well as the types of soils and landforms that develop.

The ocean is commonly considered to comprise both the beginning and the end of the global hydrologic cycle. It is sometimes referred to as the "hydrologic sink" because some 96.7 percent of all water at or near the surface of the Earth is located in the ocean at any one time, and because millennia may pass between the time a given water molecule enters the ocean as precipitation or runoff and the time it evaporates to begin another journey through the cycle.

Origin, Distribution, and Extent

Ocean water originated in the same way as did the surface and subsurface water of the land. It was released from its original solution in rock deep within the Earth as the rock moved surfaceward as magma; subsequently, it entered the atmosphere in the form of water vapor during volcanic eruptions. Cooling and condensation of the water vapor caused it to precipitate to the surface, where it accumulated in the enclosed basins that hold the ocean water.

The present geographical distribution of the ocean is closely related to the distribution of the types of rocks

The Water Hemisphere

◆ **FIGURE 9.13** This orientation of the Earth stresses the areal predominance of the oceans over the continents.

composing the Earth's crust. The entire crust "floats" on the semisolid material of the underlying mantle. Much of the crust consists of dense basaltic rock that floats at a relatively low level, producing the ocean basins. The less-dense granitic rock from which the continents are constructed is sufficiently buoyant and high-standing to remain above the ocean level (see Chapter 13). Put another way, the ocean has simply covered the lowest 71 percent of the Earth's lithospheric surface. This includes approximately 60.6 percent of the northern hemisphere and 81.0 percent of the southern hemisphere (Figure 9.13).

Several major subdivisions of the ocean exist; they also are termed oceans, and their areas and mean depths are listed in Table 9.3. As the table indicates, the mean depths of the water in the various oceans are rather consistent with the exception of the comparatively shallow Arctic Ocean. Depths range from 0 along the coasts to a maximum of 11,033 meters (36,198 ft) in the Marianas Trench in the western Pacific. The estimated mean oceanic depth of 3,865 meters (12,680 ft) is almost five times the average elevation of the world's land areas, which is estimated at 841 meters (2,760 ft). The vertical dimensions of the ocean therefore exceed those of the continents even more than do its horizontal dimensions (see Figure 9.15 on page 228).

Physical and Chemical Characteristics

Ocean water displays substantial spatial variation in a number of physical and chemical characteristics. Among the most important are chemical constituency, temperature, and the presence or absence of ice.

NONPOINT SOURCE WATER POLLUTION IN COLORADO

Nonpoint source pollution (NPS), commonly known as polluted runoff, originates from urban and construction runoff, agriculture, mining, forestry, landfills, and hydrologic modification. The main pollutants in polluted runoff are heavy metals, sediment, salts, bacteria, and nutrients. Heavy metals are toxic to aquatic life and alter the composition and nutrient uptake capabilities of wetland vegetation. Sediment from eroding soils degrades waterways and wetlands through siltation and serves as a vehicle for toxics and nutrient transport. Salinization involves the increased concentration of salts in waterways caused by irrigating saline soils in semiarid and arid climates. The presence of fecal coliform bacteria usually coincides with other bacterial organisms that cause severe gastrointestinal illness from drinking contaminated water. Nutrients (such as nitrates and phosphates) promote the eutrophication (excessive plant growth) of waterways, lowering oxygen levels and killing plant and animal life. The Environmental Protection Agency (EPA) estimates that NPS pollution constitutes 50 percent of the water pollution in the nation.

◆ **FIGURE 9.14** Waste pipe emitting chemically polluted water in Merseyside, England.
SOURCE: David Woodfall/Tony Stone Images

I live in Colorado, where 3,342 out of 14,100 miles of streams are known to be degraded by polluted runoff. It is not known to what degree groundwater has been affected. Most of Colorado's polluted runoff comes from agriculture and mining. Demonstration projects have been implemented across the state to show farmers alternative agricultural methods that add less pollutants to streams. Since these projects are voluntary and local farmers or agricultural groups have to pay 60 percent of the project costs, very few farmers are participating. Mining poses a more serious problem, however, since there are so few solutions currently available. Most of the high-pollution severity stream segments in the state are adjacent to or downstream from mining sites, some of which are or will become Superfund sites. Many abandoned mines and their waste piles contribute metals and toxic chemicals to streams with every storm due to continuous seepage of water through mined material and mine shafts. Both agriculture and mining involve relatively few citizens in the state, whereas other forms of NPS pollution are caused by the majority of Colorado citizens. My research has focused on urban and construction runoff for two reasons: First, current solutions for this type of runoff are difficult and expensive. Second, the rapid growth of Colorado's communities means that urban and construction runoff will increase and contribute a significantly larger share of the NPS pollution total.

Constituency

Seawater averages about 96.5 percent water and 3.5 percent dissolved solids by weight. Nearly every known substance is soluble to some extent in water, and about two-thirds of the 92 naturally occurring chemical elements have been found in seawater. The great majority of these solids, by weight, are ions of salts.[1] More than 99 percent consist of the six elements chlorine, sodium, magnesium, sulfur (as sulfate), calcium, and potassium. Chlorine and sodium are the two dominant elements by far and, upon evaporation of suffi-

1. An ion is an electrically charged atom or group of atoms. The dissolution of minerals in seawater causes them to dissociate into their constituent atoms, which will recombine if the water is evaporated. A salt is formed when the hydrogen atoms of an acid are chemically replaced by atoms of a metal. Salinity causes seawater to conduct electricity better than does fresh water, and electrical conductivity is normally used to measure oceanic salinity.

WHAT KINDS OF COMMON URBAN ACTIVITIES CONTRIBUTE TO POLLUTED RUNOFF?

◆ *Landscape maintenance.* Excess fertilizers and pesticides applied to yards, golf courses, and parks in urban and recreational communities are picked up by rain or snowmelt and transported to waterways.

◆ *Car maintenance.* Oil changes or car washing in driveways causes phosphates (in the detergents) or spilled oil to enter the nearest storm drain either immediately or with the next rain. This practice adds nutrients and organic chemicals to streams.

◆ *Motorized transportation.* Vehicles powered by fossil fuels contribute to polluted runoff in several ways: First, organic materials dripping from engines (gas, oil), spilled at gas stations, or emitted from exhaust pipes are washed into storm drains or streams with every rain. Second, street applications of gravel and salts to aid traction in snow cause sediments and chemical salts to be transported into waterways. Finally, paved streets create more impervious surfaces, accelerating surface runoff and increasing the transport of light or soluble pollutants to streams.

◆ *Construction.* Construction of highways, homes, shopping malls, and other businesses contributes excess sediment to streams by exposing previously stable topsoil through earthmoving activities.

HOW DO RECEIVING WATERS REACT TO THESE POLLUTANTS?

Receiving waters are disrupted not only by the presence of these pollutants, but by the sporadic and intense nature of pollutant input to the stream. Infrequent precipitation events transport pollutants to streams in "shock loadings" that can temporarily make the stream unfit for aquatic life due to toxic loadings of metals, oils, silt, and nutrients.

WHAT CAN BE DONE?

Since polluted runoff is created by multiple individual activities and can not be traced back to specific individuals, it is difficult to regulate. There are three categories of solutions: structural, policy, and educational. Structural solutions require engineering and building efforts, such as the construction of retaining walls to catch and divert runoff to storm drains that lead to water purification plants. An alternate structural solution is to build retention ponds that allow pollutants to settle out and also moderate delivery of runoff to streams. Porous pavement (now used in Sweden) reduces runoff by increasing infiltration of precipitation. The chief problem with most structural solutions in an already urbanized environment is that they are usually difficult to design and very expensive to implement.

Policy solutions include reducing atmospheric pollution, increasing the frequency of street sweeping, increasing ridership on mass transit, reducing erosion through stricter controls on construction methods, and taxation or elimination of pesticides and fertilizers within municipal boundaries.

Education is the most promising of the alternatives. The ubiquity of polluted runoff and its link with land uses negate a purely regulatory approach (it would be impossible to police everybody), and many cities cannot afford effective structural solutions. Educating the public about how common daily activities can contribute to polluted runoff would generate enough debate in the media, the legislature, and town halls that the public could start altering its behavior voluntarily. A concerted effort at public education for all ages would promote public awareness of water quality concerns. Water is the genesis of life, and we are its caretakers. Through education, we could foster a greater recognition of the honor and privilege inherent in its protection. ∎

This focus box was written by Laurel Phoenix, a resident of Lafayette, Colorado.

cient water, will combine to form common table salt, or sodium chloride (NaCl).

The origin of the salinity of the ocean is similar to that of saline lakes. Over several billion years, runoff from the land has carried to the sea tremendous quantities of dissolved salts. Because there is no natural mechanism for removing these salts until saturation occurs (and the oceans are still well below the salt saturation level), salinity continues to gradually increase. The quantity of salt in the ocean is now so great that, if removed, it would exceed in volume the portion of the continent of Africa that is above sea level.

Near-surface salinity levels vary somewhat over the open ocean because of losses of water to evaporation and gains of fresh water from rivers or precipitation. In general, salinity levels are highest in the subtropics, where large net evaporative water losses concentrate the salts (Figure 9.16). Large variations in salinity occur only in restricted seas where communication with the major ocean basins is limited. For

TABLE 9.3   Areal Extent and Mean Depths of the Ocean Basins

OCEAN	AREA		MEAN DEPTH	
	Km²	*Mi²*	*M*	*Ft*
Pacific	165,427,000	63,855,000	4,070	13,350
Atlantic	82,328,000	31,744,000	3,642	11,950
Indian	73,500,000	28,371,000	3,892	12,770
Arctic	14,060,000	5,427,000	1,204	3,950
"Antarctic"*	14,847,000	5,731,000	3,730	12,240
Other†	10,339,000	3,991,000		
Total‡	360,412,000	139,119,000	3,865	12,680

*This refers to the ocean water surrounding Antarctica.
†Includes restricted bodies of ocean water such as the Mediterranean Sea, Black Sea, and Red Sea. All figures are approximate because boundaries between adjacent ocean basins are arbitrary.
‡Estimated total volume of water is 1,192,625,000 cubic kilometers (286,230,000 mi³).

example, the Black Sea, which receives large quantities of fresh water from the Danube and other rivers, has a salinity of only 1.6 percent. The shallow Baltic Sea sometimes has salinities as low as 0.1 percent in the spring snowmelt season. On the other hand, the eastern Mediterranean and Red Seas are located in arid climates and have summer salinity levels of 4 percent or more.

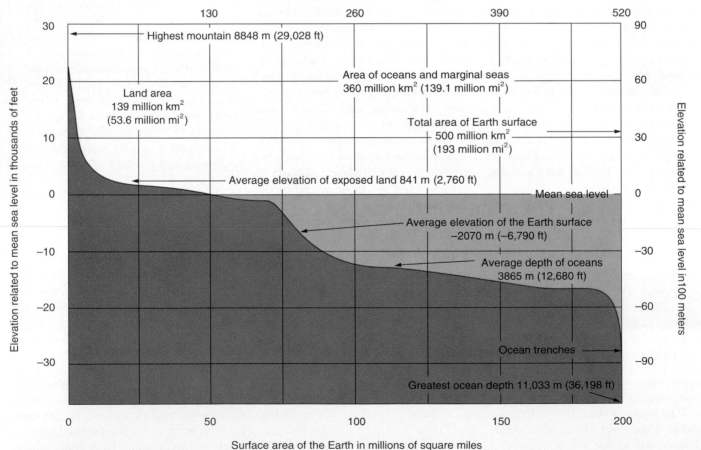

◆ **FIGURE 9.15** Hypsographic curve of Earth surface levels. The proportions of the total global surface at differing elevations and ocean depths are indicated by the slope of the curve. The two relatively level portions of the curve—near sea level and at ocean depths between 4,000 and 5,000 meters (12,000 and 16,000 ft)—indicate that these surface levels are the most widespread geographically.

◆ **FIGURE 9.16** Map of mean oceanic surface water salinity levels in parts per thousand (⁰/oo) in August. Green shaded areas have salinity levels exceeding 36 ⁰/oo, yellow areas have levels between 34 and 36 ⁰/oo, and light violet areas have salinity levels of under 34 ⁰/oo.

Temperature

The temperature of the ocean surface varies considerably in different parts of the world (Figure 9.17). Away from the continental margins, the temperature distribution pattern generally is similar to that of the atmosphere. Extensive areas of tropical ocean water have surface temperatures between 26.5° and 29.5°C (80° and 85°F), and values gradually fall with increasing latitude to readings near or slightly below 0°C (32°F) in the subpolar and polar regions. The mean temperature of the ocean surface water as a whole is about 21°C (70°F).

A basic difference in the thermal behavior of the continents and ocean is that the latter experiences smaller temperature variations, especially on a seasonal and daily basis (see Chapter 4). Annual temperature variations of the open ocean waters reach a maximum of 8 to 10 C° (15 to 18 F°) in the middle latitudes of the northern hemisphere and 5 to 6 C° (9 to 10 F°) in the middle latitudes of the more highly maritime southern hemisphere. Variations of at most 1.5 to 2 C° (3 to 4 F°) are typical of the tropics. Seasonal temperature variations decrease rapidly with depth and, in most places, do not occur at all at depths greater than 180 meters (600 ft). Daily water temperatures on the open ocean vary less than 0.5C° (1F°).

Despite the transparency of the ocean, most insolation is absorbed in the top 1.5 meters (5 ft) of water. As a

consequence, outside the polar regions, water temperatures decline everywhere with depth. The warm surface water is lighter and more buoyant than the underlying water and remains at the surface. Conversely, cold water produced in the high latitudes becomes dense enough to sink and spread outward at depth. As a result, all ocean water at depths greater than 3,000 meters (10,000 ft) is derived from the polar regions and has a constant temperature of approximately 2°C (35°F). The aggregate mean temperature of all water in the ocean is about 4°C (39°F).

It has recently been discovered that extensive portions of the ocean surface waters can vary in temperature by several degrees Celsius over periods of one or more years. Because of the constant exchange of heat, moisture, and momentum between the ocean and the atmosphere, these thermal variations can have a profound influence on weather conditions both over the oceans and on the adjacent continents. For example, during the early 1970s, water temperatures in the central Pacific became warmer than normal. This warming apparently caused the jet stream to ridge northward over the area, while a trough developed over the western United States, bringing unusually cool and wet weather to the West Coast. This pattern, in turn, apparently induced the formation of a second ridge over eastern North America, causing the East Coast to experience a string of unusually mild, dry winters. In 1982 and 1983, a large-scale warming of the eastern Pacific took place. It was caused by a temporary

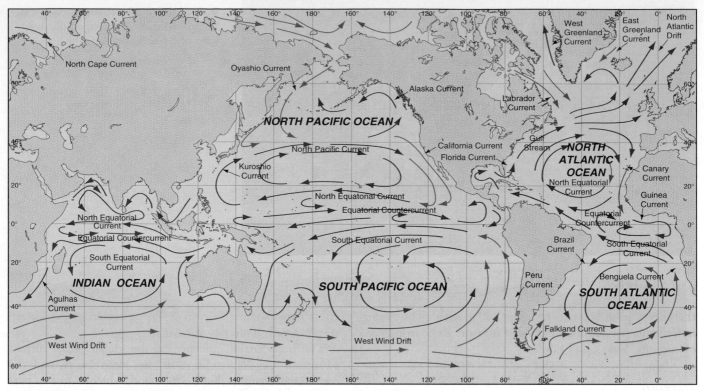

◆ **FIGURE 9.19** Global surface ocean currents. Warm currents are shown in red; cold currents in blue. The centers of the major current circulation patterns, or gyres, roughly correspond to the oceanic centers of the subtropical highs.

water eventually returns to the surface in zones of **up-welling.** Most of these zones are situated near the west coasts of continents in the subtropics, where equatorward-flowing water is redirected westward by the trade winds. The seaward push of this water causes cold water from great depths to well up to replace it.

A period of several centuries may be required for one complete cycle of convergence, descent, and upwelling. This period is estimated at about 700 years for water of the Atlantic and 1,500 years for that of the Pacific. Slow as it is, this vast vertical circulation system is crucial to sea life. The descent of surface water in areas of convergence carries dissolved carbon dioxide and oxygen needed by marine plant and animal life into the depths. Conversely, the rise of bottom water in zones of upwelling transports nutrients from the sea floor to the surface. As a result, zones of upwelling are the sites of some of the world's most productive fishing grounds (see the Case Study at the end of this chapter). Water in the vast mid-ocean basins is largely cut off from this circulation. As a consequence, it is low in dissolved gases and nutrients and nearly devoid of life.

Waves

Waves are the most familiar of the ocean movements and have the greatest influence on the characteristics of the coastline (see Chapter 19). Like currents, they are generated primarily by the wind. Unlike currents, though, waves do

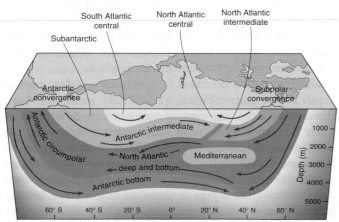

◆ **FIGURE 9.20** A profile view of the Atlantic, showing the major deep ocean currents. Most of the oceanic deep water is derived from the two zones of convergence located in the subpolar regions, because only there is the water sufficiently cold to become dense enough to sink.

◆ **FIGURE 9.21** Ocean wave terminology.

not produce a significant linear transport of water, although they are efficient transporters of energy over great distances.

Waves are produced by the frictional drag of the wind blowing over the water surface. Once started, the size and energy of waves is increased by the push of the wind against their sides. Three factors determine wave sizes: the wind speed, the **fetch** (the distance over which the wind blows), and the length of time the wind has been blowing from the same approximate direction. An increase in any of these factors increases wave dimensions. The largest wind-generated waves ever reliably reported were 17 meters (55 ft) high. One combination of conditions that would produce waves of this size is a 110 kilometer per hour (70 mph) wind blowing for 24 hours over a fetch exceeding 800 kilometers (500 mi). The largest storm waves are capable of damaging or sinking even large ships and of producing tremendous erosional and structural damage along coasts (Figure 9.22).

The waves that reach our coastlines at any given time are generated only in certain restricted areas, usually located within storms. They are able to travel hundreds or even thousands of kilometers before eventually reaching distant shores. Large waves along the California coast, for example, are often produced by storms near Alaska. On occasion, waves reaching Hawaii are generated in storms off the coast of Antarctica.

Waves lose little energy as they travel through deep water, but this situation begins to change as they approach the shore and move into shallow water. When the water depth decreases to less than half the wave length, the wave starts to "feel bottom" as its forward progress begins to be slowed by friction (Figure 9.23). The restraining influence and upward

◆ **FIGURE 9.22** Large waves such as these are generated in storms that may be centered thousands of kilometers from shore. SOURCE: H. Armstrong Roberts

push of the bottom cause the wave to rise gradually and tilt forward. When the water depth has been reduced to about 1.3 times the wave height, the upper portion of the wave pitches forward and breaks. (The process is much like that of a person who falls forward because he or she has been tripped.) The water now becomes foam-filled surf, which for the first time experiences significant net horizontal motion as its forward momentum causes it to rush up the beach as **swash** and then, its energy spent, to return as **backwash.**

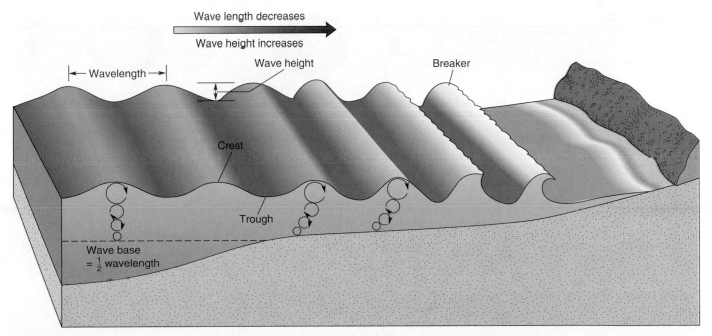

◆ **FIGURE 9.23** The water in waves moves in circular paths that gradually decrease in magnitude with depth. Waves begin to rise, tilt forward, and eventually break as they enter areas where the bottom is too shallow to permit the orbital motion to continue unimpeded.

During this process, the energy imparted to the wave days earlier in a distant storm finally is expended on the coastline. The coastal features that result are examined in Chapter 19.

Tides

The ocean **tides** are the rhythmic rise and fall of the ocean produced by the gravitational influence of the Moon and Sun. Unlike waves and surface currents, the tides affect the ocean throughout its entire vertical extent. Gravity-induced tides are not restricted solely to the ocean. They also occur in large lakes, on the continents (with displacements of up to 30 centimeters or 12 inches), and in the atmosphere (with displacements of many kilometers). Their practical significance with respect to human activities, however, is greatest by far in the ocean. The ebb and flow of the tides has been observed, pondered, and put to practical use by seafarers for thousands of years. Only in recent times, however, have sufficiently detailed records been compiled in most areas to allow for precise tidal predictions.

Although the Sun is much more massive than the Moon and has a much greater surface gravity, the tidal influence of the Moon is over twice that of the Sun because it is much closer to the Earth (Figure 9.24). The ocean water on the portion of the Earth directly beneath the Moon is pulled upward, creating a high tide when the Moon is overhead. A second high tide occurs on exactly the opposite side of the Earth because of the rotational inertia of the Earth-Moon system and because that area is farthest from the Moon, making its pull weakest there. Midway between the two areas of high tides are zones of low tides.

The combination of the Earth's axial rotation and the Moon's revolution around the Earth causes the Moon to take 24 hours and 50 minutes to return to its same longitudinal position over the Earth. The pattern of high and low tides rotates with the Moon's changing position, so that most places experience two high tides and two low tides during this period. A time span of 6 hours and 12 minutes therefore normally elapses between subsequent high and low tides.

Although the basic pattern of semidaily high and low tides is relatively simple, a number of additional influences combine to make the world pattern of oceanic tides exceedingly complex and difficult to predict. One of the most important is the gravitational influence of the Sun. Like the Moon, the Sun acts to produce a pair of high tides in those areas facing directly toward and away from it, with an intervening pair of low tides. The tides actually experienced at any one place therefore result from the combined influence of the Moon and Sun. When these two bodies are aligned either on the same side of the Earth (at the time of the new Moon), or on opposite sides (at the time of the full Moon), their gravitational effects work in concert, and tidal ranges are largest. This produces the so-called **spring tides,** whose occurrences are actually unrelated to the season of the year. When the Sun and Moon are situated at right angles with respect to the Earth (at the times of the half Moon), their gravitational influences are opposed to each other, and tidal ranges are smallest. These are called **neap tides.**

Probably the most important and certainly the most complex of the tidal influences is the geographic pattern of the ocean basins and coastal margins. Their irregular shapes

◆ **FIGURE 9.24** The ocean tides are caused by the gravitational pull of the Moon and Sun. The diagram shows the relative positions of these bodies with respect to the Earth at the times of the spring and neap tides. The more pronounced bulges around the Earth at the time of the spring tides indicate that tidal ranges are highest then because the gravitational attractions of the Moon and Sun are working in concert.

(b) Spring tides

(c) Neap tides

(a) Tidal forces

◆ **FIGURE 9.25** Tidal pattern in the North Sea region. The tidal range is given in meters, and the time of the high tide is in hours after Greenwich noon. Note the rotation of the tides around the two nodal points.

Tidal range (meters)
Over 9 | 3 – 4.9
7 – 9 | 1 – 2.9
5 – 6.9 | Under 1
— Time of high water in hours after high noon at Greenwich

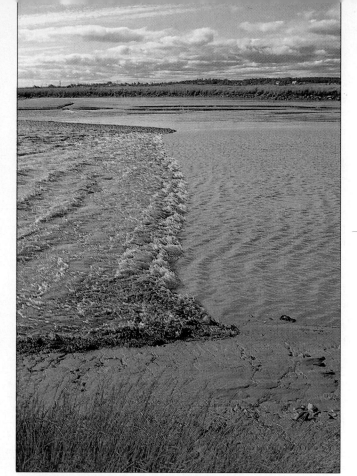

◆ **FIGURE 9.26** Tidal bore, upper Bay of Fundy, near Truro, Nova Scotia.
SOURCE: Francois Gohier/Photo Researchers, Inc.

cause the tides to be redirected or reflected in countless ways and often to be augmented, diminished, or even canceled altogether in some places. Highly uneven coastlines frequently experience tides that vary greatly both in range and in time of occurrence over small distances. In broad water bodies, including the major ocean basins, the tides display a circular pattern, with high and low tides rotating around one or more nodal points or lines that experience no tides at all (Figure 9.25).

Tidal ranges of 1 to 2 meters (3 to 6 ft) are most common for the world as a whole. The highest tides occur in funnel-shaped bays or estuaries, where the tidal influence is focused by the converging shorelines. Nova Scotia's Bay of Fundy has the highest tides of all, with one site experiencing a mean spring tidal range of 14.5 meters (47.5 ft) and an extreme range of 16.5 meters (53.5 ft).

Tides in restricted bodies of water may be so concentrated that strong **tidal currents** are formed. These consist of currents that flow landward during the period of the rising tide and seaward during the period of the falling tide. If a current is forced to advance upstream against a strong downstream flow of river water, its progress is slowed and water may be piled up at its forefront to produce a **tidal bore**

(Figure 9.26). In the lower Amazon, such a tidal bore advances upriver as a wave for a distance of 480 kilometers (300 mi). In places it forms a wall of water 5 meters (16 ft) high that travels at a rate of 22 kilometers per hour (14 mph) and produces a roar that can be heard for great distances.

The net result of the combination of tidal influences is that most localities experience alternating higher and lower high and low tides. Even a thorough understanding of all the tidal influences does not make it possible to forecast the height of the tides precisely. This is because ocean levels are also strongly influenced by constantly changing patterns of wind direction, fetch, and speed and, to a lesser extent, by variations in air pressure.

❙ SUMMARY

This chapter has examined the characteristics and geographical distribution of the Earth's surface and near-surface water. Water occurs in quantity in several different sites. It exists in a dispersed state in virtually all soils, and substantial volumes of deeper groundwater occur within porous rock and sediment layers in many regions. Surface water bodies include lakes, rivers, and the ocean. With the exception of some of the deeper groundwater, all of these sites contain water that participates in the hydrologic cycle, and all continually experience both additions and losses of water.

We now return to the Focus Questions posed at the beginning of the chapter:

1. *What factors control the movement and storage of subsurface water?*

Subsurface water enters the ground primarily through the soil pore spaces. Large quantities of water adhere to the soil particles, while the remainder makes its way to the water table to become groundwater. Groundwater normally moves slowly downhill under the influence of gravity and hydrostatic pressure, with the speed of movement controlled largely by the permeability of the subsurface rock or sediment layers through which it passes. Large volumes of groundwater are confined more or less permanently within porous aquifers.

2. *How do lakes form and disappear?*

Lakes occupy enclosed basins that are supplied with enough water to fill them at least partially. These basins form in a number of ways, but the two most important are by glacial erosion and deposition and by crustal movements, especially faulting. Lakes have relatively brief spans of existence because of the action of two processes that eventually destroy the water-holding capacity of the basin. The first is erosion of the basin rim, which may cause the lake to drain. The second is the filling of the basin by sediments and organic matter, converting it to a swamp and eventually to dry land.

3. *What factors determine geographic and temporal variations in stream discharge?*

Stream systems transport excess precipitation to enclosed interior basins or, more commonly, to the ocean. The discharge of a stream is determined primarily by climate and by the size of its drainage basin. Streams with drainage basins of similar size tend to have greater discharges in areas with heavy precipitation, cool temperatures, steep slopes, and limited vegetation. Nearly all streams display large short-term variations in flow as a result of precipitation irregularities, and many streams experience major seasonal variations in discharge. These variations are correlated with seasonal fluctuations in precipitation, temperature, and vegetative cover.

4. *What are the chief physical and chemical characteristics of ocean water?*

The ocean is the interconnected body of saline water that covers the lowest 71 percent of the Earth's lithospheric surface. It consists of approximately 96.5 percent water and 3.5 percent dissolved solids, mostly chlorine and sodium. Variations in salinity are minor, except in the vicinity of the mouths of large rivers and in basins with restricted access to the open ocean. Ocean surface temperatures are high in the tropics and decline with latitude, but the deep ocean water everywhere has a temperature near freezing. The Arctic Ocean and much of the water around Antarctica are covered with sea ice or contain ice shelves or icebergs derived from coastal glaciers.

5. *What are the causes of the three oceanic motions of currents, waves, and tides?*

Currents are continuous movements of water streams of similar temperature and density. Surface currents are set in motion by the wind and display a general similarity in flow pattern to that of the global wind currents. Deep currents are produced largely by water density variations resulting from differences in temperature and salinity. Waves are oscillations of the surface water generated by strong winds associated with storms. The three factors of wind speed, fetch, and wind duration determine the size of waves. The tides are produced by the gravitational pull of the Moon and, to a lesser extent, of the Sun. They result in a rhythmic rise and fall of the ocean surface.

▌QUESTIONS FOR THOUGHT AND REVIEW

1. What are the two categories of subsurface water? Describe the location and characteristics of each.
2. How and why does groundwater return to the surface? Where is it most likely to go?
3. Portions of the New York City area periodically experience problems with saltwater intrusion. What factors might cause this area to be especially susceptible to this problem? What factors are likely to be responsible for Phoenix, Arizona's problem of rapidly declining groundwater supplies?
4. What determines whether a lake will contain fresh water or salt water?
5. What are the two chief methods by which lake basins are formed? Give some examples of large lakes produced by each method. How else may lake basins be formed?
6. From what specific water source is your home water supply derived? How far must it travel to reach your home? Does this water supply face any problems at present with respect to quality or quantity?
7. What basic factor controls the geographic distribution of rivers? Is this same factor the major distributional control for lakes, groundwater, and the ocean? Explain why or why not for each.
8. What categories of human activities in North America use the most water? Approximately how much of this water is consumed, and how much is returned to existing water supplies for possible later use?
9. What is a flood? Explain how human landscape modifications in agricultural and urban areas can increase the likelihood of flooding.

10. In what ways are ocean temperature distributions, both vertically and horizontally, similar to those of the troposphere? In what ways are they different? Why are the deep oceanic waters cold everywhere?

11. Describe, in general terms, the pattern of surface ocean currents within the five major ocean basins. What factors are responsible for this pattern?

12. How are ocean waves formed? In what parts of the ocean are large waves produced? What factors determine the size of waves?

13. Explain the cause of the tides. Why do they vary in height in most places at regular intervals through the month? What factors complicate the global pattern of tides?

14. What is El Niño? What do the climatic consequences of recent El Niños indicate about the relationship between the ocean surface temperature and the atmosphere?

▌KEY TERMS

soil water	**drainage basin**
groundwater	**flood**
interception	**sea ice**
throughflow	**iceberg**
water table	**currents**
spring	**waves**
aquifer	**tides**
stream discharge	

10 LIFE-FORMS AND ECOSYSTEMS

OUTLINE

FOCUS QUESTIONS

1. What are ecosystems, and how do they function?
2. What are the chief environmental controls on the global distribution of plants
   and animals?

The Earth is home to a great variety of life-forms. An estimated five million species of plants and animals inhabit our planet. About 1,800,000 species have been classified, and many more—most of them small and microscopic life-forms—await discovery.

The systematic study of plants and animals comprises the subject matter of biology. The branch of physical geography devoted to the study of the biosphere is known as **biogeography.** This field is closely related to ecology, except that the geographer's emphasis is on spatial patterns. Biogeographers strive to develop a better understanding of the global distributions of associated groups of plants and animals and to better comprehend the interactions of these groups with their environment. The distribution of terrestrial life-forms is closely related to climate patterns and is also profoundly influenced by the barriers imposed by large water bodies, particularly the oceans. For this reason, our study of biogeography immediately follows the chapters devoted to these two topics.

FORMS AND FUNCTIONS OF PLANTS AND ANIMALS

Plants are organisms containing a green pigment, **chlorophyll,** that has the ability to convert water and atmospheric gases into food in the presence of light. The number of plant species currently living on Earth is estimated to exceed 300,000. Plants vary in size from microscopic single-celled algae to giants such as the California redwood and Australian gum, which can attain heights of more than 100 meters (330 ft).

Vegetation covers approximately 80 percent of the Earth's land surface and is remarkable in its diversity. Plants grow from the margins of glaciers to the heat of the equator, from areas submerged in water to almost rain-free deserts. They not only comprise the vast bulk of life on Earth, they also have made possible the evolution of animals, including human beings, by providing both food and a breathable atmosphere.

As far as we know, Earth is the only planet in our solar system (and indeed, in the universe) on which life has developed. Plants were the first life-forms on Earth. They evolved when our planet was relatively young; the earliest fossils are more than three billion years old and are apparently the remains of bacteria. Between that time and the present, evolution has produced all the plants and animals currently occupying the Earth's surface. For its first 2.5 billion years, life on Earth was apparently restricted to the oceans. About 400 million years ago, however, plants began to establish themselves on land. The conifers (gymnosperms) evolved about 350 million years ago, and the flowering plants (angiosperms) that now form the Earth's dominant plant group appeared only within the last 140 million years.

The dominant positions achieved by the conifers and flowering plants result largely from the fact that they produce seeds. Seeds consist of an embryo surrounded by a food store for nourishment during germination (Figure 10.1). Seeds are also protected by a hard outer layer, the seed coat, and can remain dormant until conditions for germination and growth are favorable. This greatly increases their chances for survival.

Photosynthesis and Energy Production

Plants produce their energy and the bulk of their solid matter directly from water and carbon dioxide in the presence of light. This process, termed **photosynthesis,** is accomplished by the organic chemical chlorophyll manufactured by the plants. The chlorophyll, present in certain tissues (mostly within the leaves) of green plants, absorbs sunlight. This energy is used in a series of reactions to synthesize adenosine triphosphate (ATP) and release oxygen. ATP provides the chemical energy to drive a second series of reactions in which atmospheric carbon dioxide is used to synthesize glucose, which is converted into carbohydrates. These include starch (the principal energy storage compound for plants) and other sugars. The plant's vascular system brings in water and nutrients and transports the sugars produced by photosynthesis to other parts of the plant.

◆ **FIGURE 10.1** The seeds of this Mexican coconut palm are protected by thick husks that not only provide long-term protection, but also are buoyant in water. This allows coconuts reaching rivers or the ocean an opportunity to be carried to other germination sites.

SOURCE: © Mark Lewis/Tony Stone Images

◆ **FIGURE 10.2** Estimated global net primary productivity in grams of carbon per square meter per year.

Only certain wavelengths of sunlight (chiefly, the orange-red and violet-blue wavelengths) are used during the photosynthetic process. Green wavelengths are not used and are mostly reflected, giving plant leaves a green appearance. Globally, only about 0.1 percent of the total insolation reaching the Earth is used by vegetation during photosynthesis.

To enable the process of photosynthesis to occur, plants absorb great quantities of water from the soil through their root systems. They also take in great volumes of atmospheric carbon dioxide through openings called stomata that are most abundant on the undersides of their leaves. The photosynthetic conversion of carbon dioxide and water to carbohydrates and oxygen removes great quantities of carbon dioxide from the atmosphere and has supplied nearly all the atmosphere's free oxygen.

The rate at which a plant produces energy by photosynthesis is termed its **primary production rate.** Plants also

consume a portion of this energy through the process of **respiration** to generate the energy for their other life processes. Respiration breaks down the carbohydrate molecules into simpler parts, usually with the help of oxygen. The energy that is produced allows the plant to conduct its life processes and is eventually released as heat to the atmosphere. A plant can only survive and grow, however, if it produces more energy by photosynthesis than it consumes by respiration. In general, somewhat over half of a plant's energy production is used during respiration. The remainder is stored as carbohydrates and is later released as the plant is consumed by animals or decomposed by microorganisms. These organisms convert the carbohydrates back to carbon dioxide, which is released into the air for possible absorption by other plants, thus repeating the cycle.

The difference between the amount of energy produced by primary production and the amount consumed by respiration is termed **net photosynthesis.** Net photosynthetic energy production depends not only on the type of plant, but also on the favorability of environmental conditions for plant growth. Particularly important are temperature, moisture, and available light.

The net photosynthesis produced by an entire biological community is termed its **net primary productivity.** This figure determines the amount of organic matter (biomass) generated by the community. Global net primary productivity is estimated at 170 billion metric tons of dry organic matter per year. The world's most productive regions occur within the humid tropics (Figure 10.2), which have a year-round growing season and receive abundant precipitation. In the oceans productivity is highest in areas where cool, nutrient-rich water upwells from the depths. This generally occurs in the upper middle latitudes and off the west coasts of the continents. These areas contain the world's most productive marine fisheries.

Animal Evolution and Characteristics

Animal life originated approximately one billion years ago in the oceans. Its origins are obscure because early animals had soft bodies that left little in the way of fossil imprints. About 400 million years ago, the first animals emerged from the ocean to occupy the land. This became possible only after the plants, their food source, had established themselves on land.

Unlike plants, animals are incapable of manufacturing their own food by photosynthesis. Instead, they rely on plants or on other animals for sustenance. Animals also are capable of independent movement, while most plants are not. In addition, most nonmicroscopic animals are aware of their surroundings and possess at least a rudimentary intelligence. These characteristics allow them to respond rapidly to stimuli, giving them a measure of self-protection from threats to their well-being.

The great majority of animal species are microscopic. The larger animals—the ones with which biogeographers are most concerned—may be classified as either invertebrates or vertebrates. **Invertebrates,** animals without backbones, comprise over 90 percent of all nonmicroscopic animal species. Important invertebrate groups include worms, mollusks, and especially insects. **Vertebrates,** which have backbones, are divided into five principal groups: fishes, amphibians (including frogs and salamanders), reptiles (especially snakes and lizards), birds, and mammals. As a group, the vertebrates are more highly advanced on the evolutionary scale than are the invertebrates.

At present, about 1.5 million animal species have been studied and named. These include roughly 8,800 species of birds, 4,070 mammals, about 4,015 amphibians, 6,600 reptiles, and more than 21,000 fish species. These vertebrates, however, are greatly outnumbered by invertebrates, with well over 1 million species of insects alone known to exist.

BIOLOGICAL COMMUNITIES

The global distribution of plants and animals is not random. Each species has specific environmental requirements, so that different biological groupings or assemblages exist in different settings. A specific site that has environmental conditions favoring its habitation by a given species or group of species is termed a **habitat.** Habitats can be identified at differing scales. **Macrohabitats** generally are climatically delineated and usually are differentiated on the basis of their vegetation. **Microhabitats** exist within macrohabitats. They may be very small, such as around the edge of a pond or under a rotting log. Microhabitats are especially important to small plants and animals, but even large animals often confine themselves to only limited portions of macrohabitats.

An assemblage of plants and animals whose members share a particular environmental setting of restricted size is referred to as a **biological community** (Figure 10.3). Biological communities of all sizes exist and are determined to a great degree by climatic characteristics at both the macro and micro levels. Each member of a biological community occupies a **niche** within the community by performing a function or occupation that contributes to the operation of the community.

Ecosystems

Because a biological community is so intricately related to its environmental setting, these two entities may be linked together by the concept of the ecosystem. An **ecosystem** is a functioning entity consisting of all the organisms in a biological community, as well as the environment in which and with which they interact. A balanced ecosystem is largely self-sustaining in its resources and has a relatively stable assemblage of life-forms. The organisms within an ecosystem are linked to each other and to the environment by transfers of energy, water, and nutrients. The quantities of energy and water and the amounts and types of nutrients available in a place largely control the numbers and types of organisms that the ecosystem can contain.

Each member of an ecosystem occupies a particular habitat and performs a specific function that contributes to the functioning of the community as a whole. The members of an ecosystem themselves act to alter the ecosystem's physical characteristics. For example, the growth of trees within a forest increases shade and adds organic matter to the soil, and the growth of vegetation along the shore of a lake increases the availability of food and habitat for the shoreline fauna. These changes, in turn, will produce changes in the types of organisms inhabiting the ecosystem.

Ecosystems exist at a variety of scales, and, in fact, the Earth itself can be considered the ultimate ecosystem. As a general rule, however, they are divided into small units in discrete environmental settings, such as a forest floor or a

◆ **FIGURE 10.3** A distinctive biological community that includes alligators inhabits this Florida swamp.
SOURCE: W. Metzen/H. Armstrong Roberts

coral reef. Ecosystems generally are treated as discrete entities that differ significantly from neighboring communities. In reality, however, they often are separated by transitional communities called **ecotones.** Ecotones share attributes of the ecological communities they border and therefore tend to exhibit relatively complex biological characteristics.

Ideally, a biological community is relatively stable. If environmental conditions are altered significantly, however, the community may undergo drastic changes. Disturbances such as high winds, floods, droughts, fires, insect invasions, or human interference can remove some species and encourage others, thereby greatly modifying the makeup of the community. If the disturbance is minor, and environmental conditions return to those that originally existed, the community will gradually tend to revert to its original composition. Reversion occurs because the original assemblage of plants and animals formed the most stable and efficient community under the existing environmental conditions, and a return of those same conditions should produce the same biological response.

The gradual change in the composition of a plant community in response to changing environmental conditions following a disturbance is referred to as **vegetative succession** (Figure 10.4). Succession normally proceeds at a relatively rapid rate when a soil cover is already present and seeds and some vegetation exist within the disturbed area. This is referred to as a **secondary succession.** In some instances, however, physical processes can produce a new surface that has no soil or vegetation cover. Examples of such processes include volcanic eruptions, landslides, glacial erosion, and the drainage of water bodies. In such cases a **primary succession** of vegetation occurs. Primary succession takes place relatively slowly because its progress is largely dependent upon the development of a soil cover.

Dissimilar environmental conditions lead to very different successional sequences. Eventually, the successional process produces the biological community best suited to survive under existing environmental conditions. This **climax community,** as it is called, theoretically continues to occupy the area until the next disturbance or environmental alteration occurs.

A climax community has several important attributes. It usually contains the greatest **biomass,** or mass of living organisms, that the area is capable of supporting. A high degree of species diversity typically exists, so that all available habitats are utilized effectively. The composition and relative distribution of species tend to be stable, and birth and death rates are balanced. Finally, most energy and nutrient needs are supplied and recycled by the organisms within the community, so that losses are minimized.

Currently, most areas of the world do not have climax biological communities. This is primarily the result of widespread human disturbance and, to a lesser extent, of geologically recent worldwide climatic changes (see Chapter 17). Nevertheless, it is possible to divide the world's plants and animals into a number of major global-scale communities, or **biomes,** that exist in relative equilibrium with their environment. We will examine these communities in the next chapter.

◆ **FIGURE 10.4** An idealized vegetative successional sequence in the southeastern United States following the abandonment of farmland.

Bare field	Grass-land	Grass-shrub	Pine forest	Oak-hickory forest climax
1	2	3-20	25-100	Above 150

Age (yr)

Biotic Ecosystem Components

Organisms within an ecosystem may be divided into three groups based on the functions they perform: producers, consumers, and decomposers. **Producers** consist of the ecosystem's vegetation. They absorb insolation, water, and carbon dioxide and manufacture carbohydrates and oxygen. The producers may be considered the foundation of the ecosystem because they provide all three groups with the energy needed to conduct their respective life processes.

Consumers are the ecosystem's animal population; they rely on the producers for their carbohydrate energy. The consumers may be divided into several groups. **Primary consumers** feed directly on the producers and are known as **herbivores.** Cattle, sheep, and deer are familiar examples of herbivores. **Carnivores,** or meat-eating animals, are **secondary consumers** and derive their food supply chiefly by feeding on the primary consumers (Figure 10.5). Familiar examples of carnivores include lions and sharks. **Tertiary consumers,** in turn, feed on secondary consumers (other carnivores) and are sometimes referred to as "top carnivores." Examples of tertiary consumers include hawks, which often feed on rodent-eating snakes, and human beings, which feed on carnivorous fish such as tuna.

The third group of organisms within an ecosystem are the **decomposers.** They consume organic waste as the principal component of their diets. In doing so, these organisms remove the last usable units of energy from the organic matter, while returning nutrients to the soil for future generations of producers to use. This group includes bacteria, worms, insects, fungi,[1] and a variety of other organisms.

As a result of this feeding hierarchy, solar energy captured by the producers passes through several links to form a **food chain.** In reality, most herbivores and carnivores have a variety of choices of food available in the environment, so that if their first choice is unavailable, they can turn to

◆ **FIGURE 10.5** A carnivore (lioness) running down a herbivore (zebra) in the East African savanna.
SOURCE: S. Purdy Matthews/Tony Stone Images

others. As a result, the food chain is more accurately described as a diffuse **food web** by which energy passes through a wide variety of producers and consumers (Figure 10.6). The web includes human beings, who are sometimes classified as **omnivores,** a group of organisms that eat both producers (plants) and consumers (animals).

The food chain or food web is perhaps the most vital component of the ecosystem, because it is the basis of the energy flow that passes through the system and maintains it.

---

[1]Fungi are organisms that can live without light because they obtain their food by absorbing dissolved organic compounds. As a result, they can live only in the presence of living or dead organic material. Because the fungi do not contain chlorophyll or engage in photosynthesis, they are not considered to be plants at all, but rather to comprise a third biological kingdom.

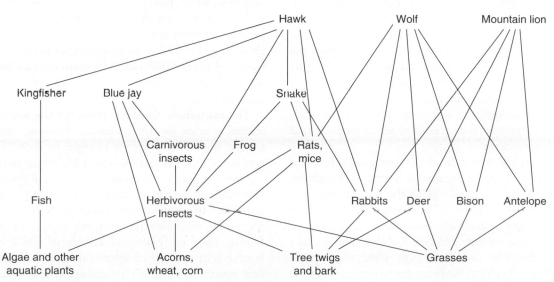

◆ **FIGURE 10.6** A portion of a food web. The entire food web of a biological community would be too complex to diagram on a single page.

Quaternary carnivores
Tertiary carnivores
Secondary carnivores

Primary carnivores

Herbivores

Photosynthetic plants

◆ **FIGURE 10.7** Diagrammatic depiction of the energy and biomass transfers through differing levels of a food pyramid. The available energy and resulting biomass at each level are approximately one-tenth that of the next lower level.

A food chain cannot extend indefinitely, however, because each of its links involves a major loss of carbohydrate energy. Only about 10 percent of biomass normally transfers from one level on the chain to the next higher level (Figure 10.7). For example, cattle fed 1,000 kilograms (2,200 lb) of corn can be expected to gain only about 100 kilograms (220 lb) of weight. Humans consuming this meat, in turn, have the potential to add only 10 kilograms (22 lb) of weight. The energy loss at successive levels on a food chain occurs mainly because animals use large amounts of energy to keep warm and to fuel their movements and body functions. In addition, some of the biomass is converted to waste products that flow to the decomposers.

Any assessment of world food resources, including the global capacity to support increased human populations, thus depends greatly on the level of the food chain from which the food is to be derived. Because ten times as much carbohydrate energy is available from the consumption of plants as from meat, many researchers feel that humans, as a group, cannot afford the continued large-scale consumption of meat. In North America, however, approximately half the total cropland acreage is planted in crops destined for consumption by domesticated animals.

Abiotic Ecosystem Components

The abiotic components of an ecosystem are the supplies of energy and nutrients needed for the ecosystem to function. As we have seen, energy is derived chiefly from insolation, which is fixed by the system's vegetation during the process of photosynthesis. The amount of available light helps determine both the kinds of life-forms that will inhabit an ecosystem and the biological processes of these life-forms. We shall examine this topic shortly in greater detail. Solar energy ultimately powers all the operations of the ecosystem, as illustrated in Figure 10.8.

Nutrients are the chemical elements needed by the life-forms of the ecosystem. The three nutrients required in greatest quantity are hydrogen, oxygen, and carbon. Plants obtain these so-called macro-nutrients in large quantities from the air and soil and convert them to carbohydrates. Together, the macro-nutrients comprise more than 99 percent of plant biomass. Other nutrients, known as micro-nutrients, are required in only small amounts. They include nitrogen, phosphorus, potassium, calcium, magnesium, and sulfur. Each of these nutrients is continuously cycled between the ecosystem and Earth's atmosphere or surface in what is referred to as a **biogeochemical cycle.** Figures 10.9 and 10.10 illustrate the carbon and nitrogen cycles.

BIOGEOGRAPHIC DISTRIBUTION CONTROLS

Plants and animals display a tremendous range of physiological and behavioral characteristics that enable them to cope with the multiplicity of terrestrial environments. Adaptation to the environment is the driving stimulus to the evolution of life-forms—both plants and animals. This adaptation is accomplished through the process of natural selection, which produces changes in the life-form's physical makeup or behavior that make it better able to cope with its physical environment. The two keys to the evolutionary process are (1) biological changes that occur through mutations and (2) the ability of individuals that are better adapted to their environment to survive while those not so well adapted perish.

The **mutations** that have produced the world's plant and animal species involve molecular rearrangements of their genetic material that create changes in their physical characteristics. Most mutations affect life-forms adversely, making them less able to survive. A very small proportion of all mutations, however, produce changes that enable a given individual to compete more successfully than before, resulting in its multiplication and the expansion of its range. During the long biological history of the Earth, literally hundreds of millions of successful mutations have produced new species of gradually increasing variety and complexity. This process has culminated in the Earth's present **biota,** or population of plants and animals.

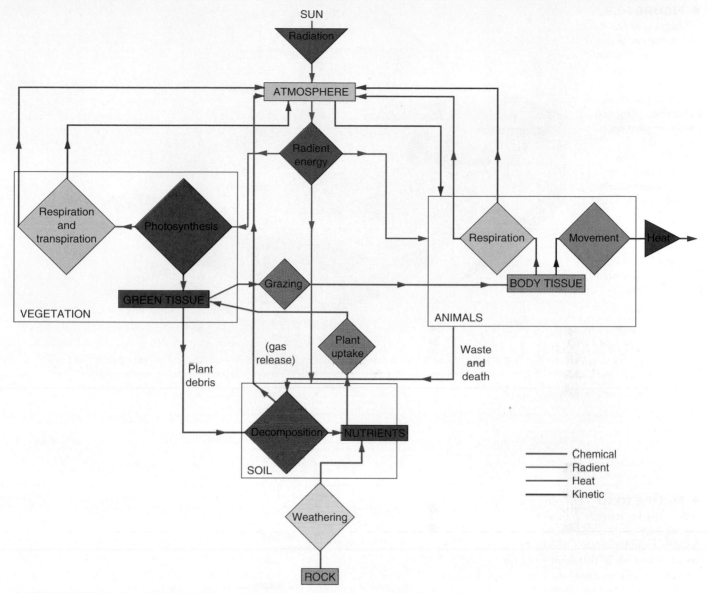

◆ **FIGURE 10.8** Energy flows and transformations in terrestrial ecosystems.

In order for a given plant or animal species to occupy a given geographic area, however, it must have had access to that area. Most species originally evolved in a restricted area, and their current ranges are a result of reproductive success and migration to other areas. Plant migration occurs primarily during the seed stage and is accomplished largely through the mechanisms of wind, water, and animal transport.

Examples of plant adaptations for seed dispersal are numerous. Orchids have extremely lightweight seeds that are easily carried by the wind. The seeds of dandelions, maple, and ash have feathery parachutes or winglike structures to aid in wind dispersal. Vetch ejects seeds explosively when its seed pods split open. Some seed, like those of the carrot, contain barbs that easily attach to the fur of animals or the clothing of humans. Other plants, like the coconut, have

seeds that are waterproof and buoyant, allowing ocean currents to transport them great distances. Still others, like cherries and blackberries, produce fleshy fruits that are attractive to animals; after being ingested with the fruit, the seeds pass through the animal's digestive system.

Animals, unlike plants, can purposefully change location to reduce stressful conditions; they are therefore less tied to the local environment. Most animals possess mechanisms for controlled movement such as wings, legs, and fins. Many animals also migrate seasonally in order to follow food sources or to avoid weather extremes.

The local abundance of an organism usually is closely related to the availability of its preferred habitat. Unfavorable areas between habitats can become **barriers** to dispersal, for the organism must cross them to reach other favorable potential habitats. Few barriers are absolute for all species,

Diagrammatic representation of the carbon cycle. Carbon dioxide enters the atmosphere by a variety of processes, particularly plant and animal respiration, decay, and combustion. It is removed chiefly by plant photosynthesis and solution in water bodies. Great quantities of carbon are stored in coal, oil, peat, natural gas, and limestone below the Earth's surface.

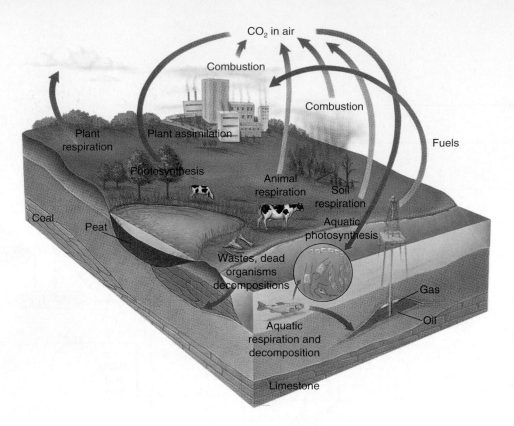

◆ **FIGURE 10.10** The nitrogen cycle. Gaseous nitrogen comprises the majority of the atmospheric gases, but cannot be used directly by plants or animals in that form. However, it can be "fixed," or converted to forms that plants can utilize, through the actions of certain microorganisms. The human manufacture of nitrogenous fertilizers is also adding vast quantities of useable nitrogen to the environment, sometimes leading to the over-enrichment of water bodies.

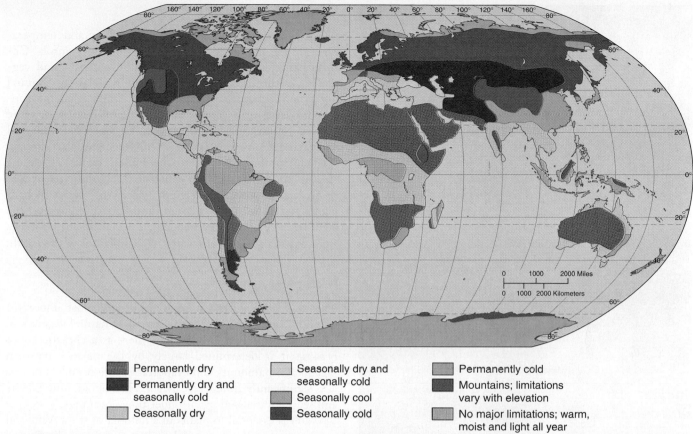

■ Permanently dry	□ Seasonally dry and seasonally cold	■ Permanently cold
■ Permanently dry and seasonally cold	■ Seasonally cool	■ Mountains; limitations vary with elevation
□ Seasonally dry	■ Seasonally cold	■ No major limitations; warm, moist and light all year

◆ **FIGURE 10.11** Global distribution of major climatic stress factors on natural vegetation.

although their effectiveness varies greatly for different plants and animals. In this respect, barriers may act as filters, allowing some species to pass through, while keeping others out. On a global scale, the three most important categories of barriers are oceans, mountain ranges, and deserts. Collectively, these three types of barriers are largely responsible for the existing geographical patterns of most terrestrial life-forms.

Despite the successful adaptations of terrestrial biota to diverse and often adverse environments, large portions of our planet are largely or entirely devoid of life. For the Earth as a whole, the leading barrier by far in restricting the areal distribution of land biota is the presence of surface water. The oceans, of course, comprise a complex biological realm that is both rich in variety and distinctively different from the land. Because of space limitations in this book, however, the subsequent discussion is limited to land biota.

Approximately 80 percent of the world's land surface contains significant plant and animal life. The remaining 20 percent is subject to environmental extremes that are prohibitive to the survival of plants and of most animals (Figure 10.11). About half of this area is too cold for plants to carry on their life processes. These regions, primarily consisting of Antarctica and interior Greenland, are covered by snow and

ice throughout the year, so that they lack soil or even a bedrock surface from which plants could extract nutrients. Most of the remaining surface area devoid of vegetation is too dry and consists of almost lifeless desert. Most notable here are portions of the central Sahara, the Arabian Peninsula, and interior Asia. Perhaps an additional 1 or 2 percent of the land surface is lacking in natural vegetation because it is too steep, contains toxic surface materials such as excessive quantities of salt, or has been subject to recent natural disturbances that have temporarily removed the vegetation cover. Where vegetation is lacking, animal life is eliminated or severely restricted by harsh environmental conditions and a lack of food.

From the preceding paragraph, it can be deduced that moisture, temperature, slope, and soil are among the factors that most strongly influence the world distribution of life-forms. These factors can be grouped into five major categories of biological controls: climate, topography, migrational access, soil, and organisms. To these five natural controls can be added a sixth, which has risen to a position of great importance within the past few centuries; this is the influence of human beings. We now examine, in turn, the influence of each on the global distributions and types of plants and animals.

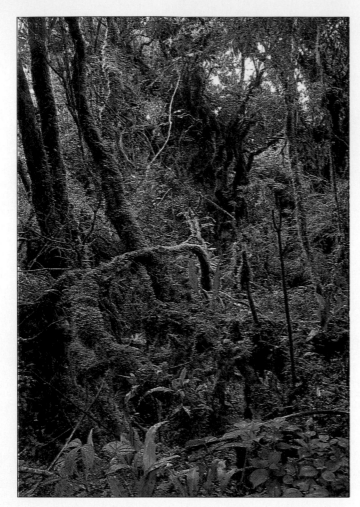

◆ **FIGURE 10.12A** Dense well watered "cloud forest" high on the slopes of Barva volcano, Braulio Carillo National Forest, Costa Rica.

SOURCE: © Greg Neise/Visuals Unlimited

◆ **FIGURE 10.12B** Sparse desert vegetation near Borrego Springs, eastern San Diego County, California.

SOURCE: Ralph Scott

Climatic Influences

Climatic characteristics, especially moisture and temperature, are the most important biogeographical controls. Climate particularly determines vegetation patterns, and vegetation, in turn, largely controls animal populations. Natural vegetation depends so strongly upon climate that the geographic patterns of climate and vegetation tend to be quite similar. Natural vegetation tends to be most abundant and diverse in regions where there are no major climatic stresses. Notable in this respect are the humid tropics. Climatic stresses, the most crucial of which are the occurrence of cold or dry seasons, greatly reduce the variety of species of both plants and animals. As a result, both the diversity and the abundance of life-forms generally decline as one proceeds from wet to dry and from warm to cold regions of the Earth.

Moisture

The availability of moisture is the single most important criterion controlling the global pattern of natural vegetation. Moisture availability is considered here as a climatic factor because it is determined largely by the balance between precipitation amounts and evapotranspiration rates, but it is also significantly influenced by soil and topographic factors and will be discussed again under those headings.

Water is essential to plants in a number of ways. Vital soil nutrients must be absorbed by plant roots and distributed throughout the plant in a dissolved state, and water is the necessary solvent. Water also is an essential reactant in the synthesis of the organic compounds (starches and sugars) manufactured by plants. In addition, the rapid transpiration of water during hot weather helps to reduce vegetative surface temperatures to tolerable levels.

The amount of soil water available is a vital factor in controlling the quantity of natural vegetation that a locality can support. Being literally "rooted to the spot," plants must rely on moisture that comes within reach of their root systems. The total biomass within an area normally increases with available moisture until a point of excess moisture is reached, after which the quantity of vegetation again diminishes. Perhaps the most striking characteristic of a desert is its relative lack of vegetation (Figure 10.12b), while the density and lushness of a well-watered tropical forest are equally obvious and impressive (Figure 10.12a).

Plants are sometimes categorized on the basis of their adaptations for coping with available moisture.[2] **Xerophytes** (from the Greek *xeros*, "dry" and *phytos*, "plant") are plants especially adapted to regions deficient in water. These regions are mostly arid deserts, but also include areas such as beaches or steep slopes, where the nature of the soil, rather than the climate, limits moisture availability. Some of the water-conserving characteristics commonly displayed by xe-

---

[2]It should be noted, though, that the definitions of these groups are not exact, that they permit considerable overlap, and that experts are not in complete agreement as to the geographic boundaries of the areas dominated by each group or even of their constituent plant members.

rophytes include extensive root systems, water storage ability, waxy leaves, thickened epidermal layers that reduce transpiration, small leaves or none at all (as in the case of some cacti), and the ability to minimize insolation receipts by turning leaves edgewise to the Sun. In addition, xerophytes tend to be small and to grow slowly.

Hygrophytes (*hygros,* "wet") are plants especially adapted to excessively moist environments. Because their environmental stress factor is the opposite of that for xerophytes, many of their physical characteristics are also directly opposed to those exhibited by xerophytes. For example, hygrophytes tend to have shallow root systems, large leaves with numerous stomata (pore spaces), thin epidermal layers, and the ability to turn leaves directly toward the Sun. In addition, they usually grow rapidly and may attain a large size when mature. Large size helps these plants compete for sunlight—an important consideration because vegetation is usually dense in areas where hygrophytes grow.

A third category of plants, the **mesophytes** (*meso,* "middle"), have physical characteristics intermediate between those of xerophytes and hygrophytes. They typically occupy well-drained habitats in areas of the low and middle latitudes with ample year-round precipitation and without excessively low temperatures.

A fourth category, the **tropophytes** (*tropos,* "to turn"), is defined somewhat more broadly than the other three. Tropophytes are adapted to alternating favorable and unfavorable climatic conditions for growth. These conditions generally consist of seasonally varying warm and cold periods, such as occur in the middle and high latitudes, or wet and dry periods, as experienced, for example, in the Tropical Wet and Dry and the Dry Summer Subtropical Climates. Because most areas experience at least one of these two types of seasonal alternations, the tropophytes are distributed more extensively than are the other three plant groups.

Most tropophytes are large, leafy plants commonly categorized as trees or shrubs. A **shrub** (or **bush**) is a woody plant with several permanent stems rising from the ground. A **tree,** on the other hand, normally is larger and contains a single primary stem or trunk that develops many branches. Many tropophytic shrubs and trees, especially those with broad leaves, are **deciduous;** that is, they lose all their foliage at the beginning of the unfavorable season. Conversely, most needleleaf bushes and trees are **evergreen;** they retain their foliage throughout the year. Tropophytes, whether deciduous or evergreen, are more versatile than xerophytes, hygrophytes, and mesophytes in that they are able to grow vigorously during favorable seasons and to protect themselves by becoming dormant during unfavorable seasons.

The availability of moisture has somewhat less direct and extensive effects on animals than on plants, but it nevertheless strongly influences faunal distributions and behavioral patterns. For example, amphibians such as frogs are much more common in humid climates than in dry climates because they must maintain moist outer skins, and most require water bodies for reproduction. Periodic alternations in moisture, such as the wet and dry seasons in tropical Africa, are largely responsible for the annual migrations of African big game animals. Wildebeest, zebras, and gazelles, in particular, follow shifting rainfall patterns as they seek fresh grazing land. By far the most important influence of moisture availability on animals, however, is its effect on plant life. Many animal species are characteristically or even exclusively associated with specific plant species or groups on which they rely for food and cover.

Temperature

The second dominant climatic control on the distribution of the Earth's biota is temperature (Figure 10.13). Plants vary

◆ **FIGURE 10.13** Contrasting vegetation assemblages growing in response to differing thermal conditions: (*left*) palm forest on a tropical beach in Sri Lanka; (*right*) Cold temperatures, a short growing season, and perennially frozen subsoils permit only a low growth of grasses and flowering plants in most subpolar regions. Shown here is a scene in northern Alaska's Brooks Range in early autumn.
SOURCE: *left:* M. Thonig/H. Armstrong Roberts; *right:* Greg Probst/Tony Stone Images

greatly in thermal tolerance, so only the coldest areas do not support vegetation. Some plant species are adapted to a broad range of thermal environments; others are restricted to a narrow range. Each plant species, though, has a maximum and minimum thermal limit beyond which it cannot survive, as well as an optimum temperature range within which it grows most vigorously.

Many plants outside the tropics require varying temperatures to initiate different crucial processes such as germination, fertilization, flowering, fruiting, and foliage growth. Because most plant growth stops when temperatures fall below 5°C (40°F), little vegetation exists in areas where temperatures never rise above this level. As temperatures increase (assuming other vegetative controls present no hindrance), growth rates and biomass concentrations increase to a maximum at about 35°C (95°F).

Of the two thermal extremes, cold has the most detrimental impact on vegetation. Tropical plants often have cold temperature limits well above the freezing point. Many plant species of the higher latitudes, conversely, can survive extended periods with temperatures far below –18°C (0°F). The most critical temperature for the survival of many plant species, however, is the freezing point. Numerous plants, including most flowers and garden vegetables that are familiar and important to residents of the middle latitudes and subtropics, are killed annually by the first period of subfreezing temperatures. In contrast, excessive heat rarely kills plants directly, and even the hottest surface locations on Earth can support vegetation. High temperatures do, however, greatly increase transpirational losses. This frequently causes plants to require more water than is available, resulting in wilting and possible eventual death.

In regions that experience the climatic stress of a cold season with subfreezing temperatures, plants must adjust their life cycles in order to survive. One group of plants, the **annuals,** do not have mechanisms that enable individual plants to survive the rigors of the cold season. Instead, they have a life cycle that is sufficiently short to enable them to mature in one warm season and produce the seeds that ensure the growth of the next generation of plants the following year. The amount of time they have to accomplish this task is termed the **growing season** (Figure 10.14). In the United States, this period generally is defined as the number of days between the last killing frost of spring and the first killing frost of fall. The length of the growing season tends to decrease with latitude and also with increas-

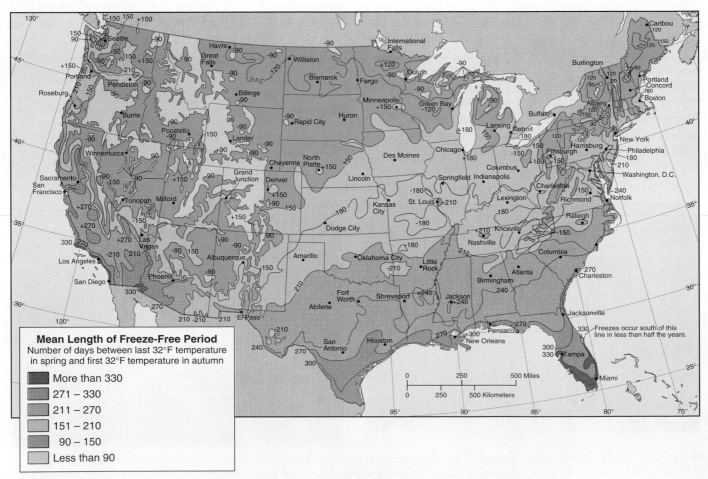

◆ **FIGURE 10.14** Variations in the length of the growing season within the coterminous United States.

ing elevation and continentality. As a result, the variety of annuals that can grow is increasingly restricted in these directions to types with short growing seasons. Most of our food crops (including cereal grains and vegetables but excluding tree crops) are annuals and must be replanted each year.

A second group of plants, the **perennials,** have multiple-year life spans. In areas with a climatic stress, they have developed protective mechanisms that enable them to survive as individuals. Often this involves a period of dormancy. Most broadleaf trees and shrubs, for example, lose their leaves and become dormant during autumn in the middle and higher latitudes, resuming their growth and producing new foliage when warm weather returns the following spring. Dormancy and a deciduous habit therefore protect plants from the stresses of a dry or cold season. Perennials have a tremendous advantage over annuals in terms of biomass, and therefore in spatial dominance, because they can add to their growth over an extended period. As a result, most natural vegetation groups are dominated by perennials; all the world's larger plant species are of this type.

Animal patterns and behaviors are also greatly affected by temperatures. Like plants, different animal species and groups are adapted to differing thermal conditions. For example, reptiles, which are cold-blooded, are numerous and varied in the lower latitudes, but decline rapidly in importance poleward. Conversely, mammals and birds, which are warm-blooded and are insulated with fur and feathers, respectively, are abundant in the higher latitudes. In general, the body size of warm-blooded animal species, as well as the relative proportion of exposed body surface area, tends to decrease toward progressively colder climate regions, apparently in order to conserve heat more efficiently.

The mobility of animals gives them a major advantage over plants in coping with environments with large daily or seasonal temperature ranges. For example, many desert animals are nocturnal, concentrating their surface activities during the night, while sleeping in underground burrows during the heat of the day. In colder climates, of course, this pattern is reversed, with the activity period concentrated in the warmer daylight hours. Other animals, especially birds, are capable of migrating long distances from wintering grounds in temperate or tropical regions to breeding grounds in higher latitudes (Figure 10.15). The longest migration route is taken by the arctic tern, which breeds in the Canadian Arctic and annually migrates to islands in the South Atlantic—a distance of nearly 20,000 kilometers (12,500 mi).

Light

Light is a third climatic necessity for most life-forms. Plants, as we have seen, use light as the energy source for photosynthesis. The competition for light also provides the stimulus for vertical growth in plants. In densely vegetated areas, the heights of the various plant species generally are associated closely with shade tolerance. Plant species frequently are excluded from a given habitat solely because of their inability to secure the required amount of sunlight. The changing availability of light is largely responsible for vegetative succession in an area recovering from some disturbance, such as a fire, that has removed the vegetation cover.

Although the amount of light available beneath a dense cover of vegetation is inadequate for most plant species, the intensity of light in an open setting is less important than its duration. The length of daylight varies significantly during the year outside the tropics. Like temperature, seasonal differences in daylight trigger biological activities such as budding, flowering, and leaf growth or shedding in many plants. They also control mating and migration patterns in many animal species.

Other Climatic Influences

Other climatic variables have a more limited, but often significant, effect on the makeup of a biological community. The relative humidity and the amount of cloudiness and fog all affect evapotranspiration rates, both on a short-term and a long-term basis. The presence of clear skies and dry air during the summer causes especially rapid water losses because these conditions are combined with warm temperatures and occur when plant and animal water requirements are at their peak.

Wind is another climatic element with both favorable and unfavorable effects on organisms. On the positive side, it increases the carbon dioxide intake of plants and is essential for pollen and seed dispersal for many species. It also aids in the movement and dispersal of many lightweight or flying animals. Conversely, the persistent high winds experienced along some exposed high latitude coasts or in mountainous areas can lead to excessive transpirational losses. Under these conditions, trees may not grow or they may be stunted or deformed (Figure 10.16). Strong winds in storms may blow migrating flocks of birds off course or blow insects out to sea.

Topographic Influences

Topographic influences on the biota range from the large scale to the extremely localized. The vegetation of the American Great Plains, for example, differs greatly from that of the adjacent Rocky Mountains largely because of the topographic diversity of these two large regions and the resulting variations in temperature and moisture conditions. On a much smaller scale, differing ecosystems frequently are encountered on different portions of a hill or between an area of sloping land and a nearby flat stream valley. The primary effect of the topography is to produce a variety of microhabitats for plants and animals. As a result, biogeographic patterns tend to be much more diverse in rugged areas than in areas of low relief.

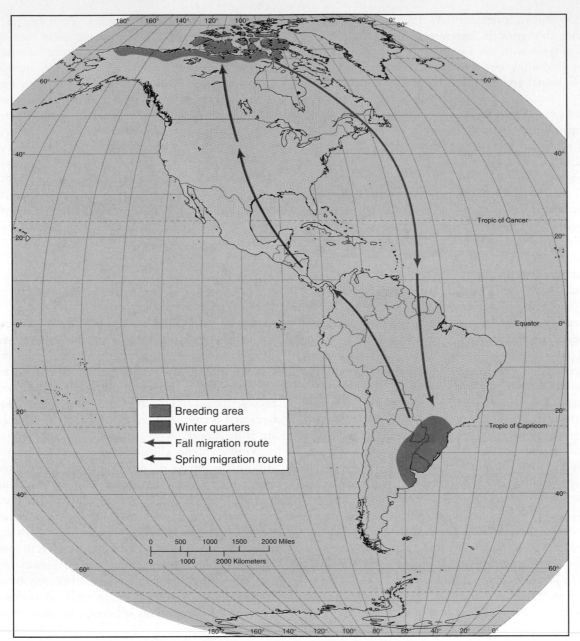

◆ FIGURE 10.15 The migration route of the golden plover extends from Arctic Canada to southeastern South America. Different routes are used for spring and fall migrations because of changing available food supplies.

The biota is influenced by a variety of topographically related factors. These include soil characteristics, insolation receipts, surface temperatures, and several factors relating to water availability such as precipitation amounts, evapotranspiration intensities, and runoff rates. Large-scale topographic influences are mostly climatic and are associated with the cooler and moister conditions that normally exist in mountain regions as compared with adjacent lowlands. The biological consequences of these climatic differences are discussed in the next chapter. Probably the two most impor-

tant localized topographic influences on vegetation are slope steepness and aspect.

Slope Steepness

Slope steepness affects vegetation primarily through its influence on water drainage and soil erosion. As slope angles increase, gravity becomes increasingly effective in causing excess water to drain rapidly downhill. The rapid runoff produces a high erosion potential for the slope, so that an

◆ **FIGURE 10.16** Stunted and wind-deformed trees at the tree line in the Colorado Rockies. Branches grow only on the downwind (east-facing) side of the trees.
SOURCE: Ralph Scott

inverse relationship exists between slope steepness and soil depth. Because they are likely to have formed recently, the thin soils covering steep slopes tend to be mineralogically immature and coarse textured. As a consequence, they typically have limited water retention abilities and may be subjected to extremes of heating and cooling. The net effect on the vegetation of steeply sloping terrain is a reduction in plant size and diversity, as well as a significant lowering of the total biomass. Steep mountain slopes have little soil and are likely to be devoid of vegetation or to contain a sparse cover of xerophytes.

Excessively flat topography, especially surface depressions, can also reduce the favorability of a habitat for vegetation. In low areas, poor drainage and waterlogging are likely. An excess of surface water leads to a lack of aeration, which hinders soil development and interferes with plant respiration. In dry climates, it may also result in the surface accumulation of salts. This subject is examined in the Focus Box beginning on page 300.

The optimum condition for plant growth generally is considered to be a very gentle slope with an inclination of about 2 or 3 degrees. This allows excess water to drain from the surface, but does not permit drainage to proceed so rapidly that it results in significant erosion.

Slope Aspect

The directional orientation or **aspect** of a slope influences its biota by controlling the duration and intensity of sunlight. Slopes inclined toward the Sun are substantially warmer during the day than are slopes inclined away from the Sun. Sunward slopes also have drier surfaces because evapotranspiration is greatly increased by the warmer temperatures and strong insolation. The presence of these contrasting microclimatic conditions may produce major differences in the nature of the habitat on different sides of the same hill (Figure 10.17).

Slope aspect has only minor significance near the equator because the Sun is high in the sky at midday and its path carries it into both the northern and southern sky at different times of the year. The significance of slope aspect increases with latitude as the angle of incidence of the Sun's rays becomes progressively lower and insolation is more concentrated on the equatorward-facing slopes. In the northern hemisphere, south-facing slopes have the warmer and drier conditions, while in the southern hemisphere this situation is reversed. East- and west-facing slopes have intermediate conditions, with west-facing slopes being somewhat warmer and drier in both hemispheres because daytime temperatures typically peak in the afternoon hours.

Accessibility Influences

Plants and animals that migrate, or are otherwise transported, to isolated locations usually will evolve within their

◆ **FIGURE 10.17** Vegetation differences resulting from variations in slope aspect in southwestern Colorado. The north-facing slopes on the left are the most densely vegetated.
SOURCE: J. K. Nakata: © http://www.sightandsound.com

restricted habitats. This causes these populations to become progressively more distinct from others of their kind, until they eventually form new subspecies and species. Thus, California, largely separated from the rest of North America by mountains and deserts, has its own distinctive flora and fauna; similarly, the biological communities that have developed on many isolated islands or island groups are unique. The number of species occupying an isolated landmass depends on several factors, including the size of the landmass, its diversity of habitats, and its accessibility to sources of plant and animal colonists, as well as the biological richness of these sources.

The Hawaiian Islands serve as a prime example of a large and extremely isolated island group that has developed a restricted and unique flora and fauna. These islands contain no true freshwater fish, no amphibians, reptiles, or mammals (except for human beings and one species of bat). Seed plants apparently were introduced largely via the digestive systems of birds; a few seeds, such as coconuts, arrived by water. Of the 1,729 known species and varieties of Hawaiian seed plants, 40 percent are of Indo-Malaysian origin, 18 percent of American origin, and the rest are of local or uncertain origin.

Many other isolated islands have developed distinctive biological assemblages. These include Madagascar, with its lemurs, and the Galápagos Islands off the coast of Ecuador, with their iguanas and giant tortoises.

Australia has also long been isolated by water bodies from the mainstream of evolutionary development and thus has its own unique, and relatively primitive, assemblage of plants and animals. More than 90 percent of the continent's native trees are varieties of a single genus, Eucalyptus. Likewise, most Australian mammals are marsupials, such as kangaroos and koalas, which carry their young in an abdominal pouch (Figure 10.18).

In similar fashion, isolated mountainous areas or even single mountain peaks may develop their own distinctive biological communities. Some of these communities are holdovers from previous colder climatic conditions. For example, dozens of plants and animals that were widely distributed in Europe during the Pleistocene epoch, when glacial and tundra conditions existed over much of the continent (see Chapter 17), are found today in only a few locations within the Alps.

Soil Influences

Soil characteristics affect the biota on a variety of scales, the smallest of which often are very localized. The influence of soil on the distribution of plants is closely related to the effects of climate and topography. Some of the numerous attributes of the soil that can affect vegetation include the size of the soil particles, the tendency of these particles to form larger aggregates, the compactness of the soil, its mineral and organic matter content, its water-retention ability, and its acidity or alkalinity. These factors are dis-

◆ **FIGURE 10.18** Marsupials such as kangaroos carry their young in abdominal pouches. Shown here is a gray kangaroo and her "joey."
SOURCE: © Mike James/Photo Researchers

cussed more fully in Chapter 12; here we briefly discuss the vegetative influence of the soil's mineral content and the effect of soil water on the availability of soil nutrients to plants.

Plant Nutrient Availability

Plants obtain the water and most nutrients they require from the soil. The nutrients are absorbed through the plant roots after being dissolved in soil water. It therefore is essential to vegetation that the soil not only contains the needed minerals, but that it contains adequate, but not excessive, quantities of water.

Mineral nutrients required by plants are derived from the two solid soil-forming constituents—bedrock and organic matter. Plant species differ in their nutrient requirements, so that soils formed in areas with different types of bedrock may encourage the growth of different vegetative assemblages. This, in turn, will help determine the composition of the local animal population. The **fertility** of the soil, or its productivity with respect to vegetation, is a somewhat relative term, because a given soil may contain adequate nutrients for one plant species but not for another. Some types of rock, however, are noted for forming soils of generally higher fertility than are others.

Assuming that the important plant nutrients are present in the soil, the ability to support vegetation is still critically dependent upon the presence of sufficient water to dissolve the minerals. Soils in arid environments are often potentially fertile, but they receive insufficient moisture to dissolve the available nutrients; hence their ability to support vegetation is highly limited. Conversely, soils in excessively wet climates frequently are infertile because most soluble minerals have been removed by the downward percolation of soil water.

Biotic Influences

The joint occupation of most areas by a diversity of plant and animal species modifies the environment in a variety of ways. The presence and activities of other life-forms increase the favorability of the habitat for certain plant species. For example, the shade provided by a forest helps conserve soil moisture and affords protection from sunlight for many shade-loving plants of the forest floor. In addition, various plants utilize the mobility of animals for pollination and seed dispersal. Organic matter produced by the decomposition of vegetation also releases nutrients to the soil for use by new generations of plants.

On the other hand, the presence of numerous plant and animal species within a region of limited size and resources necessarily results in competition for survival. Even if the climate, topography, and soil type are all favorable for a given species, it cannot survive within a given area unless it can successfully occupy a habitat despite the likely presence of competitors. Of all the species with access to a given habitat within a biological community, only those best adapted to the environment will be successful in occupying it. If a species is out-competed by other plants or animals within an area, it will be eliminated from that area. If a species is out-competed everywhere, it will become extinct. The removal of organisms less able to compete makes room for species that are better suited to survive within a given environment. It is the means by which biological diversity and ultimate advancement occur. The fossil record indicates that animal species persist on Earth for an average of about 2.7 million years. In the last 600 million years, perhaps one billion animal species have lived on Earth, but only a fraction of these—some 5 to 30 million species—are alive today. Some of the major causes of extinctions are competition, evolution into other species, and loss of food sources. There is also mounting evidence that mass extinctions have occurred on several occasions in Earth's history as a result of major natural disasters. For example, many researchers believe that the collision of a giant meteorite with the Earth some 66 million years ago produced a temporary global cooling that caused the dinosaurs and many other life-forms to become extinct.

This event, they believe, paved the way for the ascendancy of the mammals.

Human Influences

The biosphere is the most sensitive of the Earth's four environmental spheres and consequently is the one that has been altered most profoundly by human activities. The majority of the global ecosystems discussed in the next chapter are currently in a state of instability because of both direct human disturbance and climate changes resulting from global warming. As discussed in the Case Study at the end of Chapter 4, however, even the current global warming trend may be partially attributable to human actions. Ecosystem disturbances produce rapid change, altering the composition of an ecological community or even resulting in its destruction.

A wide variety of human activities can produce major ecosystem disturbances. A complete analysis of this important topic lies beyond the scope of this book, but a few of the most important categories of human disturbance of the biosphere are briefly discussed in the following paragraphs.

Habitat Destruction and Fragmentation

The nearly 6 billion human beings have the highest biomass of any life-form on Earth. We require vast amounts of space for our homes, businesses, transportation and recreation facilities, and food-producing operations (Figure 10.19). This space is taken from areas that formerly comprised the

◆ **FIGURE 10.19** Vast amounts of real estate are occupied by large cities, leaving little or no trace of the ecosystems that originally occupied these areas. Shown here is Hong Kong, one of the world's largest and busiest ports.
SOURCE: Doug Armand/Tony Stone Images

Some Conflicts in Management of National Parks in the United States and Canada

National parks in both the United States and Canada are meant to be maintained and used for human enjoyment and yet preserved for future generations. These apparently contradictory goals of use and preservation have been in conflict ever since the park systems were first formed (1872 in the United States and 1885 in Canada). National parks in the United States were originally established to preserve sites of geological wonder. But before Congress would allow a park to be formed, it had to be assured that the lands were economically useless. In the same year that Congress established Yellowstone, it passed the General Mining Law of 1872, giving anyone who discovered a hard mineral on federal land the right to stake a claim to the land and to use that land in any manner, even for nonmining uses. National park lands were not excluded from the law until its revision in 1976. Mining continues to this day in some parks. Although new claims may not be made, existing ones, numbering perhaps in the thousands, must be honored or negotiated. Formerly mined areas are hazardous, and their by-products can pollute the water and disturb animals and plants. Attitudes and laws affecting wildlife preservation have also changed. For example, hunting is now prohibited in all the parks, but had been allowed in Yellowstone in its early years.

The early Canadian national parks were established to protect natural resources and to encourage economic development of the West. Prime Minister John MacDonald recognized the importance of developing the West and used the railroads and the national parks to further this aim. For MacDonald, Canada's need was not for more wilderness, but for economic development. The railroads were active supporters of the parks; they built lines through them and frequently removed such resources as timber and minerals. From the 1880s to the early 1900s, national parks, including Banff, Yoho, Mt. Revelstoke, Glacier, and Jasper, were formed as the railroads moved west across the Canadian Rocky Mountains. Townsites were also built within some of these parks, and today remain part of the parks.

The conflict between preservation and use continues today. In both the United States and Canada, the same parks that contain vast wilderness areas also often have restaurants, hotels, gasoline stations, and ski trails and lifts. In Canada, parks containing towns have permanent residents and the associated economic activities as well.

Although wildlife is protected within the parks, problems can develop. Visitors are usually delighted to see a deer or a mountain sheep up close, and some feed the animals even though it is prohibited. Animals thus come to campgrounds and cars seeking food, but they are still wild and occasionally harm someone. In extreme cases park rangers have to shoot an animal to prevent it from hurting visitors.

In the towns within Banff and Jasper National Park, animals forage in streets and yards, generally pleasing tourists, but displeasing homeowners whose plants may be eaten or damaged (Figure 10.20). At certain times of the year, unwary visitors may be hurt by animals protecting their young.

For some species, such as wolves, wildlife management has become an extremely sensitive issue. In an effort to restore a truly balanced ecosystem, Yellowstone National Park has considered reestablishing wolves, but has encountered opposition from ranchers near the park who are concerned that the wolves will eat their livestock. The wolves had been exterminated by rangers in 1915 as a menace to such animals as elk, deer, and mountain sheep. Since those early years, the ecology of the park has changed, and the current wildlife may not be sufficient to support the reintroduced wolves. The neighboring national forests could function as wolf habitat, but only if profitable timber operations were halted.

In contrast, wolf management in Great Smoky Mountains National Park has had greater success. The red wolf was reintroduced in 1992, as part of a U.S. Fish and Wildlife project begun in 1974. The red wolf had faced extinction in the United States because of changes in land use and predator activities. Many of the remaining red wolves were believed to be interbreeding with coyotes. Great Smoky was selected because its large wilderness area makes it one of the few places in the Southeast where red wolves can

continued on next page

breed successfully. The project appears to have been effective, and the organizers foresee the release of a few additional pairs of wolves each year, until the carrying capacity of the park, 50 to 70 animals, is reached.

Great Smoky Mountains National Park has also been used to reintroduce other endangered species, including the peregrine falcon starting in 1984 and the river otter starting in 1986. Peregrine falcons have been fledged in the park, but no nests have been located. The river otter project appears to have been successful in developing a viable population.

The combined presence of automobiles and animals in both Canadian and U.S. parks results in many animals being killed and cars being damaged, even though speed limits are posted and visitors are warned about animals on the roads. Banff National Park has built fences along the portion of the Trans Canada highway that runs through the park, as well as special underpasses so that the animals may cross the road. Park personnel have also found that when visitors are made especially aware of the presence of wildlife (through intensive use of signs and other means), fewer animals have been killed.

Vegetation management in parks varies as well. The management goals for some species appear straightforward, but for others, the goals become diffuse. In general, the U.S. Park Service allows native species to flourish within the natural balances of the park and tries to protect the park from nonnative species. Some parks, such as Shenandoah and Great Smoky, have been invaded by nonnative insects that are destroying trees, including eastern hemlocks and spruce fir, along with

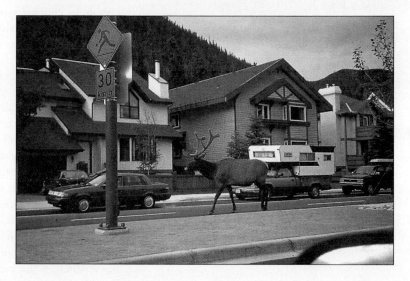

◆ **FIGURE 10.20** An elk looking for breakfast in the townsite of Banff National Park.

SOURCE: Margaret F. Boorstein

other parts of the forest ecosystem, including rare bryophytes and invertebrates. The Park Service has tried some expensive and labor-intensive techniques to protect the trees, but as funds have run out, the Park Service has decided to let nature run its course. For other trees, such as the American ash, the cause of damage may not be obvious. No one is sure whether an insect (the European mountain ash sawfly) or air pollution is responsible for the deaths of numerous trees.

In Hot Springs National Park in Arkansas, a completely different plant-management policy is followed. Exotic plants are deliberately planted and encouraged to flourish. The Park Service wants to recreate the atmosphere of a nineteenth-century mineral spa.

Over the past few decades, to protect park wilderness from being destroyed by human beings and their activities, vast areas of the parks have

been declared wilderness areas, to be kept completely free of human interference. In Canada, four mountain national parks, Banff, Jasper, Kootenay, and Yoho, now comprise the Rocky Mountain World Heritage Site, one of several areas around the world affording special natural protection. Without national parks, much of the natural beauty of Canada and the United States would not have been protected from encroachment. Management of national parks in the United States and Canada continues to require decisions regarding the proper balance between use and preservation. ∎

This box was written by Margaret F. Boorstein, Department of Earth and Environmental Science, C. W. Post College, Brookville, New York.

◆ **FIGURE 10.21** The rolling countryside of Central Washington state's Palouse region is one of many extensive areas presently almost entirely devoted to agriculture.

SOURCE: Bruce Hands/Tony Stone Images

habitats of natural ecosystems. Habitat removal and fragmentation are causing the rapid decline of many species and threatening the continued existence of many others. For example, waterfowl in North America are rapidly disappearing due to wetland reclamation. The Atlantic flyway has lost half its coastal wetlands in the last 35 years, and the number of breeding ducks and geese has dropped from 45 million to 31 million over the past 15 years.

Huge portions of the Earth's surface have been converted to agricultural production, an activity that may totally remove the original natural vegetation over extensive areas (Figure 10.21). For example, the state of Illinois once

◆ **FIGURE 10.22** Fire is being employed in this sugar cane field in Mozambique to clear vegetation and provide soil nutrients.

SOURCE: Brian Seed/Tony Stone Images

contained 145,000 square kilometers (60,000 mi²) of wild grassland. Currently, only a few patches totaling about 10 square kilometers (4 sq mi) remain. Even in areas that appear to contain natural vegetation, such as the forests of eastern North America, lumbering is likely to have occurred several times in any one locality since the time of initial European settlement. Ecosystem disruptions caused by agricultural activities are perhaps most severe in the world's less-developed countries, where human population growth rates are exceptionally high. Case studies of the resulting problems of deforestation and desertification can be found at the ends of Chapters 11 and 18, respectively.

Many societies use fire to clear land for agriculture and rapidly inject nutrients into the soil (Figure 10.22). As noted in the next chapter, this practice may have resulted in the large-scale replacement of forests by grasslands in drier environments within both the low and middle latitudes. The suppression of natural fires can also alter ecosystems. Decades of forest fire suppression in the North American West has resulted in the buildup of forest litter. This has gradually altered the composition of the forest ecosystems and has substantially increased the danger of uncontrollable fires such as those that devastated Yellowstone National Park in the summer of 1988.

Introduction of New Species or Foreign Substances

The human introduction of new life-forms or of foreign substances such as pollutants or pesticides can have a devastating impact on ecosystems that must suddenly cope with a new set of environmental conditions. Hundreds of examples of ecosystem disruptions resulting from the introduction of new or "exotic" species of plants or animals could be given. A few well-known examples from North America include the "chestnut blight" and Dutch elm disease that have largely removed the once abundant North American sweet chestnut and elm trees from the eastern woodlands. The starling and the English sparrow, two of the most common birds in North America, are introduced species. Currently, the spread of the Africanized "killer bee," the gypsy moth, the fire ant, the Japanese beetle, the Mediterranean fruit fly, and other introduced insect pests are of major concern in parts of North America.

Atmospheric pollutants such as soot, ozone, lead, and fluorides can kill sensitive plants and have greatly increased the acidity of precipitation in industrial areas, particularly in Europe and eastern North America. The Case Study at the end of this chapter deals with the problem of acid rain, while the Case Study at the end of Chapter 3 discusses the subject of air pollution.

The preceding paragraphs are not meant as a condemnation of the effects of human activities on the environment in general, for we have all benefited from the results of these activities. Indeed, it would have been impossible for humanity to have reached its current population and level of civilization without transforming the environment in a major way. In essence, we are gradually reorganizing our planet's surface to one that maximizes its ability to support human beings, and this process shows every sign of continuing to accelerate. Nevertheless, with the power over nature that we have developed, we must assume the role of stewards of the environment. We should therefore undertake future changes with due concern for the countless other life-forms that are the co-inhabitants of our environment.

During the past century, the role of national governments in preserving and protecting sensitive ecosystems, particularly in areas of great natural beauty, has greatly expanded. Both Canada and the United States, for example, have established systems of national parks that have become extremely popular tourist destinations in recent decades. The challenges North American park managers face in balancing environmental preservation with large-scale public use are explored in the Focus Box on pages 259–260.

▌SUMMARY

Plants and animals occupy about 80 percent of the Earth's land surface. Through the processes of mutation and natural selection, a great variety of life-forms adapted to a diversity of environmental conditions have evolved. Given a stable environmental setting, a given habitat will develop a biological community that will utilize the available space and resources to its maximum advantage. When a community is disturbed or removed, life-forms will generally return to the site in an organized successional sequence, eventually reaching ecological stability when a climax assemblage is achieved. The combination of human activities and recent global-scale climatic changes, however, has prevented stable climax communities from currently existing in most areas. Biogeographers are concerned with the organization of biological communities and with the factors that control the geographical distribution of the Earth's biota. In order to summarize this information, we return to the Focus Questions posed at the beginning of the chapter.

1. *What are ecosystems and how do they function?*
   Ecosystems are functioning biological communities. They consist of all the life-forms within the community as well as the environment in which and with which they interact. Each ecosystem member occupies a particular habitat and performs a specific function that contributes to the functioning of the community as a whole. The energy source upon which the ecosystem depends is derived from sunlight, which is converted to carbohydrate energy by the ecosystem's producers, or plants, through the process of photosynthesis. The net carbohydrate energy supplied by all the plants within a biological community is referred to as its net primary productivity. This energy then passes, in a food chain, to the ecosystem's animal consumers (first herbivores

and then carnivores), and eventually to the decomposers. At each step in the food chain, however, large amounts of biomass and energy are returned to the environment. As a result, a large mass of plants is required to support the carnivores at the top of the food chain.

2. *What are the chief environmental controls on the global distribution of plants and animals?*
   The chief environmental controls on the distribution of the biota are climate, topography, migrational access, soil, organisms, and human activities. Climate is probably the most critical because it influences moisture availability and temperature. Moisture availability influences such vegetative characteristics as plant size, root development, transpiration rates, leaf number and size, growth rates, and regional biomass. Temperature is especially important for its influence on the length of the growing and mating seasons, vegetative growth rates, and leaf shedding. Topographic influences often are localized. In general, areas of steep slopes support biological assemblages of reduced complexity, and poleward-facing slopes support life-forms adapted to cooler and moister surface conditions. Geographical barriers such as oceans, deserts, and mountain ranges determine the degree of migrational access to a region. Isolated habitats such as islands tend to have biological communities that are more primitive and limited in variety than those with good access to other regions. A number of soil characteristics are important to plants and to the animals that feed on them. Crucial are soil depth, mineral content, and water retention ability. Organisms must successfully coexist within a given habitat, or ecosystem alteration will occur. The interaction between the life-forms inhabiting an ecosystem is highly complex, but an overriding factor is the competition for available resources. Within the past few centuries, human beings have developed the capacity to alter and destroy ecosystems on a regional scale. As a result, ecosystems in many areas are under stress and are changing rapidly, and their futures are in doubt.

▌QUESTIONS FOR THOUGHT AND REVIEW

1. With what aspects of the biosphere is the geographer particularly concerned?
2. How do plants produce food? In what regions of the world is net primary productivity highest? Explain why.
3. What is an ecosystem? How are the various life-forms in an ecosystem related to each other?
4. What is a food chain? Explain why it is sometimes referred to as a "food web" or a "food pyramid."
5. What are the major abiotic components of an ecosystem? What vital function does each perform?
6. Explain how each portion of the world happens to be occupied by life-forms that are adapted to the local environmental conditions.

7. By what mechanisms have individual plant and animal species occupied their current ranges? Why are the ranges of some life-forms larger than others? What are the major types of barriers to the dispersal of life-forms?

8. What is a climax biological community? How does the concept of vegetative succession relate to this concept?

9. List the five major "natural" biological distribution controls. Which one has the greatest effect on world plant and animal distributions?

10. What environmental factors other than the amount of precipitation influence the quantity of moisture available to plants?

11. List several structural differences between the characteristics of xerophytes and hygrophytes.

12. Would you expect to encounter greater diversity in the flora and fauna of a level region or of a hilly region? Explain several reasons for this.

13. Discuss some of the ways human activities in your own area have influenced the distributions of plants and animals.

▌KEY TERMS

biogeography	barrier
photosynthesis	xerophyte
primary production rate	hygrophyte
net primary productivity	mesophyte
habitat	tropophyte
biological community	deciduous
ecosystem	evergreen
vegetative succession	annual
climax community	perennial
food chain	acid rain
biogeochemical cycle	

The Environmental Consequences of Acid Rain

The phenomenon of **acid rain** has become a major environmental concern in recent years. It is a complex problem that is far from completely understood but seems to be caused largely by human activities. Natural factors, including soil and bedrock characteristics and wind and precipitation patterns, also influence the severity of this problem in some regions.

The acidity of precipitation is measured by the pH scale, a logarithmic scale of acidity or alkalinity that is discussed in greater detail on page 299. Rainfall is naturally slightly acidic because it reacts with dissolved carbon dioxide to form carbonic acid. This acid, in turn, dissociates to produce the hydrogen ions that are measured by the pH scale. In addition, the atmosphere contains suspended particles from a number of sources (including volcanoes and soil bacteria) that can dissolve in rainwater to form acids. Although natural precipitation has a pH of approximately 5.6, acid rain may have pH values below 3.0, making it a corrosive substance that can be toxic to plant and animal life.

A number of categories of human activities have acted to substantially increase the acidity of precipitation, particularly in densely populated and industrialized regions. Especially important are the large-scale burning of coal by electric power–generating plants and steel mills and the combustion of gasoline by motor vehicles. These activities release large quantities of sulfur dioxide and nitrogen gases that unite with oxygen in the presence of sunlight, forming sulfur trioxide (SO_3) and nitrogen dioxide (NO_2). The sulfur and nitrogen oxides may be deposited on the Earth's surface as a thin layer of acidic dust. Alternatively, they may react with atmospheric water vapor to form sulfuric and nitric acids that are eventually carried to the surface by precipitation downwind of their sources.

In many areas, the effects of acid precipitation on the soil are not a concern because the surface is naturally buffered by the presence of lime or other alkaline substances in the rock or soil. In other areas such as the northeastern United States, southeastern Canada, and much of Europe, however, the presence of numerous coal-burning industries, coupled with naturally acidic rock and soil types, has resulted in a serious acid rain problem. Globally, an estimated 6.4×10^{10} kilograms (70 million tons) of sulfur are released each year into the atmosphere by human activities. The United States alone annually releases about 20 million tons of sulfur dioxide, mostly from coal-fired power plants and vehicles.

Several major adverse environmental consequences of acid rain have been widely reported. In large portions of Scandinavia and Germany, needleleaf forests have been slowly dying. Needles near the tops of the trees turn yellow and fall off, and eventually the entire tree may be killed. In Germany, this phenomenon was first noticed in the Black Forest in the early 1970s, and by 1990 more than half of the nation's trees were affected. In the republics of eastern Europe, where government regulations of factory emissions have long been highly limited, at best, entire forests are reported to have been destroyed. Forest dieback also has taken place recently in portions of the Appalachian Mountains of the eastern United States (Figure 10.23). Although the exact mechanism by

continued on next page

◆ **FIGURE 10.23** A spruce forest nearly destroyed by acid rain on Mt. Mitchell, North Carolina.

SOURCE: © Joseph L. Fontenot/Visuals Unlimited

◆ **FIGURE 10.24** pH values of acid rain in the midwestern and eastern United States in 1987.

which the trees are killed by acid precipitation remains uncertain, it is believed that the acidification of the soil kills organisms that participate in the soil-plant nutrient cycle and also mobilizes potentially toxic quantities of aluminum and heavy metal ions in the soil that subsequently are absorbed by the trees.

Recent reports based on forest studies in eastern Germany indicate that the growth of some trees is so inhibited by acid precipitation that their ability to absorb needed nitrogen has been greatly reduced. Nitrates are therefore entering and passing through the soil in large amounts. Because the nitrate (NO_3) ion is negatively charged, its passage through the soil removes positively charged nutrients such as calcium and magnesium that are essential for soil fertility and for forest growth. The loss of these nutrients could lead to even more rapid forest declines in the near future.

Acid rain has also long been blamed for the destruction of fish and other aquatic life in New England, upstate New York, eastern Canada, and Scandinavia. Many lakes within these areas, which previously supported large fish populations, have been virtually sterilized.

A cultural concern resulting from acid rain has been the damage suffered by numerous priceless works of art in Italy, Greece, and several other countries. Although acid rain can corrode metals, fade paints, and even etch glass, marble statues are especially affected because the acid chemically converts the marble to calcium sulfate, which is soft and soluble. This process, when combined with the effects of freezing water, causes the outer portions of the marble to be shed in layers, allowing progressively deeper penetration of acidic water.

The release of sulfur dioxide by power plants in the American Midwest has been much criticized by residents of the Northeast and adjacent southeastern Canada, who have experienced severe acid rain problems in the last few decades (Figure 10.24).

The Midwest contains a large concentration of coal-burning electric power plants. Midwestern coal is plentiful and easy to extract, but it unfortunately has a high sulfur content. In an effort to meet U.S. Environmental Protection Agency regulations, these plants are in the process of installing expensive scrubbing systems that reduce the sulfur content of the coal before it is burned and are also increasingly importing low-sulfur coal from the western United States. The complete abandonment of the productive, high-sulfur coal deposits, however, would have a devastating impact on the economy of many communities. Further reductions in the use of this coal are being vigorously resisted as long as any doubt exists about the degree to which its combustion affects acid rain contamination farther to the east.

The acid rain controversy illustrates an important environmental dilemma that societies face to some extent in the pursuit of nearly all major economic activities. It involves the "acceptable" degrees to which we should expect the natural environment to be degraded as a consequence of our actions. Some degradation is inevitable, no matter how careful we are. The minimization of the environmental consequences of human activities, however, although frequently viewed as highly desirable, can be tremendously expensive both monetarily and in terms of human livelihood.

11 DISTRIBUTION AND CHARACTERISTICS OF TERRESTRIAL BIOMES

OUTLINE

FOREST BIOMES
Tropical Rainforest
Tropical Semideciduous Forest
Tropical Scrub Woodland
Subtropical Broadleaf Evergreen Forest
Mediterranean Woodland and Scrub
Mid Latitude Deciduous and Mixed Forest
◆ FOCUS BOX: *The Gypsy Moth Invasion of North America*
Mid Latitude Evergreen Forests
Northern Coniferous Forest

GRASSLAND BIOMES
Savanna
Steppe
Prairie
DESERT BIOMES
◆ FOCUS BOX: *The Mesquite Invasion of the American Southwest*
TUNDRA BIOMES
MOUNTAIN BIOMES
◆ CASE STUDY: *The Destruction of the World's Tropical Rainforests*

FOCUS QUESTIONS

1. How are the complex global patterns of plants and animals organized for purposes of geographical analysis?
2. What are the locations and biological characteristics of the world's major biomes?

Plants and animals can be divided geographically into communities that are usually delineated on the basis of their dominant forms of natural vegetation. Four global-scale communities, or **plant associations,** can be recognized. Each of these communities—forest, grasslands, desert, and tundra—may be further subdivided into smaller units called **biomes** based on the structure and leaf characteristics of their dominant plants. The world's major biomes are listed in Table 11.1, and their locations are depicted in Figure 11.1. The discussion will give primary emphasis to the vegetative characteristics of each biome, but their faunal assemblages will be noted as well. As we saw in the last chapter, faunal distributions are much less clear-cut than are floral distributions, largely because of the greater mobility of animals. The major forest and grassland biomes are sufficiently large and important to merit individual treatment.

The global distribution of plants and animals is complex and varied when examined in detail. Such complexity makes any limited classification system such as the one used in this chapter merely a convenient abstraction of reality. In actuality, there is no set number of biomes, and the character of an area's life-forms normally does not suddenly change when "boundaries," such as those shown in Figure 11.1, are crossed. It can be argued that biogeographical regions are devised more to fit our intellectual limitations than they are to delimit natural boundaries. Moreover, human activities have profoundly altered the biological composition of many areas. Indeed, in discussing the natural vegetation and wildlife of large parts of the world, we are often dealing with past biological conditions, not present-day reality.

However inadequate global-scale biomes may be in describing the complexity and rapid change that currently exists, their formulation does serve useful purposes. It indicates that life-forms differ greatly on a regional basis and facilitates a generalized discussion of these differences. In addition, a geographic analysis of biomes emphasizes the crucial fact that the global distribution of plants and animals is not random, but instead is closely related to other environmental parameters, especially patterns of temperature and available moisture.

In the following discussion of life-forms, we attempt to provide an element of geographical continuity by proceeding in a general poleward direction and by stressing the factors responsible for the basic changes in characteristics from one grouping to the next. We begin with the forest association, which not only dominates the equatorial regions, but is the most areally extensive and important association from both a biological and a human occupancy standpoint.

FOREST BIOMES

A **forest** is an assemblage of trees that grow sufficiently close together for their foliage to overlap and largely shade the ground for at least a portion of the year. Trees are not the only type of vegetation in a forest, but they are the dominant form.

Trees far exceed all other terrestrial life-forms combined in total biomass. The key to their success in the competition with other forms of vegetation is their size, which has enabled them to secure the sunlight they need and to shade out rivals. In essence, forests dominate wherever the soil, the topography, and especially the climate allow their growth. They are, however, subject to two important climatic limitations that exclude them from large areas and thereby allow other forms of vegetation to dominate. The first is that trees need a warm growing season of reasonable length. Even the most cold-tolerant forests cannot grow where the warmest month has a mean temperature below 10°C (50°F). The second and more geographically important limitation is the large water requirement of forests. Because they are large perennials, trees need substantial moisture during the growing season and at least some water even when they are dormant. This restriction excludes forests from extensive areas with perennial or seasonal soil moisture deficiencies, including most of the Low Latitude and Mid Latitude Dry Climates, the Tropical Wet and Dry Climate, and portions of the Dry Summer Subtropical Climate (see Figure 11.2 on page 270).

Trees commonly are classified on the basis of leaf characteristics. **Broadleaf** trees, as the term implies, have leaves of

◆ **TABLE 11.1** Major Terrestrial Biomes[1]

I Forest
   A. Tropical rainforest
   B. Tropical semideciduous forest
   C. Tropical scrub woodland
   D. Subtropical broadleaf evergreen forest
   E. Mediterranean woodland and scrub
   F. Mid latitude deciduous and mixed forests
   G. Mid latitude evergreen forest
      1. West coast coniferous forest
      2. Southern pine forest
   H. Northern coniferous forest
II. Grassland
   A. Savanna
   B. Steppe
      1. Low latitude steppe
      2. Mid latitude steppe
   C. Prairie
III. Desert
IV. Tundra

[1]Roman numerals precede plant associations; letters precede biomes. Because of their diverse nature, mountain biomes do not fit into this classification.

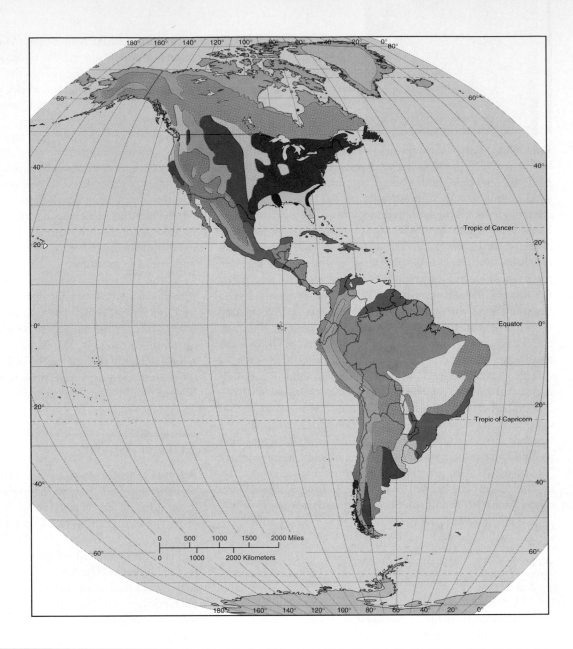

considerable width, while **needleleaf** trees have leaves that are very narrow and sometimes pointed. Many species of needleleaf trees bear seeds in cones and therefore are described as **conifers.** Familiar middle latitude examples of broadleaf trees include oak, maple, and hickory, and well-known needleleaf trees include pine, spruce, and fir. The broadleaf trees, as a group, are more complex and biologically more advanced than the needleleaf trees and, under favorable environmental conditions, hold the competitive edge because of their higher shade tolerance. As a result, they are the more widely distributed of the two groups. On the other hand, the needleleaf trees are more tolerant of cold climates, drought, and infertile soils; therefore, they gain a competitive advantage when any of these conditions prevail. Trees and shrubs are also classified as either deciduous or evergreen, depending on whether they undergo a seasonal loss of foliage. The use of these two pairs of antonyms permits trees and shrubs to be separated into four groups in

a simple classification system. These groups are broadleaf evergreen, broadleaf deciduous, needleleaf evergreen, and needleleaf deciduous.

For the purposes of this book, we divide forests into eight biomes based on their structure and leaf characteristics. Most of the forest biomes exist in several widely separated areas of the world that have similar climates. This underscores the dominance of climate as the major natural vegetation distribution control. The following descriptions are of climax or near-climax biological assemblages in areas undisturbed by human activities.

Tropical Rainforest

The tropical **rainforest** is the most equatorward and the most extensive forest biome. It covers approximately 6 percent of the Earth's surface and contains up to four million

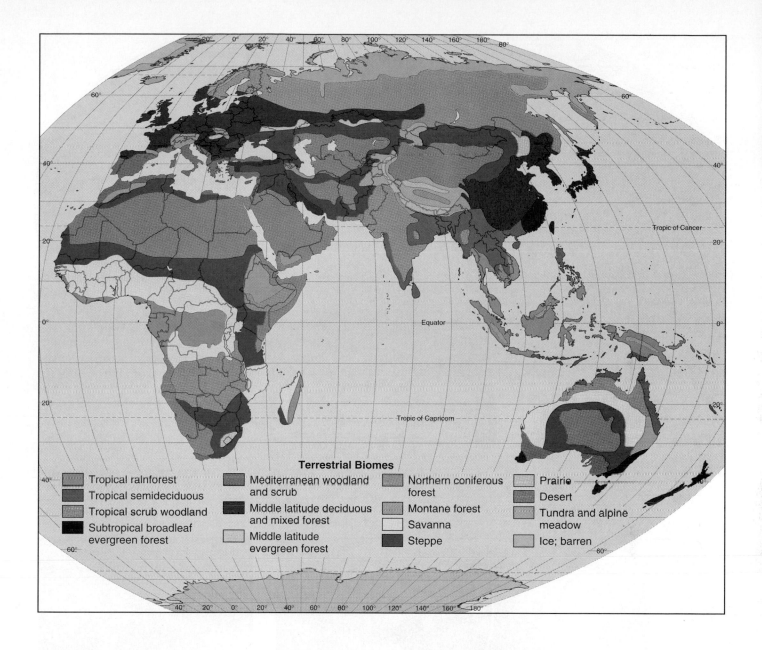

Terrestrial Biomes

- Tropical rainforest
- Tropical semideciduous
- Tropical scrub woodland
- Subtropical broadleaf evergreen forest
- Mediterranean woodland and scrub
- Middle latitude deciduous and mixed forest
- Middle latitude evergreen forest
- Northern coniferous forest
- Montane forest
- Savanna
- Steppe
- Prairie
- Desert
- Tundra and alpine meadow
- Ice; barren

plant and animal species, which collectively comprise the majority of the world's total biomass.

The geographical distribution of the tropical rainforest coincides closely with that of the Tropical Wet Climate (compare Figures 8.3 and 11.1). The largest areas with this forest type are the Amazon Basin of northern South America, the Congo (Zaire) Basin of central Africa, and Southeast Asia and the islands of the western Pacific.

A highly favorable climate for vegetative growth is responsible for the lushness of the rainforest (Figure 11.3). The absence of both a cold season and a lengthy period of drought means that no major climatic stresses are present. As a result, a twelve-month growing season exists, vegetation is dense, and the structure of the forest is determined primarily by the competition for light.

The tropical rainforest is characterized by an amazing variety of broadleaf evergreen trees. While an acre of land within the forests of the middle latitudes may contain a

dozen or so tree species, an area of similar size within the rainforest may contain a hundred or more species. Some of the better known varieties include mahogany, rosewood, and ebony, all prized sources of wood for furniture; ironwood, which has wood so dense that it cannot float; balsa, the lightest wood in commercial use and familiar to makers of model planes; and cinchona, the source of quinine used to treat malaria. The numerous tree species are well integrated, so that "stands" of a single type do not occur. This characteristic has made human exploitation of the forest for specific tree species difficult and has been a factor in the preservation of extensive tracts of virgin rainforest.

Despite the great diversity of types of wood produced by the trees of the rainforest, they are surprisingly similar in structure. The forest is tall and very dense. Trees are straight and relatively slender, branching only near their crowns and typically producing an interlacing canopy of leaves and branches sufficiently dense to shade the ground completely.

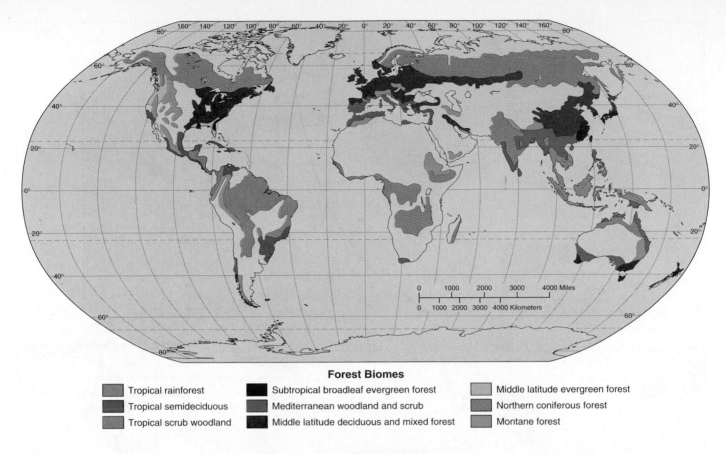

Forest Biomes

- Tropical rainforest
- Tropical semideciduous
- Tropical scrub woodland
- Subtropical broadleaf evergreen forest
- Mediterranean woodland and scrub
- Middle latitude deciduous and mixed forest
- Middle latitude evergreen forest
- Northern coniferous forest
- Montane forest

◆ **FIGURE 11.2** World distribution of forest biomes. Areas in light tan did not have forest or woodland as their natural vegetation.

(a)   (b)

◆ **FIGURE 11.3** (a) View of the canopy of virgin tropical rainforest in the upper Amazon Basin near Iquitos, Peru. (b) Dense tropical rainforest in the Danum Valley, Malaysia.

SOURCE: a: J. L. Medeiros; b: Daniel J. Cox/Tony Stone Images

Larger trees often have developed buttress roots for support in the soft, damp soil. The root systems of these trees are also shallow because of the ready availability of near-surface moisture and because most of the limited supply of soil nutrients is concentrated very near the surface.

Because of the density of the vegetation, success in the competition for sufficient sunlight to carry on photosynthesis becomes the key to survival for the plants of the rainforest. The forest typically exhibits a layered appearance caused by the tendency of large numbers of trees to attain similar heights. This produces canopies of leaves and branches at two, three, and sometimes four distinct levels. This vertical zonation of tree heights results not so much from age differences of the trees as from variations in shade tolerance between different groups of species. A typical three-tiered rainforest contains an upper story of solitary giant trees, consisting of species that require the most sunlight, and averaging perhaps 55 meters (180 ft) in height. An intermediate story forms a continuous dense canopy 9 to 18 meters (30 to 60 ft) above the ground. A third discontinuous layer of highly shade-tolerant trees such as palms and tree ferns constitutes the lowest story.

The floor of the tropical rainforest is surprisingly open, allowing for reasonably easy passage on foot. This openness results from the density of the vegetative canopy, which normally allows less than 1 percent of the available light to penetrate to the surface. The forest floor is a still, damp, and dim world, permeated with the moldy odor of decomposing vegetation and containing a scattering of only the most shade-tolerant plants. Light, life, and growth are concentrated in the treetops. Here resides a tremendous assortment of plants and animals, representing many more species than are found in the biota associated with any other biological assemblage. Many thousands of the plant and animal denizens of the tropical rainforest have not as yet even been biologically classified and cataloged. Among the plants are giant woody vines called **lianas,** which festoon the trees and whose trunks can reach diameters of up to 20 centimeters (8 in) at ground level. **Epiphytes,** or "air plants," are very common. These small plants typically grow in hollows or on the upper surfaces of horizontal branches and derive their moisture and nutrients from the air. Included in this category are mosses, bromeliads, ferns, and thousands of species of orchids whose flowers add color that helps relieve the monotony of the sea of dark green rainforest vegetation.

The tropical rainforest contains the greatest variety of animal life of any biome because of the richness of its food resources and the relative constancy of environmental conditions throughout the year. Most animals live in the treetops. Large mammals are relatively rare, but birds and monkeys fill the air with their raucous chatter, snakes slither silently along the branches, and a wide variety of other small mammals, reptiles, and amphibians are to be found. While many of the mammals are arboreal (tree-dwelling), there also are many ground-dwelling forms such as deer, wild pigs, and a variety of rodents. Most numerous of all, however, are the insects in their countless millions, exhibiting tremendous variety and far outnumbering all other species of animals combined.

As already noted, most of the tropical rainforest is free of dense undergrowth because of insufficient light. Where breaks in the canopy permit sunlight to reach the surface, though, nearly impenetrable thickets of leafy vegetation, commonly known as **jungle,** are likely to occur. Often patches of jungle result from human activities. Subsistence farmers in many rainforest areas practice a pattern of shifting cultivation in which they clear small plots of land, raise crops on them for several years, and then abandon them, allowing them to turn to jungle. Settlements often are surrounded by extensive tracts of jungle that were previously farmed. Strips of dense jungle are also common along the banks of the numerous rivers that flow through the rainforest. They form because the rivers create breaks in the forest that allow sunlight to reach each bank for at least a portion of the day. Riverside strips of jungle are readily apparent to travelers on river craft—still the predominant means of transportation in many rainforest regions—and have helped foster the erroneous impression that impenetrable jungle is widespread throughout the rainforest.

The continued existence of tropical rainforest ecosystems has been jeopardized by the recent rapid encroachment of human activities. Large portions of the rainforests are being cleared for roads, settlements, pasture, and lumber. The potential consequences of rainforest destruction are explored in the Case Study at the end of the chapter.

Low-lying coasts in the tropics frequently have forests of mangrove trees, which have a tolerance for brackish water. Their stiltlike prop roots trap sediment and gradually extend the land in a seaward direction. Higher and drier coasts with sandy beaches are typically backed by forests of palm trees, which are tolerant of salty air and occasional coastal storms.

Tropical Semideciduous Forest

Progressing poleward from the equatorial regions dominated by rainforest vegetation, a climatic stress begins to appear in the form of a winter dry season of increasing length. Trees have sufficiently extensive roots to withstand relatively short periods of dry weather by tapping soil moisture at lower levels; as a consequence, rainforest vegetation extends into areas with as much as two months without rainfall. Where the dry period increases to between two and five months, however, the vegetation begins to exhibit important differences in its characteristics and becomes tropical semideciduous forest.

This forest represents what might be considered a deterioration in vegetative quantity and quality from that of the rainforest as a result of less favorable moisture conditions. The mean size of the trees diminishes, with maximum tree heights generally of 15 to 30 meters (50 to 100 ft). The variety of plants also is reduced, with perhaps 30 or 40 tree species per acre and an increasing tendency for large stands of a single type to occupy some localities. As the dry season

◆ **FIGURE 11.4** Tropical semideciduous forest in Sao Paolo State, Brazil.

SOURCE: © Jacques Jangoux/Tony Stone Images

increases in length in a poleward direction, the forest undergoes a gradual transition from broadleaf evergreen to broadleaf deciduous, as progressively more trees lose their foliage and become dormant in order to conserve moisture. The term **semideciduous** indicates that this forest contains a mixture of both evergreen and deciduous trees.

One of the most important differences between the tropical rainforest and the tropical semideciduous forest is that the latter has a significantly reduced vegetative density and a much lower biomass. Trees are not crowded so closely, their average size is smaller, and the canopy of leaves and branches is much less dense and continuous (Figure 11.4). Branches tend to start at relatively low levels, and tree crowns are much more rounded than those of the rainforest.

◆ **FIGURE 11.5** A tropical scrub woodland/savanna transition area in the Lubombo Mountains of Swaziland.

SOURCE: Sinclair Stammers/Science Photo Library Photo Researchers

A vertical zonation of tree heights is often still apparent, but usually only two foliage layers are present. The larger trees are most frequently deciduous because of their greater transpirational demands, while the smaller trees are predominantly evergreen. The decreased density of the forest, coupled with the deciduous habit of many of the tree species, enables a considerable amount of sunlight to reach the forest floor, especially during the dry season. This in turn supports substantial amounts of undergrowth.

Animal life within the tropical semideciduous forest is diverse and generally is similar to that of the tropical rainforest. It is rich in birds, insects, and arboreal reptiles and mammals, and many areas also contain a considerable variety of ground-dwelling vertebrate species, including large cats such as jaguars and panthers.

Tropical semideciduous forests are scattered in a number of widely separated regions of the tropics. In theory, this vegetation type should be found in continuous strips on both poleward margins of the tropical rainforests. The impact of dry season fires, of natural and increasingly of human origin, however, has considerably reduced the extent of this vegetation type. Where fires have been numerous, the trees have largely been destroyed and the growth of grasses encouraged. In many areas, therefore, the rainforest is bordered by the savanna grasslands discussed later in this chapter. Currently, large expanses of tropical semideciduous forest are located in Southeast Asia, India, some of the islands of the western Pacific, portions of northern South America and Central America, and southern Brazil.

Tropical Scrub Woodland

Progressing into tropical environments with a still longer dry season, the natural vegetation is further modified. Many areas with a dry season four to seven months long are vegetated with a mixture of small trees, bushes, grasses and other herbaceous (nonwoody) plants; these areas are termed tropical scrub woodland (Figure 11.5). The term **woodland** indicates that, although woody trees and shrubs are the dominant forms of vegetation, no continuous tree canopy exists, and it would therefore be improper to characterize this plant assemblage as a forest. The term **scrub** refers to small, stunted trees and shrubs growing thickly together.

The vegetation within this biome varies greatly in composition and appearance in different areas. In some localities it consists of small individual trees 6 to 12 meters (20 to 40 ft) in height surrounded by tall grasses and other low-growing plants. Elsewhere it takes the form of dense thickets of thorny shrubs separated by grassy openings. In still other instances, it consists of a variable mixture of small deciduous trees, shrubs, and grasses. In all cases, the vegetation consists predominantly of tropophytes that grow actively during the rainy season and become dormant during the long, often hot, dry season. Because of the striking alteration in appearance between the lush greens of summer and the sere buffs and browns of winter, these areas are often described as having "rain-green vegetation" (see Figure 8.12 on page 184).

The geographic distribution of the tropical scrub woodland biome has also been complicated and generally reduced by the destructive effects of fire so that several large but detached areas currently exist. One of the largest is in portions of South and Southeast Asia, where thickets of bamboo often are encountered. In northern Australia, species of eucalyptus trees dominate the landscape. A large section of southern and eastern Africa is dominated by species of acacia trees. In portions of eastern Brazil, Paraguay, and northern Argentina, a mixture of thorny bushes, cacti, and trees such as the quibracho make some areas almost impassable.

Most fauna in the tropical scrub woodland are ground-dwelling. The extensive grasses and low-growing trees encourage the presence of groups of large herbivores such as zebras, elephants, and antelope in Africa, and kangaroos in Australia. These animals are in turn preyed upon by carnivores such as lions and scavengers such as hyenas.

Subtropical Broadleaf Evergreen Forest

The occurrence of a dry season of increasing length, as the subtropical high pressure belt is approached, is responsible for the transition (at least ideally) of tropical rainforest to tropical semideciduous forest and then to tropical scrub woodland. Progressing still farther poleward, the climate eventually becomes so dry that in most places trees cannot grow. A major climatically imposed gap thus divides the world's forests into tropical and nontropical segments in most areas (Figure 11.2). The subtropical high, however, is not strong enough everywhere to produce this vegetative transition. Especially along the east coasts of continents paralleled by warm ocean currents, enough precipitation falls that trees are able to continue to flourish, resulting in a link between the tropical and nontropical forest biomes. This forest, termed the subtropical broadleaf evergreen forest, differs in several important respects from those found at either higher or lower latitudes.

Within the areas occupied by this forest type, precipitation is moderate to abundant and is distributed evenly enough throughout the year that no period of severe soil moisture depletion occurs. A moderate climatic stress is produced by occasional subfreezing winter nights in most areas, but most trees are sufficiently cold tolerant to retain their foliage. In contrast to the broadleaf evergreen forests nearer the equator, the subtropical broadleaf evergreen forest contains a limited number of tree species within any given locality, and large stands of a single species are common. The forest is not particularly dense, and average tree sizes are modest. Leaves are smaller and more leathery than those of the rainforest. Some vertical zonation is normally apparent, with an upper story of larger trees that require substantial sunlight and a lower story of small, shade-tolerant trees, shrubs, and sometimes grasses. Although the exact vegetative composition varies considerably from place to place, the dominant tree species in many areas include live oaks, magnolia, and laurel. The trees often are draped with lianas and epiphytes. One epiphyte familiar to residents of the southeastern United States is Spanish moss (Figure 11.6).

The most extensive subtropical broadleaf evergreen forests grow in eastern Asia, particularly in eastern China, southern Korea, and southern Japan. Other areas of significance include portions of the extreme southeastern United States, New Zealand, and southeastern Australia.

Mediterranean Woodland and Scrub

On the poleward sides of the subtropical highs, the climate again becomes moist enough to support the growth of trees,

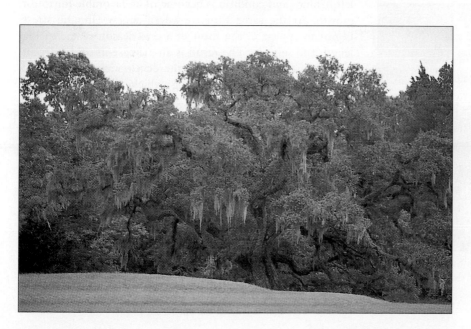

◆ **FIGURE 11.6** Live oak trees draped with Spanish moss are a common sight in the subtropical broadleaf evergreen forest of the southeastern United States.
SOURCE: © John D. Cunningham/Visuals Unlimited

and forests return as the dominant vegetation type. Once again, the transition from dry-land vegetation to humid forest is not abrupt, but proceeds through a series of intermediate stages. These stages are well illustrated within the Mediterranean woodland and scrub biome.

This biome occurs mostly on the west sides of continents and is closely associated with areas having a Dry Summer Subtropical Climate. Included are southern and central California, the Mediterranean Basin region, central Chile, the southern tip of Africa, and much of southern Australia. These areas have dry, warm summers dominated by the poleward-shifting subtropical high and mild, humid winters influenced by traveling cyclonic storms within the westerly wind belt. The dry summer season, usually lasting from four to seven months, is the major climatic stress factor. It is largely responsible for producing a vegetative grouping dominated by shrubs and small trees that somewhat resembles the tropical scrub woodland.

The composition of the constituent plant species varies greatly between the widely separated regions of the Mediterranean woodland and scrub biome. Despite their biological differences, however, a remarkable physical similarity in vegetative forms and community organizations is exhibited in response to similar environmental conditions. Most trees and shrubs are classified as **sclerophylls** (from the Greek *sclero*, or "hard," and *phyllos*, or "leaf"). They have small, hard, thick, leathery leaves that minimize transpirational losses. Other common xerophytic adaptations include deep root systems, viscous fluids, waxy leaf coatings, and thick bark. Many plants also have thorns as a protection from grazing animals.

Better vegetated regions typically display one of two biological assemblages. **Mediterranean woodland** consists of a parklike combination of scattered small- to medium-sized oak trees surrounded by grasses and flowering herbaceous plants. An assemblage known as **Chaparral** in California and **Maquis** in the Mediterranean region is somewhat more common (Figure 11.7). It consists of an often dense cover of woody shrubs and small trees ranging up to 6 meters (20 ft) in height. Species of oak most frequently predominate, although in Australia eucalyptus and acacia are most frequent and in California manzanita shrubs are abundant.

Climax vegetation is extremely rare within this biome because of a long history of human occupancy and exploitation. The removal of trees for lumber and firewood and overgrazing by sheep and goats, especially in extensive areas within the Mediterranean Basin, have produced a serious decline in both the quality and quantity of the vegetation. Fire, too, has played a major role. Brush fires in the tinder-dry vegetation are an annual late summer occurrence in many regions, but are probably most notorious in California and Australia. Today, those areas most extensively affected by overgrazing and fire are covered by a dense growth of thorny shrubs commonly less than a meter (3 ft) in height.

The Mediterranean woodland and scrub biome at one time supported a varied fauna with numerous herbivores, such as ground squirrels, deer, and elk, and predators, such as wolves and mountain lions. The larger animals, however, have largely disappeared as a result of human occupancy. Smaller animals that coexist well with humans, such as birds and rodents, are now dominant.

Mid Latitude Deciduous and Mixed Forest

Progressing poleward from the subtropics into the middle latitudes, the dryness produced by the subtropical highs is left behind, and conditions become more favorable for forest growth. At the same time, however, a new climatic stress begins to appear in the form of a cold season of increasing length and severity. The result is an increase in cold-tolerant needleleaf evergreen trees and the development of a period of vegetative dormancy, during which the cold-sensitive broadleaf trees that dominate the forest must reduce their moisture content and shed their leaves to survive.

The mid latitude deciduous and mixed forest, as this biome is often termed, dominates areas with warm summers, subfreezing periods in winter, and moderate to abundant precipitation well distributed throughout the year. It is the most extensive biome in the middle latitudes. The greatest expanses occur within the northern hemisphere, which, unlike the southern hemisphere, has large landmasses with substantial annual temperature ranges in the middle latitudes. Three large regions of this forest type within the northern hemisphere are the eastern United States, much of Europe and a portion of west-central Asia, and east-central mainland Asia. In the southern hemisphere, its only signifi-

◆ **FIGURE 11.7** This portion of California's Santa Monica Mountains contains a dense vegetation cover of shrubs and small trees known as chaparral.

SOURCE: © John D. Cunningham/Visuals Unlimited

◆ **FIGURE 11.8** Autumn begins to color the trees in a rural portion of the mid latitude deciduous and mixed forest in the eastern United States.

SOURCE: Richard Jacobs/JLM Visuals

cant area of occurrence is in the southern Andes of Chile and Argentina.

During the summer season, the closed canopy of broadleaf trees bears some similarity to that of the tropical rainforest (Figure 11.8). Important differences, though, exist between the characteristics of the two forest types. One difference is the much lower vegetative biomass of the middle latitude forests. Mature trees reach only intermediate heights of perhaps 30 to 45 meters (100 to 150 ft). Individual trees are quite variable in size, and size is related more to age than to species type. No clear-cut vertical zonation of heights is normally observable, although there frequently is an understory of shade-tolerant trees and shrubs. The variety of tree species is rather limited in any one area, and often only three or four are locally dominant. Likewise, fewer vines and epiphytes use the trees for support, although certain climbing plants such as the honeysuckle, poison ivy, and Virginia creeper of the eastern United States may be locally abundant. Although the forest is dense enough to produce a closed canopy, even in summer patches of sunlight penetrate to the forest floor, which is not dark and gloomy like that of the rainforest. The intensity of shade does, however, inhibit the growth of near-surface vegetation.

Probably the major difference between the tropical rainforest and the mid latitude deciduous and mixed forest is that the former is evergreen and the latter experiences a seasonal loss of most foliage. While the rainforest is constant and rather timeless in appearance, the deciduous forest undergoes a seasonal transformation, in visual terms, between life and death. Because the trees spend much time in a dormant state, growth is much slower than in the tropics.

Diversity in the mix of tree species is sufficient to give a distinctive character to the forests of different areas. In the southeastern United States, oak, hickory, and poplar are among the dominant tree types. Farther north, maple, birch, and beech are dominant. Many of these trees are noted for their brilliant autumn colors and are an important tourist attraction in New England, the Canadian Maritimes and the Great Lakes region. In Eurasia, species of oak, beech, and ash are most common, and in southern Chile, beech trees are dominant.

Coniferous needleleaf evergreen trees are scattered throughout many portions of the mid latitude deciduous and mixed forest. These trees generally are out-competed by the broadleafs, but are more tolerant of adverse conditions such as cold temperatures and rocky, sandy, or highly acidic soils. They therefore tend to be more numerous and even locally dominant in mountainous areas within the higher Appalachians and the Alps and in sandy areas such as the Pine Barrens of southern New Jersey and the Landes district of southwestern France. Needleleaf trees also become increasingly abundant near the northern border of the forests in

THE GYPSY MOTH INVASION OF NORTH AMERICA[1]

The gypsy moth, *Lymantria dispar* (L.), was introduced to North America when M. Henri Trouvelot accidentally released the insects in Medford, Massachusetts, sometime in 1868 or 1869. Trouvelot, an amateur entomologist experimenting with silk production, was aware of the dangers inherent with nonnative species but he could not have fully anticipated the scale of the environmental, economic, and sociological consequences that followed his mishap. The gypsy moth is a destructive pest of temperate mixed forests with a strong preference for oak leaves as a food source. Perhaps because of a lack of predatory controls, the insect's population waxes and wanes in boom and bust cycles. Boom years, such as 1981, see the outbreak of billions of caterpillars, the destruction of thousands of hectares of forest, and inconvenience and losses suffered by many people (Figure 11.9).

Although insect life stages have been recorded across the continent, it is in the Northeast where populations currently reach destructive proportions

◆ **FIGURE 11.9** The upper portions of these trees in Tuscola County, Michigan, have been largely defoliated by gypsy moth larvae, while the understory remains nearly intact. This photo was taken in July, when the trees would normally be in full leaf.
SOURCE: © John Sohlden/Visuals Unlimited

(Figure 11.10). Two agents of diffusion can be identified as contributing to the expansion of the gypsy moth's range: the insect itself and human beings. As North American gypsy moth females are flightless, spread occurs primarily in the first caterpillar stage when it floats on a silk thread in search of suitable food. This method of transport usually restricts movement

North America and Eurasia, as these areas are transitional to the adjacent northern coniferous forest.

The mid latitude deciduous and mixed forest has been greatly altered and in many cases completely removed by humans. Most regions containing this forest type have dense human populations and high agricultural and industrial productivities, conditions that are incompatible with extensive forest coverage. Large areas now consist of farmland, roadways, and urban sprawl. The removal of the forests is most complete in eastern Asia, where only the most rugged and inaccessible regions contain substantial numbers of trees. Only in the United States is the majority of the surface still forested, and nearly all has been logged on several

occasions since the time of initial European settlement. Selective cutting has resulted in inferior stands of trees in many areas. In other areas, the planting or encouragement of more "desirable" species, such as pines, has altered the composition of the forest. Introduced diseases and insect pests have also altered the composition of the forest. At present, the rapid areal expansion of the gypsy moth, which has defoliated oak trees in large portions of the eastern United States, is causing concern. This topic is examined in the Focus Box above.

Animal life in the mid latitude deciduous and mixed forest has also been influenced greatly by human occupancy. Animals well suited for coexistence with humans, such as

◆ FIGURE 11.10   The expansion of the North American range of the gypsy moth.

to a few kilometers at most. Long-distance movement occurs when fertile females or egg masses are inadvertently transported on vehicles by human beings.

Available historical data have revealed irregular pulses in the invasion process: a high rate of expansion from 1900 to 1915, a period of slow expansion from 1916 to 1965, and a greatly accelerated rate since 1966. Explanations for the different rates may be related to geographical and temporal changes in mean minimum temperature, the varying distribution of forest species composition, and differing levels of human intervention. Increased recreational use of forests may account for the increased rate of expansion since 1965.

Suppression and control have been attempted since the time of the initial release. Methods have varied from the manual removal of caterpillars and egg masses to widespread aerial spraying with broad-spectrum insecticides. In the 1940s an attempt was made to halt the spread at the Hudson River. The intensive eradication campaign used the pesticide DDT and may have slowed, but eventually did not prevent, the expansion of the insect's range into Pennsylvania, New Jersey, and Maryland. Today, specific pesticides are applied to carefully selected target areas. The development and use of geographic information systems have been beneficial in monitoring insect populations, selecting treatment areas, matching treatments to environmental conditions, and assessing the effectiveness of alternative suppression methods.

The gypsy moth currently infests only one-fifth of its potential range, and there is no evidence that suppression activities have reduced the rate of expansion. As the insect moves south, it encounters more favorable climatic conditions and prime habitat, much of it of great economic value. Humans will continue to have to adapt to the dynamics of the insect rather than attempt to remove the threat from our environment. ■

[1]This focus box was written by Gregory Elmes, Professor of Geography at West Virginia University.

deer, squirrels, rodents, and a great variety of birds, are still abundant in many areas. Other animals that once inhabited the forests, such as bears, wild boar, wolves, and several species of wild cats, are now rare or absent in most regions.

Mid Latitude Evergreen Forests

Broadleaf trees predominate over needleleaf trees in most of the middle latitudes. Two large areas, both in North America and primarily within the United States, however, contain needleleaf evergreen forests. The reasons for the existence of these forests are unclear, because they do not represent the vegetative assemblages that would be expected within their respective areas based on climatic conditions. It is believed that fire may contribute to their maintenance.

One of these two forests is the **west coast coniferous forest** (Figure 11.11a). The only fully developed forest of this type extends along the west coast of North America from northern California to southern Alaska. In other mountainous Marine Climate locations such as southern Chile and western Norway, mixed broadleaf and needleleaf forests exist that appear to represent a transitional phase between this and the mid latitude deciduous and mixed forest.

The west coast coniferous forest contains the world's most impressive assemblage of needleleaf evergreen trees. The

(a)

(b)

◆ **FIGURE 11.11** (a) Giant redwood trees in Redwood National Park, California, a spectacular portion of the west coast coniferous forest. (b) An open stand of longleaf pine trees within the southern pine forest in Florida.

SOURCE: a: Redwood Empire Association; b: JLM VISUALS

forest has a high biomass, probably second only to the tropical rainforest. Trees are very large, with tall, straight trunks. Heights can reach well over 60 meters (200 ft), with some coastal redwoods more than 90 meters (300 ft) in height and 4.5 meters (14 ft) in diameter. The size of these trees exceeds that of all other living organisms on Earth.

The largest and densest stands occur on the moist western (windward) sides of the Cascades and Coast Mountains of British Columbia. In these areas Douglas fir, western hemlock, and Sitka spruce are among the dominant species. In the drier lands to the east of the mountains are less dense forests of smaller Douglas fir, Ponderosa pine, and white pine. Near the coasts, where hard freezes are rare, there is some admixture of broadleaf deciduous and broadleaf evergreen trees.

An extensive portion of the southeastern United States stretching from eastern Virginia south to Florida and west to Texas is covered by another forest composed mostly of species of pines (Figure 11.11b). Known as the **southern pine forest**, it is closely associated with the sandy, drought-prone, and rather infertile soils of the Atlantic and Gulf coastal plain. Its existence probably can be attributed to the resistance of pine trees to droughty soil conditions, their low

soil nutrient demands, and their ability to withstand ground fires.

The pine forests occupy well-drained sandy surfaces that are situated between stream valleys. The most common tree species include longleaf, shortleaf, pitch, and loblolly pines. These trees are valuable sources of lumber, pulpwood, and naval stores (turpentine and resin). Within the swampy stream valleys, hardwoods such as live oak, bald cypress, tupelo, and red gum predominate.

Northern Coniferous Forest

The most poleward forest biome is the **northern coniferous forest,** also known by its Russian name **Taiga.** It is second only to the tropical rainforest in total areal extent. The northern coniferous forest primarily occupies regions having a Subarctic Climate with short mild summers and long frigid winters, but the forest's southernmost margins enter Humid Continental Climate regions. It exists within two giant areas, both in the northern hemisphere between approximately 47°N and 70°N. These are the northern interior of North America, extending from Alaska to Labrador, and the even larger interior of northern Eurasia, reaching from Scandinavia to the Pacific Ocean. Logging activities have removed or modified the forest over portions of its southern range. The central and northern portions of the forest, however, have been influenced relatively little by human activities so that this forest biome as a whole remains more nearly in a natural state than any other.

The northern coniferous forest is dominated by cold-tolerant needleleaf evergreen trees (Figure 11.12). These trees have an advantage over deciduous varieties because they need not grow new foliage each spring, but can instead begin adding to their growth as soon as warm weather arrives. The needleleaf trees also are sufficiently drought resistant to withstand the long winter period of physiological dryness imposed by the freezing of soil moisture.

Only a small number of tree species have developed the ability to cope with the severe climatic stresses imposed by the short growing season and extremely low winter temperatures of these areas. As a result, the northern coniferous forest is biologically the least complex forest biome. It is common for only one or two tree species to dominate over large areas, and the forest usually is locally uniform in height and density. Spruce and fir trees are most abundant for the region as a whole. In the northern United States, these trees are mixed liberally with pine trees, especially white pine, red pine, and jack pine. In western Eurasia, Scotch pine and stone pine are locally abundant. Eastern Siberia, the coldest region of all, is unique in being dominated by two species of needleleaf deciduous larch trees.

The trees of the northern coniferous forest are moderate in size near their southern margins, but become progressively smaller poleward. Most attain heights of 15 meters (50 ft) or less. Near their northern limits, they are dwarfed by the shortness of the growing season, and mature specimens as much as a century old may be only a meter or two tall. The density of the forest also tends to diminish in a poleward direction. The forest floor typically is covered with a mat of lichens and mosses. Especially dominant in northern Canada is a pale-colored species of lichen referred to as "reindeer moss."

During dry periods in summer, the forest can become highly susceptible to lightning-set forest fires. Forest openings produced by fires normally are occupied by certain broadleaf deciduous trees, including birch, aspen, alder, and poplar. These deciduous patches are conspicuous from even great distances because of their distinctively lighter color. They eventually are replaced by the climax needleleaf trees, but the pace of change is very slow in this cold environment, and the transition can take centuries.

In addition to forest fires, breaks in the continuity of the forest have resulted from the lasting effects of the continental glaciers that covered parts of the area as little as 5,000 years ago (see Chapter 17). Their legacy is a landscape dotted by myriad lakes of all sizes. Lakes are relatively ephemeral features on the landscape; as they gradually fill with sediment, they are occupied by a succession of hygrophytic vegetation assemblages. Many small lakes of glacial origin within this region already have been converted to **muskegs**—swampy areas covered with mosses, hygrophytic grasses, shrubs, and scattered water-tolerant trees such as larches (Figure 11.12a). In Canada and Scandinavia, some severely glaciated areas still have patches of exposed bedrock on which little vegetation grows.

The diversity of fauna in the northern coniferous forest is limited by severe winters, a lack of food, and a limited variety of habitats. Most evident are large herbivores, especially species of deer, which are often found in large herds. Carnivores include wolves, weasels, mink, and bears. Birds are most numerous in the summer, when the swarms of insects that breed in the numerous lakes and marshes provide a plentiful supply of food. Most bird species migrate south for the winter.

(a)

(b)

◆ **FIGURE 11.12**   (a) Taiga forest with muskeg in the foreground. Gore Range, Colorado. (b) This taiga forest near the northern limit of tree growth in Central Alaska is dominated by black spruce, with a scattering of birch and cottonwoods.

SOURCE: a: John S. Shelton; b: © Steve McCutcheon/Visuals Unlimited

GRASSLAND BIOMES

The **grasses** are herbaceous (nonwoody) plants with blade-like, opposing leaves connected to a jointed stem by a sheathlike attachment. A great number of species of grasses exist, including many that are commercially important. Both directly and indirectly, members of the grass family are largely responsible for producing the basic foods on which humanity depends. All of the cereal grains, including rice, corn, wheat, oats, and barley, are grasses. In addition, most animals raised for meat and dairy products feed primarily or exclusively on grasses.

Grasses grow on virtually all vegetated portions of the Earth's surface and are the dominant form of natural vegetation on more than a third of the world's land area (Figure 11.13). Their widespread geographical distribution results both from their hardiness and their diversity. Grasses are abundant from the equator to the tundra and from swamps to deserts. A key to their ability to survive in environments with severe periodic climatic stresses is the fact that many grasses are annuals and can simply regrow from seeds during favorable times. Numerous other grass species, unlike most trees, can grow new shoots from their subsurface stem sections after the above-ground portions have been killed by fire or adverse weather conditions, or have been removed by grazing.

The chief competitive disadvantage of grasses is their relatively small size. Most grasses attain heights of only a meter or two. Grasses are not sufficiently xerophytic to dominate the vegetation in most arid regions, although they are well represented in many semidesert areas. As a result, most of the world's grasslands are situated on the margins of forests—in those areas either too wet or too dry for trees. These areas generally include regions with subhumid, semiarid, and seasonally dry climates, as well as more restricted areas that experience poor drainage or periodic flooding.

Over the last several millennia, the areal extent of the world's grasslands has gradually been expanding as they have encroached upon their forest margins. This trend may be partially the result of post-Pleistocene climatic changes, but is believed to be caused primarily by human activities such as the gathering of firewood, overgrazing by domesticated animals, and especially the use of fire. Expansion has been occurring for such a long time that researchers are unsure about the original extent of the world's "natural" grasslands. It is certain that, especially in the tropics and subtropics, both grasslands and deserts will continue to expand at the expense of forests in the near future.

The grassland vegetation association is divided into three biomes: the savanna, steppe, and prairie. The savanna grasslands occur in the low latitudes, mostly within drier portions of the Tropical Wet and Dry Climate. They will be discussed first. Next to be examined are the steppe grasslands, which are associated with semiarid climates in both the low and middle latitudes. The prairie, which is restricted largely to subhumid portions of the middle latitudes, is covered last.

Savanna

The **savanna** is unique among the plant groupings discussed in this chapter in that it typically contains a mixture of two greatly different types of vegetation. It may be described as a tropical grassland with scattered trees or shrubs (Figure 11.14). In theory, the savannas are situated on the dry (normally poleward) margins of the tropical scrub woodland subassociation, with the boundary between the two located where grass becomes areally dominant over trees and shrubs. They generally occupy the drier portions of the Tropical Wet and Dry Climate, where the winter dry season lasts between six and eight months. The actual boundary between the

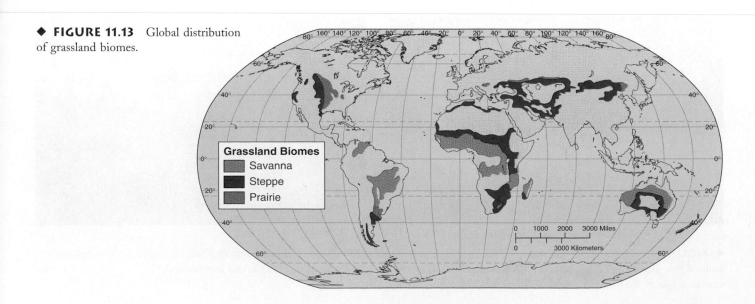

◆ **FIGURE 11.13** Global distribution of grassland biomes.

Grassland Biomes
- Savanna
- Steppe
- Prairie

◆ **FIGURE 11.14** The East African savanna during the dry season. Scattered acacia trees and a solitary giant baobob stand above a ground cover of dry bunch grasses.

SOURCE: J. L. Medeiros

tropical scrub woodland and the savanna is, of course, very difficult to define precisely because it is subject to individual interpretation and is determined not only by the length of the dry season but also by the effects of fire, grazing animals, and human activities. The dry margins of the savannas occur where the climate becomes semiarid and the mixture of tall grasses and woody plants is replaced by short tropical steppe grasses.

The most extensive savannas are located on the continents of Africa, South America, and Australia. The African savannas are the largest and best known. Most occur in a broad arc surrounding the tropical rainforest of the Congo Basin to the north, east, and south. These savannas, especially those of eastern Africa, are frequently seen on "big game" movies or television documentaries because of their great herds of large animals. In South America, extensive savannas are located in Venezuela and Colombia as well as in southern Brazil and Paraguay. Large portions of northern, eastern, and western Australia are also often considered to be savanna, although the mix of plants here is somewhat different. Other smaller but noteworthy savannas exist in much of western India, western Madagascar, and the Everglades of southern Florida.

Savanna grasses thrive because they consist of species that become dormant during the dry season. The above-ground portions of the plants die, but new shoots grow with the onset of the summer rains to turn the landscape lush and green. Great variations in grass heights exist in different areas, ranging from 0.3 meters (1 ft) to more than 4 meters (12 ft). Most savannas are dominated by coarse, short-bladed bunch grasses that grow in dense tufts, with intervening patches of bare earth. They have well-developed root systems that support rapid growth during the rainy season.

Trees may be scattered throughout the savanna as individuals to produce a parklike appearance, or they may exist in groups. They consist of a mixture of stunted broadleaf deciduous and broadleaf evergreen species, which, with their thick bark, lack of lower branches, and flattened or umbrella-shaped crowns, are highly similar to those of the tropical scrub woodlands. Trees do not dominate because the seasonal rainfall distribution is too uneven to readily support large perennials with year-round moisture requirements. In general, the number of trees declines toward drier areas, as does grass height.

Narrow strips of dense tree growth commonly occur along the banks of streams. Their presence indicates that when water is available, trees will assert dominance over grasses. Often, the trees along either bank are sufficiently large that their foliage intermeshes, enclosing the stream within a tunnel of vegetation. These strips of forest are consequently termed **galeria,** the Italian word for tunnel.

In some savanna areas, the woody vegetation takes the form of shrubs rather than trees. The shrubs are fire-resistant and often thorny, and tend to grow in dense thickets. Most shrubs are xerophytic broadleaf evergreens with small, hard leaves.

The dominant fauna of the savanna are the large herbivores. These animals are most numerous and best known in Africa, where they include elephants, zebras, wildebeests, antelope, giraffes, and many others (Figure 11.15). Large carnivores, of which the lion is best known, also thrive here. A diversity of smaller animals, especially birds and rodents, occupy the savanna as well.

◆ **FIGURE 11.15** Impalas and zebras are two of the large herbivores native to Kenya's savanna.
SOURCE: © Will Troyer/Visuals Unlimited

Steppe

The steppe comprises the short grasslands of the low and middle latitudes. It is closely associated with semiarid climates. Many geographers use the term **steppe** to refer only to the short grasslands of the middle latitudes, and, indeed, they attain their greatest degree of distinctiveness as a separate biome there. Large areas of the tropics, however, also are covered primarily by short grasses because of similarly marginal amounts of rainfall; for this reason, it seems justifiable to include them in this category.

The low latitude steppes are located in regions with a dry, hot climate typically punctuated by a brief, often intense, rainy season lasting two to four months. Ideally, they occupy

◆ **FIGURE 11.16** Short steppe grassland in the High Plains of eastern Colorado.
SOURCE: © Rod Planck/Photo Researchers, Inc.

the semiarid margins of the subtropical deserts. Larger areas include the borders of the Sahara and Kalahari Deserts in Africa and portions of north-central Australia. Smaller tropical steppes occur in several localities, including northwestern Argentina and northern Venezuela.

The vegetation of these regions consists predominantly of a short grass cover that lies dormant for much of the year. The grass is similar to that of the savanna in that it is tufted, with patches of bare earth between. Varying amounts of often thorny broadleaf deciduous shrubs are interspersed with the grasses and may be locally dominant.

The steppes of the middle latitudes, like their low latitude counterparts, generally occupy the semiarid margins of deserts. On their humid sides, they grade into either tallgrass prairies or forests. Two principal areas exist: one occupies the Great Plains of the United States and south-central Canada, the other extends more than 7,200 kilometers (4,500 mi) from the north shore of the Black Sea east through central Asia to northern China. Additional steppe areas occur in interior portions of the Middle East and southwestern Asia. Only limited expanses of steppe exist in the southern hemisphere; these include eastern New Zealand and the South African veldt, both of which are transitional to prairie.

Most steppe lands of the middle latitudes consist of vast expanses of flat or gently rolling terrain covered almost exclusively with short grasses less than half a meter (2 ft) high (Figure 11.16). In some areas the grass is dense and continuous, and the combination of soil and matted grass roots forms a rough sod that protects the surface from erosion. Important sod-forming species in the American Great Plains include buffalo grass and bluestem wheatgrass. In other areas, especially toward the desert margins, the vegetative cover consists predominantly of bunch grasses such as feather grass. Most grasses are perennials that lie dormant during the winter and sprout vigorously in the spring. During the summer, high temperatures and evapotranspiration rates produce severe soil moisture deficits, and grass growth is greatly reduced.

Along with the grasses are various flowering annuals as well as some shrubs. Shrubs are especially common on rocky hillsides where the decreased density of grasses reduces the risk of destructive fires. An increase in xerophytic shrubs also is evident toward the dry margins of the steppes and in regions overgrazed by domesticated animals. In most cases, trees are lacking in the middle latitude steppes except along the banks of streams, where such water-tolerant species as cottonwoods and willows are found.

Prairie

The **prairie** is one of the smallest and most distinctive of the world's biomes. It has also been more nearly destroyed by human activities than any other. The underlying soils have proven so fertile that virtually all prairie areas have been plowed up and converted to cropland. Remaining small patches of prairie along roadsides, the edges of fields, and on steeper slopes can provide an idea of the original appearance of the vegetation, but they certainly cannot convey an adequate impression of the original immensities of shoulder-high grasses extending to the horizon (Figure 11.17).

The prairie regions are located in the middle latitudes, mostly near the dry (subhumid) margins of the Humid Continental Climate. The largest area is centered in the midwestern United States, extending in a rough triangle from Indiana northwestward to southern Alberta and southward to eastern Texas. Smaller patches within the United States include the "Black Belt" of western Alabama and the Palouse country of eastern Washington, which is transitional to steppe. A second extensive area of prairie is the pampa of South America, located within northeastern Argentina, Uruguay, and southern Paraguay and Brazil. A small but important prairie region also occupies the Danube plain in Hungary, and vegetation transitional to prairie is found along the humid northern margin of the steppe lands of central Eurasia in a narrow discontinuous band extending from Ukraine to Manchuria in northern China.

The predominant vegetation of the prairie was a mixture of tall grasses that were sufficiently dense to form a continuous sod cover over large areas. The grasses were mostly 0.6 to 1.2 meters (2 to 4 ft) high, but were reputedly as tall as 3 meters (10 ft) near the eastern margin of the prairie in Indiana and Illinois. They diminished in both height and density toward the drier steppe margins. A variety of annual and perennial species contributed to the grass cover, and in the American Midwest, species of bluestem were predominant. Interspersed with the grasses were a large number and variety of herbaceous flowering plants that produced a colorful landscape, especially in late spring. Particularly common in the American Midwest were the sunflower and black-eyed Susan.

On the original prairie, trees existed only in scattered patches near the forest margins and, somewhat in galeria fashion, in strips along streams. With settlement, large numbers of trees were planted for decoration and shade and as windbreaks; these generally have thrived without special care and often have attained large dimensions. The reason for the original lack of trees therefore poses a rather perplexing problem. Most theories concerning the absence of trees revolve around the following four ideas:

1. Fires of both natural and human origin may have been sufficiently frequent to destroy tree seedlings while doing little harm to the herbaceous vegetation.
2. The high lime content of most prairie soils may have encouraged the growth of grasses while suppressing tree growth. This factor is known to have been instrumental in the dominance of grasses in some areas, such as the Alabama Black Belt.
3. The large herds of bison and other grazing animals that ranged across these areas may have trampled and eaten vulnerable tree seedlings.

◆ **FIGURE 11.17** Tucker Prairie, a protected tall grass prairie in central Missouri.

SOURCE: Marbut Memorial Slide Collection/American Society of Agronomy

4. The climate may have been significantly drier during the Pleistocene epoch, becoming humid enough to support trees so recently that a relict grassland cover still persisted. In support of this idea, it is known that forests were gradually encroaching on the eastern margins of the American prairie at the time of settlement.

It is, of course, quite possible that the presence of a grass cover resulted from a combination of these factors.

At present, the prairies consist largely of a checkerboard of fields of crops. It should not be surprising that cereal grains such as wheat and corn are currently most important, because these domesticated tall grasses are biologically quite similar to the wild tall grasses that once grew so profusely in the prairie.

The dominant fauna of both the steppe and the prairie are large migratory grazing animals. In North America, vast herds of bison once roamed these grasslands, and deer and pronghorn antelope also were numerous. In Eurasia, saiga antelope and wild horses were among the dominant animals, and in Australia, kangaroos were numerous. These herbivores have now been largely replaced by domesticated animals such as cattle and sheep.

DESERT BIOMES

Desert biomes are closely associated with the world's arid climate regions. These regions are located mostly in the subtropics, but also occur in portions of the middle latitudes (Figure 11.18). The largest desert area by far extends across the Sahara of North Africa, through the Arabian Peninsula, and well into southwestern Asia. Other important subtropical deserts include the Kalahari of southwestern Africa, the Mojave and Sonoran Deserts of the southwestern United

◆ **FIGURE 11.18** Global distribution of desert biomes.

States and adjacent Mexico, the Atacama of western South America, and the Australian Desert. Within the middle latitudes, deserts occupy drier portions of the American

◆ **FIGURE 11.19** Desert vegetation can assume a variety of diverse forms, as illustrated in this photo from Organ Pipe Cactus National Monument in Arizona.

SOURCE: Larry Ulrich/Tony Stone Images

Great Basin, southern and western Argentina, and much of Central Asia.

Despite their aridity, most deserts contain a variety of vegetation (Figure 11.19). The one-quarter, or less, of all deserts that are essentially devoid of plants occur mostly in areas that are extremely dry, have shifting sands, or have steep, rocky slopes. Desert vegetation typically consists of a rather diverse mixture of small, predominantly xerophytic plants that are especially adapted to survive in an arid environment. Desert plants must also be tolerant of extremes of both heat and cold. Species vary greatly in different areas, but the environmental similarity of these areas leads to the development of plant groups that are similar in general appearance and characteristics.

Perhaps the most basic attribute of desert vegetation is its wide spacing. Most other plant assemblages produce a more or less continuous ground cover. The lack of surface moisture does not permit this in deserts, despite the xerophytic nature of the plants. In general, the spacing between plants increases with aridity, but even in better watered areas, most of the surface is bare. Two other basic characteristics of desert plants are small size and slow growth. Most plants attain heights of only a meter (3 ft) or less, and many take years to reach maturity. The vegetation of the desert can almost be viewed as progressing through life in slow motion, so that the desert normally projects a somewhat timeless image.

Desert plants can be divided into several botanical groups. Most widespread and important are shrubs of both evergreen and deciduous varieties. These are very common in the southwestern United States, where familiar species include sagebrush, creosote bush, and mesquite. Desert shrubs, particularly mesquite, are currently increasing rapidly in number and areal extent in parts of the Southwest. The causes and consequences of this phenomenon are discussed in the Focus Box on the next page.

The best-known desert plants in North America are the leafless evergreen herbs, which include most cacti. It is

THE MESQUITE INVASION OF THE AMERICAN SOUTHWEST

When the early explorers and settlers entered the southwestern United States during the sixteenth through the nineteenth centuries, they encountered a region covered largely by native grasses. These grasses provided sustenance for a variety of animals, including deer, antelope, and bison. The early farmers and ranchers also used these grasses as forage for their livestock. Woody plants were present, but in most places they were widely scattered and were often concentrated in stream valleys.

During the past century, however, there has been an almost explosive increase of trees and shrubs throughout the Southwest. This has not only considerably reduced the grazing potential of most areas, but has altered the basic vegetative composition of the entire region.

The plant that has most greatly increased its range during the past century is mesquite. Mesquite is a spiny, deciduous tree or shrub with an extensive root system that makes it highly drought resistant (Figure 11.20). Originally confined largely to draws and washes, it has invaded extensive areas of flatter surfaces from Texas and Oklahoma west to central Nevada. Today, a low, scrubby covering of mesquite occupies some 17 million hectares (70 million acres) of rangeland in the southwestern United States.

A number of researchers have attempted to determine the cause of mesquite's rapid expansion in range. They have concluded that it is largely a culturally induced phenomenon resulting from changes in land-use patterns following European and American settlement of the region. Two factors are believed to be crucial:

1. *Overgrazing and trampling.* The wild herbivores of the Southwest did not overly tax the recuperative ability of the grasses, because their numbers were kept in check by natural predators and Native American hunters. The grasses therefore were able to hold their own against the encroachment of mesquite or other woody plants. The cattle and sheep ranchers of the nineteenth and early twentieth centuries, however, fenced in their lands and often overstocked them, so that the grass cover was reduced greatly by overgrazing and trampling. This opened up large areas to colonization by mesquite and other unpalatable shrubs such as sagebrush and creosote bush.

2. *Suppression of fires.* Summer fires set by lightning were once common on the dry grasslands of the Southwest. The Native American inhabitants also set fires both to promote fresh grass growth and to drive game during hunts. These fires did little permanent damage to the grasses, which quickly sprouted new shoots from their undamaged root systems, but they were deadly to small trees and shrubs. The European and American farmers and ranchers, however, viewed fire as their enemy and did their best to suppress it. In addition, overgrazing in many areas reduced the supply of dry vegetative matter on which wildfires fed. The reduction in the frequency of fires

◆ **FIGURE 11.20** A mesquite tree near Phoenix, Arizona.
SOURCE: Ralph Scott

generally enables mesquite and other woody plants to reach maturity.

The expansion of mesquite shrubland at the expense of grassland is one of the major problems facing ranchers in many parts of the Southwest today. Vigorous efforts are underway in many areas to eradicate or control mesquite growth. How successful these efforts will be, only time will tell. ■

commonly believed that cacti are the dominant plants in deserts. Actually, they are native only to the Americas and are abundant in only a relatively few localities. Their striking appearance and often rather large size, however, cause them to stand out from the more mundane desert plants, and this probably accounts for their wide renown.

A third group of desert plants are the sclerophylls, broadleaf evergreen trees or shrubs with small, hard, leathery leaves. These plants are most closely associated with the Mediterranean woodland and scrub forest biome, and they are most frequently found near the poleward margins of the subtropical deserts that border the Mediterranean Climate. The dominant sclerophyll of the American West is the creosote bush.

A fourth group, the **halophytes** (from the Greek *halos*, or "salt"), are salt-tolerant plants that normally occupy lowland basins of interior drainage where the soils are highly saline. Because these plants have developed an ability to cope with saline surface conditions, they are able to grow where no other plants can survive and are therefore free from competitors. Important examples of halophytes in the western United States are the salt bush and shadscale scrub.

A fifth group of desert plants are the **ephemerals.** They are small, short-lived plants that go through their life cycles quickly following a substantial rainfall. The ephemerals take advantage of those brief, rare periods when the desert is not dry, and they therefore cannot be classified as xerophytes. They include a wide variety of plants; most are grasses or flowering herbs whose seeds or bulbs have lain dormant for months or even years. The abundance of brightly colored flowers among the ephemerals has led to the frequent observation that the desert "blooms" after a rainstorm. Within a month or so, however, the moisture is gone, the flowers have produced their seeds and died, and the desert reverts to its normal stark appearance.

Most desert localities, of course, contain a mixture of several of these plant groups in varying proportions. The somewhat better watered semidesert regions usually are dominated by scattered shrubs with a considerable number of smaller herbaceous plants such as cacti and grasses interspersed. Within the more sparsely vegetated drier deserts, widely spaced individual shrubs, cacti, and bunch grasses are common.

Trees are not found in most desert areas because their moisture requirements are too high. On occasion, though, scattered xerophytic trees do occur, especially near the desert margins. This is especially true in Australia, where occasional drought-resistant eucalyptus trees are found in many desert areas. Other localities where trees can secure enough moisture for growth are **oases,** where the water table intersects the surface to produce a spring, and riverbanks, where dense **riparian** (riverbank) vegetation often flourishes. Riverbanks in parts of the southwestern United States have been invaded recently by the tamarisk, an imported tree whose dense foliage has choked out the native vegetation.

The desert fauna consist mostly of small, nocturnal animals that remain hidden from the often intense heat of the day under stones or in burrows so as to conserve water. Rodents are common, as are reptiles such as snakes and lizards. Arachnids such as scorpions and tarantulas also are well represented.

TUNDRA BIOMES

The **tundra** is the most cold-tolerant plant association and consequently is found at higher latitudes and elevations than any other. Because the tundra has been influenced relatively little by human activities, it is the least modified of all the major plant groups.

Most tundra areas are located within the northern hemisphere (Figure 11.21). The three largest regions are northern North America from Alaska to Labrador, the Arctic coast of

◆ **FIGURE 11.21** Global distribution of tundra vegetation. Small, scattered patches of tundra at high elevations in mountain ranges such as the Rockies and the Alps are not depicted due to the small scale of the map.

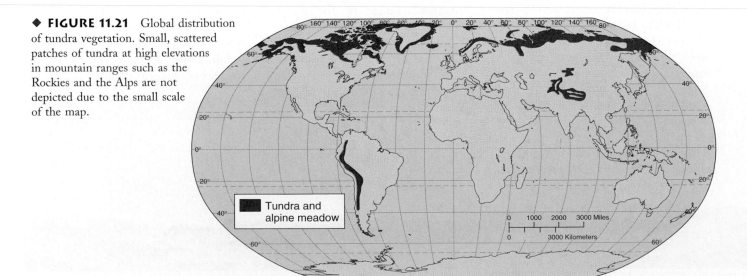

Eurasia from Norway to the Bering Sea, and the coastal rim of Greenland. These areas all have a Tundra Climate characterized by long cold winters, summers with monthly means below 10°C (50°F) and occasional frosts, and a growing season of only about two months. Only during the summer, extending from late June or early July into early September, does the snow melt and the vegetation appear.

Tundra vegetation consists of a mixture of small plants, most of which are herbaceous perennials (Figure 11.22). Two layers of plants normally cover most of the surface. The upper layer of grasses, grasslike sedges, flowers, and sometimes small shrubs often produces a meadowlike appearance. Beneath the upper layer exists a well-developed surface layer of lichens and mosses.

Probably the two most important plant groups are the grasses and the lichens. Grasses are so common that some biogeographers consider the tundra a type of grassland rather than a distinct plant association. Several species are common, but perhaps the best known is cotton grass, a tall grass that grows in poorly drained areas and is surmounted by a cottony tuft. Lichens consist of algae and fungi growing together in a symbiotic relationship to produce a plant form unlike either one alone. The best-known example is the so-called "reindeer moss," which grows to heights of 5 to 10 centimeters (2 to 4 in).

Near the southern margin of the tundra, dwarf trees and shrubs are found on better drained sites. These trees are of the same species as those of the neighboring northern coniferous forest and include spruce, willow, birch, and alder. They grow extremely slowly and at maturity may attain heights of only a meter or two. On occasion, they display a tendency to grow along the ground more like vines than trees. The very small size of the trees and shrubs is believed to result largely from the alternate freezing and thawing of the surface, which causes the ground to heave, thus tearing and crushing the roots. In addition, an underlying layer of **permafrost,** or perennially frozen soil, acts as a barrier to root development. In better drained river valleys, the depth to permafrost is greater, and larger trees are able to grow. As a result, narrow strips of coniferous forest extend up river valleys well into the tundra.

Localized topographic variations exert perhaps a greater influence on the vegetation of the tundra than on any other plant group. Permafrost and past glaciation have caused large areas to be poorly drained, and lakes and marshes abound. The marshes are sites where hygrophytic grasses grow. Slightly higher and drier sites a short distance away are likely to be covered with completely different types of vegetation, including patches of colorful flowers such as buttercups, anemones, and poppies. With the low Sun angles that always exist in the high latitude tundra, slope orientation also is crucial; south-facing hillsides are sunnier, warmer, and better vegetated.

Proceeding poleward through the tundra, the density, size, and biological sophistication of the plants gradually decline. At its poleward edge, the tundra surface consists

◆ **FIGURE 11.22** Tundra dominated by cottongrass in the Richardson Mountains, Northwest Territories, Canada.
SOURCE: © Steve McCutcheon/Visuals Unlimited

largely of bare rock, with only widely scattered patches of moss and lichen.

Despite its harsh climate, the tundra is home to several species of large migratory herbivores, including caribou, reindeer, and musk ox. Smaller mammals include snowshoe hares and lemmings, while carnivores are represented by wolves, Arctic fox, owls, and hawks. Numerous birds summer in the Arctic, feeding on the huge insect population, and migrate south to warmer climates for the winter.

MOUNTAIN BIOMES

Within mountainous regions, all the factors that control the distribution of plants and animals are subject to large variations over small distances. The climate rapidly becomes colder and moister with increasing elevation, and, in addition, large differences in precipitation usually exist between windward and leeward slopes. Rugged topography is the hallmark of a mountainous region; slope steepness and orientation vary greatly. Soils may be deep and well developed in valleys but completely absent on the steepest slopes; because of the complex geology typical of many mountainous areas, soils frequently have formed from a variety of parent materials. As a result, great differences in life-forms are encountered within small distances, especially in higher mountains.

The resulting biogeographical patterns are too complex to discuss in detail except at the local level. At any specific site, however, within the constraints imposed by topography and soils, the flora and fauna may be expected to be associated closely with the local climate. This correlation again indicates that climate is the most powerful of the biological distribution controls.

◆ **FIGURE 11.23** A zonation of vegetation types based on elevation is apparent in this view of Wyoming's Grand Teton Mountains. Prairie vegetation in the foreground gives way to coniferous forest, which in turn merges with tundra and, finally, with ice and bare rock.

SOURCE: Richard Jacobs/JLM Visuals

◆ **FIGURE 11.24** The treeline occurs at about the 3,500 meter (11,000 ft) level in the Colorado Rockies.

SOURCE: Albert Copley/Visuals Unlimited

Local climatic conditions may cause the vegetation within a highland area to be completely unlike that of the surrounding lowlands. Areas of montane forest therefore can be found in the midst of deserts, and mountain tundra (or **alpine meadow**) can exist at the equator. Patterns of mountain vegetation are by no means random, however, especially when examined at the broader scale. Because climatic influences are dominant, and elevation largely determines mountain climates, a vertical zonation of biomes that represents a compression of latitudinal life-form gradients is normally well developed (see Figures 8.51 and 11.23). A very important natural vegetation boundary in mountain areas is the **treeline,** or uppermost limit of tree growth. It is located approximately at the elevation at which the warmest month of the year averages 10°C, or 50°F (Figure 11.24). In general, the biological assemblages of mountain areas are those that thrive under cooler and moister conditions than those of the surrounding lowlands.

▮ SUMMARY

The geographic distribution of the Earth's planet and animal life results from the interaction of a number of environmental variables, the most important of which is climate. In general, life is most abundant and diverse in regions where the climate is warm and moist, while the existence of cold or dry conditions introduces climatic stresses that reduce the variety of fauna and flora. Because Earth's temperature and precipitation patterns are broadly associated with zones of latitude, areas containing similar groupings of life-forms generally occupy specific latitudinal settings. In order to

further summarize important information on the global distribution of plants and animals, we now return to the Focus Questions.

1. *How are the complex global patterns of plants and animals organized for purposes of geographical analysis?*

The world's plants and animals are generally divided into global-scale communities that are differentiated on the basis of their dominant forms of natural vegetation. The four major vegetative groupings, or plant associations, discussed in this chapter are the forest, grassland, desert, and tundra. Each of these associations, in turn, is divided into biological communities, or biomes, that are associated with a specific range of climatic conditions.

2. *What are the locations and biological characteristics of the world's major biomes?*

Forests cover the largest geographical area of all the plant associations and comprise the great majority of the total biomass of land areas. They generally dominate in areas with at least a five-month period of annual soil moisture surplus and with a warmest monthly mean temperature of 10°C (50°F) or higher. Most of the tropics, middle latitudes, and subpolar regions meet these criteria and are forested. Broadleaf evergreen trees dominate near the equator, while broadleaf deciduous trees are dominant in forested portions of the subtropics and middle latitudes. Needleleaf trees are most common in subpolar and alpine environments. Grassland regions typically are encountered on the dry margins of forests. Savanna grasslands, which are tropical grasslands with scattered trees, occupy large portions of the low latitudes with dry seasons lasting five to eight months. Short steppe grasslands are found in semiarid regions of both the low and middle latitudes. Tall prairie grasslands formerly occupied subhumid areas within the middle latitudes, but most have been removed for agriculture. The driest deserts are unvegetated, but most deserts are covered by a variety of xerophytic herbs and shrubs. Deserts are widespread in the subtropics, but also exist in portions of the middle latitudes. Tundra vegetation consists of a mixture of small, cold-tolerant herbaceous perennials, which often present a meadowlike appearance, along with a moss and lichen understory. The plants of the tundra are only visible, and biologically active, for approximately three months in summer. They dominate subarctic regions with warm season monthly means between 0°C and 10°C (32°F and 50°F).

▌QUESTIONS FOR THOUGHT AND REVIEW

1. What competitive advantage do trees have over other types of vegetation? Why don't trees dominate the vegetation of all areas?
2. Discuss the basic vegetation differences between the tropical rainforest and the tropical semideciduous forest. What factor causes these differences?
3. Why are the world's forest areas geographically separated into tropical and nontropical segments? Are these two segments connected? Where and why?
4. Through what natural vegetation groupings would you pass on a trip from New York City to San Francisco? Explain briefly the environmental factors responsible for each of the changes in vegetation you would encounter.
5. What basic similarities and differences exist between the vegetation of the tropical rainforest and the mid latitude deciduous and mixed forest?
6. Name two regions of the coterminous United States where needleleaf evergreen trees dominate. Explain why they exist in these areas.
7. In what respects is the northern coniferous forest different from all the other forest biomes?
8. Describe the basic environmental conditions that favor each of the four plant associations.
9. Describe the typical appearance of the savanna grasslands. In what ways do they differ from the grasslands of the middle latitudes?
10. Discuss the controversy over the origin of the prairie. What factors may have contributed to its formation?
11. How do the high temperatures and lack of moisture in desert environments influence the types and behaviors of the plants and animals that live there?
12. Discuss the major environmental limitations of the tundra, and explain how they have influenced the tundra flora and fauna.
13. Imagine a supercontinent extending from the equator to the North Pole. Starting with the tropical rainforest, list in order the hypothetical sequence of vegetation groups that a person traveling poleward would encounter.
14. In what biome is your own area located? Explain how local environmental conditions have resulted in the existence of this biome.
15. What are the chief causes for the current destruction of the world's tropical rainforests? What would be the major consequences of the destruction of these forests?

▌KEY TERMS

plant association	grass
biome	savanna
forest	steppe
broadleaf	prairie
needleleaf	tundra
rainforest	alpine meadow
jungle	treeline
woodland	

The Destruction of the World's Tropical Rainforests

Reference was frequently made in the last two chapters to the impact of human activities on global vegetation patterns. In some areas, this impact has taken place over many centuries, and little now remains of the original surface cover. In other areas, only minor modifications have occurred, and the original vegetation remains largely intact. In at least one instance—that of the tropical rainforests—humans are currently destroying a major vegetation assemblage. This destruction is proceeding with unprecedented speed and is likely to produce major adverse consequences for the tropics and perhaps even for the entire planet.

CAUSES AND EXTENT OF DEFORESTATION

Until a few decades ago, most portions of the tropical rainforest were untouched and largely unexplored. Little lumbering took place because of the lack of roads, the long distances to markets, and the great diversity of tree species. Beginning in the 1950s and 1960s, however, a number of factors combined to alter the economic conditions that previously had protected the rainforests. This initiated the large-scale forest destruction that continues at an accelerating pace to the present. One of the chief factors in this turnabout was population pressure caused by the tre-

mendous growth rates of the developing countries that contain tropical rainforests. Many landless families in Asia, Africa, and South America are willing to settle and farm rainforest areas despite the land's low fertility and are burning rainforest to clear land for agriculture.

Commercial lumbering is increasingly profitable, despite the inherent difficulties already mentioned (Figure 11.25). Modern machinery has greatly increased the efficiency of operations, and uses have been found for more and more tree species. Clear-cutting methods are often employed, and even if they are not, uncut trees frequently are damaged and later die from insect infestation and disease. Settlers also need wood for lumber and fuel. Exports of tropical rainforest lumber, much of it to Japan, bring in several billion dollars a year. This income is desperately needed by the developing countries, most of which are burdened with huge foreign debts. Other important causes of rainforest destruction include cattle ranching, mining, and the flooding of vast areas behind dams built for hydroelectric power generation.

In 1990, the remaining area of tropical rainforest totaled some 7.8 million square kilometers (3 million mi^2), distributed among 70 countries. This is an area nearly the size of Europe. Each year, approximately 113,000 square kilometers (44,000 mi^2) of rainforest are destroyed by land clearing for crop production, fuelwood gathering, and cattle ranching, and an additional 45,000 square kilometers (17,000 mi^2) are damaged or destroyed by commercial lumbering operations. A study released in June 1990 estimated the rate of tropical rainforest destruction to be 0.6 hectares (1.5 acres) each *second*. In

total, an area slightly larger than New York and Vermont combined is lost each year. Within a few decades, only two large rainforest areas will remain; in the western Amazon Basin and in the central Congo Basin.

The destruction of the tropical rainforests after some 60 million years of continuous existence is a major event in the Earth's biological history and will considerably alter the face of our planet, even as viewed from far out in space. The scope of the clearing operations is larger in magnitude than the earlier removals of the temperate forests of North America and Europe and is being accomplished in a much shorter time span.

CONSEQUENCES OF DEFORESTATION

The removal of the rainforests is likely to produce a number of adverse consequences, the exact nature and magnitude of which are difficult to forecast. One is the permanent deterioration of the vegetative cover. The removal of the trees will not, of course, produce a bare surface for long. Replacement by a dense secondary growth of weeds, shrubs, vines, and saplings is rapid. If undisturbed, the area may eventually return, after several centuries, to its original state. In most cases, though, cutover areas are settled or grazed, thus inhibiting reforestation. In addition, if cutting is too extensive, few seeds are locally available for the return of the trees. The tendency is for the area to revert to a coarse savanna grassland.

A second important concern is soil deterioration. Soils in the humid tropics are naturally low in fertility because of the leaching of soluble plant nutrients by downward-moving soil water from heavy rains. (This process is discussed further in the next chapter.) The rainforest trees, however, are highly efficient at recycling nutrients by absorbing them into their root systems before they can be leached away. At any one time, a large proportion of the total nutrients are locked up in the trees themselves, to be released gradually as they die and decompose. Lumbering activities therefore directly remove many of these nutrients and also destroy the mechanism for protecting the rest from the leaching process. In addition, erosion is greatly accelerated in these rainy regions by the absence of both a protective canopy of leaves and branches and the soil-binding influence of tree roots. Flooding also becomes more common when the trees, with their large water-absorbing capabilities, are gone, and the soils become more compact. Finally, some tropical soils tend to bake brick-hard upon exposure to sunlight.

Concern has also been expressed that rainforest destruction may increase global temperatures because the rainforests currently are responsible for the absorption of more atmospheric carbon dioxide than the rest of the world's vegetation combined. Without this continuing CO_2 removal and storage in the form of trees, the greenhouse effect may intensify. The rainforest areas also may become considerably drier because vegetative transpiration provides a substantial proportion of the atmospheric water vapor content for rainfall received in rainforest areas.

The greatest tragedy from a biological standpoint is the impending extinction of countless plant and animal species. The tropical rainforests contain literally millions of species of plants and animals that exist nowhere else on Earth and that would not survive if their habitat were destroyed. Many of the denizens of the rainforests have not yet been studied scientifically; extinction would eliminate them before we have a chance to learn about them or, in some cases, even to become aware of their existence. In addition, these varied life-forms represent a great storehouse of genetic diversity. Their destruction would considerably diminish the planetary pool of genetic resources, probably altering future evolutionary history.

PARTIAL SOLUTIONS

The fate of the tropical rainforests as vast expanses of virgin forest appears to be sealed. The human transformation of the face of the Earth, which is an inevitable consequence of the scientific revolution and the global population explosion, is well underway. Sadly, the maintenance of large uninhabited and unutilized wilderness areas such as the rainforests appears to be incompatible with this transformation.

If the rainforests themselves cannot survive, at least some of the adverse consequences attending their removal can be minimized. For instance, trees can be replanted immediately after logging operations within an area are completed. Replanting would protect the soil and would likely reduce the climatic consequences of deforestation. Little replanting is now taking place. A second action, currently being undertaken by several countries including Brazil, is the establishment of large natural rainforest reserves. These reserves should save many plant and animal species from extinction and preserve for future generations at least some segments of a unique ecosystem.

12 SOILS

FOCUS QUESTIONS

1. What physical and chemical characteristics of the soil most affect its ability to support vegetation?
2. What are the five factors in soil development, and what influence does each have on soil characteristics?
3. What are the major world soil orders, and what are the chief attributes of each?

Soil is one of our most essential natural resources. We depend on it directly or indirectly for nearly all our food, and without it, advanced life-forms could not exist. Humankind has had an intimate relationship with the soil for thousands of years. About 11,000 years ago in Mesopotamia (located in present-day Iraq), agriculture began to replace hunting and gathering, thereby making possible the establishment of large permanent settlements and the rapid growth of human populations.

Most of the world's population still lives close to the land—indeed, more people are supported by agriculture than by all other occupations combined. Both Canada and the United States were settled chiefly by farmers. North American agriculture has become the most advanced and productive in the world, allowing most of the population to pursue other means of earning a livelihood. Despite the advances of the twentieth century, however, a variety of problems raise questions about the ability of North American agriculture to continue its past successes. Two of these problems—erosion and increasing soil salinity—are examined in this chapter.

Unlike the vegetation it supports, the soil is not a living entity, but it is intimately associated with life in several major respects. In the first place, the soil contains a great deal of life. Soil organisms include a tremendous variety of both plants and animals. Much of this life is microscopic, but some, like the root systems of large trees, dwarf all but the largest animals living on the surface. Second, the soil supports nearly all plant life, and this, in turn, supports animal life. Even aquatic plants and animals largely depend upon minerals derived from soil particles eroded from the land by water or wind. A third association with life is the soil's life-like ability to adjust to its environment. The soil is sensitive, to some degree, to almost every aspect of its natural environment. Soil characteristics at any given site therefore are determined by the interplay of past and present environmental factors at that site. The soil's response to its environment is dynamic. Not only does the soil develop characteristics that are in equilibrium with the environment in which it forms, but it also responds to environmental changes such as variations in climate or vegetation. Any such change produces a slow alteration of soil characteristics until a new equilibrium is achieved. The geographic distribution of soil types, then, is not random, but is closely related to other components of the environment.

SOIL ORIGIN AND CHARACTERISTICS

The soil is so complex and variable in composition, and means so many different things to different people, that it is difficult to define. In general, though, **soil** can be described as a loose mixture of weathered rock material, organic matter, water, and air that can support plant growth. Over extended periods, soils tend to develop distinctive layers that differ in their physical, chemical, and biological characteristics. By volume, a typical soil might contain 45 percent inorganic material, 5 percent organic matter, 25 percent water, and 25 percent gases (Figure 12.1). The last two components fill the voids or **pore spaces** between the solid soil particles. The proportions of these four components vary greatly among soils, but nearly all soils contain at least a small quantity of each. Each component plays a vital role in the ability of the soil to perform its most important function—the support of plant life.

The most abundant constituent of most soils is weathered rock material. This material is a major source of the minerals required by plants and animals living in the soil. Weathering, which is discussed in more detail in Chapter 15, involves the action of a variety of physical and chemical processes that break down rock and may eventually convert it to soil-forming materials. Weathering alone does not produce soil, although it certainly is a major and necessary preliminary step. It does, however, form a **regolith** (weathered rock) layer above the unaltered bedrock. The regolith is transformed gradually into soil through physical, chemical, and biological processes. These processes, discussed in subsequent sections of this chapter, involve the movement of some materials, the formation of new materials, the addition of organic matter, and the development of distinctive layers within the soil.

The organic components of the soil consist of the remains of plants and animals. The organic matter undergoes gradual decomposition, which slowly releases its mineral content.

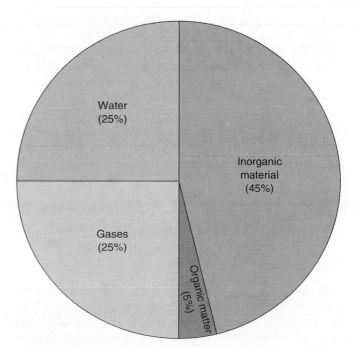

◆ **FIGURE 12.1** Relative proportions of the four components of a typical soil.

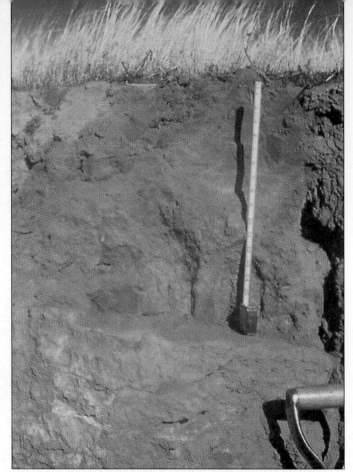

◆ **FIGURE 12.2** Soils with contrasting colors. The soil on the left is a dark, humus-rich South Dakota mollisol that contains a white layer of calcium carbonate ($CaCO_3$) within the E horizon. The soil on the right is an alfisol from California that has been colored orange by iron and aluminum oxides.

SOURCE: *left:* Marbut Memorial Slide Collection/American Society of Agronomy; *right:* Soil Taxonomy AH 436, Soil Conservation Service

Many of these organically derived minerals are required by the soil's living population of plants and animals. Organic matter therefore is an important source of plant nutrients. In addition, chemical reactions involving organic material release nutrients bonded to inorganic mineral particles so that they become available for plant use.

Soil water and air both are derived primarily from the atmosphere. The proportions of these two components are to some extent inversely related, since both share the soil pore spaces. Following a rainstorm, water floods the pore spaces, expelling much of the air. As the water gradually drains away, evaporates, or is used by plants, air refills the pores.

We now examine several important soil characteristics.

Soil Color

Soils exist in a great variety of colors (Figure 12.2). One group varies from white through all shades of gray to black. Other soils range through yellows, reds, and browns. Colors often differ considerably with depth at a single site. Color is not important for its own sake; its significance derives from the information it provides about the mineral content and the physical and chemical condition of the soil. Because color is the most easily observable soil characteristic, it usually is the first noted during an examination. Soil color charts, such as the Munsell chart, which contains 175 color gradations, are used for soil identification and description.

Most soil color is imparted by just two substances—humus and iron. As the humus content increases, soils tend to darken through shades of brown to nearly black. Oxides of iron usually are responsible for imparting a reddish or yellowish color. These two colors are especially typical of the soils of humid tropical and subtropical areas, because iron tends to form relatively insoluble compounds with oxygen under warm, moist conditions. With good drainage and aeration, oxidized iron compounds produce a bright rust red color. If moderately poor drainage conditions prevail, yellow colors are produced by hydrated iron compounds. Under very poor drainage conditions, reduced iron compounds can give the soil a greenish or gray-blue color.

In drier environments, the soil may contain whitish flecks or layers due to the presence of calcium carbonate ($CaCO_3$) or salts. These substances are soluble in water and do not collect in the soils of humid climates. In cool humid climates, nearly white subsoils sometimes exist where iron

◆ **TABLE 12.1** **Soil Texture Grades**

SIZE GRADE	DIAMETER (MM)	DIAMETER (IN)
Gravel	greater than 2.0	greater than 0.08
Very coarse sand	1.0–2.0	0.04–0.08
Coarse sand	0.5–1.0	0.02–0.04
Medium sand	0.25–0.50	0.01–0.02
Fine sand	0.10–0.25	0.004–0.01
Very fine sand	0.05–0.10	0.002–0.004
Silt	0.002–0.05	0.00008–0.002
Clay*	below 0.002	below 0.00008

*The term *clay* denotes the grade, or size, of the soil particles and does not mean that they consist of clay minerals.

compounds and other coloring agents have been removed by downward-percolating water.

Soil Texture

In contrast to soil color, which is important largely for the information it provides about the composition and condition of the soil, the **soil texture,** or size of the individual soil particles, has important direct effects on the soil. Excluding gravel and larger rocks, soils consist of three basic sizes or "grades" of particles. The largest or coarsest grade is **sand,** while **silt** particles have an intermediate texture, and **clay** particles are the smallest or finest. Each texture grade has been assigned a specific range of particle diameters, as indicated in Table 12.1. Only inorganic materials are included when texture measurements are made, organic matter being removed beforehand. Many soils contain gravel and stones exceeding the maximum sand-size diameter of 2 millimeters (0.08 in). Such materials are not considered part of the soil, as they require further weathering to reduce them to soil-sized particles.

A factor complicating the description of soil textures is that nearly all soils contain a mixture of soil particles of differing sizes. Various proportions of particles of different sizes can therefore produce a variety of texture classes. The classes adopted by the U.S. Department of Agriculture, which are in general use in the United States and many other countries, are shown in Figure 12.3 and are explained in the accompanying legend. In general, a soil containing 50 percent or more silt-sized particles has "silt" or "silty" in its name. Clay-sized particles exert a strong influence on the soil, so soils containing more than about 27 percent clay have this term in their names. The finest clay particles are so small that they can remain suspended in water indefinitely and are soft and gelatinous. They are termed **colloids** and, as will be seen, are essential to the biochemical reactions that allow plants to extract nutrients from the soil. Soils that are significantly influenced by all three particle grades, and therefore have an intermediate texture, are referred to as **loam** soils.

The soil texture in any given area is controlled largely by the climate and by the chemical stability of the material from which the soil was derived. Regions with warm, moist climates typically have deep, fine-textured soils because chemical weathering processes are exceptionally active under these conditions. Conversely, cold, dry areas usually have thin, coarse-textured soils, if a soil cover has developed at all. Likewise, soils developing from chemically stable rock or sediment, especially if it has a high quartz (SiO_2) content, tend to be coarse while those developing from parent material with unstable mineral assemblages weather rapidly to produce deeper, finer-textured soils.

The texture of the soil is of practical significance in two major ways. First, it affects the looseness and "workability" of the soil from an agricultural standpoint. Coarse-textured, sandy soils are "light" and easily worked. Plant roots penetrate readily, but the soil may lack the firmness to keep plants from being bent or uprooted during floods or windstorms. Fine-textured soils, in contrast, tend to be dense, sticky when wet, and "heavy." They may be difficult to till, but provide strong physical support for plants.

Second, the soil's ability to absorb and retain water is greatly influenced by soil texture. Coarse-textured soils have relatively large intergranular pore spaces because sand-sized particles generally do not fit together as closely as do silt and

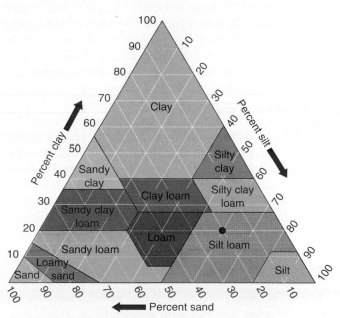

◆ **FIGURE 12.3** The soil texture triangle is used to determine the textural class of the soil, based on the proportions of sand, silt, and clay it contains. Each side of the triangle indicates the percentage content of one of the three texture grades. The three lines that represent the proper proportions of sand, silt, and clay in a soil sample are traced inward to their point of juncture. For example, a soil consisting of 20 percent sand, 60 percent silt, and 20 percent clay is classified as a silt loam. (Note the location of the dot in the diagram for this soil type.)

clay particles. As a result, sandy soils have high water infiltration rates but low water retention capacities because water passes readily through the pore spaces and relatively little adheres to the soil particles. The situation is reversed for fine-textured soils, which have large water retention capacities, but absorb water very slowly. This may lead to poor drainage and excessive runoff, resulting in rapid erosion and in sediment pollution of streams near cultivated fields. Loamy textures generally are considered most favorable for agriculture because they provide a balance between the problems resulting from excessively fine- and coarse-textured soils.

Soil Structure

Individual soil particles normally do not remain detached from one another. If they did, fine-textured soils, when dry, would behave like the finest dust, to be easily washed away by rainwater or blown away by the slightest wind. Instead, soil particles tend to adhere to one another, forming larger aggregates called **peds.** The shape, size, and organization of these peds constitute the soil's **structure.**

Peds are held together largely by the clay-sized particles and by the soil colloids of organic origin. These fine particles develop strong electromagnetic charges that bind the soil particles together. As a result, coarse-textured soils, with little colloidal material, tend to have weakly developed structures or even a complete lack of structure. As a general rule, the coarser the texture of a soil, the weaker its structure. The type and strength of the structure also are influenced by the chemical nature of the clay mineral particles, the type and abundance of soil microorganisms, the frequency and extent of wetting/drying and freezing/thawing cycles, and cultivation practices.

Although a wide variety of specific structural forms can develop, soil structures are divided into four primary groups (Figure 12.4). **Blocky** structures contain peds that are compact and relatively flat-sided. Soils with **platy** structures consist of a large number of thin platelet-shaped peds paralleling the surface. **Prismatic** or **columnar** structures (Figure 12.5) sometimes are associated with excessively drained soils, because the pore spaces between the peds offer water a direct path downward through the soil. They are the most massive of the four types. **Granular** or **crumb** structures are common in soils with a moderately high humus content. The degree of structural development and even the type of structure may vary with soil depth.

One of the most important practical effects of a soil's structure is the influence it exerts on the ease of absorption and movement of soil air and water. A soil with a well-developed structure typically is less compact and has greater permeability and porosity than does a structureless soil. Some types of structure, however, may restrict the absorption of water. For example, a soil with a strongly developed platy structure may be poorly drained because each "plate" acts as a barrier to water infiltration.

Blocky

Crumb

Columnar (Tops rounded)

Prismatic (Tops level)

Platy

◆ **FIGURE 12.4**   Appearance of peds in five important soil structural groups.

The strength with which the individual peds are internally bound influences both the soil's resistance to erosion and the ease of cultivation. A strong structure holds the soil together and causes it to resist erosion. This same characteristic, though, makes the soil more difficult to plow. Plowing tends to alter and usually to weaken soil structure, and the passage of farm machinery leads to soil compaction. As a result, the soils of agricultural areas typically have poorly developed structures and lower water-infiltration capacities than do nearby unfarmed areas.

Soil Air and Water

Although soil water and air come mostly from the atmosphere, their chemical constituencies are modified by their residence in the soil. These chemical changes have a significant effect on the soil biota.

The soil atmosphere is critical to soil organisms because it supplies the oxygen needed for animal and plant respiration and the carbon dioxide required for plant photosynthesis. A lack of oxygen in a saturated soil can suffocate both plants and animals. In addition to air from the atmosphere that enters the soil through its pore spaces, the soil also contains gases released from chemical reactions and biological activities. As a result of evaporation and plant respiration, soil air contains slightly less oxygen and nitrogen and somewhat more water vapor and carbon dioxide than does the atmosphere.

Soil water, of course, is derived primarily from precipitation. During a heavy rainfall, large amounts of water typically enter the soil. Most passes downward through the pore spaces toward the water table in response to gravity. If the soil is freely drained, the excess water takes about two days to pass completely through. The soil at this point retains its

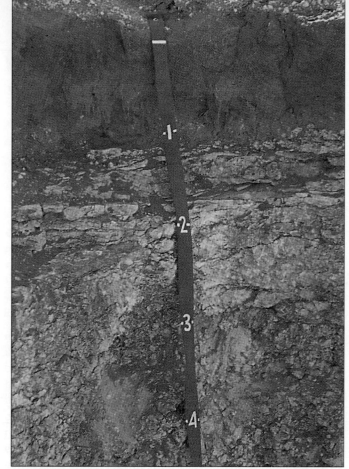

◆ **FIGURE 12.5** A duricrust layer appears between the 1- and 2-foot levels in this aridisol from southern New Mexico. The horizontal structure of this layer contrasts markedly with the vertical prismatic structure of the overlying A horizon.

SOURCE: Marbut Memorial Slide Collection/American Society of Agronomy

field capacity for water, which is the maximum amount of water the soil is capable of holding against the pull of gravity. This water is vital to the soil's plant and animal population because it is retained until used or evaporated.

Soil water can be held against the pull of gravity because it is attracted to the surfaces of the soil grains by **capillary tension.** This property results from the electromagnetic charges developed by the soil particles. Because fine-textured soils have a large total soil surface area per unit of volume, they have higher water-retention abilities than do medium- or coarse-textured soils. Plants are able to withdraw much of the soil water for their own use, but finally, at the **wilting point,** the remaining water is bound so tightly to the soil particles that plants are unable to extract it.

Water in the soil dissolves soil minerals, making them available to plants as they absorb water through their root systems. As water passes through the soil, dissolved minerals and fine-textured particles are carried, or **translocated,** from one place to another. These materials often are deposited a considerable distance from their points of origin.

In humid climates, an excess of downward-percolating water removes minerals from the upper soil in a process called **leaching.** Leaching involves the removal or **eluviation** of dissolved and fine-textured solid materials from one portion of the soil, followed by the translocation and deposition (or **illuviation**) of these materials in another, usually deeper, layer of the soil. Leaching, where strongly developed, often results in soluble plant nutrients being deposited so deeply in the soil that they become inaccessible to most vegetation; under very wet conditions, many of these materials are carried completely out of the soil. The leaching process is the primary reason for the general lack of fertility of soils in humid climates. Fertilizers may be used to replace leached or otherwise unavailable plant nutrients temporarily, but these, too, are susceptible to leaching so that reapplications on a periodic basis are necessary.

In arid climates, the drying of the surface soil causes capillary water to be drawn toward the surface. This water transports minerals dissolved from the parent material or lower soil upward to be deposited near the surface in a sort of reverse leaching process. Although the upper soil's supply of plant nutrients may be increased by this process, some of the substances brought up are toxic to most plants. A prime example is provided by salts of various types. In many arid areas where water is available for agriculture, the high salinity of the soil makes the raising of crops impossible or impractical. The use of irrigation water has in many cases intensified the problem by dissolving large quantities of salts in the lower soil layers and carrying these salts to the surface. The problem is worsened if the irrigation water itself has a significant dissolved salt content. Approximately one-third of the irrigated farmland in the western United States has been damaged by increasing salinity, and some once-productive land has had to be abandoned. The problem of soil salinization resulting from irrigation is discussed in the Focus Box on pages 300–301.

The surfaceward movement of water also may deposit substances that form a resistant **duricrust layer** on or just beneath the surface (Figure 12.5). When well developed, a duricrust layer may resemble a buried layer of concrete in its ability to block the penetration of water or plant roots, and under extreme conditions, dynamite has been employed to break it up.

As a general rule, the most naturally fertile soils exist in subhumid or semiarid climates, where precipitation approximately equals evaporation. Under these conditions, vertical translocations of soil water and dissolved minerals in either direction are minimized.

Organic Matter in the Soil

Organic matter is incorporated into most soils in the form of humus. **Humus** is finely divided, decomposed organic matter with a brownish black color. Humus is formed from the remains of both plants and animals, but for the most part, it has a vegetative origin, being derived largely from leaves,

FOCUS BOX

SOIL SALINIZATION—A GROWING PROBLEM IN ARID LANDS

The soils of many desert regions are inherently fertile, and the use of irrigation water can make these regions "bloom," creating some of the most productive agricultural lands on Earth. Irrigated cropland, which constitutes about 15 percent of all cultivated land, produces up to 30 percent of the world's food supplies. The extensive irrigation of California's Imperial Valley by water from the Colorado River, for example, has helped make that state the nation's leader in value of agricultural production. Unfortunately, the irrigation of arid lands often causes a gradual increase in soil salinity. This problem, unless controlled, eventually can ruin even the most productive agricultural lands. Approximately one-fourth of the world's irrigated land suffers from excessive salinity, and that figure is increasing.

◆ **FIGURE 12.9** Salt-encrusted farmland in California's Imperial Valley that has been severely damaged by salinization caused by the surfaceward movement of irrigation water.

SOURCE: Science VU/Visuals Unlimited

Salinization of soils caused by irrigation is probably the most ancient of all factors promoting desertification (see the Case Study at the end of Chapter 18). In the fourth millennium B.C., the Sumerians occupied the southern Tigris-Euphrates Valley and built a prosperous civilization based on irrigation agriculture. By the end of the second millennium B.C., this region, often referred to as one of the "cradles of civilization" in world history texts, was so badly salinized that agriculture was largely abandoned. Even today, many parts of this valley are semidesert wastelands.

The rapid spread of irrigation agriculture has greatly increased the geographical scope of the problem during the past century. For example, extensive areas of Pakistan have been affected, as have large portions of the cotton-producing lands of southwestern Asia. Salinization is becoming a

Soils develop horizons for two basic reasons. First, as we have seen, the soil is sensitive to its environment, and nearly all **pedogenic** (soil-forming) environmental controls vary rapidly in a vertical direction. Second, vertical translocations of materials occur in all soils because of the movements of soil water. As a result, selected materials are eluviated from some portions of the soil and illuviated into others.

The soil horizons are identified by a lettering system, as shown in Figure 12.10. From the surface downward, five major horizons, designated the O, A, E, B, and C horizons,

are recognized. To provide greater precision, transitional horizons often are identified. For example, the subhorizons EB and BE are sometimes recognized between the E and B horizons. The characteristics and extent of development of horizons vary greatly in differing pedogenic environments, and some of the subhorizons, or even horizons, may not be present in a given soil sample.

The uppermost horizon, referred to as the **O horizon,** contains a substantial proportion (20–30 percent or more) of fresh or partially decomposed organic matter. This horizon problem in Egypt's Nile Valley now that the Aswan High Dam has ended the annual flood that once flushed the salts from the soil. Extensive portions of the American West also suffer from this problem, and some areas have already been abandoned (Figure 12.9).

THE SALINIZATION PROCESS

Soil salinization is a problem because salts in large quantities are toxic to most crops. Common table salt (NaCl) is the worst offender. As it dissolves, the sodium cations increase the pH of the already alkaline soil to 9 or 10. Sodium attacks the soil structure, disintegrating the structural peds that provide channels for the downward percolation of moisture. Precipitated salts clog the soil pores, making the soil impervious to air and water and producing a white crustal layer.

Most irrigation systems use canals to transport water from a river or well to the fields. These fields are periodically flooded, either wholesale or along furrows. The fields often are not leveled, so that excess water is needed to flood the higher portions. Frequently, two or three times the amount of water actually needed by the crops is used. This excess water percolates down to the water table. If natural groundwater drainage is poor and no artificial drainage system is provided, the water table rises. The rising water carries with it dissolved salts from both the lower soil and the irrigation water (which in arid regions often is slightly saline). As capillary action causes the water to approach the surface, the water evaporates and the salts are left behind. Salinization and waterlogging increase in severity until further farming becomes uneconomical and the land is abandoned.

POSSIBLE SOLUTIONS

There are a number of existing and potential solutions to the soil salinization problem that provide a measure of optimism for the future. In areas where fresh water is abundant and the water table is not too high, the application of large quantities of water can flush away the accumulated salts. If the irrigation water itself contains a large quantity of dissolved salts, it can be desalinated before use. Unfortunately, this is a costly process because of the large amounts of energy required.

Irrigated fields in some regions now employ "tile" drainage—a system of perforated plastic pipes under the surface that conducts away excess irrigation water and dissolved salts. This system is expensive to install, but is widely used in California's Imperial Valley and in the Central Asian republic of Uzbekistan.

The process of drip irrigation, pioneered by the Israelis in the 1960s, currently is considered one of the most promising methods of conserving water and combating salinity in arid lands. It employs a system of aboveground plastic pipes containing perforations that drip water into the soil directly beside each plant.

Potentially, the most promising solution of all involves the genetic engineering of crops to produce salt-tolerant varieties. Irrigation water containing up to 3,000 ppm (parts per million) of dissolved salts is now being routinely used in some areas for a wide variety of salt-resistant crops, including wheat, cotton, potatoes, and sugar beets. Some crops are now being irrigated with water containing up to 10,000 ppm dissolved salts—close to one-third the salinity of seawater. Researchers dream of developing strains of crops that can be irrigated with seawater itself, thereby assuring humanity of an unlimited supply of irrigation water. This dream may someday become a reality; already salt-resistant varieties of barley have been grown with seawater irrigation under laboratory conditions. ∎

develops only where a cover of natural vegetation exists and the climate is sufficiently cool to allow organic matter to accumulate. It therefore is absent in arid and some tropical regions.

The second, or **A horizon,** often is referred to as the **topsoil.** It contains a mixture of inorganic and organic matter (much of the latter in the form of humus) and typically is low in density and dark in color. The topsoil generally is the most fertile soil horizon and is the horizon in which most seeds germinate.

Below the A horizon is the **E,** or **eluvial horizon.** It is a mineral horizon that is lighter in color than the overlying horizons, and is best developed in high rainfall areas, especially under a forest cover. Clay-sized particles and minute organic substances have been washed (eluviated) from it by downward-percolating water.

The underlying **B horizon,** sometimes termed the **subsoil,** often is a zone of illuviation. It typically is denser than the A and E horizons and may develop zones of clay accumulation. In some cases, iron oxides give the B horizon

Zone of eluviation
- O — Loose and partially decayed organic matter
- A — Fine-textured humus
- E — Light-colored mineral particles
- EB — Transition to B; more like E than B
- BE — Transition to B; more like B than E

Zone of illuviation
- B — Colloid accumulation; duricrust likely
- BC — Transition to C

C: either a deposit or material weathered from bedrock

R — Unweathered bedrock

◆ **FIGURE 12.10** Idealized soil profile from a humid climate region in the middle latitudes. The relative positions of the various horizons are indicated.

a reddish color. Unless humus has been translocated downward from the A and E horizons, the B horizon contains little organic material. The A, E, and B horizons together comprise what is sometimes referred to as the "true soil," or **solum.**

The **C horizon** consists of weathered rock material that as yet has been little affected by pedogenic processes. This material currently is undergoing changes that eventually will convert it to soil. Rapid weathering in the tropics can form a C horizon more than 50 meters (150 ft) thick. In the middle latitudes, thicknesses of 1 to 3 meters (3–10 ft) are typical, but much greater thicknesses occur in some places. Below the C horizon lies unweathered bedrock or other parent material, sometimes labeled the **R horizon,** that represents the raw material for future soil.

Considerable time is required for soil horizons to develop. For this reason, the presence of well-developed horizons indicates that the soil is mineralogically mature and that environmental conditions have remained relatively constant over a prolonged period.

Pedologists study soils by examining **soil profiles,** which are vertical slices of the soil in which the various horizons are exposed. A number of photographs of soil profiles appear in this chapter. Each soil type exhibits a unique profile that can be used to identify it. In this respect, a soil profile is much like a fingerprint, no two of which are exactly alike. A person's fingerprint, though, tells little or nothing about the characteristics of its owner, whereas a soil profile can provide a great deal of information about a soil's physical and chemical characteristics. In many places soil profiles have been destroyed by human activities such as plowing and the construction of buildings and highways.

FACTORS IN SOIL DEVELOPMENT

A mature soil exemplifies a system in **dynamic equilibrium** with its environment. The term "dynamic equilibrium" may seem paradoxical, since dynamic implies constant change and equilibrium implies balance or changelessness. The two terms, however, each refer to a different attribute of the soil. The soil is dynamic because it undergoes a constant progression of soil-forming activities. Each soil mineral particle passes, in turn, through a sequence of horizons, beginning as unweathered material and eventually being eroded at the surface as A horizon material. On the other hand, the soil can achieve and maintain equilibrium with its environment. If the environmental controls in a given area remain constant long enough, the soil type will remain the same even though different material is constantly passing through the soil "assembly line" from initial weathering to erosion.

To this point, our discussion has centered on the physical and chemical characteristics of soils. Although we have briefly alluded to the reasons, or controls, for some of these characteristics, we have not covered this subject in any systematic fashion. It is the geographic distribution of the pedogenic controls, however, that has determined the global pattern of soil types. Before proceeding to a discussion of world soil types, then, we need to examine the five major factors controlling the type of soil that will develop in any given locality. These five factors are summarized by the so-called clorpt equation, devised by H. Jenny, which can be written as follows:

$$\text{Soil} = f\,(\text{CL, O, R, P, T})$$

This equation indicates that the soil at any given site is a function of climate, organisms, relief, parent material, and time.

Parent Material

The first factor to influence a newly developing soil is its **parent material,** the material from which the soil develops. The parent material is the "raw material" of the soil, which undergoes progressive physical and chemical alterations as the soil forms. As might be expected, the parent material has the greatest influence on recently formed soils. Its influence diminishes, but does not entirely disappear, as the soil matures. Pedogenic processes operating in differing environments can cause greatly differing soils to form from the same parent material.

Nearly all parent materials formed from combinations of the most abundant elements in the Earth's crust. Table 12.2 lists the eight elements that individually comprise 2 percent or more, by weight, of the crust. Differing parent materials vary greatly in the numbers and proportions of the elements they contain. In addition, some of these elements are essential for plants, while others are not. As a result, the type of parent material is an important factor in determining soil

◆ TABLE 12.2   Abundant Elements of the Earth's Crust

ELEMENT	PERCENTAGE OF CRUST BY WEIGHT
Oxygen	46.6
Silicon	27.7
Aluminum	8.1
Iron	5.0
Calcium	3.6
Sodium	2.8
Potassium	2.6
Magnesium	2.1
Total	98.5

fertility. For example, soils formed from sandstone or quartzite generally are much less fertile than those formed from limestone or basalt because the latter two rock types contain more of the minerals that provide important plant nutrients.

The mineral content of the parent material also plays a part in determining how rapidly the soil matures. If chemically unstable minerals predominate, their rapid weathering may soon produce a soil bearing little similarity to its parent material. Conversely, a predominance of stable minerals results in slower weathering rates, slower soil formation, and less variation of the soil from its parent material.

Areas with complex geologic histories may contain a variety of parent materials for soils to develop upon. Soil types often change abruptly as boundaries between different parent materials are crossed. Conversely, the soils over a large area may be made more similar if the entire area is underlain by the same type of parent material.

From the standpoint of origin, soil parent materials often are designated as either residual or transported. **Residual** parent materials consist of weathered or unweathered bedrock that remains in place before and during the pedogenic process. **Transported** parent materials, which are geographically widespread, have been brought in from other sites through the actions of gravity, water, ice, or wind before being converted to soils. These transport mechanisms and the chief areas in which they operate are discussed in some detail in Chapters 15 through 19.

Relief

Topographic relief, like parent material, is a pedogenic control that can vary greatly over a very short distance, leading to a variety of localized soil types. As noted in Chapter 10, both slope steepness and aspect (directional orientation) influence the soil.

Most differences in soil characteristics relating to slope steepness occur because of its effects on soil erosion and drainage. Because erosional processes increase in effectiveness as slope angles increase, soils on steep slopes tend to be thin, coarse textured, and poorly developed. Soil depths decrease as steepness increases until, at slope angles exceeding about 45°, loose material washes, slides, or falls off slopes as soon as it forms, leaving bare rock surfaces. Steep slopes are largely responsible for the stark, rocky surfaces of the mountains of much of western North America as opposed to the more gently sloping soil-covered and forested Appalachians.

The combination of thin, coarse-textured soils and steep slopes frequently results in excessive drainage. This may cause soils in mountainous areas to be droughty despite precipitation amounts that would normally be considered adequate. On the other hand, flat valley areas, which receive both water and soil from adjacent slopes, tend to have deep and often poorly drained soils. Globally, the best-developed and most agriculturally productive soils are located in low relief areas.

Slope aspect affects soil for the same reasons that it affects vegetation. North-facing slopes in the northern hemisphere and south-facing slopes in the southern hemisphere are inclined away from the Sun. As a result, they develop under conditions of lower temperature and greater moisture (resulting from decreased evapotranspiration rates) than do slopes oriented toward the Sun. East- and west-facing slopes, as well as all slopes in the tropics, are not so greatly influenced, because the directional orientation of sunlight is less constant in these settings.

Organisms

Organisms play an essential role in soil formation. The numerous plants and animals living in the soil release minerals from the parent material, supply organic matter, aid in the translocation and aeration of the soil, and help protect the soil from erosion. The types of soil organisms present greatly influence the soil's physical and chemical characteristics. In fact, for mature soils in many parts of the world, the predominant type of natural vegetation is considered the most important direct influence on soil characteristics. For this reason, the description of a soil as a "prairie soil" or "tundra soil" conveys to the pedologist a great deal of information about the soil's attributes.

The quantity and total weight of soil flora generally far exceed that of soil fauna. By far the most numerous and smallest of the plants living in the soil are the **bacteria.** Under favorable conditions, a million or more of these tiny, single-celled plants can inhabit each cubic centimeter of soil. It is the bacteria, more than any other organisms, that enable rock or other parent material to undergo the gradual transformation to soil. Some bacteria produce organic acids that directly attack soil parent material, breaking it down and releasing plant nutrients. Others decompose organic litter to form humus. A third group inhabits the root systems of plants called legumes. Many important agricultural crops, including alfalfa, clover, soybeans, peas, and peanuts, are legumes. The bacteria they host within their root nodules

Litter fall

Humus-rich A horizon

Downward leaching of dissolved organic materials

Absorption and recycling of organic materials

◆ **FIGURE 12.11** Trees help to maintain soil fertility by drawing up nutrients from the deeper soil layers through their root systems. These materials are subsequently added to the surface soil through the fall of organic litter.

resentatives of the animal kingdom, and like the bacteria, a million or more can sometimes inhabit each cubic centimeter of soil. Protozoa feed on organic matter and hasten its decomposition.

Of the larger animals, the earthworm probably is the most important. Up to a million earthworms, with a total body weight exceeding 450 kilograms (1,000 lb), may inhabit an acre of soil under exceptionally favorable conditions. As they feed, earthworms ingest large quantities of soil, chemically alter it, and excrete it as casts. The casts form a high-quality natural fertilizer. In addition, earthworms mix soil both vertically and horizontally, improving aeration and drainage.

Insects such as ants and termites also can be exceedingly numerous under favorable climatic and soil conditions. In addition, mammals such as moles, field mice, gophers, and prairie dogs sometimes are present in sufficient numbers to have a significant impact on the soil. These larger animals primarily work the soil mechanically. The soil is aerated, broken up, fertilized, and brought to the surface, hastening soil development.

Climate

While natural vegetation may be considered the most important single control determining the characteristics of a mature soil in a direct sense, climate certainly is the most important control in an indirect or ultimate sense. This is because climate is the major controlling factor for the world distribution of natural vegetation. Indeed, climate has at least some effect on all of the factors in soil development discussed in this section. Only the nature of the residual parent material (if present) and, in some cases, the amount of time the soil has had to develop are independent of at least some climatic influence. The same two climatic elements that most strongly affect human activities—temperature and precipitation—exert the greatest effect on the soil (Figure 12.12). Extremes of either element have an unfavorable effect on soil characteristics.

Soil formation begins when minerals in the parent material combine chemically with water from precipitation. If this water is not available, as in an arid climate, most physical and chemical weathering reactions cannot occur, and a mature soil will not form. In addition, a lack of precipitation means that little moisture will be available to support biological activity and humus formation. Conversely, an excess of precipitation may lead to conditions of poor drainage, low aeration, leaching, the formation of organic acids, and possible excessive erosion. Many of the soil-forming weathering reactions again will be inhibited, especially by a lack of oxygen, and most plant organisms will not be able to survive. An intermediate amount of precipitation therefore is most favorable for the formation of fertile soils.

The precise amount of precipitation needed for optimal soil development varies greatly from place to place, depending on factors such as evapotranspiration rates, slope steep-

have the unique ability to extract nitrogen from the atmosphere and change it to a form available to plants. This process, termed **nitrogen fixation,** can actually make the soil more fertile. Other microscopic plants also are important in soil development. For example, in highly acidic soils where few bacteria can survive, fungi frequently become the chief decomposers of organic matter.

Higher vegetation plays several vital roles with respect to the soil. Trees, grass, and other large plants supply the bulk of the soil's humus. The minerals released as these plants decompose constitute an important nutrient source for succeeding generations of plants as well as for other soil organisms. In addition, trees in particular are able to reach and recycle nutrients located deep within the soil, returning them to the surface (Figure 12.11). Finally, plants perform the vital function of slowing runoff and holding the soil in place with their root systems, thus combating erosion. The accelerated erosion that often accompanies agricultural use of sloping land is principally caused by the removal of its protective cover of natural vegetation.

The faunal counterparts of the bacteria are the **protozoa.** These single-celled organisms are the most numerous rep-

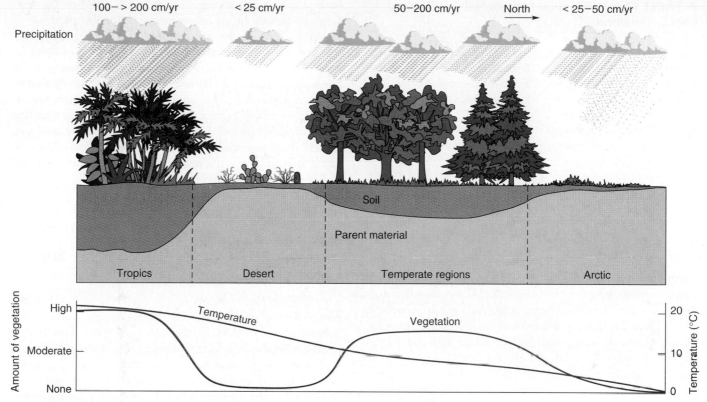

Precipitation

100- > 200 cm/yr   < 25 cm/yr   50-200 cm/yr   North   < 25-50 cm/yr

Soil

Parent material

Tropics   Desert   Temperate regions   Arctic

◆ **FIGURE 12.12** Latitudinal representation of soil depths in relation to temperature, precipitation, and vegetation type and density. Soil-forming processes operate most vigorously where temperatures and precipitation are high and vegetation is abundant. The left margin of the diagram represents the equator, while the right margin represents the North Polar region.

ness, and the nature of the parent material. As a general rule, optimal conditions exist in areas where precipitation amounts roughly equal evapotranspiration rates. When this balance occurs, no large-scale translocations of soil water and minerals take place, and neither significant leaching nor duricrust formation occurs. Table 12.3 summarizes the differences in soil characteristics between excessively wet and dry environments. It should be kept in mind that the most productive soils develop under moisture conditions approximately midway between those represented in the table.

In the absence of inhibiting factors such as insufficient moisture, the rate of soil formation is positively correlated with the mean temperature of an area. The tropics therefore contain the deepest soils, and soil depths generally decrease poleward. In fact, soil formation rates in the humid tropics average about 9 or 10 times those of the subpolar regions because the rate of chemical weathering is greatly accelerated by warm temperatures and ceases entirely when the soil is frozen. The soils of warm and humid environments also tend to display an increased rate of differentiation from the characteristics of their parent materials. This is because they pass through a greater variety of chemical weathering reactions in a shorter time than do the soils of colder climates.

Another important difference between the soils of warm and cold environments is that soil microorganisms are much more numerous and active in warm environments. This factor, in turn, affects the rate of humus formation and consumption. In cold, heavily vegetated regions such as the

◆ **TABLE 12.3** Differences between Soils in Wet and Dry Environments

WET	DRY
1. Deep (unless formation inhibited by standing water)	1. Shallow
2. Intermediate to fine texture	2. Coarse texture, stony
3. Chemical weathering predominant	3. Physical weathering predominant
4. Leached of soluble minerals	4. Near-surface accumulation of soluble minerals
5. Acidic condition	5. Alkaline condition
6. Variable humus content (depending on temperature and abundance of vegetation)	6. Low humus content due to lack of vegetation

WARM	COLD
1. Deep	1. Shallow
2. Fine textured	2. Coarse textured
3. Chemical weathering predominant	3. Physical weathering predominant
4. Rapid rate of soil formation	4. Slow rate of soil formation
5. High degree of mineralogic maturity	5. Low degree of mineralogic maturity
6. Soil microorganisms abundant	6. Soil microorganisms sparse
7. Humus lacking	7. Humus abundant
8. Less acidic	8. More acidic

northern coniferous forests, humus forms relatively slowly. Once formed, though, it may remain for many years before being consumed by microorganisms. As a result, humus in the high latitudes can accumulate to excessive amounts, leading to peaty soils and highly acidic soil conditions. Within the tropics, conversely, organic materials on or in the soil are rapidly consumed by microorganisms, producing soils deficient in humus. As a consequence, intermediate temperatures produce soil conditions that generally are most favorable for agriculture. As at least a partial result of this fact, the most fertile soils are concentrated chiefly in the middle latitudes. Table 12.4 summarizes the major characteristics of soils developing in warm and cold environments.

Time

Each of the four factors in soil formation just discussed needs time to influence the soil. Therefore, a fifth and final essential factor determining the characteristics of the soil in any locality is the amount of time that has elapsed since soil formation began.

The amount of time needed for soil formation and maturation varies geographically, making it impossible to cite precise figures that will hold true everywhere. A rough approximation of 2.5 centimeters (1 in) per century as a world average rate of soil formation is sometimes given, but actual rates vary so greatly in different areas that a world average is of limited value. In general, soil development is fastest where the climate is warm and moist, where soil microorganisms are abundant, and where the soil parent material is composed primarily of chemically unstable minerals that weather quickly. Specifically, average rates of development are most rapid in the humid tropics and slowest in polar and arid environments. Rates of soil formation range from almost zero in extremely cold and arid environments to measured rates of 1 centimeter (0.4 in) per *year* on recent volcanic ash in humid tropical locations.

Even at a particular site, rates of soil formation and development do not remain constant over time. A recently formed soil is not in equilibrium with many of the soil-forming factors in its environment. Change therefore is relatively rapid. As the soil matures, it gradually approaches equilibrium with its environment, and the pace of development slows. Eventually, a steady-state condition may be reached in which soil characteristics remain essentially unchanged with time. This condition of equilibrium can be attained only if all factors affecting the soil remain constant for a prolonged period. In the face of major recent geologic, climatic, and especially human alterations, few places on Earth currently have the necessary degree of environmental stability for this situation to exist.

SOIL CLASSIFICATION AND WORLD SOIL ORDERS

The geographic pattern of world soil types is highly complex. As is true of all natural Earth phenomena, however, the pattern is not random; it displays an organization that is closely related to the distribution of the pedogenic factors just discussed. Because the two most important pedogenic controls are climate and natural vegetation, the pattern of world soil types is more closely correlated with these two factors than with the other pedogenic influences. It should be stressed, though, that each of the five factors controlling soil development is capable, under certain conditions, of exerting a dominating influence on local soil characteristics, and that each will exert at least some influence on the soil at any given site. Some of these factors, such as parent material and slope, can produce large differences in soil characteristics over very short distances, especially in areas of rugged topography.

Soil Classification Systems

Numerous attempts have been made to classify soils. The problems involved in devising an "ideal" soil classification system, however, are even more imposing than those encountered in developing classification systems for climate and for natural vegetation. Among the major problems are the following:

1. Soils display numerous physical, chemical, and biological characteristics. Which ones should be stressed in a classification system? A key factor is the relative importance of the various soil characteristics. As already noted, the importance of each characteristic varies spatially, depending on the soil's environmental setting.
2. Significant variations in soil characteristics exist on a much more localized geographic scale than they do for climate and for natural vegetation. Any detailed soil classification system therefore must be exceedingly complex, and at its most detailed level of classification will contain thousands of soil types. There is, in fact, no set

limit to the number of soil types that exist, and any soil classification scheme will, to some extent, be imposing arbitrary divisions on what is actually a continuum of soil types.

3. In many areas soils have been altered greatly by human activities. Should the classification system recognize these modifications, or should it be based only on natural factors?

Differences in soil types have been recognized from antiquity. Early soil classification systems, however, were primarily descriptive and differentiated soils on the basis of physical characteristics such as color, structure, and depth. European researchers in the early and middle 1800s assumed that regional soil differences were produced chiefly by variations in parent material and used this factor, along with the texture of the topsoil, as the primary basis of classification.

Russian soil scientists first developed modern concepts of soil formation and classification in the late 1800s. They discovered that the soil is not a static medium produced merely by the weathering of parent material, but rather that it evolves through time as the result of energy and material gains and losses from climate and biological activity. Emphasis on soil classification therefore shifted from parent material to the process of soil development and evolution as determined by the analysis of soil profiles. Further work, especially that undertaken by pedologists in the U.S. Soil Survey of the Department of Agriculture, led to the publication in 1938 of a genetic (formation-oriented) soil classification system that came into widespread use around the world. This system, which was modified slightly in 1949, was in official use in the United States for 27 years.

The Soil Taxonomy

During the 1950s, growing dissatisfaction with various shortcomings of the Russian-American system led the U.S. Soil Conservation Service to appoint a group of pedologists to devise a completely new soil classification system for official use. This system, which was called the "U.S. Comprehensive Soil Classification System," was first published in 1960 and was adopted for official use in the United States in 1965. The system, often referred to as the **"Soil Taxonomy,"** has been expanded and refined and remains in use today. It is the system used in the inventory of world soil types appearing in this section. It should be noted that although a number of countries use the Soil Taxonomy, others, including Canada, have their own national soil classification systems. As a consequence, no one soil classification system has a clear dominance in world usage.

The major characteristics of the Soil Taxonomy are as follows:

1. Soils are classified on the basis of their characteristics rather than their method of formation. It is thus a descriptive, rather than genetic, system.

2. Soils modified by human activities are classified along with "natural" soils. (The earlier 1938 system did not recognize the profound influence of human activities on soil characteristics.)

3. The names of soil types convey exact information concerning their characteristics. Every syllable has a meaning, so that each name describes the soil in almost the manner of a chemical formula.

The system is organized into a hierarchy of six levels of classification, with individual soil types forming the lowest level, as follows:

> Orders
> Suborders
> Great groups
> Subgroups
> Families
> Soil series

Eleven soil orders exist in the system. The number of entries increases at each lower level; for example, over 13,000 soil series are recognized in the United States alone. (Several times as many series exist for the entire world, but they are as yet incompletely classified.) The chief characteristics of the 11 soil orders are described in the next section and summarized in Table 12.5. Their geographical distribution is mapped in Figure 12.13. As this map shows, some soil orders are much more geographically extensive and important than others.

World Soil Orders

Oxisols

The **oxisols** (*oxi* for "oxide") are the deepest and most chemically altered of the soil types. Almost all minerals other than quartz have been weathered to form oxides and hydroxides, and virtually all soluble bases have been leached from the upper soil layers. Humus concentrations are very low despite the typical dense vegetation cover, because the warm, moist climate promotes the rapid consumption of organic material by bacteria and insects. Oxisols occur on ancient, geologically stable surfaces within the humid tropics, especially in South America and Africa.

The intense leaching that characterizes these soils results from heavy precipitation, which at least seasonally greatly exceeds evapotranspiration. Not only are dissolved soil constituents transported downward, but even clay-sized particles frequently are removed as they are flushed through the pore spaces separating the larger soil grains. Much of the leached soil material is carried completely through the soil profile and is lost permanently from the soil. Because there is no soil layer in which any substantial illuviation takes place, soil horizons are rather weakly developed and difficult to recognize (see Figure 12.14 on page 310).

Intensely leached soils usually are highly acidic because bases are especially susceptible to removal. The acidity of the

◆ **FIGURE 12.13** Global distribution of the soil orders of the U.S. Comprehensive Soil Classification System (the Soil Taxonomy).

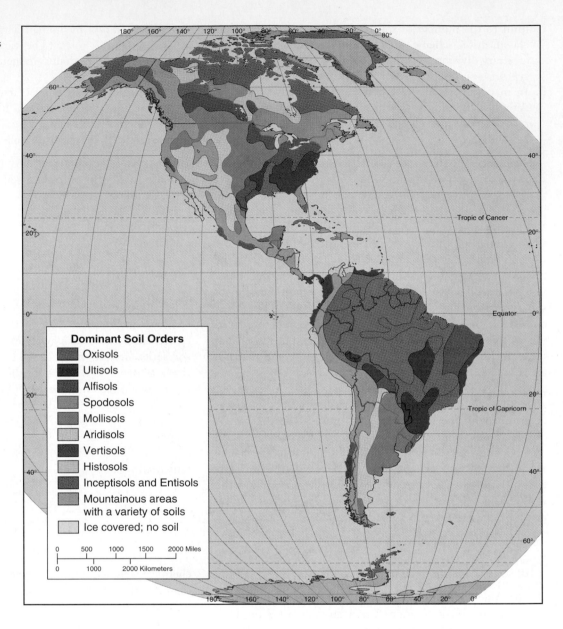

Dominant Soil Orders
- Oxisols
- Ultisols
- Alfisols
- Spodosols
- Mollisols
- Aridisols
- Vertisols
- Histosols
- Inceptisols and Entisols
- Mountainous areas with a variety of soils
- Ice covered; no soil

◆ **TABLE 12.5** U.S. Soil Taxonomy Soil Orders

ORDER	EQUIVALENT SYSTEMS: 1938 GENETIC SYSTEM (CANADIAN SYSTEM)	GENERAL LOCATION AND CLIMATE	WORLD LAND AREA (%)	U.S. LAND AREA (%)
Oxisols	Latosols, lateritic soils	Warm, humid tropical regions	9.2	0.02
Ultisols	Red-yellow podzolic	Humid subtropical forest regions	8.5	12.9
Alfisols	Gray-brown podzolic, degraded chernozem (luvisol)	Humid temperate forest regions	14.7	13.4
Spodosols	Podzol, brown podzolic (podzol)	Cool, humid regions; needleleaf forest cover	5.4	5.1
Mollisols	Chernozem, chestnut (chernozemic)	Subhumid and semiarid temperate grasslands	9.0	24.6
Aridisols	Red desert soils, gray desert soils, sierozems	Tropical, subtropical, and temperate deserts	19.2	11.5
Vertisols	Grumusols, tropical black clays	Restricted semiarid grassland regions	2.1	1.0
Histosols	Peat, muck, bog (organic)	Poorly drained lowlands and depressions	0.8	0.5
Entisols	Azonal soils, tundra soils	Recently formed or disturbed soils in all climates	12.5	7.9
Inceptisols	Subarctic brown forest soils, lithosols (brunisol, cryosol, gleysol wet)	Immature soils mostly within mountain areas, floodplains, or tundra regions	15.8	18.2
Andisols		Active volcanic regions	<1.0	<1.0

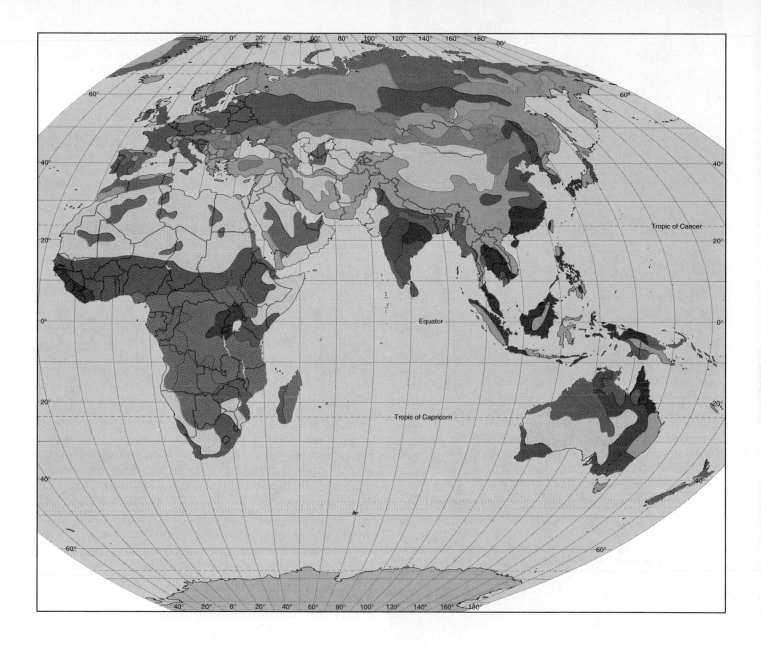

oxisols, however, is kept in check by their lack of humus and its associated organic acids. An only slightly acidic pH, usually in the 5 to 7 range, results. At this acidity level, silica, which usually is a highly stable and inert soil component, enters into weathering reactions that result in its solution. Large amounts of silica therefore are leached from the soil. Iron and aluminum, conversely, weather to form insoluble oxides that accumulate in the soil. Iron compounds typically impart an orange-red color that is one of the most striking characteristics of the soils of the humid tropics.

The low natural fertility of the oxisols presents imposing problems for agricultural use. Despite the presence of year-round warmth and often abundant moisture, large amounts of fertilizer are required for any sustained agricultural use of these soils due to their lack of humus and base nutrients. The removal of the trees from forested areas often causes further deterioration of soil quality. Trees draw up a modest supply of soluble bases, which are placed on the surface as

the plants die and decompose. A nutrient-cycling system is thus established, and a limited amount of soil fertility is maintained. Tree removal causes these bases to be leached, increases the likelihood of erosion, and in some places encourages the formation of a rock-hard subsurface layer of iron oxide called **laterite.** The current large-scale removal of tropical forests, discussed in the Case Study at the end of Chapter 11, is causing the general deterioration of oxisols over large areas.

Ultisols

The **ultisols** (*ult* for "ultimate") are highly chemically weath-ered soils that have developed under warm, moist climatic conditions. They are somewhat similar to the oxisols, but display less mineral alteration.

Ultisols have a distinctly reddish A horizon that has been colored by iron and aluminum compounds and is low in

◆ **FIGURE 12.14A** An oxisol developed from weathered basalt on the island of Kauai, Hawaii.

SOURCE: Marbut Memorial Slide Collection/American Society of Agronomy

◆ **FIGURE 12.14B** Oxisol exposed on an eroding bank in Sri Lanka.

SOURCE: JLM Visuals

humus because of rapid bacterial decomposition. The illuvial B horizon has a considerable accumulation of clay minerals; it is composed of highly weathered minerals with a relatively low cation exchange capacity (Figure 12.15).

Most ultisols have developed under a broadleaf forest cover where the trees maintain some soil fertility by recycling soluble bases from the B horizon to the A horizon. The clearing of the forests in many areas, though, has led to the rapid leaching of these plant nutrients. The fine texture of the soil also inhibits water infiltration and surface compaction has resulted in severe erosion in many areas that were cleared for agriculture.

The ultisols are especially associated with humid subtropical climate regions, but also occur in portions of the humid tropics. They are the dominant soil type of the southeastern United States, where in many areas they have suffered from nutrient depletion and erosion resulting from excessive cotton cultivation. They also are important in much of China, southern Brazil and Uruguay, the northern Andes, West Africa, northeastern Australia, and the islands of the tropical western Pacific.

Alfisols

The **alfisols** (*alf* for "aluminum and iron") are fine-textured soils with a yellowish brown A horizon that has been colored by iron and aluminum compounds (hence the name) and is relatively low in humus (Figure 12.16). They have a clay-rich B horizon with a moderately high cation exchange capacity.

The alfisols are well-developed soils that are distributed widely throughout the low and middle latitudes. They tend to develop in areas that contain clay-forming parent materials and that have a season in which evapotranspiration exceeds precipitation. They therefore are especially typical of forest/grassland transitional environments. In the United States, they occur mostly in the southern Great Lakes, Ohio Valley, Texas, and Mississippi River Valley regions. Other middle latitude sites of importance include eastern Europe, southern Asiatic Russia, and southern Australia. Within the tropics, they are especially important in southern and eastern Africa and in the African Sahel, as well as in large areas of India and Pakistan.

Their widespread geographical distribution and moderate to high natural fertility have caused the alfisols to rival the mollisols in worldwide agricultural potential. The wide variety of crops raised in these soils constitute a substantial proportion of total world food production.

Spodosols

The **spodosols** (*spodos* for "wood ash") are acidic soils with a light-colored eluvial E horizon and an illuvial B horizon (Figure 12.17). Spodosols generally develop from a quartz-rich and often sandy parent material in a cool, humid climate. They usually occur in association with a needleleaf

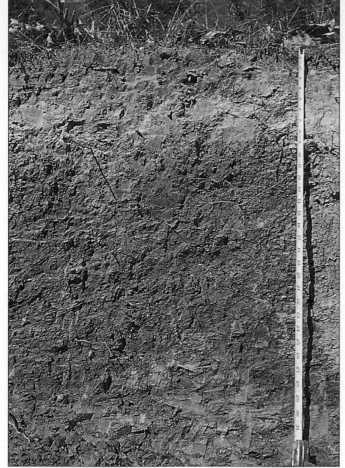

◆ **FIGURE 12.15A** An ultisol developed on metamorphosed sandstone in the Ouachita Mountains of western Arkansas.

SOURCE: Marbut Memorial Slide Collection/American Society of Agronomy

◆ **FIGURE 12.16A** An alfisol developed under an oak-hickory forest in southern Michigan.

SOURCE: Marbut Memorial Slide Collection/American Society of Agronomy

◆ **FIGURE 12.15B** Rows of tobacco growing in an eastern North Carolina ultisol.

SOURCE: Will & Deni McIntyre/Photo Researchers

◆ **FIGURE 12.16B** Harvesting cabbage grown in an alfisol in Estonia.

SOURCE: Jeff Isaac Greenberg/Photo Researchers

◆ **FIGURE 12.17A** This spodosol, developed on sandy glacial outwash in northern New York, has a well-defined whitish silica-rich E horizon.

SOURCE: Soil Taxonomy Alt 436, Soil Conservation Service

◆ **FIGURE 12.17B** Early autumn scene in a rural Vermont area containing spodosols. The mixture of deciduous and evergreen trees typical of this region is evident on the hillside.

SOURCE: © Chromosohm/Joe Sohm/Photo Researchers

evergreen forest cover. The most extensive areas with this soil type contain the northern coniferous forests of North America and Eurasia.

The acidity of the spodosols results from the presence of a well-developed O (organic) horizon, as well as abundant humus in the A horizon and surplus moisture. The soil pH is especially low in needleleaf forests because these trees have low nutrient requirements and do not cycle many bases from the subsoil, where soluble bases tend to collect. The leaching of soluble constituents is promoted by the abundant precipitation and by the acidity of the soil. Soil acidity restricts the removal of silica in solution, while promoting the leaching of iron and aluminum compounds. As a result, silica is the primary mineral constituent in the E horizon, giving it a pale gray, ashy appearance. This characteristic led to the name given this soil type.

Dissolved materials and clay-sized particles, as well as some humus, are deposited in the B horizon. The B horizon is therefore an illuvial horizon that is denser, finer in texture, and higher in soluble bases than the E horizon. It generally has a reddish brown coloration resulting from the presence of iron oxides.

The combination of low natural fertility, high acidity, a short growing season, and sometimes a lack of soil depth presents serious drawbacks to agricultural development in regions containing spodosols. These regions are most noted for the production of root crops, such as potatoes, that have modest soil nutrient requirements and a tolerance for acidic conditions.

Mollisols

The **mollisols** (*molli* for "soft") are soils that contain a thick, dark, humus-rich A horizon and have a soft consistency when dry (Figure 12.18). They exhibit a high CEC and a B horizon accumulation of calcium carbonate. Mollisols are associated chiefly with the prairie and steppe grasslands of the middle latitudes. Seven suborders are recognized, based largely on moisture conditions. Nearly all mollisols are developed on transported parent material. This material usually consists of alluvium, or of glacial or **loessal** (wind-blown silt) deposits.

The amount of soil moisture is perhaps the most important influence on the mollisols. These soils form in regions that have a rough balance between precipitation and evapotranspiration. As a result, there is little translocation of soil materials and no significant leaching or accumulation of soluble minerals.

Climatic and vegetative conditions generally combine to allow the mollisols to maintain a high humus content. Water from winter snowmelt and spring rains promotes rapid early luxuriant grass growth. Drier, hotter conditions in mid and late summer as well as cold winter temperatures greatly reduce vegetative decomposition during these periods, allowing humus to accumulate. The soils of the tall-grass prairies have the highest humus content, which is generally well mixed by the upper-soil fauna throughout a thick, black-colored A horizon. In the drier steppe areas, the humus content is reduced by the smaller quantity of available organic matter, and the soils are lighter brown in color. Soil depths also decrease toward drier areas because less water is available for chemical weathering.

Soil pH values are conducive to the growth of a wide variety of vegetation. The better watered areas are almost neutral or even slightly acidic, but pH increases with dryness, producing soils that are moderately alkaline and causing a subsoil concentration of calcium carbonate to develop—a process known as **calcification.**

The mollisols have a high degree of natural fertility and are associated with some of the world's most productive agricultural regions. The better watered areas support major grain-producing economies. The American Midwest, South American pampas, Ukraine, and a part of Chinese Manchuria, which are considered the major "breadbasket" regions of the world, produce huge surpluses of corn, wheat, and oats. The drier steppe areas generally receive too little precipitation to support intense cash crop economies (except on irrigated lands) and are used primarily for raising livestock.

Aridisols

The **aridisols** (*arid* for "dry") are soils lacking in available water for extended periods. They are the world's most extensive soil type, occupying large areas of the low and middle latitudes that receive less than about 25 centimeters (10 in) of liquid precipitation annually. In all, they are estimated to cover 19 percent of the world, largely within the subtropical deserts.

Chemical weathering processes are considerably hampered by a lack of water in arid regions. As a result, the aridisols are shallow, stony, and mineralogically immature, with poorly developed horizons (Figure 12.19). Soil textures generally are coarse and sandy, leading to a low water-retention ability even when water is available. The humus content is low to completely absent because of the sparseness of the vegetation cover.

An important characteristic of the aridisols is their high alkalinity. Evapotranspiration exceeds precipitation, producing a surfaceward movement of groundwater and dissolved minerals. Well-drained soils typically experience an accumulation of calcium carbonate and other soluble bases at the site of water evaporation, which is normally several centimeters below the surface. This frequently produces a duricrust layer. In poorly drained depressions, the deposition of salts, a process termed **salinization,** results in toxic conditions for most vegetation. The salinization of dry land soils is often worsened by irrigation practices that cause salts to accumulate in the near-surface soil layers. This problem is discussed in the Focus Box on pages 300–301.

Despite their shortcomings, aridisols often are potentially fertile and have proven highly productive under certain conditions. The chief problem is a lack of water. Irrigation is

◆ **FIGURE 12.18A** This mollisol, developed on glacial drift in central Iowa, has a thick, humus-rich A horizon and a granular structure.

SOURCE: Marbut Memorial Slide Collection/American Society of Agronomy

◆ **FIGURE 12.18B** Farm scene in a mollisol soil area near Bellevue, Iowa.

SOURCE: © David R. Frazier Photolibrary/Photo Researchers

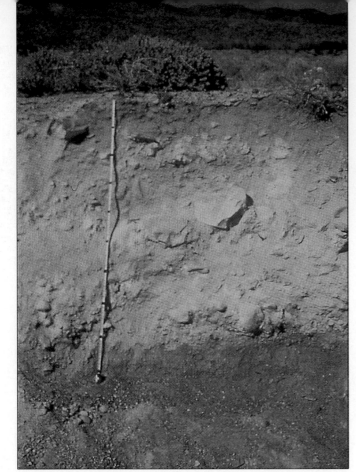

◆ **FIGURE 12.19ᴀ** A stony Utah aridisol is covered by semidesert vegetation dominated by sagebrush.

SOURCE: Marbut Memorial Slide Collection/American Society of Agronomy

◆ **FIGURE 12.19ʙ** This aridisol in Arizona's Sonoran Desert supports a sparse vegetative cover of cacti and perennial shrubs.

SOURCE: © John D. Cunningham/Visuals Unlimited

essential not only to support plant growth, but often initially to flush excess salts from the soil. Most agricultural activity occurs in enclosed basins where water is available, the surface is flat, and soils are deep enough for farming.

Vertisols

The **vertisols** (*vert* for "invert") contain large amounts of montmorillonite clay that swells when wetted and shrinks when it dries. These soils are located in areas that experience alternating periods (often seasonal) of moisture surplus and deficiency. During dry periods, vertisols shrink and develop cracks that may be as large as 2.5 centimeters (1 in) wide and 75 centimeters (30 in) deep. Surface soil crumbles and is blown, or washes, into the cracks. When moistened, the soil swells, closing the cracks and covering the soil that has entered the cracks from above. The vertisols thus possess the unique property of mixing themselves and even, to some extent, of turning themselves "inside out." As a result of this natural churning action, they are almost totally lacking in horizons. Vertisols generally develop in semiarid regions under a grass cover, have a moderately high cation exchange capacity and organic matter content, and form fertile and agriculturally productive soils (Figure 12.20).

The vertisols are relatively restricted in geographical extent. Most are located in areas with Tropical Wet and Dry climates in western India, eastern Australia, and parts of northeastern Africa. They occur to a lesser degree in portions of the southern and western United States, notably in Texas and California.

Histosols

The **histosols** (*histos* for "tissue") are dark organic soils of poorly drained areas. They consist primarily of decomposed vegetation, often forming a peat or muck. Lack of oxygen inhibits organic decay, so these soils can attain considerable depth through the continued addition of vegetative matter at their surface. Most histosols are highly acidic and low in fertility (Figure 12.21).

The histosols are one of the least extensive soil orders. They can occur in almost any nonarid environment where drainage is poor, but are most common in the tundra regions of North America and Eurasia, where the underlying permafrost prevents adequate drainage. They also are found in lowland areas to the south that once were covered by continental glaciers. Glacial erosion and deposition in these areas disrupted the previously existing drainage systems. Significant sites of histosol formation in the coterminous United States include the Florida Everglades and the Mississippi River Delta in Louisiana.

The artificial drainage of histosol areas leads to improved aeration and often to a major temporary improvement in their fertility and agricultural potential. Oxidation of the organic matter, though, leads to eventual compaction and subsidence and to fire danger in dry periods. In recent years, smoky fires in the Everglades, resulting from a decline in the

◆ **FIGURE 12.20B** The profile of this Texas vertisol shows that, in places, subsoil is being drawn to the surface by the soil's natural churning action.

SOURCE: Natural Resources Conservation Service

◆ **FIGURE 12.20A** A vertisol consisting largely of montmorillonite clay in the Lajas Valley of Puerto Rico.

SOURCE: Marbut Memorial Slide Collection/American Society of Agronomy

(a)

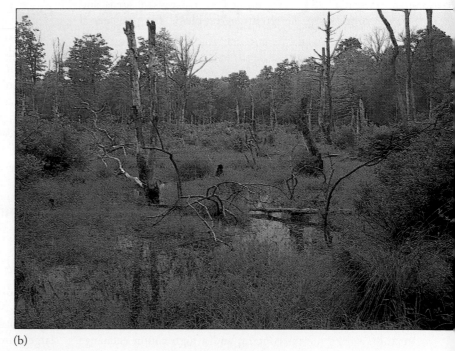

(b)

◆ **FIGURE 12.21** (a) A histosol occupying a poorly drained depression in southern Michigan. (b) The Parker Brook Swamp in High Point State Park, New Jersey, is a prime site for histosol formation.

SOURCE: a: Marbut Memorial Slide Collection/American Society of Agronomy; b: Michael P. Gadomski/Photo Researchers

◆ **FIGURE 12.22A** A New Zealand inceptisol that has developed on loess deposits.

SOURCE: Soil Taxonomy Alt 436, Soil Conservation Service

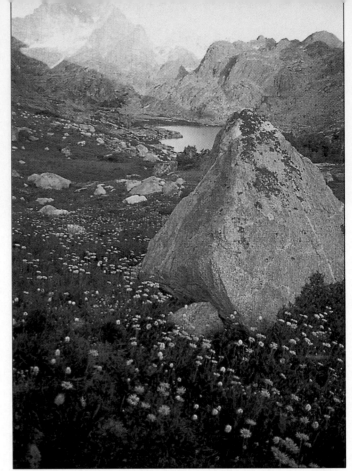

◆ **FIGURE 12.22B** An inceptisol supports this rocky tundra meadow in the Wind River Range, Wyoming.

SOURCE: © Charlie Crangle/JLM Visuals

region's water table, have burned over extensive areas and have caused severe air pollution problems for the coastal cities.

Inceptisols

The **inceptisols** (*incept* for "inception or beginning") are recently formed, immature soils that have some profile development, but contain weatherable minerals. They are primarily eluvial soils with little indication of illuvial layers or the vertical translocation of materials (Figure 12.22).

The inceptisols are found largely in three types of environmental settings, each of which lacks the physical stability for the development of a mature soil. The first is in moderately steep mountain areas, such as the Appalachians, where erosion and landslide activity are relatively common. The second is in tundra areas, where soil development is inhibited by cold temperatures, poor drainage, and permanently frozen subsoils (**permafrost**). The third type of setting, and the most important from a human standpoint, is in floodplains or deltas. These soils typically have a loamy texture, a high humus content, and a high cation exchange capacity. The alluvial soils of large floodplains, such as those of the Mississippi, Nile, and Ganges Rivers, are intensively farmed and highly productive.

Entisols

The **entisols** (*ent* for "recent") are poorly developed soils with little or no profile development (Figure 12.23). As their name implies, most have formed very recently.

Entisols develop in a number of highly dissimilar and often localized environments; consequently, they vary greatly in their physical and chemical characteristics. Their most widespread areas of occurrence are in sandy deserts and in areas of regolith in dry environments. They also exist in tundra areas, in areas that have been recently glaciated, and in areas recently covered by **alluvium** (deposits of river sediments) or volcanic ash. Soils at construction sites as well as other sites at which the soil has been extensively eroded or disturbed by human activities also are classified as entisols.

Andisols

The **andisols** (*ando* for "volcanic ash") are a new soil order, recognized in 1990 from a group of soils previously classified as inceptisols and entisols. Their parent material is derived largely from volcanic ash and volcanic glass erupted from volcanoes (Figure 12.24). They may contain buried soil profiles (paleosols) above which the current soils have formed following major volcanic eruptions. Andisols contain

◆ **FIGURE 12.23B** Sea oats and beach grass partially cover an entisol formed from wave-deposited sand at Cape Hatteras, North Carolina.
SOURCE: Breck P. Kent/JLM Visuals

◆ **FIGURE 12.23A** This southwestern Wisconsin entisol developed on a river floodplain and contains many horizons. The development of each horizon was interrupted by an overlying deposit of fresh alluvium during a flood.
SOURCE: Marbut Memorial Slide Collection/American Society of Agronomy

(a)

(b)

◆ **FIGURE 12.24** (a) An andisol near Pahala, Hawaii, developing from volcanic ash (horizontal layers) that overlie weathered basalt. (b) Plantation of irrigated sugarcane growing in an andisol on the Hawaiian island of Kauai. The steep slopes in the distance are typical of the deeply eroded mountains of the western Hawaiian islands.
SOURCE: a: Bernard W. Pipkin, University of Southern California; b: Marbut Memorial Slide Collection/American Society of Agronomy

large amounts of iron and aluminum oxides and have high cation exchange capacities. They are generally fertile and under natural conditions typically support a dense vegetation cover that gives them a high humus content and dark coloration, especially in cooler climatic settings.

Andisols form over a wide range of latitudes, but since they exist in patches surrounding volcanoes, they are too localized to appear on our map of global soil types. Notable sites of andisol development include the Hawaiian Islands and the vicinity of many of the volcanoes that surround the Pacific Basin.

▌SUMMARY

Soil is the loose mixture of weathered rock material and organic matter that mantles most of the Earth's land surface. It is the medium in which most plants and many animals live, and it provides the water and nutrients they need in order to survive. The global distribution of soils is complex, and the type of soil at any given site is determined by the interplay of a variety of environmental controls.

In order to summarize systematically the most important information contained in the chapter, we return to the Focus Questions:

1. *What physical and chemical characteristics of the soil most affect its ability to support vegetation?*

The soil characteristics of greatest importance to vegetation include its texture, structure, organic matter content, and base saturation. The soil texture is determined by the mix of sand-, silt-, and clay-sized particles. Texture especially affects the soil's density and ease of cultivation, as well as its ability to absorb and retain water. The soil structure is produced by the tendency of individual soil particles to adhere to one another, forming peds of distinctive sizes and shapes in differing environments. The structure notably influences the ability of the soil to absorb water and to resist erosion. Soil organic matter supplies a variety of plant nutrients, plays an essential role in the transfer of the nutrients to plants, and improves the soil's water-retention ability. The base saturation of the soil relates to its ability to obtain base cations that are essential for plants. It therefore is a quantifiable measure of soil fertility. The soil's base cation exchange capacity (CEC) is influenced by a variety of factors, including its humus content and its pH.

2. *What are the five factors in soil development, and what influence does each have on soil characteristics?*

The five factors in soil development collectively determine the type of soil that will develop at a given site. An unlimited number of combinations of these five factors is possible. The first factor to influence the soil is its parent material, which may be either residual or transported, and may consist of bedrock, sediment, organic matter, or some combination of these materials. A second factor is the topographic relief of the land. Steep slopes usually are adversely affected by erosion, and their soil cover typically is shallow, coarse textured, and often relatively infertile. A third factor is biological organisms. Microscopic plants and animals, notably bacteria and protozoa, are crucial in decomposing soil organic matter to form humus. Larger plants and animals help to loosen and aerate the soil and add organic matter. The characteristics of mature soils in well-vegetated regions are often more a direct product of the natural vegetation than of any other factor. A fourth factor, the climate, plays both a direct and an indirect role in soil characteristics because of its influence on vegetation. The two climatic elements with the greatest direct influence are moisture and temperature, and moderate conditions with respect to each are generally associated with the most fertile soils. Excessively moist conditions promote poor aeration and leaching, while dry conditions are associated with shallow and immature soils, alkaline conditions, a lack of humus, and insufficient moisture to support vegetation. Soils developing in high temperature regions often lack organic matter, while cold regions often have shallow, acidic soils. The final factor in soil development is time. All soils need time to develop, and all gradually evolve with time, so that a variety of different soils will form from the same parent material in diverse environments. Time also allows soil horizons to form because of the differential operation of pedogenic processes at varying depths.

3. *What are the major world soil orders, and what are the chief attributes of each?*

Pedologists have long been concerned with developing ways of logically and effectively classifying the great variety of existing soils. The Soil Taxonomy, which divides all soils into 11 major soil orders, currently is in official use in the United States and was used in this chapter.

Oxisols are the deepest and most chemically weathered of the soil types, but are relatively infertile. They are closely associated with the humid tropics. High bacterial decomposition rates and intense leaching have removed most humus and plant nutrients from these soils and have left concentrations of insoluble oxides of iron and aluminum that give them a distinctive red color. **Ultisols** are reddish soils that are highly chemically weathered, have a high aluminum and iron content and limited natural fertility. They typically are associated with humid subtropical climates and are common in the southeastern United States. **Alfisols** are mature soils that are fine textured and relatively low in humus; they also have a high aluminum and iron content. Alfisols are usually rather fertile, are typical of forest/grassland transitional environments, and are widely distributed in North America. **Spodosols** are acidic soils with a silica-rich eluvial E horizon and an illuvial B horizon. They are relatively low in fertility and generally develop under a needleleaf forest cover in the higher latitudes. **Mollisols** typically develop in middle latitude grassland regions. They have a thick A horizon that is rich in humus and has a high cation exchange capacity. Mollisols often are considered the most naturally fertile of the soil orders. **Aridisols** develop in arid regions and usually are shallow, mineralogically immature, and alkaline. **Vertisols** are clay-rich soils that swell when wetted, but shrink and crack as they dry. The resulting churning action allows for little profile development. Vertisols are of limited global extent, but most are associated with grasslands. **Histosols** are composed largely of organic matter that has accumulated in swampy areas. In their natural state, most are highly acidic and infertile, but they can be productive when drained and treated for acidity. **Entisols** and **inceptisols** are immature soils with poorly developed profiles. They have formed in areas where pedogenic processes are disturbed by natural or human-induced erosion or deposition. The final order, **andisols,** are iron- and aluminum-rich soils that have formed largely from volcanic ash. They are often fertile, but are limited geographically to small areas that are volcanically active.

▌Questions for Thought and Review

1. From what four constituents is the soil formed? What proportion of each comprises a typical soil? What alterations must occur to the soil's two solid constituents before they can be incorporated into the soil?

2. What is soil texture? What practical effects does it have on the soil?

3. What is soil structure? What relationship exists between texture and structure? What practical influences does structure have on the soil?

4. What is humus? What favorable effects does humus have on the soil? What environmental conditions maximize humus formation?

5. By what process do plants withdraw nutrients from the soil? What is soil pH, and how does this factor relate to the availability of soil nutrients to plants?

6. Why do most soils develop horizons? In what major respects do the A, E, and B horizons differ from each other?

7. What are the basic differences in the characteristics of soils that have developed in wet versus dry environments? In warm versus cold environments?

8. Which of the 11 major soil types is the deepest? Why? Explain what physical and chemical characteristics limit its agricultural potential, and what environmental factors have resulted in the development of these characteristics.

9. Which of the 11 major soil types is generally considered to have the highest natural fertility? What soil characteristics produce this fertility?

10. Which of the 11 major soil types dominates in your area? Describe its basic characteristics and agricultural potential in a short paragraph.

11. What are the chief agricultural products in your local area? What characteristics of the soils in your area support these products?

12. Why do soils in arid and semiarid regions become salinized? How does irrigation affect this process?

13. How does soil erosion occur? How can agricultural practices greatly accelerate soil erosion? What agricultural activities can reduce soil erosion rates?

▌Key Terms

soil	Soil Taxonomy
soil texture	oxisol
loam	ultisol
ped	alfisol
soil structure	spodosol
translocation	mollisol
leaching	aridisol
duricrust	vertisol
humus	histosol
cation exchange capacity	inceptisol
soil horizons	entisol
soil profile	andisol
parent material	

North American Agriculture and Accelerated Soil Erosion

The North American continent has been endowed with vast expanses of fertile soils. The presence of these soils, coupled with the enterprise of American and Canadian farmers and the availability of modern technology, has allowed North American agricultural productivity to far surpass that of any other region of the world. The successes of North American agriculture have not been without cost, however. The greatest problem—one that raises serious questions about our ability to continue to expand our agricultural production—is the alarming rate of soil erosion from agricultural land (Figure 12.25).

Soil erosion is a natural and unavoidable phenomenon that becomes detrimental when it occurs more rapidly than new soil can form. Such an unbalanced situation results in a net loss of soil and eventually of agricultural potential. Most human activities that alter the landscape disrupt the delicate natural balance between soil formation and removal and greatly increase the rate of erosion. Globally, since World War II, more than 12 million square kilometers (3 billion acres) of agricultural land—an area larger than China and India combined—have been damaged by human actions and some of this land may prove impossible to reclaim.

By far the leading cause of accelerated erosion throughout the world is the clearing of land for agriculture and grazing land. Removal of the protective vegetation cover exposes the soil to rapid erosion by water in wet periods and by wind in dry periods. Both agents of erosion remove topsoil, which in most cases comprises the most fertile segment of the soil.

A 1992 United Nations study of global soil erosion found that approximately 35 percent of erosion was caused by livestock overgrazing. This problem appears to be most serious in Africa and Oceania. Another 30 percent of erosion resulted from deforestation. This factor, which is most serious in Asia and South America, is the subject of the Case Study at the end of Chapter 11. Another 28 percent of soil degradation results from such harmful agricultural practices as overfertilization and ignoring fallow periods, which the study cited as the primary causes of soil erosion in North America.

In the United States, the estimated average rate of soil formation is 2.5 centimeters (1 in) per century, or approximately 1,360 kilograms (1.5 tons) per acre per year. Estimates of the annual loss of topsoil from agricultural land in the United States range from 5,500 to 12,500 kilograms (6 to 14 tons) per acre per year. Our farmland therefore is being eroded at several times the rate at which it is being formed. As a result, the sediment loads of rivers draining into the Atlantic are estimated to be four or five times greater than before European settlement.

Approximately 20 percent of the land area of the United States is currently under cultivation; about 75 percent of the 4.5×10^{12} kilograms (5 billion tons) of soil lost annually to water and wind erosion in the 48 contiguous states is derived from agricultural land. During the past two centuries, at least one-third, and perhaps as much as one-half, of the topsoil on U.S. farmland has been lost. For every kilogram of grain produced in the fertile Mississippi Valley,

◆ **FIGURE 12.25** Severe soil erosion in an agricultural field in Arkansas resulting from storm runoff.

SOURCE: Garry D. McMichael/Photo Researchers

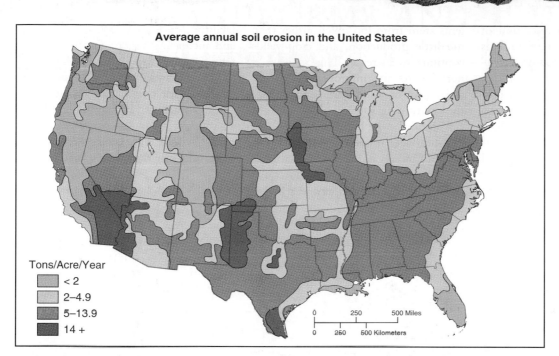

Average annual soil erosion in the United States

Tons/Acre/Year
- < 2
- 2–4.9
- 5–13.9
- 14 +

0   250   500 Miles

0   260   500 Kilometers

◆ **FIGURE 12.26**   Soil erosion severity map of the United States.
SOURCE: Information provided courtesy of the U.S. Department of Agriculture.

4 kilograms of soil are eroded from the grain fields. Of the 3.6×10^{12} kilograms (4 billion tons) of topsoil removed annually by water erosion, 0.9×10^{12} kilograms (1 billion tons) is carried to the sea, where it is irretrievably lost. The remaining 2.7×10^{12} kilograms (3 billion tons) settles in lakes, reservoirs, and river floodplains, where it causes serious sedimentation problems.

Sedimentation can be viewed as the "other side of the coin" in the soil erosion problem. In contrast to erosion, where the effects often are gradual and sometimes unnoticeable over the short term, the results of greatly accelerated sedimentation rates often are painfully apparent. For example, river channels can be raised so greatly by deposition that they become highly susceptible to flooding. Reservoirs are filled so rapidly by siltation that their useful life spans often are reduced to only a few decades. Harbors and river channels must be

dredged constantly to permit continued shipping.

Many factors affect the severity of soil erosion. The most important is slope steepness, while others include length of slope; soil texture, structure, and organic content; cultivation practices; types of crops grown; and climatic characteristics—especially the frequency of heavy rainfalls and strong windstorms. The U.S. Department of Agriculture estimates that approximately 70 percent of American farmland has experienced moderate to severe erosion, while 30 per cent—mostly on very flat surfaces—has experienced slight or no erosion. Figure 12.26 shows the distribution of soil erosion in the United States produced by the combined actions of water and wind. It can be seen that the most severe problems are occurring in the Southeast, the western Midwest, the Great Plains, and in parts of the Southwest. These areas include the most intensively farmed and agricul-

turally productive areas of the country. Within portions of these areas, considerable farmland has been abandoned because of erosion and gullying. Some 200 million acres of land in the United States have been removed from agricultural availability due to erosion, coupled with rapid urbanization and highway construction. This is more than half the total acreage still being farmed.

Some reduction of soil erosion rates in the United States has occurred over the past few decades, although erosion still remains unacceptably high. Conservation measures such as contour plowing and strip cropping (shown in the chapter opening photo on page 292), and minimum tillage, mulching, as well as government incentives and technical assistance, are all playing a part in reducing the problem. Unfortunately, the implementation of many of these conservation practices results

continued on next page

◆ **TABLE 13.1**   Percentage of Land Area Occupied by Major Topographic Categories

	NORTH AMERICA (INCLUDING GREENLAND)	SOUTH AMERICA	EURASIA	AFRICA	AUSTRALIA AND NEW ZEALAND	ANTARCTICA	WORLD
Flat plains	8	15	2	1	4	0	4
Rolling and irregular plains	23	30	30	44	51	0	30
Plateaus	9	10	3	5	1	0	5
Plains with hills or mountains	18	18	11	22	19	0	15
Hills (60–300 m relief)	8	5	10	11	12	0	8
Low mountains (300–900 m relief)	10	11	21	13	12	1	14
High mountains (900+ m relief)	16	11	23	4	1	1	13
Ice caps	8	0	0	0	0	98	11
Percentage of world area	16	12	36	20	6	10	100

SOURCE: Glenn T. Trewartha, Edwin H. Hammond, and Arthur H. Robinson, *Fundamentals of Physical Geography*, 2d ed. (New York: McGraw-Hill, 1968), p. 231.

example is the Colorado Plateau, in the southwestern United States, whose level surface originally formed on the ocean floor. In some instances, however, plateaus are produced by outpourings of tremendous volumes of lava, as in the cases of the Columbia Plateau of the northwestern United States and the Deccan Plateau of India. The high elevations of plateaus cause streams draining them to have a large amount of gravitational energy, frequently enabling them to carve deep, steep-walled **canyons** or **gorges** into the plateau surfaces (Figure 13.3e). The Grand Canyon, which has been eroded by the Colorado River into the rock surface of the Colorado Plateau to a mean depth of 1.6 kilometers (1 mi), is undoubtedly the world's best-known example.

◆ **FIGURE 13.5**   An example of the effect of tectonic forces: intense folding of sedimentary rock strata exposed in a California roadcut.
SOURCE: JLM Visuals

Genetic Approach to Landform Study

The genetic approach to the study of landforms emphasizes the causal factors in the development of the topography rather than its physical attributes. The chief subdivisions within this system are the various landform-producing forces outlined in Table 13.2. As indicated by this table and the earlier text, two basic groups of geomorphic forces exist (Figures 13.5 and 13.6). Each is powered by a different energy source; one influences the Earth's surface from below and the other from above.

The first group of forces, **tectonic forces,** originate within the Earth. They are powered primarily by the nuclear decay of unstable isotopes of the elements uranium and thorium. The general tendency of the tectonic forces is to increase the elevation and relief of the surface. Consequently, they are often described as being constructive forces in landform development. From a global standpoint, tectonic forces are largely responsible for raising land above sea level and for initially producing mountains and plateaus.

I. Tectonic forces
   A. Diastrophism
      1. Folding
      2. Faulting
   B. Volcanism
      1. Intrusive
      2. Extrusive

II. Gradational forces
   A. Gravity
   B. Rivers
   C. Ice
   D. Wind
   E. Ocean waves and currents

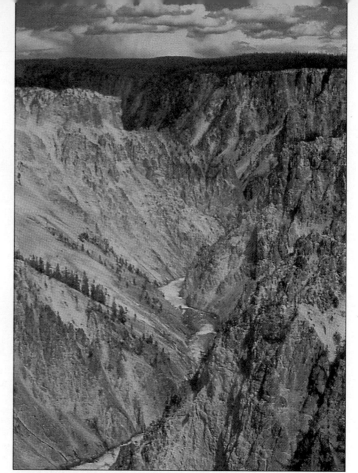

◆ **FIGURE 13.6**   An example of the effect of gradational forces: the gorge of the Yellowstone River in Yellowstone National Park, Wyoming.
SOURCE: JLM Visuals

Tectonic forces are, in turn, divided into two categories. **Diastrophism** involves solid-state movements of the Earth's crust. This displacement may occur either through the **folding** (bending) of the rock material or through **faulting,** which involves rock fracturing followed by the movement of the two sides of the fracture relative to one another. **Volcanism** involves the transfer of molten rock material, either from one point to another beneath the surface or, more rarely but spectacularly, its expulsion onto the surface. The mechanisms by which tectonic forces are generated and the nature and distribution of the resulting diastrophic and volcanic landforms are examined in the next chapter.

The second group of landform-producing forces—the **gradational forces**—originate above the Earth's surface. They are powered by solar energy and are given a downward directional component by gravity. The overall tendency of gradation is to lower the surface and to reduce topographic relief. The application of these forces is very uneven, however, so that a sequence of changing features is produced during the gradational process. The ultimate effect of gradation would be to reduce all land surfaces to smooth surfaces at or slightly below sea level.

Three different substances accomplish most of the actual gradational reduction of the surface. These are water, ice, and air (as wind), often collectively referred to as the "three tools of gradation." It is noteworthy that one is a liquid, one a solid, and one a gas. These substances accomplish their work by being heated, evaporated, or both, largely by solar energy. This causes them to expand, become more buoyant, and rise into the atmosphere. When the energy is later lost, gravity causes them to return to the surface in the forms of liquid or solid precipitation or wind. Earlier chapters have described the processes by which this is accomplished. Once on the surface, they tend to flow downhill or, in the case of air, to areas of lower pressure. Frictional drag exerted on the surface by this movement accomplishes the work of gradation. The force of gravity acting alone can also transport materials downhill in areas with loose or weak surface materials.

Examined in a temporal sequence, there are four steps in the gradational process, regardless of which "tool" is involved. The first step is **weathering.** Weathering actually is a preliminary process that makes gradation possible. It involves the breakup of solid rock or large rock fragments into pieces small enough to be moved by the three tools of gradation. This is necessary because the forces of gradation are not as powerful as the tectonic forces. For example, although a mountain range may be raised tectonically as a single unit, it must be carried off by water, ice, or wind piece by piece.

Once weathering (discussed further in Chapter 15) has prepared the surface, the weathered materials are in relatively quick succession picked up or **eroded,** transported, and finally deposited by one or more of the three tools of gradation. Because erosion involves the removal of material, it lowers the surface. Deposition, conversely, increases surface elevations. Erosion occurs where the tools of gradation have excess energy and are able to perform the work involved in picking up and transporting weathered material. Deposition usually occurs at relatively low elevations and in areas of low relief. It takes place where the gradational tools lack sufficient energy to continue to transport the materials they have been carrying. For the world as a whole, erosion must in the long run equal deposition, since everything that is picked up must eventually be put back down. The geographical distributions of sites in which erosion and deposition dominate differ greatly, though, because the environmental conditions that favor these sets of processes directly oppose one another. As a basic geographical generalization, erosion exceeds deposition on land areas, resulting in their net gradational reduction, while deposition exceeds erosion on the ocean floor.

Tectonic and gradational forces are in essence diametrically opposed to one another in terms of their effects on the surface and can be envisioned as being engaged in a constant struggle for a controlling influence on the topography. As soon as tectonic forces produce surface uplift, forming hills, mountains, and plateaus, gradational forces attack these uplands, carrying them bit by bit to lowlands or to the sea. Tectonic forces are much more powerful and fast-acting than gradational forces; they are capable, for example, of thrusting up a volcanic peak in only a few years—an amount of time so short that the gradational effect is minimal. Tectonic forces, however, operate very sporadically, with periods of intense activity followed by long interludes of quiescence. A good example was provided by the violent eruption in 1980 of Mt. St. Helens after 123 years of dormancy. The gradational sequence, conversely, is generally slower in its speed of operation, but it is continuous and inexorable. As soon as a tectonic feature is produced—even as it is being uplifted—gradational forces attack it and begin to wear it back down.

Each year, gradational forces erode some 1.0×10^{13} kilograms (11 billion tons) of material from the land and transport it to the oceans. If erosion proceeded unopposed at this rate, the continents would be reduced to sea level in a span of only 40 million years. A comparable mass of material, however, is lifted above the oceans annually by tectonic activity, so that the total mass of land tends to remain relatively constant.

Although the net gains and losses of mass from the Earth's land areas resulting from tectonic and gradational forces may be roughly equal, the geographical distributions of these opposing forces and the types of surface features they produce differ greatly. In a basic sense, the tectonic forces can be considered as primarily responsible for the large-scale patterns of surface features, such as the arrangements of the continents and ocean basins and the locations of the major mountain ranges. Gradational forces, on the other hand, can only operate where, and when, the tectonic forces have first created topographic features to erode.

The sizes, shapes, and locations of the continents, ocean basins, and mountain ranges are among the most basic and important aspects of our planet's physical geography. They form the largest units of the Earth's surface features. In addition, because of the complex interactions that exist among the various Earth surface phenomena, they exert a vital influence on the global patterns of climate, plant and animal distributions, and human activities.

Insofar as astronomers and geomorphologists are able to tell, the Earth's topography is unique among the planets and satellites of our solar system. Detailed surface observations of many of these astronomical bodies have been made in recent years, and none has been found to have a landform assemblage even remotely similar to that of the Earth's. The uniqueness of the Earth's topography results more from the gradational forces than from the tectonic forces. All the planets and larger satellites of the solar system apparently either currently have, or at one time had, tectonic forces roughly comparable to those of the Earth. (Some bodies, such as the Moon, seem at present to be tectonically inactive, apparently due to the depletion of their internal energy supplies.) Not all these astronomical bodies, however, have gradational forces other than the force of gravity. The existence of these forces requires both an atmosphere and sufficient proximity to the Sun to supply an effective quantity of insolation. Those bodies without an atmosphere, such as the Moon, therefore have no way of erasing tectonic features or meteorite impact craters. They contain ancient landscapes that are nearly frozen in time. The Earth's surface, in contrast, is dynamic and is unique even when compared with the planets and satellites that have an atmosphere and a reasonable supply of solar energy. This is because the surface of the Earth contains an abundance of life-forms, is largely covered by water, and has an overlying atmosphere containing water vapor that is engaged in a constant exchange with the water at the surface. The presence of life-forms makes possible a large number of biochemical reactions that greatly accelerate weathering processes, while the cycling of surface and atmospheric water permits the erosion of the surface by streams, ocean water, and ice. Without a hydrologic cycle, only the wind, the weakest of the three tools of gradation, would be available to erode the surface. This would greatly reduce both the speed of gradation and the variety of surface features produced.

If one looks at the workings of the tectonic and gradational forces from a long-term global standpoint, a complex geological cycle can be discerned (Figure 13.7). The same sequence of activities occurs over and over again, and the same Earth materials are continually used and reused. The sequence can be envisioned as beginning with the tectonic uplift of a portion of the surface. The material forming the uplifted region is then subjected in turn to weathering, erosion, transport to the ocean floor or some other basin, and deposition. The deposited sediment is eventually buried, compacted by the weight of the overlying materials, and gradually reconstituted as rock. At some later time, this rock is once again uplifted by another episode of tectonic activity, and the sequence begins anew.

EARTH'S INTERIOR

The geographer is concerned primarily with the Earth's surface. The internal composition and structure of our planet are relevant to physical geography largely because they shed

◆ **FIGURE 13.7** The geological cycle. In the top view, a rugged mountainous surface is formed by tectonic uplift. The gradational forces of water, ice, and wind, in conjunction with gravity, gradually erode the mountains (lower right). The eroded material is transported, largely by streams, to the ocean, where it is deposited to form layers of sedimentary rock (lower left). At some later time, renewed tectonic activity lifts these rock layers to form a new mountainous landmass, and the cycle is repeated.

light on surface processes and features. Events deep within the Earth produce the tectonic activity responsible for the formation of the continents, the oceans, and the atmosphere.

All knowledge of the Earth's interior must be obtained indirectly because humans have, almost literally, barely scratched the surface in their mining and drilling activities. Most information on the composition and structure of the interior has been gained through the analysis of shock waves produced by earthquakes and explosives.

The Earth's interior has a layered structure produced by density differences resulting from variations in composition and temperature. It is dominated by a small group of chemical elements consisting mostly of iron, silicon, and oxygen, along with lesser quantities of magnesium, aluminum, and calcium. These are the heavy elements produced in greatest quantities by the fusion reactions occurring in massive stars, which later spew them throughout the universe in supernova explosions (see the Focus Box on pages 12–13 in Chapter 2). The interior is differentiated into an iron-dominated central region, called the **core,** and a larger silicon- and oxygen-dominated periphery containing the **mantle** and **crust** (Figures 13.8 and 13.9).

The development of a layered interior apparently occurred during the Earth's formation. The process of planetary accretion was associated with very high temperatures, which, it is believed, were sufficient to partially melt the developing planet. The melting of the Earth's constituents allowed them to settle at different levels according to their respective masses to form the present layers of the interior. Fortunately, the separation process was not complete, or none of the heavy elements that are so essential to us today would exist at the surface.

The Core

The core is the innermost layer of the Earth. It occupies one-fifth of our planet's total volume, but contains a third of its mass. The core forms a dense metallic sphere with a radius of about 3,460 kilometers (2,160 mi). Its uppermost margin is located about 2,900 kilometers (1,800 mi) beneath the surface. Temperatures and pressures within the core are higher than anywhere else inside the Earth. All values are estimates, but temperatures are believed to range from 4,000° to 5,000°C (7,200° to 9,000°F), and pressures from 1.4×10^6 to 3.5×10^6 kilograms per square centimeter (20 to

Earth's continental crust	Earth's oceanic crust	Earth's mantle
SiO₂ (60.1)	SiO₂ (49.8)	SiO₂ (47.5)

SiO₂ (60.1)

Al₂O₃ (15.6)

CaO (5.2)

FeO (3.9)

Na₂O (3.9)

MgO (3.6)

Fe₂O₃ (3.1)

K₂O (3.4)

TiO₂ (1.1)   P₂O₅ (0.3)

SiO₂ (49.8)

Al₂O₃ (16.7)

CaO (11.8)

MgO (8.0)

FeO (7.0)

Na₂O (2.8)

Fe₂O₃ (2.0)

K₂O (0.2)   TiO₂ (1.5)   P₂O₅ (0.2)

SiO₂ (47.5)

MgO (28.8)

FeO (15.5)

Al₂O₃ (3.1)

CaO (2.5)

Na₂O (1.2)   K₂O (0.2)   TiO₂ (0.1)   P₂O₅ (0.2)

Relative chemical abundances, by percent

◆ **FIGURE 13.8** Proportional chemical constituencies of the mantle, the oceanic crust, and the continental crust.

inner core with a radius of approximately 1,200 kilometers (750 mi). Temperatures here undoubtedly are even hotter than those of the outer core, but pressures are so high that liquefaction cannot occur.

The core, despite its remoteness, appears to play an important indirect role in tectonic processes affecting the surface. Apparently, it contains a large proportion of the radioactive elements that power tectonic activity. The two most important of these elements, uranium and thorium, have high masses and would have tended to sink into the core during the period when the Earth was molten. Their gradual atomic decay through time has released great quantities of heat that has slowly made its way to the surface to help power tectonic processes. Fluid movements within the core are also believed to generate the Earth's magnetic field, which shields our planet from high-energy particles that stream from the Sun.

The Mantle

The mantle forms an extensive layer of rocky material extending from the top of the outer core almost to the surface of the Earth. It comprises about four-fifths of the Earth's total volume. In contrast to the metallic core, the mantle is composed primarily of rock. The rock type apparently is peridotite, which is dense, dark in color, and consists largely of oxides of silicon, iron, and magnesium (Figure 13.8). Recent research indicates that the boundary between the core and mantle is highly complex and irregular in shape. The mantle seems to penetrate several kilometers into the core in some places, and pockets of molten rock the size of small seas exist along the boundary zone in others.

The mantle, like the core, can be subdivided on the basis of its physical state. The lower mantle, below a depth of 700 kilometers (450 mi) beneath the surface, consists of immobile crystalline rock. Most of the upper mantle, in contrast, seems to exist in a plastic state and is capable of undergoing a slow creeping flow in response to forces acting on it. This plastic layer, termed the **asthenosphere,** plays a critical role in tectonic processes at the Earth's surface. The extreme upper mantle, above the asthenosphere, is a rigid solid. The uppermost portion of the mantle and the overlying crust to which it is attached are sometimes collectively termed the **lithosphere.**[1]

Heat from the decay of radioactive elements in both the core and mantle is conducted very slowly surfaceward through the mantle. When it reaches the asthenosphere, the heat is sufficient to partially melt this layer, generating slow convection currents within the rock material. The radioactive elements and the heat they produce are distributed unevenly within the Earth, just as they are at the surface. In places where radioactive materials are relatively abundant, more heat is generated, and the mobile areas within or above

50 million lb/in²). The composition of the core has not yet been ascertained, but because of its mass, it is thought to consist largely of iron, perhaps combined with a small quantity of other elements.

Seismic data indicate that the core consists of two distinct layers. The larger **outer core** apparently is liquid, since it cannot transmit certain types of earthquake waves that pass only through solids. Temperatures here must be high enough in relation to pressure to have caused melting to occur. The outer core is the only major portion of the Earth's interior to exist in the liquid state. Recent research indicates that the upper surface of the outer core may be highly irregular, perhaps forming a mirror image of the Earth's surface features that lie far above. Inside the outer core is a solid

---

1. Note that this use of the term "lithosphere" is much more restrictive than its earlier application to denote the entire interior of the Earth.

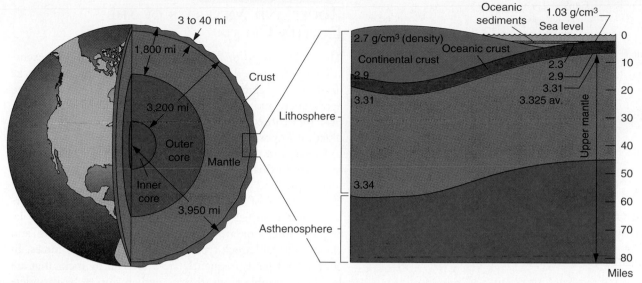

◆ FIGURE 13.9 The internal layers of the Earth. On the right is an expanded view of a portion of the upper mantle and crust, with representative rock densities indicated.

them in the asthenosphere expand, increase in buoyancy, and become the sites of rising convection currents. In places where they are not as abundant, less heat is produced, and the affected mobile areas are likely to be the sites of descending currents of somewhat cooler and denser material. A continuous circulation system is thus established (Figure 14.4). It has been found that the regions of mantle upwelling and sinking are not fixed, but instead slowly shift in various directions at mean rates of several centimeters a year. The movements of currents within the asthenosphere and their influence on the Earth's surface are explored in greater detail in the next chapter.

The Crust

The crust is the outermost layer of the Earth. Although it contains only 0.01 percent of the Earth's mass, its characteristics are overwhelmingly important to humans and to life in general and are closely interrelated with all the topics covered in this text. Landforms are merely the surface irregularities of the crust.

The crust consists of a relatively thin "skin" of rigid low-density rock that covers the underlying mantle to a depth of 5 to 70 kilometers (3–40 mi). Nearly three-fourths of the crust consists of compounds formed from the elements oxygen and silicon. It is separated from the mantle by a rather sharply defined boundary zone known as the **Mohorovičić discontinuity** after the Yugoslavian physicist who first identified it in 1909 from seismic data (Figure 13.10). Generally termed the **Moho,** this one-kilometer (0.6 mi) thick boundary zone tends to mirror the surface topography by lying deepest beneath high-elevation continental areas and most closely approaching the surface beneath deep ocean basins.

Like both the core and mantle, the crust is divided into two segments. The lowermost is the **oceanic crust,** a thin,

continuous layer of relatively dense basaltic rock that reaches the surface beneath the ocean basins. Above this, and forming the continental landmasses, is a discontinuous layer of very low density granitic rock referred to as the **continental crust.**

Approximately 65 percent of the Earth's surface, virtually all located beneath the ocean, is covered only by oceanic crust. This crustal layer generally ranges from 5 to 8 kilometers (3–5 mi) in thickness. It has a mean specific gravity of about 3.0 grams per cubic centimeter, a value some 10 percent greater than that of the continental crust.[2] In addition to oxygen and silicon, the oceanic crust contains a higher proportion of the elements magnesium, iron, and

2. **Specific gravity,** commonly stated in grams per cubic centimeter, is a frequently used measure of density. Fresh water has a specific gravity of exactly 1.0 g/cm³; therefore, a given volume of oceanic crust weighs three times as much as the same volume of water. By comparison, the mean specific gravity of the Earth as a whole is about 5.5 g/cm³, and that of the inner core averages 12.5 g/cm³.

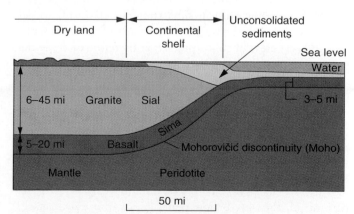

◆ FIGURE 13.10 Cross-sectional view of a continental plate margin. The oceanic crust is shown in green and the continental crust in violet.

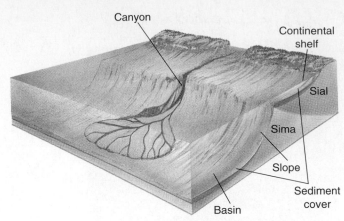

◆ FIGURE 13.11 The tapered margins of the continental crust are overlain by shallow ocean water covering the continental shelves. As the continental crust ends completely, ocean depths increase dramatically.

calcium than does the continental crust (Figure 13.8). The oceanic crust is sometimes referred to as the **sima,** a term derived from the words *si*licon and *ma*gnesium. Its thinness and high density cause it to rest at a relatively low level on the underlying mantle; for this reason it forms the ocean basin floors.

The continental crust is the uppermost of the solid layers of the Earth and is the only one that is discontinuous. It covers approximately 35 percent of the Earth's surface. In most places where it occurs, the continental crust extends from the surface to a depth of 10 to 70 kilometers (6–45 mi), making it much thicker than the oceanic crust. The continental crust contains a relatively high proportion of the elements aluminum, potassium, and sodium, which often combine with oxygen and silicon to form granite and related rock types. The continental crust is sometimes referred to as the **sial,** a term derived from the words *si*licon and *al*uminum.

Because its thickness and low density of about 2.7 grams per cubic centimeter give the continental crust enough buoyancy to form the highest portions of the Earth's surface, it is the material of which the continents are composed. The underlying oceanic crust or sima stretches elastically to sag downward beneath the weight of the sial and the continents. The thickest areas of continental crust therefore underlie the highest mountains while the thinnest areas are beneath shallow ocean water, usually along the continental margins. Ocean depths increase dramatically as the blocks of continental crust taper off and end (Figure 13.11).

While all of the oceanic crust is believed to have formed within the past 200 million years, much of the continental crust is very old. The most ancient portion of the crust to be dated so far is located in Canada's Northwest Territories and has an age of approximately 4.0 billion years. This is 87 percent of the 4.6 billion–year age of the Earth itself. Because of its great age, many areas of continental crust have been subject to multiple episodes of tectonic activity and currently exhibit complexly deformed rock structures.

ROCKS AND MINERALS OF THE EARTH'S CRUST

The solid surface of the Earth is composed almost exclusively of rock materials and their weathering products. Rocks in different regions vary greatly in origin, composition, appearance, and in both physical and chemical characteristics. Two consequences of this distribution that are of great practical importance to humanity are the global patterns of soil types and the patterns of mineral deposits. Of greatest relevance to this section of the text, however, is the fact that rock types and structures are crucial in determining the characteristics of landforms. In general, high-standing portions of the surface either have been uplifted recently by tectonic forces or are composed of physically strong and chemically stable rock types that resist gradational forces. In contrast, lowlands frequently are underlain by rocks that are highly susceptible to weathering and erosion. A basic understanding of the major categories of surface rock types is therefore an important prerequisite to the study of landforms.

The Earth's crust contains only a small number of chemical elements in abundance. Only eight, in fact, individually account for more than 1 percent of the weight of the crust (Table 13.3). Oxygen and silicon are about three times as abundant as all the rest combined. These two elements usually occur in chemical combination with metals to form the **silicate minerals,** the dominant mineral group of the crust and mantle. A **mineral** is a naturally occurring, solid, inorganic compound. It has a specific chemical composition and usually exhibits a crystalline structure that results from a distinctive atomic arrangement of its constituent elements. The 2,000+ known minerals are divided into several major mineral groups dominated by the silicates, which comprise 92 percent of the Earth's crust.

Minerals can be considered the building blocks of rocks. A **rock** is simply a solid aggregate of minerals. Most rocks contain several different minerals, the individual crystals of which are sometimes readily visible (Figure 13.12). In most areas the solid surface rock, or **bedrock,** is overlain by

◆ TABLE 13.3 Abundant Elements of the Earth's Crust

ELEMENT	PERCENTAGE OF CRUST BY WEIGHT
Oxygen	46.6
Silicon	27.7
Aluminum	8.1
Iron	5.0
Calcium	3.6
Sodium	2.8
Potassium	2.6
Magnesium	2.1
Total	98.5

◆ **FIGURE 13.12** This photo of a coarse-grained granitic rock clearly displays its crystalline structure.

SOURCE: Photo courtesy of Sue Monroe

comparatively thin layers of weathered rock and soil. Most of the continental crust is composed of rocks with a high content of silicon and aluminum. These rocks are normally low in density and light in color. The oceanic crust, in contrast, consists almost entirely of denser, dark-colored rock containing a greater abundance of iron, magnesium, and calcium. This rock is sometimes brought to the surface by volcanic eruptions or is exposed by the long-term erosion of a landmass.

Major Categories of Rocks

From the standpoint of the Earth scientist, the most meaningful way of categorizing rocks is by their method of formation. Three basic rock categories are recognized: igneous, sedimentary, and metamorphic. Within each category, a large variety of individual rock types exists. Differences among rock types are determined by mineralogical makeup and mode of origin as well as by the presence, size, and orientation of their constituent mineral crystals or rock particles.

Any rock type in any one of the three major rock categories can be converted by Earth processes to one or more types in either of the other two categories. These conversions occur because of changes in the environmental conditions that influence the rocks. A given rock type can form only in an environment where physical and chemical conditions favor its formation. If the rock-forming environment changes, or if the rock is transported to a different environmental setting, it is no longer in equilibrium with existing conditions. It therefore begins to change toward a condition of equilibrium with its new environment.

The continuous cycle of rock formation and transformation is known as the **rock cycle.** As in the hydrologic cycle, the same materials are used over and over again. A diagram of the cycle appears in Figure 13.13. The energy that powers the rock cycle is derived from both the Sun and the interior of the Earth. As long as these two energy sources continue to operate, the materials of the Earth's crust will continue to be involved in a constant cycle of change.

Igneous Rocks

Igneous rocks (from the Latin *igneus,* "from fire") are formed directly from the solidification of molten rock and are produced by tectonic activity. If the Earth at one time existed in a molten state, as is believed to have been true during the period of planetary formation, it must have consisted entirely of igneous rock when it solidified. This means that the constituents of all rocks must originally have been derived from igneous rock. The great majority of the Earth, including all of the mantle and oceanic crust and even most of the continental crust, still consists of igneous rock. Only within a few kilometers of the surface are sedimentary and metamorphic rocks found.

Most igneous rocks, including many now at the surface, formed in the lower crust or upper mantle. Although both of these segments of the Earth's interior consist primarily of solid rock, localized **magma** (molten rock) pockets form in the vicinity of zones of weakness where pressures are not sufficiently high to keep the rock from expanding and melting. Such masses of magma may be injected into overlying rock layers before solidification occurs, forming bodies of **intrusive igneous rocks.** Where bodies of hot or molten igneous rock exist near the Earth's surface, they may heat large volumes of groundwater that can be withdrawn by humans for heating or electricity. The subject of **geothermal energy** is examined in the Focus Box beginning on page 342.

The most visible characteristic of intrusive igneous rocks is the large size of their constituent mineral crystals. These crystals, which have time to form because of the very slow rate of cooling and solidification of intrusive igneous rock bodies, normally are easily visible, and their sharp edges give the rock a rough, granular texture. The best-known intrusive igneous rock type is **granite,** which is composed mostly of the minerals quartz, mica, and feldspar (Figure 13.14a).

Intrusive igneous rock bodies (known as **plutons**) often serve as sites for the formation of veins of economically important metallic minerals. These veins form along fractures or other zones of weakness in the slowly cooling plutons. Mineral-bearing fluids enter and move along the fractures, depositing dissolved minerals because the reduction of pressure in these zones of weakness reduces the ability of the fluids to carry minerals in solution. Richly mineralized igneous rock bodies are the sources of much of the world's supplies of metals such as gold, silver, copper, and tin. The Case Study at the end of the chapter examines the global distribution and major methods of extraction of gold—historically the most sought-after of all the world's metals.

On occasion, magma makes its way to the Earth's surface, where it flows out as **lava** or is thrown out as **volcanic ejecta**

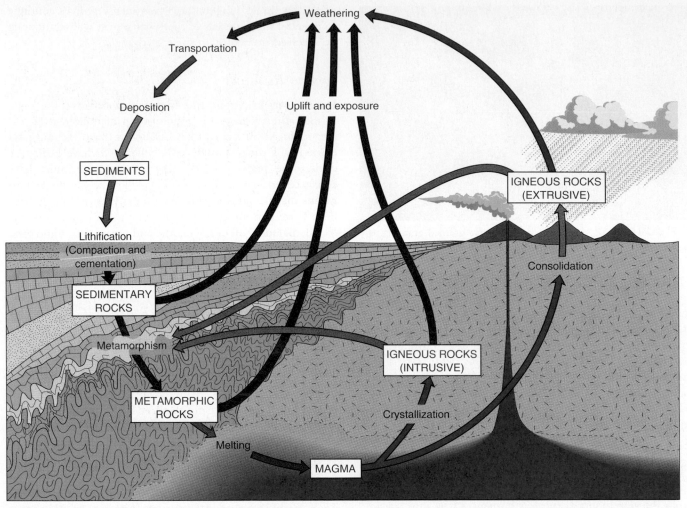

◆ FIGURE 13.13 The rock cycle. Under proper conditions, a rock in any one category can be transformed into a rock in any other category.

(Chapter 14). The solidification of this material forms what is termed **extrusive igneous rock** because it is extruded onto the surface while still molten. Because of the rapidity with which it cools, extrusive igneous rock generally has microscopic mineral crystals and, unless it contains air pockets, is relatively smooth. **Basalt,** a dense, black rock containing substantial quantities of iron and magnesium, is the best-known extrusive igneous rock type (Figure 13.14b). The Hawaiian Islands and the Columbia Plateau of the northwestern United States consist largely of basalt. A less dense, lighter-colored extrusive igneous rock containing larger quantities of silicon and aluminum is termed **rhyolite,** while a third common rock type, **andesite,** is intermediate in composition. Some igneous rock types, including **obsidian** and **pumice,** have cooled so rapidly that mineral crystallization did not occur at all, and they are glassy in texture (Figures 13.14c and d).

Figure 13.15 depicts the global distribution of surface rock types. It can be seen that intrusive and extrusive igneous rocks together cover about one-third of the Earth's land surface, and that the intrusive rocks are the more common of

the two groups. The distributional pattern of intrusive igneous rocks is complex, but the largest areas consist of the cores of ancient plutons, such as those in eastern Canada, Scandinavia, and Africa. Only the erosion of great thicknesses of overlying rock masses has allowed these ancient rocks, which originally formed at great depth, to exist at the surface.

Extrusive igneous rocks are more commonly associated with oceanic crust than with continental crust. On land, they occur mostly within active volcanic zones located along the margins of the continents. They are quite common, for example, all along the borderlands of the Pacific. Because volcanic eruptions usually are rather localized phenomena, the resulting extrusive igneous rocks are scattered in numerous patches, many of which are too small to appear in Figure 13.15. In a few instances, though, large-scale fissure outpourings of lava have covered extensive areas. Notable in this regard are the Columbia Plateau of the western United States, the Paraná Plateau of southern Brazil, and the Deccan Plateau of India. These outpourings are discussed further in the next chapter.

◆ **FIGURE 13.14** Important igneous rocks. (a) Granite specimen with individual mineral crystals that are readily apparent. (b) Hand specimen of basalt. (c) Obsidian, or volcanic glass, is an extrusive igneous rock that solidified so quickly that mineral crystals did not have time to form. (d) Pumice specimen. This rock type contains so many gas pockets (or vesicles) that it can float on water.

SOURCE: a: Photo courtesy of Reed Wicander; b: JLM Visuals; c–d: Bernard W. Pipkin, University of Southern California

Sedimentary Rocks

Weathering processes frequently produce loose surface materials sufficiently small and light to be picked up and transported to other sites by gradational agents. Any such solid materials, of either organic or inorganic origin, that have been transported by water, ice, wind, or gravity are collectively termed **sediments.** Familiar examples include mud or gravel deposits left by streams, beach sand and shells washed ashore by the ocean, rock debris deposited by retreating glaciers, and windblown leaves and dust.

Sediments are deposited where the energy of the transporting agents is insufficient to carry them farther. Because water is by far the most effective gradational tool, most depositional sites are associated with bodies of slow-flowing or standing water. Examples include valley floors, lake bottoms, and especially the ocean floor. The ocean can be thought of as the ultimate sediment trap. Once sediments reach it, as they do in tremendous quantities, there is normally no way for them to be removed because most sediments are denser than water and sink to the seafloor. The great majority of land-derived sediments reach the ocean by way of rivers, so that the greatest rates of sediment accumulation occur along the continental margins, especially near the mouths of large rivers.

Where conditions remain favorable for the deposition of sediments over extended periods of time, they gradually accumulate to form horizontal layers termed **strata.** If accumulation is rapid, great thicknesses of material eventually collect, resulting in the deep burial of the lowermost sediments. This generates a large amount of pressure, compacting these sediments. Frequently, water circulating through the compacted sediments precipitates chemicals that cement the deposits together. The processes of compaction and cementation, along with chemical changes in the sediments themselves, result in their gradual conversion to **sedimentary rock.**

The boundaries or **bedding planes** between adjacent strata mark periods in which local environmental conditions at the time of deposition changed, resulting in changes in

◆ FIGURE 13.15 World distribution of rock types. Sedimentary rocks are most abundant, covering approximately two-thirds of the Earth's land surface and nearly all of the sea floor.

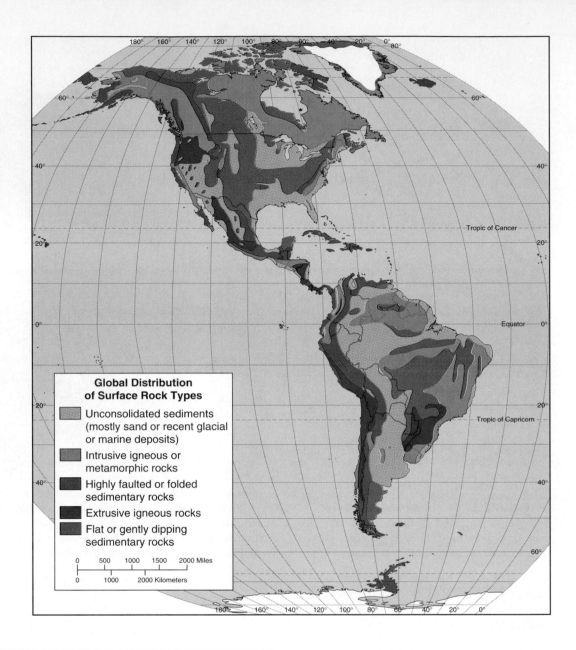

Global Distribution of Surface Rock Types

- Unconsolidated sediments (mostly sand or recent glacial or marine deposits)
- Intrusive igneous or metamorphic rocks
- Highly faulted or folded sedimentary rocks
- Extrusive igneous rocks
- Flat or gently dipping sedimentary rocks

0   500   1000   1500   2000 Miles

0   1000   2000 Kilometers

the types of sediments being deposited. Thick sequences of sedimentary rocks containing many hundreds of individual strata occur in various parts of the world (Figure 13.17). Where deposition was continuous, these strata may provide information about past environmental conditions over a lengthy geological time span. They bear eloquent testimony to the great age of the Earth and to the inconstancy of surface environmental conditions.

SEDIMENTARY ROCK TYPES. Sedimentary rocks are categorized by their method of formation. Three basic categories are recognized: clastic, organic, and chemical. Each contains a number of individual rock types, the most important of which are listed in Table 13.4.

Clastic sedimentary rocks are composed of particles of preexisting rocks. These particles were derived from the weathering of the parent rock and were subsequently trans-

ported, deposited, and **lithified,** or converted to rock. Clastic sedimentary rock types are differentiated on the basis of the sizes of their constituent particles. **Conglomerate** contains the largest particles, which must be at least gravel size (Figure 13.18a). **Sandstone,** one of the most abundant sedimentary rock types, consists of sand-sized particles. Most sandstones contain quartz sand grains because the great chemical stability of this mineral allows it to remain in rock form long after other minerals have weathered away. **Siltstone** and **shale** are composed of silt-sized and clay-sized particles, respectively. These sediments were nearly always deposited on the bottom of a standing body of water such as a lake or ocean. Shale is the chief sedimentary rock produced on the deep seafloor and comprises approximately half of all sedimentary rock. Both siltstone and shale, although chemically stable, are physically weak and susceptible to erosion so that they are frequently associated with

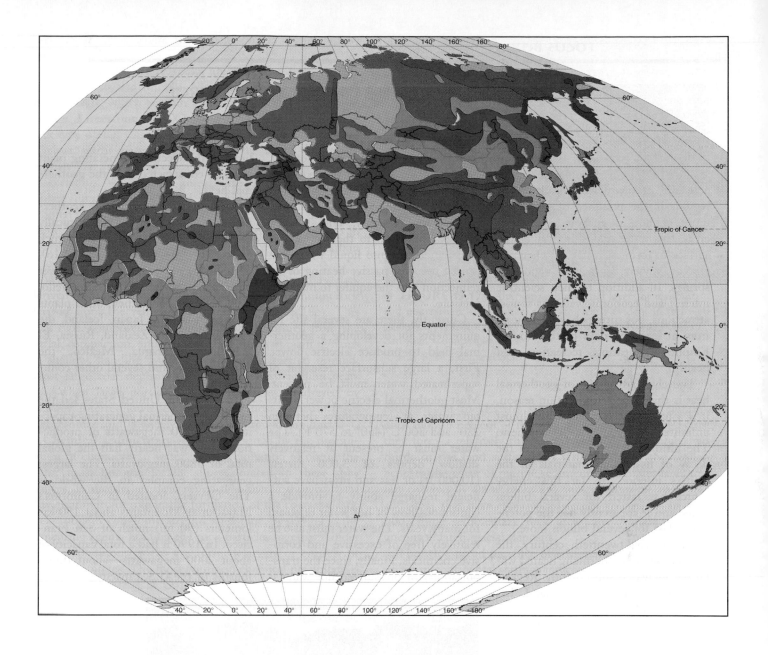

◆ **TABLE 13.4**　**Major Sedimentary Rocks and Their Sediment Sources**

ROCK CATEGORY	ROCK TYPE	SEDIMENT SOURCE
Clastic	Conglomerate	Rounded pebbles, cobbles, boulders
	Sandstone	Sand (usually quartz or feldspar)
	Siltsone	Silt
	Shale	Clay, mud
Organic	Limestone ($CaCO_3$)	Marine organisms
	Coal	Plants
Chemical	Limestone ($CaCO_3$)	Direct precipitation
	Dolomite [$CaMg(CO_3)_2$]	Direct precipitation; chemical alteration of limestone
	Chert (SiO_2)	Direct precipitation
Evaporite	Halite (NaCl)	Direct precipitation
	Gypsum ($CaSO_4 \cdot 2H_2O$)	Direct precipitation

(a)

(b)

(c)

◆ **FIGURE 13.19** Important metamorphic rocks. (a) Slate, which is formed from metamorphosed shale. (b) Gneiss, which probably formed from granite. (c) Marble, formed by the recrystallization of limestone. The application of heat and pressure during tectonic activity caused the alignment of the mineral crystals in each of these rocks, giving them foliated textures.

SOURCE: a: Photo courtesy of Sue Monroe; b–c: Bernard W. Pipkin, University of Southern California

◆ **TABLE 13.5** Major Metamorphic Rocks and Parent Rock Types

METAMORPHIC ROCK	PARENT ROCK(S)
Quartzite	Quartz conglomerate, sandstone, or siltstone
Slate	Shale
Phyllite	Shale, slate
Schist	Shale, slate, phyllite
Gneiss	Granite, diorite
Marble	Limestone, dolomite
Anthracite coal	Bituminous coal

uplift with little accompanying disturbance to the rock layers. The nearly horizontal sedimentary rocks of the Atlantic and Gulf coastal plain of the eastern United States serve as good examples. In other places, though, major tectonic upheavals have lifted sedimentary rock sequences great distances above sea level, often complexly deforming them while doing so. Most of the Earth's major mountain ranges, including the Rockies of northern North America, are composed chiefly of sedimentary rocks that formed on the ocean floor.

Metamorphic Rocks

Metamorphic rocks (from the Greek *metamorphosis*, or "transformation") are rocks altered from their original form by heat, pressure, and/or chemical activity, while remaining in the solid state. Metamorphic changes in rocks produce textures, structures, mineral compositions, and general appearances that differ from those of the original rocks. The constituent chemical elements remain the same, but they are rearranged into different mineral combinations. Metamorphism results from the high temperatures and pressures generated well beneath the surface by tectonic processes. Most metamorphic rocks are produced by large-scale or **regional metamorphism** that occurs during mountain-building activity.

All types of rock are subject to metamorphism with sufficient applications of heat and pressure; this includes igneous, sedimentary, and even metamorphic varieties. Each rock type has one or more metamorphic equivalents. Varying degrees of metamorphism exist, from slight alterations to complete recrystallization (Figure 13.19). The type of metamorphic rock produced at any given site depends on the original type of rock and on the nature, intensity, and duration of the metamorphic processes. Most metamorphic rocks are classified as one of seven important types, listed in Table 13.5.

Metamorphic rocks are by far the least abundant of the three major surface rock categories. They are much more common deep within the crust where they have been produced by tectonic movements. Because great thicknesses of overlying materials must be eroded before rocks subjected to regional metamorphism appear at the surface, they are exposed mostly in ancient plutons. Most metamorphic rocks comprised the root zones of major mountain ranges that were thrust up by collisions between blocks of crustal material.

SUMMARY

This chapter has examined the composition and internal structure of the Earth, as well as the nature of the forces that mold Earth's surface materials into landforms. Our planet contains only a small number of chemical elements in abundance, but those elements have been acted upon by a variety of processes in differing environmental settings. This has produced a large number of materials with diverse physical and chemical properties. This diversity is partially responsible for the great variety of landforms that exist.

The branch of Earth science devoted to the systematic study of landforms is called *geomorphology*. Until about a century ago, it generally was believed that large topographic features such as mountain ranges and canyons were created suddenly as a result of cataclysmic events. It is now known that, although catastrophes such as earthquakes and floods can quickly produce small-scale landscape modifications, large-scale surface features form and evolve very slowly. As a result, most landscapes exhibit features of widely varying age and often contain relict landforms produced by forces no longer regionally active.

We turn now to the Focus Questions posed at the beginning of the chapter:

1. *What are the important landform-producing forces, and where do they originate?*

Landform-producing forces may be divided into two major groups. The first group, the tectonic forces, originate from heat released by the nuclear decay of radioactive elements deep within the Earth. The general tendency of the tectonic forces is to produce surface uplift and increase topographic relief. As a result of tectonic activity, new landmasses are produced and mountain ranges are formed. The second group of landform-producing forces is the gradational forces. Powered largely by solar energy and given a downward directional component by gravity, these forces lower surface elevations and generally reduce topographic relief. Gradation operates chiefly through the erosive action of water, ice, and wind.

2. *What are the characteristics and locations of the internal layers of the Earth?*

The Earth is divided into three major concentric layers on the basis of composition. The innermost layer, the core, is a dense metallic sphere composed largely of iron. The sur-

rounding mantle is an extensive layer of dense rock containing large amounts of iron and magnesium. The lower and extreme upper portions of the mantle exist in a rigid solid state, but an intermediate portion, the asthenosphere, is plastic and contains slow-moving currents. The Earth's surface layer, the crust, consists of low-density rock composed largely of oxygen and silicon. A continuous lower portion, the oceanic crust, reaches the surface in the ocean basins. The uppermost portion, the continental crust, is discontinuous and forms the continents.

3. *What are the major categories of rocks and minerals comprising the Earth's surface?*

Most rocks comprising the Earth's crust consist of silicate minerals, which are composed largely of oxygen and silicon. This is especially true of the continental crust, which has a predominantly granitic composition. The oceanic crust consists largely of basaltic rock. It is denser than the continental crust and contains a greater proportion of ferromagnesian minerals.

Rocks commonly are divided into three categories according to their mode of formation. Igneous rocks have solidified directly from a molten state. Intrusive igneous rocks form below the surface and may be exposed by the erosion of the overlying materials. Extrusive igneous rocks are formed at the surface by volcanic eruptions or outpourings of lava. Sedimentary rocks are formed from the accumulation and lithification of sediments. Clastic sedimentary rocks, including conglomerate, sandstone, and shale, consist largely of particles of preexisting rocks. Organic sedimentary rocks, including limestone and coal, are formed from the remains of plants or animals. Chemical sedimentary rocks result from chemical precipitation. Metamorphic rocks are altered from their original states by mountain-forming tectonic stresses.

QUESTIONS FOR THOUGHT AND REVIEW

1. Explain the differences between the geomorphic principles of catastrophism and uniformitarianism. Which principle is viewed as being more accurate today? Why?
2. Explain how the population distribution pattern within your own state or local area is related to its topographic features.
3. What are the two basic approaches to the systematic classification of landforms? What are the advantages and shortcomings of each?
4. What are the two groups of geomorphic forces, and what is the ultimate energy source for each? How are the two groups of forces opposed to each other in terms of their influence on the topography?
5. Why are the Earth's surface features different from those of the other planets and larger satellites in our solar system?
6. Name the internal layers of the Earth and briefly describe their physical and chemical characteristics.

7. Explain how the continental and oceanic crusts differ in terms of composition, thickness, density, age, and surface distribution. How does this distribution relate to the current global distribution of continents and ocean basins?

8. List and define the three major categories of rocks. Describe the environmental settings in which each is formed. What types of rocks dominate in the area in which your college or university is located?

9. What is the rock cycle? Give an example of the operation of the cycle, beginning with igneous rock, and passing through the other two rock categories before returning to igneous rock. Explain how and why each change in your example occurs.

10. What geological and geographical conditions are currently required for the establishment of an economically competitive geothermal electricity plant?

11. What are the two major categories of gold deposits? Explain how each is produced. In what areas of North America has gold mining been chiefly concentrated? What adverse impacts do mines have on the environment?

▌KEY TERMS

topography	crust
geomorphology	asthenosphere
catastrophism	lithosphere
uniformitarianism	oceanic crust
plain	continental crust
plateau	mineral
tectonic forces	rock
diastrophism	igneous rock
volcanism	sediment
gradational forces	strata
weathering	sedimentary rock
core	metamorphic rock
mantle	

Gold Distribution and Mining

Gold has historically been the most desirable and sought-after of all metals. Despite its relative rarity, it has been mined for thousands of years and has played an important role in human history. Among the qualities that have long made gold attractive are its unique color, its widespread distribution, and the fact that it is easily mined and refined because it usually occurs in the pure metallic state, requiring no complex separation techniques. Its major uses are for decoration (especially in jewelry) and as an instrument of monetary exchange. During the nineteenth century, in fact, most countries used gold as the sole basis for their currencies. Although the United States stopped minting regular gold coins in 1934, it, along with many other countries, including Canada, now mints special gold coins, and gold is accepted by all nations as a medium for international payment.

TYPES OF GOLD DEPOSITS

Gold occurs in limited quantities in all igneous rocks, but large concentrations are unusual. Commercially extractable gold ore bodies are of two types: hydrothermal veins and placer deposits. **Hydrothermal veins** are produced by the surfaceward movement of heated mineral-rich solutions that are squeezed from cooling magma bodies as they solidify. The superheated water, which contains dissolved metallic minerals such as gold, silver, and copper, is forced into cracks, fissures, and pores in the cooling rock body. As the solution cools, it loses its ability to hold minerals in solution, causing the metals to crystallize. The resulting gold deposits usually occur in the form of veins or masses (Figure 13.20).

Placer deposits are derived from the weathering and erosion of gold-bearing hydrothermal veins and are usually located in stream gravels. The gold typically occurs in the form of invisible grains, but more rarely exists as visible flakes or nuggets. Large gold nuggets weighing up to 90 kilograms (200 lb) have been found in Australia. Prospectors who discover placer deposits often work their way upstream, hoping to encounter the "mother lode"—the vein deposit from which the placer deposits were derived.

HISTORY OF GOLD MINING

The details of the earliest gold mining operations are lost in the mists of antiquity. Gold mines in the Nubian Desert operated by the early Egyptians were the chief source of gold for the ancient world and played a key role in financing the Egyptian rulers.

The greatest early surge in world gold supplies followed the European discovery of the New World. The looting of gold from the Aztecs and Incas by the Spanish conquistadores provided Europe with a large influx of the metal during the sixteenth century. In the seventeenth century, gold mines in Colombia, Peru, Ecuador, Panama, and Hispaniola accounted for 61 percent of the world gold production; in the eighteenth century, they supplied 80 percent.

Following the discovery of gold in California's Sierra Nevada in 1848, North America became the world's chief supplier, and in 1851, additional major deposits were found in Australia. As a result of these discoveries, more gold was mined between 1850 and 1875 than in the previous 350 years. By 1890, gold production from new fields in Alaska and the Canadian Yukon surpassed that of California, and during the early twentieth century, the discovery of huge new deposits in South Africa's Transvaal made that region the world's largest gold producer.

continued on next page

◆ **FIGURE 13.20** Native gold in quartz veins can appear in a variety of forms.
SOURCE: John D. Cunningham/Visuals Unlimited

◆ **FIGURE 13.21** Pie chart showing the proportional gold production of major gold-mining countries in 1992. Values indicate the percentage of total global production.

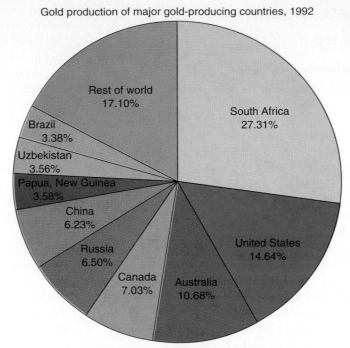

Gold production of major gold-producing countries, 1992

Rest of world 17.10%
Brazil 3.38%
Uzbekistan 3.56%
Papua, New Guinea 3.58%
China 6.23%
Russia 6.50%
Canada 7.03%
Australia 10.68%
United States 14.64%
South Africa 27.31%

World total gold production in 1992:
2,247,840 kilograms (4,955,590 lb)

At present, the Republic of South Africa produces more than a quarter of the world's gold supply, followed by the United States, Australia, and Canada (Figure 13.21). The quantity and distribution of proven reserves (like those of most mineral resources) are constantly changing because of new discoveries, improved processing techniques, and large price fluctuations. In 1991, however, more than 95 percent of global reserves were located in South Africa (81.7%) and Zimbabwe (13.7%). Only 0.1% of world reserves were located within the United States.

GOLD MINING AND PROCESSING METHODS

Gold vein and placer deposits are mined by different techniques, and important quantities of gold are also obtained as by-products of the refining of other metals. Placer mining methods are employed when the gold occurs in unconsolidated sands and gravels. A variety of mining techniques are used; all are based on the fact that gold is much heavier than the rock materials with which it is mixed. (A cubic centimeter of gold weighs 19.3 grams, while a cubic centimeter of most rock types weighs about 2.7 to 3.3 grams.) The simplest method of placer mining is panning. This method, which was widely employed by the Gold Rush prospectors in California, Alaska, and the Yukon, involves swirling a mixture of gold-bearing gravel and water in a large, gently sloping pan about 5 centimeters (2 in) deep. The mixture is swirled rapidly enough to carry the water and lighter rock over the edge of the pan, while the gold remains on the bottom. Using this method, an experienced miner can work through 0.7 cubic meters (1 yd³) of sediments in a 10-hour period. A more rapid method uses a sluice box—a U-shaped trough with a series of transverse bars along its bottom. Sand and gravel are shoveled onto the top of the trough, and a gate to a water supply is opened. The lighter rock washes out at the far end, while the gold gets caught behind the bars. A more modern placer mining technique is hydraulic mining. Here, a strong stream of water is directed at sand and gravel banks, washing away the sand and gravel, while leaving the heavier gold behind.

Lode or vein mining uses conventional mining techniques, which nearly always involve sinking a shaft to follow the deposit. This is currently the leading gold mining method, accounting for more than half of all gold production. Extensive waste (or tailings) piles result, because approximately 100,000 units of ore must be processed to produce one unit of gold.

◆ **FIGURE 13.22** (left) Aerial view of the Jamestown Mine in Jamestown, California. The tailings pond is at the left, with the waste rock piled behind it. Settling ponds below the mill structures trap suspended sediments so that the water can be reused in the milling process. (right) Flowchart of the Jamestown Mine's operation. The final leaching and gold pour is at the company's Buckskin plant near Yerington, Nevada. Only 0.06 ounce of gold is produced from each ton of ore mined.
SOURCE: D. D. Trent

Gold extracted from vein ores must be refined to remove the large quantities of unwanted rock with which it is mixed. A variety of refining procedures are used; the system employed by the Jamestown Mine in Jamestown, California, is depicted in Figure 13.22.

A method developed in the 1980s that has increased rapidly in use to become one of the major gold refining methods in the United States is cyanide heap leaching. In this method, a dilute cyanide solution is sprayed on the surface of a pile of crushed gold ore. The solution passes down through the pile, dissolving the gold, and eventually accumulates in a pond. The contents of the pond are then pumped to a processing plant, where the gold is removed by chemi-

cal and electrical techniques, melted in a furnace, and poured into molds.

ENVIRONMENTAL CONCERNS

Although the mining of gold and other metals has played a key role in the settlement and development of western North America, metal mining and refining operations produce a variety of environmental problems. Two of the most serious are the adverse visual impact of the huge piles of rock tailings and the contamination of groundwater supplies. Each year global mining activities generate twice as much solid waste as do all other industries and municipalities combined. In the future, the rate of waste generation will increase further

as high-grade ore supplies are depleted and progressively lower grade ores are processed.

After they have been exposed at the surface, the remaining metals in abandoned mine tailings are leached by downward-percolating rainwater to form highly acidic and often toxic solutions. This water then frequently flows into nearby streams and pollutes adjacent groundwater supplies. To prevent a further deterioration of mined areas, it will be necessary for technological developments to provide the means, and for environmental legislation to provide the incentive, to return the landscape as nearly as possible to its original condition.

14 TECTONIC PROCESSES AND LANDFORMS

FOCUS QUESTIONS

1. How do plate tectonic movements occur, and how do they create and destroy crustal material?
2. What causes rocks to fold, and what topographic features result from folding?
3. What are the major types of fault motions and their resulting topographic features?
4. What causes volcanic eruptions, and what types of surface features do they produce?

On the morning of May 18, 1980, Mt. St. Helens volcano in Washington State ended a 137-year period of dormancy with a spectacular eruption. The northeast side of the mountain collapsed, releasing a huge cloud of ash and superheated gases that soon spread over thousands of square kilometers, plunging areas within 390 kilometers (240 mi) of the volcano into total darkness. The eruption produced major ash falls over much of the Pacific Northwest, took 60 human lives, and caused $600 million in damage to forests, crops, and personal property. It also helped to make the North American public more aware of the hazards associated with volcanically active areas.

In this chapter we examine the nature of tectonic forces such as those responsible for the Mt. St. Helens eruption (see Figure 14.33) as well as the impact these forces have on the Earth's surface. As we shall discover, the Earth's crust is both formed and destroyed by tectonic forces, and their continued operation is slowly changing the sizes, shapes, and locations of both the continents and the oceans.

PLATE TECTONICS

Until a few decades ago, the processes by which tectonic forces operated were largely unknown. It had long been realized that these forces were powered by energy from our planet's interior, but the exact sources of this energy and the processes by which it produced tectonic features at the surface were matters of speculation. It is now known that the lithosphere, for all its seeming solidity, is divided into a number of segments, or **plates,** that are in constant motion. This motion is powered by the atomic decay of radioactive elements in the core and mantle. As the lithospheric plates move, collisions occur, and existing plates are distorted or even broken apart. In addition, new crustal material is formed from magma rising from the mantle, and old crust is destroyed as it plunges back into the mantle. Each of these motions is associated with its own group of tectonic landforms.

The entire process of plate formation, movement, and destruction is known as **plate tectonics.** First proposed during the 1940s, plate tectonic theory gained widespread acceptance during the late 1950s and 1960s. Continuing research has provided a great deal of supporting evidence for the theory and has enabled us to refine our understanding of the various processes involved.

The development of plate tectonic theory has been the landmark occurrence in modern Earth science. It has undoubtedly resulted in the greatest increase in our understanding of landform-producing processes since the concept of uniformitarianism was accepted during the nineteenth century. It has shown that the Earth's surface does not consist merely of a random assortment of land and water bodies, but instead forms an integrated system that is intimately related to materials and processes deep within the interior.

Evidence for Continental Drift

Many of us have at some time been struck by the thought that a map of the Earth resembles a jigsaw puzzle, with the continents representing the puzzle pieces separated by areas of ocean. On closer inspection, a number of the continental "pieces" do seem to fit together remarkably well. Especially striking is the match between the east coast of South America and the west coast of southern Africa. In a few other easily seen match-ups, the island of Madagascar fits neatly onto the east coast of Africa, the Arabian Peninsula into northeastern Africa, and New Guinea into northern Australia. Scholars as early as the sixteenth century noticed these coastal similarities and wondered whether they could be coincidental or if, indeed, the landmasses had somehow separated and moved long distances over the Earth's surface.

The matter remained largely in the realm of speculation until 1912, when German meteorologist and geophysicist Alfred Wegener presented a detailed theory of **continental drift** that not only traced past positions and subsequent motions of the continents, but also for the first time suggested a means by which this movement could have been accomplished. According to Wegener, all the larger landmasses were at one time assembled into a single supercontinent that he called **Pangaea.** This supercontinent later separated into a number of smaller blocks that subsequently drifted to their present positions (Figure 14.1). The mechanism that Wegener proposed for the motions of the continental blocks was soon proven to be incorrect, however, and the idea of continental drift fell into disrepute for a time. Wegener had nonetheless succeeded in focusing attention on the possibility that the positions of segments of the crust could somehow have shifted.

A revival of research into continental drift took place following World War II. By this time, more sophisticated geologic instrumentation was available, and Earth scientists soon began amassing additional evidence in support of large-scale crustal movements. It was discovered, for example, that certain rock sequences extending to the coast of Brazil reappeared on the African coast precisely where they would be expected if the two continents were once joined. Fossil remains indicated that similar or identical plants and animals appeared during the same geological periods on landmasses currently located great distances from one another. Evidence of a past ice age was unearthed in currently subtropical portions of South Africa and South America, and rocks containing fossils of tropical plants were discovered in Antarctica and Britain. In addition, the present-day locations of active volcanoes, earthquake foci, and the Earth's

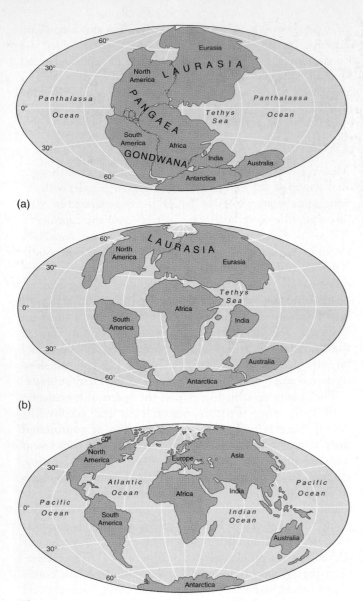

(a)

(b)

(c)

◆ **FIGURE 14.1** Map sequence showing the breakup of Pangaea and the movement of the continental blocks to their present positions. Map (a) shows the positions of the continents 180 million years ago. Map (b) shows their positions 65 million years ago. Map (c) shows the present positions of the continents.

Depth contour, m

Ridge axis

◆ **FIGURE 14.2** Magnetic rock patterns along a portion of the Mid-Atlantic Ridge southwest of Iceland (see top map for location). The shaded areas display normal polarity; they are separated by areas of reversed polarity. Note the symmetry and the continually increasing age of the magnetic bands outward from the ridge axis.

SOURCE: Robert A. Phinney, ed., *The History of the Earth's Crust.* Copyright © 1968 by Princeton University Press. Fig. 6, p. 83, reprinted with permission of Princeton University Press.

high mountain belts in general are all strongly aligned with the apparent boundaries of the crustal plates.

Some of the strongest evidence for continental drift came from the analysis of the orientation of crystals of the mineral magnetite (Fe_3O_4) in oceanic basalts. When free to move, these crystals act like tiny compass needles by aligning themselves along the axis of the Earth's magnetic field. When lava containing magnetite crystals solidifies, the crystals are frozen into place with their orientation recording the direction toward the North and South Magnetic Poles at the time that solidification occurred. While recent lava flows contain magnetite crystals aligned toward the present poles,

the alignment in progressively older flows gradually rotates away from the present polar orientations. When magnetite crystal data are compared with the theorized patterns of continental drift since the breakup of Pangaea, a good correlation is obtained, providing almost conclusive evidence that these movements did in fact take place.

In the 1960s, researchers also discovered that the magnetic North and South Poles on occasion actually "flip," or undergo a reversal in polarity. This causes the magnetic North Pole to temporarily become the magnetic South Pole, and vice-versa. Any magnetite crystals in basalt that solidified during a period of reversed polarity have an opposite

■	Pleistocene to Recent (0−1.6 M.B.P.)		■	Paleocene (58−66 M.B.P.)
■	Pliocene (1.6−5 M.B.P.)		■	Late Cretaceous (66−88 M.B.P.)
■	Miocene (5−24 M.B.P.)		■	Middle Cretaceous (88−118 M.B.P.)
■	Oligocene (24−37 M.B.P.)		■	Early Cretaceous (118−144 M.B.P.)
■	Eocene (37−58 M.B.P.)		■	Late Jurassic (144−161 M.B.P.)

◆ **FIGURE 14.3** The age of the rocks forming the ocean floor as established from magnetic anomalies. The youngest oceanic crust is adjacent to the oceanic spreading ridges, and the age increases away from the ridge axes. Dates are given in millions of years before the present.

alignment from those of the present time. We now know that the Earth's magnetic field has reversed polarity dozens of times over the past few tens of millions of years, with the latest reversal occurring approximately 700,000 years ago.

In 1964, aircraft equipped with magnetic sensing devices flew over the Mid-Atlantic Ridge and discovered that the basalt on both sides exhibited a mirror-image pattern of stripes of normal and reversed magnetic polarity (Figure 14.2). The Mid-Atlantic Ridge is part of a system of oceanic ridges extending through the Atlantic, Pacific, and Indian Oceans that reaches heights of several thousand meters above the surrounding deep ocean floors. These ridges are the sites of active and nearly continuous submarine volcanism, which erupts great quantities of lava that solidifies into basalt containing magnetite crystals. The inescapable conclusion was that these ridges are the boundaries between newly forming portions of two crustal plates that are diverging at rates of several centimeters a year. As the plates diverge, the patterns of normal and reversed magnetic polarity are frozen into the magnetite crystals of the solidi-

fying basalt, producing the symmetrical bands of magnetic polarities.

Additional evidence for the divergent movement of the crust from the oceanic ridges is provided by the progressive increase in the age and thickness of the overlying cover of sediments (Figure 14.3) and by the gradually lowering crustal temperatures outward from the ridges. Because most new crustal material seems to form and diverge along these oceanic ridges, continental drift appears to be attributable largely to the process of **seafloor spreading.**

Plate Tectonics Processes

It is now known that the moving plates are not composed exclusively of crustal material, but also include the rigid uppermost portion of the mantle. Taken together, the crust and uppermost mantle comprise the lithosphere, which averages about 100 kilometers (60 mi) in thickness. It therefore is more accurate to refer to these plates as **lithospheric plates** than to use the older term "crustal plates."

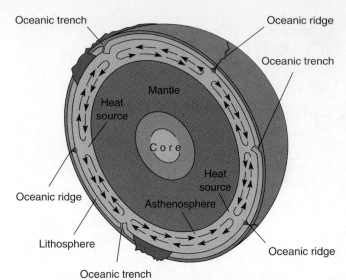

Oceanic trench

Oceanic ridge

Oceanic trench

Mantle

Heat source

Core

Heat source

Oceanic ridge

Asthenosphere

Lithosphere

Oceanic ridge

Oceanic trench

◆ **FIGURE 14.4**   Diagrammatic view of the Earth's interior, showing the likely cause of plate tectonic movements. Convection currents in the asthenosphere, powered by the decay of radioactive elements in the core and mantle, drag the overlying lithospheric plates with them. The resulting plate collisions produce mountain ranges, while seafloor spreading zones in areas of plate divergence are the sites of new oceanic crust formation.

The asthenosphere is the layer upon which the lithospheric plates float. As noted earlier, the asthenosphere is plastic in constituency and contains a complex pattern of slow-moving convection currents produced by uneven heating from the decay of radioactive elements in the mantle and core (Figure 14.4). Heat from the decay of these elements is conducted very slowly surfaceward through the mantle. When it reaches the asthenosphere, this heat generates slow convection currents within the rock material. The radioactive elements and the heat they produce are distributed unevenly within the Earth, much as they are at the surface. In places where these materials are relatively abundant, more heat is generated, and the mobile areas within or above them in the asthenosphere expand, increase in buoyancy, and become the sites of rising convection currents. In places where radioactive elements are not as abundant, less heat is produced, and the affected mobile areas are likely to be the sites of descending currents of somewhat cooler and denser material. A continuous circulation system thus is established. The lithospheric plates, which float on the upper portions of the convection currents, are carried slowly along like pieces of driftwood on a stream. Atomic energy, manifested in the form of convection currents in the asthenosphere, therefore provides the long-sought mechanism for continental drift.

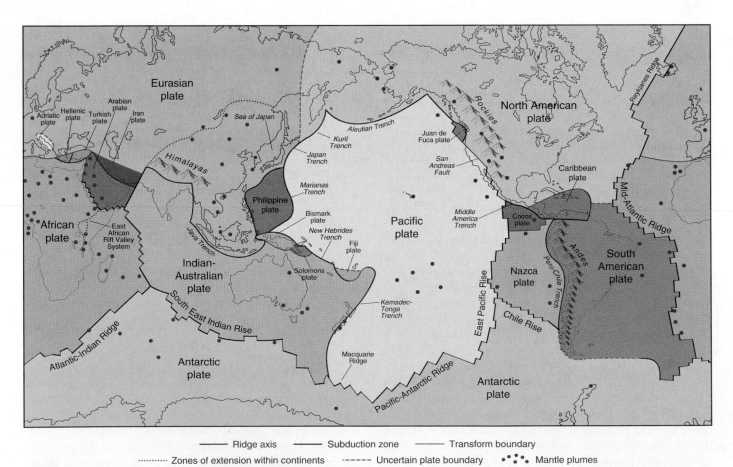

—— Ridge axis   —— Subduction zone   —— Transform boundary
········· Zones of extension within continents   - - - - - Uncertain plate boundary   •∴•∴ Mantle plumes

◆ **FIGURE 14.5**   Global pattern of lithospheric plates and types of plate boundaries. Red dots indicate the sites of known plumes.

◆ **FIGURE 14.6** The Emperor Seamount-Hawaiian Island chain formed as a result of the movement of the Pacific plate over a mantle plume. The line of seamounts and volcanic islands indicates the direction of plate movement. The numbers indicate the ages of the islands in millions of years.

Continental drift in turn, results in adjacent lithospheric plates either colliding, moving laterally with respect to each other, or separating, and these occurrences produce tectonic landforms.

The lithosphere is divided into seven major plates as well as a number of smaller plates, as shown in Figure 14.5. The plates are moving in various directions at rates of up to 20 centimeters (8 in) a year. Movement in most cases is away from the oceanic ridge along which the plate is being manufactured. For example, the North American Plate, which contains the continent of North America as well as the western half of the Atlantic Ocean, is being manufactured along the Mid-Atlantic Ridge. As new crustal material is extruded along the ridge, the plate is carried westward by convection currents in the underlying asthenosphere. This westward motion causes the North American Plate to collide with the Pacific and the Juan de Fuca Plates, which are located to its west. This collision has helped to crumple the forward edge of the North American Plate to produce the Rocky Mountains. Plate motions are also gradually increasing the width of the Atlantic Ocean, which began forming about 150 million years ago when North and South America first began to separate from Europe and Africa as part of the breakup of Pangaea. In similar fashion, the African and Indian-Australian Plates are moving northeastward, thrusting up a great mountain chain extending from the East Indies westward to North Africa and Spain as they collide with the Eurasian Plate.

Plate Formation and Movement

Convection currents within the asthenosphere are responsible not only for the motions of the lithospheric plates, but for their formation and destruction as well. Internal heat seems to be carried surfaceward largely in the form of upwelling "plumes" that form in various locations within the asthenosphere. As these plumes approach the upper margin of the asthenosphere, they gradually change direction from vertical to horizontal. More than 120 active plumes have been tentatively identified. They are most numerous along and near the oceanic ridges, but a large number are scattered in seemingly random locations beneath other parts of the oceans and the continents. Many are associated with surface volcanic activity, and most mid-oceanic volcanic islands or island groups such as the Hawaiian Islands, Azores, Iceland, and Réunion have formed on the sites of active plumes.

Isolated mantle plumes such as the one forming the Hawaiian Islands are able to protrude through a lithospheric plate without seriously disrupting it (Figure 14.6). In places where several nearby plumes break through the crust, however, they may weaken it so much that it splits apart to produce two smaller plates. (By analogy, a single nail usually can be driven into a board with no adverse effects on its integrity. When too many nails are placed too closely together, however, the board is likely to split.) This is one mechanism by which the lithosphere is believed to have been originally broken into a variety of individual plates.

The splitting of a plate causes a substantial reduction in pressure on the underlying asthenosphere. The reduced pressure lowers the melting point of the rock comprising the upper mantle, causing the selective melting of its lighter constituents, which have the lowest melting points. The zone of selective melting extends from a depth of some 50 to 70 kilometers (30 to 40 mi) below the surface upward to the base of the crust. Only 10 to 20 percent of the rock is able to melt, and it eventually collects in a magma chamber beneath the new crustal break. Eventually, part or all of this magma makes its way to the surface through a system of vertical passageways lying along the fracture and is expelled onto the ocean floor during submarine volcanic eruptions. As it solidifies, this material forms new oceanic crust of basaltic composition. The combination of upward expansion due to heat and the expulsion of magma produces an oceanic ridge and apparently forces the divergence of the newly forming lithospheric plates on either side. Plate divergence may also be aided by gravity as the plates slide downslope from the elevated oceanic ridge systems toward the deep oceanic trenches that usually border the far sides of the plates.

In some instances, a plume-induced break occurs in a portion of the lithosphere overlain by continental crust. In this case, subsequent divergence of the sides of the break will first stretch and weaken the crust, producing a deep, linear, steep-sided trough called a **rift valley.** An excellent example of such a trough, produced by the early separation stages of two segments of continental crust, is the East African Rift Valley (Figure 14.7). If divergence continues, the two continental plate segments will become completely separated as ocean water fills in the gap that is being covered by newly manufactured oceanic crust. This stage of the divergence process is well illustrated by the Red Sea, located between Africa and the Arabian Peninsula. Continued divergence eventually may produce a full-sized ocean, as in the case of the Atlantic.

At present, three major zones of plate formation and divergence marked by oceanic ridges exist on the Earth's surface: the Mid-Atlantic Ridge, the East Pacific Rise and Pacific-Antarctic Ridge system, and the Y-shaped Indian Ocean Ridge system (Figures 14.5 and 19.22). They total some 64,000 kilometers (40,000 mi) in length.

The rate of seafloor spreading varies considerably in different areas. Along the Mid-Atlantic Ridge, it averages about 4 centimeters (1.5 in) a year. The fastest rate of spreading is along the East Pacific Rise, where it averages about 15 centimeters (6 in) a year. Even though spreading rates of a few centimeters a year may seem very slow, these movements can total thousands of kilometers in a few tens of millions of years. They therefore are capable of completely rearranging the surface of our planet over relatively short geologic time spans.

Plate Collisions and Destruction

Based on their relative motions, lithospheric plate boundaries are classified as divergent, transform, or convergent. Along **divergent boundaries,** the adjacent plates move away from one another. Along **transform boundaries,** the motions are lateral, so that the two plates slide past each other while moving in opposite directions relative to each other. Neither of these two types of boundary movements is directly responsible for the production of a large proportion of Earth's continental landforms. Diverging boundaries do produce volcanic islands and rift valleys, and, as separation

◆ **FIGURE 14.7** The East African Rift Valley, shown here, is apparently being formed by the gradual separation of eastern Africa from the rest of the continent as a result of divergent plate movements. The earlier separation of other blocks formed Madagascar and the Arabian Peninsula, and opened up the Red Sea and Gulf of Aden.

continues, seas and even oceans may eventually form. The chief material manufactured along these boundaries, though, is oceanic crust, which normally is located on the seafloor or is buried beneath continental crust. Discussion of the various types of seafloor features that form on this layer can be found in Chapter 19.

Convergent boundaries have by far the greatest influence on the production and distribution of continental tectonic features. Not only are they responsible for producing most tectonic landforms, these movements actually form continental crust in the process of destroying oceanic crust. Without the convergence of lithospheric plates, no major mechanism would exist for the production of light, buoyant, and therefore high-standing masses of continental granitic rock, and the Earth would be almost completely covered by ocean.

Because the Earth's size remains constant, the rate of lithospheric plate destruction must equal its rate of formation. This destruction occurs at the convergent plate boundaries, and the nature of the surface features produced depends primarily on whether the plates are covered by oceanic or continental crust. Convergent boundaries can be formed by the collision of two oceanic plates, by a collision between an oceanic and a continental plate, or by a collision of two continental plates.

When two oceanic plates collide, one of the plates is bent downward by the other so that it is pressed gradually into the asthenosphere at an angle of 30° or more in a process called **subduction** (Figure 14.8). As the subducted plate reaches the base of the asthenosphere at depths exceeding 600 kilometers (400 mi), it is heated to the point that it loses its rigidity. The entire process from the time of initial plate contact to the loss of rigidity of the subducted plate may take some 10 million years.

The portion of the oceanic plate that is being subducted is likely to be covered by a great thickness of sediments. These sediments have accumulated on the seafloor since the time the plate was manufactured along an oceanic ridge. The sediments contain a large amount of water, and as they are subducted along with the plate, this water reduces the melting point of the lighter rock-forming materials both within the subducted plate and in the portions of the mantle with which the plate comes into contact. This results in the selective melting of these materials, which rise as magma toward the surface. The magma may either solidify beneath the surface to form large granitic plutons known as **batholiths** or reach the surface in volcanic eruptions. In either case, the resulting igneous rock material becomes continental crust.

Because this type of plate convergence occurs between two oceanic plates, the initial tectonic features are produced on the seafloor. Among the most striking features are **submarine trenches** that form along the line where the denser oceanic plate is being subducted (Figure 14.8). These narrow chasms contain the deepest points on the ocean floor. Challenger Deep in the Marianas Trench reaches a depth of 11,033 meters (36,198 ft).

Several hundred kilometers from a trench, and typically paralleling it above the zone where the subducted plate extends deep into the asthenosphere, an **island arc** is produced near the margin of the overriding plate. This occurs when the volcanic eruptions associated with the rising continental crust–producing magmas have deposited a sufficient quantity of igneous material to protrude above sea level. Examples of such island arcs include the Marianas of the western Pacific, the Aleutian Islands of Alaska, and the Lesser Antilles that separate the Atlantic from the eastern Caribbean Sea.

◆ **FIGURE 14.9** Diagrammatic view of the collision between a continental plate and an oceanic plate.

The gradual accumulation of lighter rock from both surface volcanism and subsurface granitic intrusions eventually forms continents. This process is accomplished largely by the collision and joining of numerous island arcs or other small segments of continental crust, each of which is termed a **terrane**, by the sweeping action of the lithospheric plate movements. The west coast of North America from California to Alaska, for example, has been formed by the accretion of numerous terranes onto the western margin of the American Plate. In addition, some geologists believe that the Piedmont region of the eastern United States consists of several crushed-in layers of ancient island arcs that were collected at a time when a plate boundary existed along the axis of the Appalachians.

When an oceanic plate collides with a continental plate, the denser oceanic plate is always subducted (Figure 14.9). A marginal volcanic zone forms as magma from the subducted oceanic plate rises into and sometimes through the adjacent margin of the overriding continental plate to form new continental crust. This category of plate collision is the most common of the three and is primarily responsible for the fact that the majority of the world's actively growing mountain belts are located along the margins of the continents. The Pacific Plate is currently being subducted by continental plates on all sides, producing its surrounding ring of volcanoes. They include the Cascade Mountains of Oregon and Washington, as well as the volcanoes of south-central Alaska.

Behind the coastal volcanic zone, the overriding continental plate may experience extension, compression, or both. If it is being extended, the plate is stretched and downwarped, sometimes producing an inland sea. The Sea of Japan, between volcanic Japan and mainland Asia, is a good example. In the Great Basin region of the western United States, crustal extension has broken the crust into numerous north/south-oriented fault blocks, some of which have been raised and tilted to form mountain ranges. If compression occurs, a broad mountain area dominated by folding and faulting is formed. The Rocky Mountains to the east of the Great Basin region serve as a prime example.

When two continental plates collide, neither one is subducted, and little volcanism normally occurs. Instead, a broad zone of compression and diastrophism, often many hundreds of kilometers wide, is formed (Figure 14.10). Frequently, one of the plates is thrust for great distances over the other. This produces a double thickness of continental crust, which, in combination with its light weight and buoyancy, results in the formation of a high-elevation mountainous landmass.

The crumpling of the forward edges of the two colliding continental plates produces extremely complex geological structures. The plate boundaries often lose their identities, and instead a broad and poorly defined zone of compression is formed. Mountains may form along a zone of weakness rather than on or near the actual line of plate convergence. Continued convergence eventually results in the fusion, or **suturing**, of the two plates into a single larger plate. This mechanism balances the process of plate separation and keeps the crust from being divided into progressively smaller units.

The premier example of a continental plate collision is occurring between the Indian subcontinent and southern Asia. This collision, which began 50 million years ago because of the northward movement of the Indian-Australian Plate, has produced the Himalayas—the world's loftiest mountain range (Figure 14.11). Asia has proven to be the weaker of the two plates and has deformed the most. Faulting in Asia extends nearly 4,800 kilometers (3,000 mi) north of the collision zone, and the continent has been shortened latitudinally and stretched longitudinally by the collision. A similar type of collision at an earlier stage is occurring to the west, where the northward-moving African

Deformed and metamorphosed
subduction complex

Mountain
range

Oceanic crust
fragments

Continental
crust

Magma

Asthenosphere

Upper
mantle

Oceanic crust

◆ **FIGURE 14.10** Diagrammatic view of a collision between two continental plates. The Himalayas have been formed by this type of collision (see Figure 14.11).

Plate is beginning to run into the European portion of the Eurasian Plate. The intervening Mediterranean Sea is the remnant of an ocean that once separated these two plates, and continuing plate movement should eventually compress it into oblivion. These plate collisions are major generators of earthquakes—a topic explored later in this chapter.

The process of continental plate collision results in the formation of great thicknesses of continental crust in the zones of convergence. For this reason, ancient plate boundaries, which now generally are located within the interiors of sutured lithospheric plates, are exceptionally strong and resistant to further deformation. These areas are termed **shields** because of their strength and broadly convex surface profiles.

Shields comprise the stable geologic cores of the continents. They have remained near or above sea level for long periods of geologic time, causing them to be subjected to slow erosion that has exposed the granitic and often metamorphosed "roots" of the ancient mountain systems. In many areas, the surfaces of the shields, which contain the oldest rocks on Earth, are exposed. In other localities, though, these surfaces lie beneath a veneer of much younger sedimentary rocks deposited when these shield areas were temporarily covered by shallow seas. At present, shields occupy more than a third of the continents. The Canadian Shield, for example, extends over most of eastern Canada, while the African Shield comprises the majority of that large continent. The locations of the major shields are shown in Figure 14.12.

Isostatic Uplift

The lithospheric plates float on the asthenosphere to just the depth needed to support their weight and are said to be in **isostatic equilibrium** (from the Greek *isos*, "equal" and *stasia*, "standing") with their surroundings. When erosion removes material from the surface, the weight of the plate in that location is reduced. The reduction in weight results in the **isostatic uplift** of the region. This is a broad, gentle upwarping that, like the gradual rise of the underwater portion of an iceberg when its top melts, allows the plate to remain in a state of isostatic equilibrium.

Isostatic uplift occurs most rapidly in mountainous regions, where the reduction of mass by erosional processes generally is more rapid than it is on flatter surfaces. As erosion proceeds, the surface lowering of a region resulting from the continuing removal of weathered materials is

◆ **FIGURE 14.11** View of the Himalayas. These mountains are being formed by the northward movement of the Indian subcontinent into the Eurasian Plate. Mt. Everest, the world's highest peak, is on the right.

SOURCE: Gary J. James/Biological Photo Service

Horst formation

Graben formation

Linear escarpment

Shortly after uplift

Shortly after downdropping

Retreating escarpment

Sediment cover

After erosional modification

After erosional modification

◆ **FIGURE 14.21** Block diagram depictions of a horst and graben shortly after formation (top views), and after erosional modification (bottom views).

Great Basin is bounded by the Sierra Nevada of eastern California. This mountain system consists of a giant tilted block of igneous and metamorphic rock that originally formed as an igneous intrusion (or batholith) about 150

◆ **FIGURE 14.22** The Great Basin of Nevada and western Utah is in an area of crustal extension and consists of tilted and block-faulted mountain ranges, with broad intervening basins.
SOURCE: A.J. Copley/Visuals Unlimited

million years ago. The fault motion producing the Sierra Nevada began 35 million years ago, as the eastern portion of the batholith was gradually lifted. This has produced an asymmetrical mountain range, with a gentle west-facing slope and a precipitous east-facing escarpment. Active uplift continues at a rate of approximately a centimeter (0.4 in) a year. Glaciation within the Sierras has largely destroyed their linearity as an upthrust fault block and has carved a series of individual peaks.

Earthquakes

The earthquakes that fault movements frequently generate can greatly influence the inhabitants of an affected region. **Earthquakes** are sudden movements of the surface. Most major earthquakes are produced when the slowly accumulated strain along the opposing sides of an active fault is released as the two sides are torn apart. Individual fault motions are small and involve relative displacements ranging from a fraction of a centimeter up to 5 to 6 meters (15 to 20 ft). As a general rule, larger displacements produce more violent earthquakes.

The same faults experience slippage and produce earthquakes repeatedly, as the accumulated stress periodically exceeds the shear strength of the rock. The rocks on the

CALIFORNIA'S LOMA PRIETA EARTHQUAKE—A HARBINGER OF GREATER EARTHQUAKES TO COME?

The best-known fault in North America and the one most feared for its earthquake potential is California's San Andreas Fault (Figure 14.25). Minor slippages occur frequently along this great fault and its numerous offshoots, but the last large-scale movement took place in 1906, when the opposing sides of the northern portion of the fault were displaced by 3 to 6 meters (10–20 ft). This produced the great San Francisco earthquake and fire that leveled much of the city and claimed 3,000 human lives.

Much more recently, on October 17, 1989, another slippage on the San Andreas Fault caused the Loma Prieta earthquake. This earthquake, which lasted less than 15 seconds, had its epicenter in the Santa Cruz Mountains about 100 kilometers (60 mi) south of San Francisco (Figure 14.23). It was responsible for 67 deaths and 3,500 injuries, left 12,000 homeless, and caused property damage estimated at nearly $6 billion (Figure 14.24). The Loma Prieta earthquake had a Richter scale magnitude of 7.1, making it only one-tenth as powerful as

the great earthquake of 1906. Future earthquakes along the fault are inevitable, and Californians are increasingly concerned that one of catastrophic proportions may take place in the near future.

Two lines of evidence point to the possibility of a powerful earthquake along the San Andreas Fault within the next few decades. In the first place, 90 years have now elapsed since the San Francisco earthquake of 1906. Prior to that time, major slippages had

continued on next page

(a)

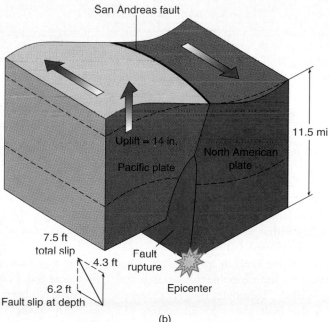

(b)

◆ **FIGURE 14.23** (*a*) Mainshock and aftershocks of the Loma Prieta earthquake in relation to the surface trace of the San Andreas fault. Size of the symbols is relative to the magnitudes of the aftershocks. (*b*) Inferred slip along the San Andreas fault during the Loma Prieta earthquake. The Pacific plate moved upward and northward in relation to the North American plate.

CHAPTER 14 Tectonic Processes and Landforms **369**

◆ **FIGURE 14.24**   San Francisco Marina district fire, caused by gas lines ruptured by the Loma Prieta earthquake.

SOURCE: Matin E. Kimek, *Marin Independent Journal*

been taking place at approximately 50-year intervals, so the next movement is now overdue. A second line of evidence comes from measuring instruments and geological observations that indicate that a large amount of strain has developed between the opposing sides of the fault in several areas. One ominous indication of this strain is the so-called "Palmdale Bulge," centered on the fault about 55 kilometers (35 mi) northeast of Los Angeles, where the Earth's surface has risen about 0.3 meters (12 in) in recent years. The southern section of the San Andreas Fault apparently became tightly locked

after its last large-scale movement, which produced the Fort Tejon earthquake of 1857; to date it has resisted any further movements of consequence. Many experts believe that the southern portion of the fault is more likely to be the site of the next major earthquake than is the northern portion near San Francisco.

There has been a significant increase in earthquake activity throughout California in the past decade. Recent scientific estimates of chances for a major earthquake in the near future have been raised for southern California, and for the San Francisco area as

well, as new faults have been discovered and as smaller fault movements have occurred that are believed to have increased the strain on the San Andreas Fault. Both the Lowa Prieta earthquake and the 1994 Northridge earthquake, which caused $20 billion in damage and claimed 61 lives, are indications that even more powerful and destructive earthquakes are likely to occur in the near future.

The tremendous development that has taken place in western California during the twentieth century has greatly increased the potential for severe loss of life and property in a major

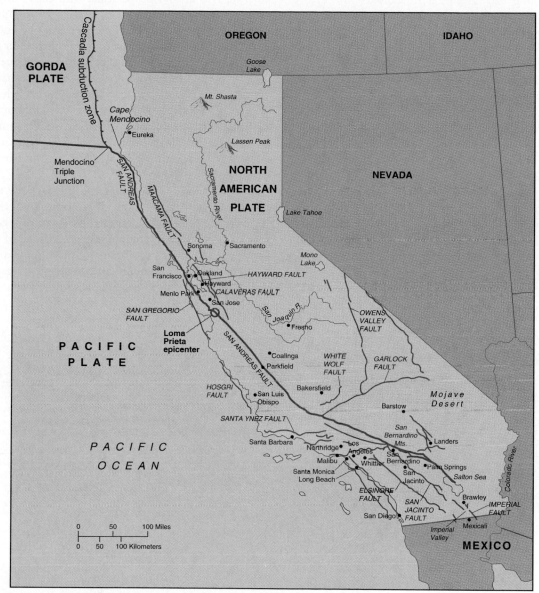

◆ FIGURE 14.25 Map of California showing the location of the San Andreas fault and its subsidiary faults.

earthquake. Most at risk are older, multistoried masonry buildings that predate 1934 antiseismic construction laws. Thousands of these structures still exist in the state.

On the positive side, major advances in engineering and construction methods, as well as earthquake protection retrofitting of older structures, are now causing earthquake casualty estimates to be substantially reduced. It is hoped that similar advances will be made in earthquake prediction techniques in the near future. ■

opposing sides of a fault deform elastically in much the same fashion as a rubber band, which stretches in response to a buildup in strain, then suddenly snaps when the strain becomes too great. Although stress tends to accumulate at a constant rate along any given fault, elapsed time spans between successive earthquakes often vary because the two sides of the fault may be more tightly locked together on some occasions than on others. In some cases, the opposing sides of a fault are not locked together at all, and strain is released gradually and continually by a slow **fault creep** that may be detectable only by the gradual offsetting of features on opposing sides of the fault. Several hundred thousand earthquakes occur annually around the world, but most are undetectable except to delicate recording instruments called **seismographs.** Thousands of perceptible earthquakes, however, occur each year somewhere on Earth.

The shaking of an earthquake is actually produced by the passage of a shock wave resulting from the sudden fault movement. A variety of different types of shock waves are produced, but the two most important are the **P waves,** or primary waves, which result from compression, and the **S waves,** or shear waves, which exhibit a side-to-side motion. The analysis of these waves has provided scientists with much of their information on the interior layers of the Earth.

Earthquake magnitudes are commonly measured by use of the **Richter scale,** devised by seismologist Charles F. Richter in 1935. It is a logarithmic scale, with each number having 10 times the earthquake wave amplitude (amount of side-to-side motion) and about 31.5 times the total energy release of the preceding number. A Richter value greater than about 2.5 is detectable without special instruments, one above 5 is considered damaging, and one of 8 or above is catastrophic. The great San Francisco earthquake of 1906 had an estimated magnitude of 8.3, and the Alaskan earthquake of 1964 was measured at 8.6. The effects and future implications of central California's 1989 Loma Prieta earthquake, which had a magnitude of 7.1, are discussed in the Focus Box beginning on page 369.

The distribution of earthquakes is essentially identical to that of active faults—they are concentrated along lithospheric plate boundaries, especially those where convergence and active subduction are taking place. A map of earthquake locations clearly outlines the major plate boundaries (Figure 14.26). Because these boundaries frequently are the sites of active tectonic uplift, earthquakes on land generally are

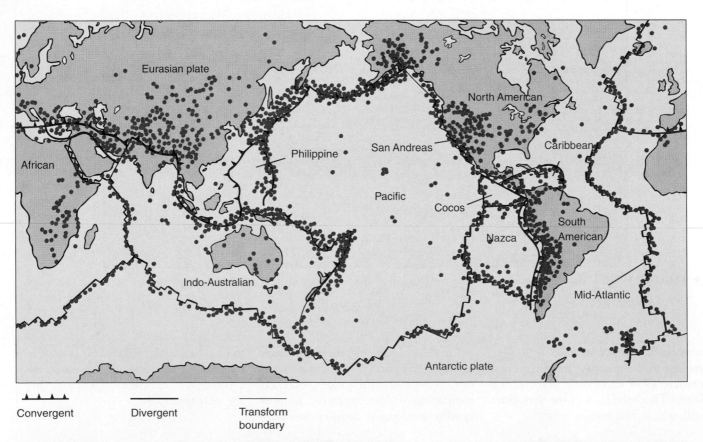

Convergent   Divergent   Transform boundary

◆ **FIGURE 14.26** Global distribution of the epicenters of earthquakes between 1963 and 1977. (The **epicenter** is the surface position directly above the earthquake location, or **focus.**) Approximately 80% of earthquakes occur within the circum-Pacific belt, 15% within the Mediterranean-Asiatic belt, and the remaining 5% within plate interiors or along seafloor spreading zones. Note that the dots collectively outline the lithospheric plate boundaries depicted in Figure 14.5.

centered in mountainous regions. Nearly 95 percent of recorded earthquakes occur around the rim of the Pacific and from southern Asia westward through southern Europe. Most other earthquakes take place along generally less active faults located within the lithospheric plates. Some of the movements along these faults can also produce devastating earthquakes on occasion. For instance, much of the city of Charleston, South Carolina, was destroyed by an 1886 earthquake. A series of quakes centered near New Madrid, Missouri, in 1811 and early 1812 sent huge waves rolling down the Mississippi River, changing its course in several locations. Although some regions are notably less prone to earthquakes than others (Figure 14.28), few if any parts of the world can be considered immune to a major quake.

Individual earthquakes generally have only a minor impact upon landforms. They may offset surface features by a few meters, produce small fault scarplets (Figure 14.20), and trigger landslides that may on occasion block streams to produce lakes. By far the most important effect, though, is on people, especially those occupying buildings constructed of massive materials and those living along coasts that may experience earthquake-produced tsunami (see the Focus Box on page 374). Major earthquakes rank with hurricanes and volcanic eruptions as producers of the most catastrophic natural disasters in recorded history (Figure 14.27 and Table 14.1). Cities are especially vulnerable, since buildings may be shaken to the ground, and ruptured gas pipes and downed electric wires cause fires that cannot be reached to be extinguished. The greatest loss of life resulting from an earthquake in the twentieth century occurred in Tangshan, China, in 1976, where an estimated 230,000 persons lost their lives. A 1923 earthquake in Japan destroyed most of Tokyo and claimed 140,000 victims. More recently, on December 7, 1988, an earthquake in Armenia took 30,000 lives, and another on January 17, 1995, claimed 5,300 lives

◆ **FIGURE 14.27** The Northridge Fashion Center, near the epicenter of the 1994 Northridge earthquake. The earthquake caused the second and third floors of Bullocks Department Store to collapse onto the bottom story.
SOURCE: NOAA

in and near Kobe, Japan. As already noted, the growing potential for a major earthquake along the San Andreas Fault is currently of great concern to the residents of California.

Volcanism and Volcanic Landforms

Volcanism is the other major category of tectonic activity. In contrast to diastrophism, which involves solid-state movements of the Earth's crust, **volcanism** involves the surfaceward movement of fluid magma. The magma, which may be derived from either the lower crust or upper mantle,

◆ **FIGURE 14.28** Map of estimated U.S. earthquake potential. The numbers are proportional to the estimated relative intensity of horizontal ground movement in a major earthquake.

TSUNAMI

Most ocean waves are produced by the frictional effects of the wind. The largest wind-generated storm waves are more than 15 meters (50 ft) high. The largest and most destructive waves of all, however, are seismic waves caused by sudden tectonically induced movements of the seafloor. These movements result most frequently from earthquakes, but they may also be caused by volcanic eruptions or large underwater avalanches. In the past these waves were commonly referred to as "tidal waves," but they have nothing to do with the tides and are more properly termed **seismic waves** or given their Japanese name **tsunami** ("harbor wave"). They are most numerous in the Pacific, which is ringed by active fault lines and volcanoes. Like wind-generated waves, they are capable of traveling great distances from their points of origin.

Tsunami are essentially shock waves produced by underwater displacements. These disturbances of the seafloor cause the overlying ocean water to be displaced and, since water is incompressible, this produces oscillations at the sea surface. Tsunami travel much faster than wind-generated waves, and their speed increases with water depth. For example, in water 300 meters (1,000 ft) deep, they travel at 185 kilometers per hour (115 mph), but in water 4,600 meters (15,000 ft) deep, they travel at 770 kilometers per hour (480 mph). This rapid movement provides relatively little time for threatened coastlines to make preparations.

A tsunami quickly loses height as it travels outward from its point of origin, and it passes unnoticed beneath ships in deep water. Upon approaching a coastline, however, the frictional drag of the increasingly shallow ocean bottom on the rapidly moving wave slows the wave and causes it to increase in height to as much as 30 meters (100 ft) or more. Its approach is often first indicated by a marked fall of water levels along the coast over a period of several minutes. This is followed by a rapid rise of water that soon inundates the entire shore and may advance well inland.

Tsunami effects are most extreme on low-lying coasts and especially in funnel-shaped bays, where the force of the rising water is magnified. A dramatic example of the magnification of a tsunami in a narrow bay occurred in Lituya Bay, Alaska. On July 10, 1958, a tsunami generated by a nearby undersea earthquake destroyed trees to an elevation of 525 meters (1,720 ft) at the head of the bay.

One of the most destructive tsunami on record was produced by the explosion of the East Indian volcano Krakatau on August 27, 1883. This tsunami devastated portions of the densely populated coasts of Java and Sumatra, taking 36,000 lives (see the Case Study at the end of this chapter). The Hawaiian Islands also have been visited on occasion by destructive tsunami. For example, in 1946 an earthquake near Alaska's Aleutian Islands generated a tsunami 17 meters (56 ft) high that took 173 lives in Hawaii (Figure 14.29). Following this event, the U.S. Coast and Geodetic Survey established the Pacific Tsunami Warning System—a cooperative early warning network jointly operated by 23 Pacific Rim nations and headquartered in Honolulu. ■

◆ **FIGURE 14.29** A tsunami, generated by an earthquake in the Aleutian Islands, destroying the pier at Hilo, Hawaii, in 1946. The man indicated by the arrow was never seen again.
SOURCE: NOAA

TABLE 14.1 Major Earthquakes of the Twentieth Century

DATE	LOCATION	RICHTER MAGNITUDE	DEATHS
April 18, 1906	San Francisco, California	8.3	3,000
Sept. 1, 1923	Tokyo, Japan	8.2	143,000
Dec. 27, 1939	Erzincan, Turkey	8.0	40,000
May 21–30, 1960	Central Chile	8.5	5,700
March 28, 1964	Southern Alaska	8.6	131
May 31, 1970	Northern Peru	7.8	66,000
Dec. 23, 1972	Managua, Nicaragua	6.2	5,000
July 28, 1976	Tangshan, China	7.6	230,000
Sept. 16, 1978	Iran	7.7	25,000
Sept. 19, 1985	Mexico City, Mexico	8.1	7,000
Dec. 7, 1988	Armenia	6.9	30,000
Oct. 17, 1989	Loma Prieta, California	7.1	67
Oct. 20, 1991	Uttar Pradesh, India	6.1	1,700
Jan. 17, 1994	Northridge, California	6.7	60
Jan. 17, 1995	Kobe, Japan	7.2	5,300

frequently solidifies into igneous rock before reaching the surface. Volcanism occurs where the lithosphere contains deep fractures or faults, a common situation along and near plate boundaries. These breaks in the lithosphere serve two essential purposes in promoting volcanic activity. First, they may relieve pressures sufficiently to permit the localized melting of the rock that forms the magma; second, they provide pathways for its surfaceward movement.

In the discussion of rock types in Chapter 13, intrusive igneous rocks, which solidify below the surface, were differentiated from extrusive igneous rocks, which reach the surface in a molten state before rapid solidification occurs. This section covers the processes by which these movements take place, the geomorphic features they produce, and the geographical distributions of volcanic processes and features.

Extrusive Volcanism

Extrusive volcanism consists of processes by which molten rock reaches, and is extruded upon, the Earth's surface. Although it occurs less frequently than intrusive volcanism and generally involves smaller quantities of molten rock, it is much better known, is more spectacular, and has a more direct influence on the topograpy.

Two basic groups of extrusive volcanic features exist— volcanoes and lava flows. The difference between the two is determined largely by the fluidity of the magma and the force with which it is expelled onto the surface. A comparison of volcanoes and lava flows provides an excellent illustration of the fact that features with similar modes of origin may differ greatly in their topographic expressions.

VOLCANOES. A **volcano** is a hill or mountain constructed of materials from the Earth's interior that have been ejected under pressure from a vent (Figure 14.30). Large volcanic peaks are among the most beautiful and distinctive types of landforms. They typically have a conical and sometimes nearly perfectly symmetrical shape and may extend to great heights above the surrounding landscape. They often are surmounted by a cap of snow and ice and, if active, may extrude steam, smoke, and sometimes ash or lava. A volcanic eruption, especially when viewed at night, may be the most spectacular and awe-inspiring of all geomorphic events (Figure 14.31d). Humanity's fascination with volcanoes also is increased by their unpredictable nature and demonstrated potential for dealing death and destruction on a large scale.

It is difficult to overstate the impact that volcanic activity has had on the Earth's surface. Most of the Earth's crust has been produced by the solidification of magma or lava, while

◆ **FIGURE 14.30** Cutaway view of a composite volcano, showing the magma chamber, system of vents, and layers of lava and ash comprising the cone.

◆ **FIGURE 14.31A** Geologically recent cinder cone south of Mono Lake, California.

SOURCE: Ralph Scott

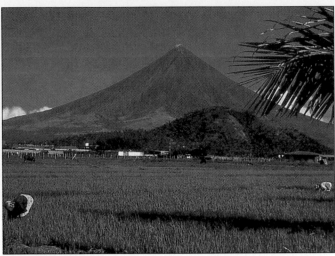

◆ **FIGURE 14.31B** Mt. Mayon in the Philippines is a composite volcano with an almost perfectly symmetrical profile.

SOURCE: Lawrence R. Solkoski, consulting geologist, Vancouver, B.C. Canada

◆ **FIGURE 14.31c** Oregon's Mt. Hood is a dormant composite volcano whose outlines have been greatly modified by glacial erosion.

SOURCE: D.R. Crandell/USGS

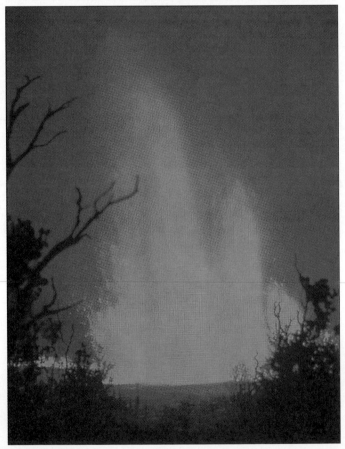

◆ **FIGURE 14.31D** Lava fountains, such as this one in Hawaii, are particularly spectacular at night.

SOURCE: USGS

both the Earth's atmosphere and surface water bodies originated from gases released during volcanic eruptions. Major volcanic eruptions have apparently played key roles in altering global climates, thereby affecting our planet's biological history. Even the 1991 eruption of Mt. Pinatubo in the Philippines is believed to have lowered Earth's mean temperature by about 0.6C° (1F°) over a one-to-two-year period.

On the other hand, perhaps because of their spectacular nature, it is tempting to overrate the present-day importance of volcanic landscapes. Only about 1,300 volcanic peaks occur on land, and these often exist in clusters, so that volcanoes are relatively restricted in geographic extent. In

Spreading ridges   Convergent plate margins   Volcanoes

◆ **FIGURE 14.32** Global distribution of recently active volcanoes, and their geographic relationship to the lithospheric plates. Most active volcanoes surround the Pacific Ocean.

North America, there are only about 70 large volcanic peaks. Even in areas where they do occur, volcanic peaks generally are relatively localized, and other geomorphic processes, such as stream erosion, are likely to exert a much greater geomorphic influence on a regional scale.

At present, approximately 540 active volcanoes are scattered throughout the world's land areas (Figure 14.32). These volcanoes have erupted within the last few centuries and are considered likely to experience future eruptions. In any given year, approximately 50 volcanic eruptions take place. Volcanic eruptions constitute one of the Earth's best-known natural hazards and have taken approximately 200,000 human lives during the past four centuries. The greatest known loss of life during that period occurred in the 1815 eruption of Tamboro volcano in the East Indies, which killed an estimated 92,000 people. Table 14.2 provides the

TABLE 14.2 Selected Major Volcanic Eruptions

YEAR	NAME AND LOCATION	DEATHS	VOLUME OF MATERIAL ERUPTED IN KM³
79	Mount Vesuvius, Italy	20,000	3
1815	Tamboro, Indonesia	92,000	80–100
1883	Krakatau, Indonesia	36,400	18
1902	Mont Pelée, Martinique	28,000	?
1980	Mount St. Helens, Washington State	60	4
1985	Nevado del Ruiz, Colombia	23,000	1
1991	Mount Pinatubo, Philippines	800	12

At some sites beneath divergent plate boundaries and above mantle plumes, temperature and pressure conditions allow the partial melting of large quantities of rock within the upper mantle, producing reservoirs of basaltic magmas.

(a)

(b)

(c)

Wizard Island   Crater Lake

(d)

◆ **FIGURE 14.34** The proposed sequence of events leading to the formation of Crater Lake, Oregon. Figures show the collapse of prehistoric Mt. Mazama into its magma chamber following the eruption that emptied the chamber. Crater Lake formed from rainwater that filled the crater, while Wizard Island is a cinder cone produced from a later eruption. The bottom photo shows Crater Lake as it appears today.

SOURCE: Photo courtesy of James S. Monroe

If the magmas reach the surface, extensive outpourings of highly fluid **flood basalts** are produced. This is the case for the lava flows produced by Hawaii's Kilauea Volcano, pictured in Figure 14.35. Most flood basalt flows occur on the ocean floor along the axes of the oceanic ridges. The movement of the lithospheric plates away from these ridges has spread their eruption products over the Earth to produce the lower crust, or sima.

Of greater direct significance to the topography of the continents are the large outpourings of flood basalts on land above "hot spots" apparently associated with mantle plumes. These outpourings typically occur as **fissure eruptions** along a major fracture or fault rather than from a single vent, as in a volcano. Huge outflows of flood basalts have occurred in a number of widely separated locations, including the Columbia River Plateau and Snake River Plain of the northwestern United States (Figure 14.36), the Deccan Plateau of northern India, the Paraná Plateau of southern Brazil, and the Drakensberg Plateau of South Africa, as well as in portions of Antarctica, Bolivia, Ethiopia, Iceland, Russia, and northwestern Arabia. Much older sites of flood basalts have been identified in other locations such as the Piedmont of the eastern United States and in the British Isles. Iceland is the only location where such eruptions have taken place during historic times, and these eruptions occurred on a relatively limited scale. The last major Icelandic fissure eruption took place in 1783, when approximately 12.5 cubic kilometers (3 mi^3) of lava were released along the 25-kilometer (16 mi) long Laki fissure.

Fissure eruptions may largely or completely bury the original surface, producing a gently rolling landscape that, depending on its elevation, may be termed a **lava plain** or a **lava plateau.** In some cases, successive flows may accumulate to depths far greater than 1,000 meters (3,000 ft). The Columbia Plateau, for instance, consists of approximately 260,000 square kilometers (100,000 sq mi) of lava-covered surface, and in some locations the lava reaches depths of 3,000 meters (10,000 ft). The lavas of the Columbia Plateau have been extruded geologically recently in a large number of separate eruptions. The eruptive cycle probably is not yet completed, and future outpourings of flood basalts may one day again cover large portions of this region.

Intrusive Volcanism

Intrusive volcanism occurs when rising magma solidifies into igneous rock before reaching the surface. If solidification takes place far below ground, any effect on the surface is indirect and occurs at a much later geologic period when the deep-seated pluton is exposed by the erosion of the overlying rock. Near-surface intrusions as well generally have their greatest influence on the topography when exposed by erosion. Unlike plutonic (deep-seated) intrusions, however, near-surface intrusions may also have a direct and immediate effect on the surface if they deform the overlying rock layers during their upward movement. The surrounding rock may

◆ FIGURE 14.35 Advancing lava flow from Hawaii's Mt. Kilauea volcano.
SOURCE: NOAA

be domed up and may also be metamorphosed by the heat and pressure of the molten rock. When intrusive igneous rock masses are exposed at the surface, their greatest effect on landforms results from differential erosion rates. In most but not all cases, the igneous rock is more resistant to erosion than the surrounding rock and produces high-standing features.

By far the largest of the plutonic rock features are batholiths. These huge plutons with diameters usually exceeding 50 kilometers (30 mi) form by the slow cooling and solidification of magma deep within the crust (Figure 14.38). Rock crystallization is sufficiently slow to allow mineral differentiation to occur. If only the upper portion of the batholith is exposed at the surface, as is normally the case, this differentiation causes it to consist largely of granite. The slow cooling and solidification also allow time for mineral-bearing fluids to deposit veins of metallic minerals in fractures, so that today valuable metal deposits are being mined within exposed batholiths in a number of regions. Batholiths currently are exposed in the cores of many mountain ranges, and it is believed that they may underlie, at depth, all major mountain regions. Within North America, batholiths are exposed in the Sierra Nevada, the Salmon River Mountains of central Idaho, the Coast Mountains of British Columbia, the Front Range of the Colorado Rockies, the Adirondacks of upstate New York, and the Black Hills of South Dakota (Figure 14.17).

Laccoliths are similar, but much smaller bodies of relatively acidic magma that work their way surfaceward through vents and solidify as dome-shaped intrusions between flexible sedimentary strata (Figure 14.38). If they approach the surface closely enough, they are capable of bulging up the overlying strata to create **laccolithic domes.** In time, the igneous core of the laccolith may be exposed by

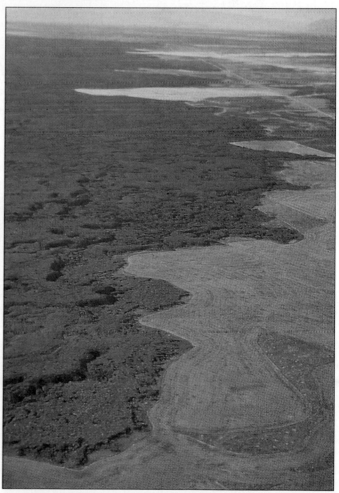

◆ FIGURE 14.36 The margin of a geologically recent basaltic lava flow on Idaho's Snake River Plain within the Columbia Plateau.
SOURCE: John S. Shelton

Yellowstone—A Volcanic "Hot Spot" with an Explosive Past and an Uncertain Future

The Yellowstone region, located mostly in the northwestern corner of Wyoming, is an area whose scenic beauty and unique geothermal features caused it to be designated in 1872 as the first national park in the United States. Yet, from a geological standpoint, the Yellowstone is one of the most unstable and dynamic segments of the North American lithospheric plate. Like the island of Hawaii, it is situated above a "hot spot" in the mantle and has been the site of repeated violent volcanic eruptions. Unlike the Hawaiian hot spot, however, the Yellowstone hot spot is located beneath continental crust, and the eruptive material is viscous and chemically acidic (that is, it has a high aluminum and silicon content). This type of magma is associated with explosive volcanic eruptions.

Large volumes of magma apparently are being generated continuously in the upper mantle and lower crust and are accumulating in one or more huge chambers deep beneath the Yellowstone Plateau. Geologic evidence indicates that similar magma accumulations were released in three huge, but short-lived volcanic eruptions during the past two million years. Each eruption emptied the subterranean magma chambers; this caused the overlying surface to collapse, forming large calderas, or volcanic depressions. Recent studies have indicated that the erupted material was not primarily lava, but instead consisted largely of ash and hot gas. The resulting ash flows were larger in scale than those known anywhere else in the world, and upon cooling they formed extensive deposits of rhyolite and tuff.[*]

Yellowstone's most recent phase of volcanic activity began about 1.2 million years ago, and the climactic eruption that last emptied the magma chambers occurred about 600,000 years ago. This eruption expelled about 1,000 cubic kilometers (250 mi^3) of ash. (By comparison, the 1980 eruption of Mt. St. Helens released about 4 cubic kilometers, or 1 mi^3, of ash.) The expulsion of this material created a caldera that was 45 kilometers (28 mi) wide and 75 kilometers (47 mi) long. The caldera has been partially filled by sediments and by material from subsequent smaller eruptions.

The floor of the Yellowstone Plateau currently is being uplifted again at the relatively rapid rate of 1.4 centimeters (0.5 in) per year. This uplift apparently is being caused by the injection of new magma into the magma chambers underlying the plateau. At some future time, the volcanic cycle will undoubtedly be repeated as the chambers are emptied by another violent eruption. ■

[*]Robert B. Smith and Robert L. Christiansen, "Yellowstone Park as a Window on the Earth's Interior," *Scientific American*, 242, no. 2 (1980): 104–17.

◆ **FIGURE 14.37** Boiling geothermal pools, like Fireside Lake shown here, are common in Yellowstone National Park. Their heat is derived from a subterranean magma chamber.
SOURCE: M. Schneider/H. Armstrong Roberts

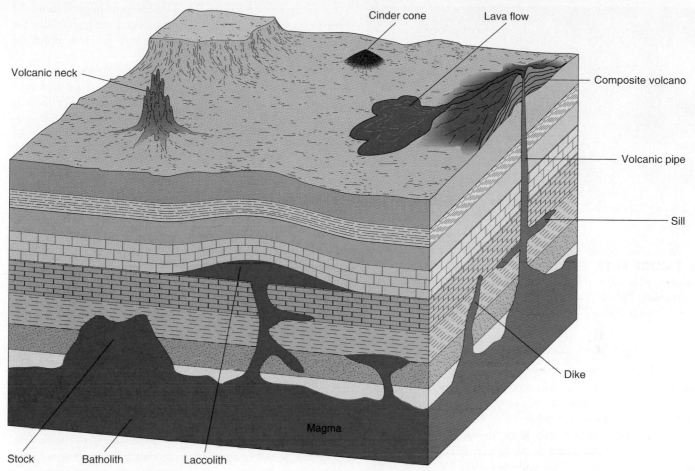

Volcanic neck
Cinder cone
Lava flow
Composite volcano
Volcanic pipe
Sill
Dike
Magma
Stock
Batholith
Laccolith

◆ **FIGURE 14.38** Block diagram showing various intrusive and extrusive volcanic features.

erosion, forming a central mass of hills or mountains. Several mountain masses in southern Utah, including the Henry and LaSal Mountains, represent the cores of exposed laccoliths.

Sills consist of sheets of magma injected between sedimentary rock strata relatively near the surface. They are rather similar to laccoliths, but are flatter because of the greater fluidity of the magma involved. As a result, they do not noticeably dome the overlying rock layers. Sills vary greatly in breadth and in thickness, depending on the quantity of magma involved. Larger sills may have horizontal dimensions of tens of square kilometers and thicknesses of more than 30 meters (100 ft). Exposures of the edges of large sills often display the columnar jointing typical of basalt, as in the well-known examples of the Giant's Causeway on the Irish coast and the Palisades along the lower Hudson River in New York.

Dikes are igneous rock masses formed from sheets of magma that have been injected through preexisting rock fractures. They are very much like sills, except that they cut across the structures into which they are intruded while sills parallel the structures. Dikes are extremely narrow compared to their length. They may extend for many kilometers, but usually are only a few centimeters to several meters thick. They are distributed widely, being found in "swarms" be-

neath some volcanoes or volcanic remnants, where they often radiate outward from the central vent that supplied their magma.

Volcanic Erosion and Remnants

Individual volcanic peaks are relatively short-lived features on the landscape. Once eruptions cease, the forces of gradation quickly lower and eventually remove them completely. Most symmetrical volcanoes currently on the Earth's surface have been active within the past few thousand years. Cinder cones and composite volcanoes, because of their steep slopes and loose pyroclastic composition, erode especially rapidly. The uneven distribution of more resistant lava flows on composite volcanoes typically results in an irregular erosional profile because the softer pyroclastic portions are removed more rapidly than the harder portions composed of solidified lava. Shield volcanoes, with their great masses, gentle slopes, and predominant composition of solid basalt, are more durable.

A volcano composed primarily of loose pyroclastic materials may leave as its last erosional remnant a **volcanic neck,** a solid spire of hardened lava that was in the vent of the volcano at the end of its final eruption. The rapid removal of the unconsolidated material that once surrounded it and

◆ **FIGURE 14.39** Ship Rock is a volcanic neck with radiating dike ridges that stands conspicuously above the surrounding New Mexico desert.

SOURCE: John S. Shelton

formed the volcanic cone may leave the volcanic neck standing conspicuously alone above the neighboring landscape. A well-known example of a volcanic neck in the western United States is Ship Rock, in New Mexico (Figure 14.39). On occasion, volcanic necks may have exposed **dike ridges** radiating from them and appearing like stone walls. The degree of topographic expression of such volcanic remnants depends largely upon their resistance to erosion compared to that of the surrounding rocks. They stand out most conspicuously in areas where they have been intruded into weak sedimentary strata.

Flood basalts contain much more massive quantities of volcanic materials than do individual volcanic peaks and remain as landscape features for correspondingly longer time spans. Their basaltic composition, however, causes them to weather rapidly at the surface, so that in most areas they soon become covered by fertile soils. If the accumulation of basalt has formed a high-standing plateau, over time it is likely to become increasingly dissected by steep-sided river valleys or canyons. This has been the case in the Columbia Plateau where streams like the Snake River have eroded steep gorges.

▌SUMMARY

This chapter has examined the methods by which tectonic processes operate and the landforms they produce. Tectonic forces originating deep within the interior are responsible for the continual alteration of the Earth's surface. New crustal material is manufactured, while old material is destroyed. At the same time, the positions of the continents and oceans are constantly rearranged by the slow creep of huge volumes of plastic rock in the mantle. These motions are responsible for the formation of the continents and for the uplift of mountain ranges.

Tectonic processes are divided into two general categories based upon the physical state of the rock material undergoing movement and deformation. Diastrophism involves solid-state motions and encompasses the processes of folding and faulting. Volcanism involves the movement of magma within the Earth and its expulsion onto the surface. These processes produce a variety of tectonic landforms. Most of these landforms are created along and near the boundaries between lithospheric plates and are associated with the differential movements of these plates.

We now return to the Focus Questions posed at the beginning of the chapter:

1. *How do plate tectonic movements occur, and how do they create and destroy crustal material?*

Lithospheric plate movements apparently are caused largely by the frictional drag of convection currents in the underlying asthenosphere. These currents are produced by heat released by the nuclear decay of radioactive elements in the mantle and core. Upwelling mantle plumes result in the formation and divergence of new oceanic crust along the oceanic ridge systems. A collision between two plates having differing densities results in the subduction and destruction of the denser plate as it is forced downward into the mantle. As some of the lighter and more volatile components of the subducted plate melt and move surfaceward, they form continental crustal material, which is volcanically erupted onto the surface. The collision of two continental plates often produces a high-standing mountain belt composed of greatly thickened and complexly deformed continental crust.

2. *What causes rocks to fold, and what topographic features result from folding?*

Most folding is caused by the linear compression of sedimentary rock strata resulting from lithospheric plate collisions. This results in the formation of elongated anticlinal and synclinal rock structures whose long axes, or strikes, lie perpendicular to the axis of compressive stress. Subsequent weathering and erosion of rocks of differing erosional resistance typically will produce shifting patterns of ridges and valleys. Rounded fold structures in the form of domes and basins are caused by the subsurface injection or removal of plastic or fluid substances such as magma.

3. *What are the major types of fault motions and their resulting topographic features?*

Faulting involves the fracturing of rock, followed by the subsequent offsetting of the two sides. Individual fault movements are small, but they may be repeated at frequent intervals, eventually producing major topographic features. Two of the four basic fault types chiefly involve vertical motions and may eventually produce mountain ranges or large escarpments. These are normal faults, which have an expansionary component, and reverse faults, which result from compression. Faults primarily involving horizontal offsetting typically produce lower relief topographic features, but are feared for their earthquake potential. They include transcurrent (strike-slip) faults, in which lateral motion of the two sides of the fault predominates, and thrust faults, which result from extreme compression. Multiple faults often occur in areas subjected to tectonic stresses, and the intervening blocks of land may be raised, lowered, or tilted.

Sudden movements of the Earth's surface resulting from fault displacements produce earthquakes. Earthquakes occur as the rocks on opposing sides of a fault plane are torn apart and then rebound elastically after being subjected to gradually increasing strain.

4. *What causes volcanic eruptions, and what types of surface features do they produce?*

Volcanic eruptions involve the surface expulsion of molten rock in the form of lava or pyroclastic fragments. The violence of many volcanic eruptions results from the explosive release of pressurized gases originally dissolved in the rock. Ideally, a volcano takes the form of a symmetrical conical peak with a central vent. Three basic types of volcanoes are recognized. Cinder cones typically are small and steep and are composed primarily of loose pyroclastic material. Shield volcanoes, conversely, are very large, gently sloping, and composed of accumulated flows of basaltic lava. Composite volcanoes consist of alternating layers of ash or other pyroclastic material and of lava. Lava flows associated with volcanic eruptions usually are of limited extent, but extensive fissure outpourings of fluid basaltic lava in the geologic past have covered large regions.

Rising magma often solidifies before reaching the surface, resulting in the formation of intrusive volcanic features such as batholiths, laccoliths, sills, or dikes. These may be exposed by erosion at a later time, and they often form high-standing surface features because of their resistance to erosion.

∎ QUESTIONS FOR THOUGHT AND REVIEW

1. What are lithospheric plates? Name the six largest plates.
2. Briefly explain four types of evidence that support the concept of continental drift.
3. Explain how and why lithospheric plate movements occur.
4. How are lithospheric plate motions responsible for the major physical features of the Earth's surface, including the ocean basins and major mountain systems?
5. What are the three types of plate motions, based on the relative movements of the plates involved? Describe the current motion of the plate on which you live and relate its movement to the major topographic features of the plate.

6. Describe the process of subduction. Why is oceanic crust destroyed by the subduction process, while continental crust is not? How does the subduction of oceanic crust create new continental crust?
7. What will be the ultimate fate of the process of plate tectonic movements? What impact will this have on the Earth's surface?
8. Under what tectonic and geologic conditions does folding take place? What is the relationship between the direction in which stress is applied and the orientation of the fold axis? Cite three examples of folded mountain belts and relate them to particular lithospheric plate movements.
9. Why is it rare to find a good accordance (correlation) between fold structures and topography within a folded region? What is the chief factor responsible for the existing patterns of ridges and valleys within a folded mountain region such as the Appalachians?
10. What is faulting? List and describe each of the four basic fault types. Describe the types of lithospheric plate boundaries with which each normally is associated.
11. What causes earthquakes? Why do many experts expect a major earthquake in western California in the near future?
12. What is volcanism? What is the source of the magma involved in volcanic activity, and how does it reach the surface?
13. Explain, in plate tectonic terms, why a zone of active volcanoes (including Mt. St. Helens) extends from northern California to southern British Columbia.
14. Describe the general characteristics of each of the three major types of volcanoes, and give two examples of each type. Which type has produced the most catastrophic eruptions from a human standpoint, and why?
15. What are flood basalts? In what major respect do they differ from lava flows produced by volcanoes? What effects do flood basalts have on the topography? What portion of North America is covered by flood basalts?
16. What causes tsunamis? Why do they pose a major hazard to coastlines, but not to ships at sea? Are they a greater threat to the east coast or the west coast of North America? Why?
17. What characteristics of the 1883 eruption of Indonesia's Krakatau volcano have helped make it one of the best-known historical eruptions?

15 WEATHERING, MASS WASTING, AND KARST TOPOGRAPHY

OUTLINE

FOCUS QUESTIONS

1. How do the weathering processes operate, and in what environmental settings are they most effective?
2. What causes mass wasting, and what types of mass wasting are most widespread and important?
3. What are the causes and characteristics of karst topography, and how is it distributed around the world?

On the night of August 19, 1969, torrential rains from the remnants of Hurricane Camille triggered hundreds of landslides in rural Nelson County, Virginia. Within the space of a few hours, much of the centuries-old layer of soil and weathered rock that mantled the county's hillsides was catastrophically released onto the agriculturally productive valleys below, causing great destruction and loss of life. This event is discussed in greater detail in the Case Study at the end of the chapter. While most gradational activities are not so sudden or dramatic, the process of gradation itself is of utmost importance in molding Earth's surface features and in aiding the weathering processes that are essential to our planet's ability to support complex life-forms.

Gradation, unlike tectonism, is continuously active everywhere on land. As a consequence, gradational features dominate a much larger total surface area than do tectonic features. Gradation primarily involves the transport of materials from higher to lower elevations by the three "tools of gradation": water, ice, and wind. Chapters 16 through 19 are devoted to the actions of these gradational agents and to the landform features they produce. In order for gradation to operate at a geomorphically effective rate, however, surface materials must first be reduced in size and erosional resistance through weathering processes. Earth materials, especially if weakened by prior weathering, do not always require transportation by water, ice, or wind, but can be eroded directly by the force of gravity. In addition, differential weathering processes can selectively dissolve some rock types to produce a distinctive landscape assemblage called **karst.** The three topics of weathering, gravity movements, and karst are covered in this chapter.

WEATHERING

Weathering is the combined action of physical and chemical processes that disintegrate and decompose bedrock. The term is used because agents associated with the weather—specifically, air and water derived from precipitation—perform weathering operations. Weathering processes are of vital importance because they produce essential weathering products, and because they greatly increase the effectiveness of subsequent gradational processes. Weathered rock material is the major component of most soils, a commodity without which advanced plant and animal life on land would never have evolved. Weathering also produces loose sediments that are later converted to the sedimentary rocks that comprise the majority of the Earth's surface rock cover.

Weathering is an essential prerequisite for gradation because it reduces the size or alters the chemical composition of rock materials in such a way that erosional agents can remove and transport them much more readily. The tools of erosion, even ice, are largely ineffective when operating on solid bedrock. The gradational process therefore would be slowed immeasurably if bedrock were not first prepared for removal by weathering activities. Weathering sets the pace of gradation; because its progress is slow and steady, the gradational processes also operate in a slow and steady manner.

Weathering, by itself, involves no significant material transport. The products of weathering remain at or near their original sites awaiting transport by water, ice, wind, or gravity. Weathering is normally, but not necessarily, followed by erosion by one of these agents.

Because weathering involves contact between rock materials and constituents of the atmosphere, it proceeds most rapidly and effectively on near-surface materials. In general, the nearer rock materials lie to the surface, the more they have been altered by weathering. Weathering at any given site should not be thought of as a single process, but rather as a series of processes, each of which alters the rock further and generally weakens or disassembles it. Waves of sequential weathering processes work their way slowly down through the rock material; the position of the lowermost wave sometimes is called the **weathering front.**

The speed and types of weathering at any given site are controlled by five factors: rock type, rock structure, topography, vegetation, and, most important, climate. The significance of rock type is related primarily to the rock's physical strength and chemical stability. Physically weak and chemically unstable rocks are most readily attacked by weathering agents. Rock structures such as joints or bedding planes may provide avenues for attack by increasing the amount of surface area within reach of the weathering agents. The influence of topography on weathering primarily involves the slope angle and the resulting depth of soil cover. A deep soil layer, such as often covers areas with gentle slopes, may insulate the underlying bedrock from some types of weathering, but may also supply chemicals that promote other weathering reactions. Vegetation generally promotes weathering both through physical stresses produced by the growth of roots and by the production of organic acids. Climate exerts a strong control on weathering by determining temperature and moisture as well as by influencing the topography and vegetation.

The mix of these five factors can vary significantly over small distances, resulting in differing weathering rates. This, in turn, is likely to lead to differing erosion rates and to the production of a variety of topographic features.

Types of Weathering

For purposes of classification and discussion, weathering activities are commonly divided into the two categories of physical weathering and chemical weathering. Although in this section the various processes in each category are

◆ **FIGURE 15.1** This granite outcrop in Wisconsin displays two clearly defined joint systems. They will serve as pathways for attack by subsequent physical and chemical weathering activities.
SOURCE: © JLM Visuals

discussed individually, it should be realized that multiple processes normally act in conjunction or in sequence at a given site.

Physical Weathering

Physical weathering involves processes that reduce the size of rock masses without altering their chemical composition. Often described as producing rock disintegration, these processes convert masses of rock into more numerous smaller pieces of the same kind of rock that may be more readily removed by erosion. The most important physical weather-

◆ **FIGURE 15.2** Exfoliation on granite of the Sierra Nevada Batholith in Sequoia National Park, California.
SOURCE: Gerald Brothen

ing processes are joint and fracture formation, frost wedging, salt crystal growth, and biological forces.

JOINT AND FRACTURE FORMATION. The first weathering process to affect many rocks, and the deepest acting, is the formation of joints or fractures. Such breaks often are produced by tectonic stresses or result from cooling contractions. The formation of joints or fractures by any cause facilitates subsequent weathering by greatly increasing the amount of rock surface exposed to physical and chemical attack (Figure 15.1).

Fracture systems paralleling the surface also may be produced by the process of **unloading.** This refers to rock breakup caused by the gradual decrease in gravitational pressure as erosion removes overlying rock material. The resulting vertical expansion causes the rock to **exfoliate,** or to separate into exfoliation sheets (from the Latin *folium,* "leaf") paralleling the surface. Exfoliation sheets may vary in thickness from less than a centimeter to several meters; when the rock is exposed at the surface, they tend to peel or break off somewhat like the layers of an onion. When this process operates on large masses of granite, mountain-sized *exfoliation domes* can be produced (Figure 15.2). Half Dome in Yosemite National Park, California, and Stone Mountain near Atlanta, Georgia, are well-known examples of granite exfoliation domes.

FROST WEDGING. **Frost wedging** results from the growth of ice crystals within rock fractures or hollows. This process can generate pressures of more than 100 kilograms per square centimeter (1,400 lb/in^2) of rock surface. The force is concentrated at the apex of the fracture and often is sufficient to cause further splitting and widening.

Frost wedging can be highly effective in regions where daily freeze and thaw cycles occur during a large part of the year and where little soil cover exists to insulate the solid bedrock from the effects of atmospheric temperature fluctuations. In particular, locations within subpolar regions and at high elevations above the tree line but below the zone of perennial ice and snow are susceptible to this process. Within the middle latitudes, frost wedging may be effective during the winter in areas where bedrock outcrops occur at the surface. In the low latitudes, it is ineffective, except at high elevations, because of a lack of subfreezing temperatures.

Active frost wedging in mountainous regions can litter the surface with angular rock fragments of all sizes, producing **boulder fields** (Figure 15.3). Most of the rocks typically break off nearby cliffs. The lower portions of large, steep-sided cliffs may be buried beneath cone-shaped accumulations of such rocks. These accumulations, known as **talus cones,** are common in mountainous portions of western North America (Figure 15.4).

SALT CRYSTAL GROWTH. In arid regions, saline groundwater is often drawn surfaceward, and the salts are eventually

◆ **FIGURE 15.3** A boulder field produced by active frost wedging in Greenland.
SOURCE: R. B. Colton/USGS Photo Library

◆ **FIGURE 15.4** Talus cones produced chiefly by frost wedging lie beneath towering peaks in the Canadian Rockies near Moraine Lake in Banff National Park, Alberta.
SOURCE: Judi Scott

deposited at or near the surface as the water evaporates. Runoff from precipitation can then dissolve the salt and carry it into bedrock fractures, intergranular voids, or other openings. The subsequent evaporation of the water allows salt crystals to form and grow, generating expansionary pressures similar to those associated with frost wedging. Fractures can thus be enlarged and rocks eventually split apart. Crystalline rocks, such as granite, may have individual mineral crystals pried loose by growing salt crystals, and granular rocks such as sandstone can have their constituent grains detached one by one. In portions of the southwestern United States, this process has occurred on a large enough scale to produce hollows in sandstone cliffs that became the habitation sites of cliff-dwelling Amerindian tribes.

BIOLOGICAL FORCES. Biological forces are of limited importance as a cause of physical weathering. Probably the most significant effect is produced by the growth of plant roots, especially those of trees, in rock crevices. The buckling of concrete sidewalks by tree roots is a familiar illustration of the power that growing plants can bring to bear on rock material.

If the actions of human beings are included in the category of biological forces, however, it attains substantially more importance. Humans, especially during the present century, have developed the technological capacity to alter landforms on a major scale (Figure 15.5). Examples of human activities resulting in the mechanical breakup of bedrock include drilling and blasting associated with mining and quarrying, building and road construction, and warfare.

Chemical Weathering

Chemical weathering involves chemical changes in the rock material that result in the formation of new substances. In general, the new substances are softer, more finely divided (i.e., they consist of smaller pieces), more soluble, and larger

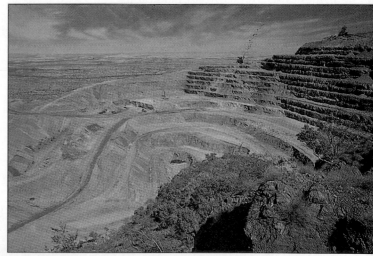

◆ **FIGURE 15.5** Argyle diamond mine, the world's largest producing diamond mine, provides a graphic example of the large scale of some human-produced landform modifications. This mine is located 120 kilometers (70 mi) southwest of Kununurra, Western Australia.
SOURCE: Bill Bachman/Photo Researchers

in volume than the original rock. These changes make the rock material more susceptible to further weathering and erosion. The breakup of rock by chemical weathering often is referred to as rock decomposition.

Chemical weathering often, but not always, follows physical weathering processes such as joint formation and frost wedging that open the way for chemical attack. As a result, weathering advances most rapidly along fractures or other lines of weakness such as bedding planes. Whereas rocks that have been influenced primarily by physical weathering tend to be angular, those dominated by chemical weathering typically are more smoothly rounded. Rounding occurs because rock corners and edges have large surface areas in proportion to their masses, allowing chemical attack to alter them relatively rapidly (Figure 15.6).

As a rule, rocks formed under the highest temperatures and pressures are the least stable chemically. In particular, the chemically basic igneous rocks derived from constituents of the upper mantle and lower crust tend to be chemically unstable. Sedimentary rocks, which form much nearer the surface, are chemically more stable but are physically much weaker, and have a tendency to separate into their original constituents when acted upon by weathering agents. From a broad perspective, then, weathering processes occur because rock materials tend to undergo alterations that increase their physical and chemical stability under the environmental conditions that exist at or near the surface. Most areas therefore display a weathering "profile" ranging downward from highly altered materials, approaching an equilibrium with surface conditions, to completely unweathered deeply-lying bedrock in equilibrium with the interior environment in which it was formed (Figure 15.7).

Several important categories of chemical weathering activities exist. Most involve chemical reactions with water or with gases dissolved in water; consequently, the availability of water is critical to chemical as well as physical weathering processes. The five most important categories of chemical weathering processes are hydration, hydrolysis, oxidation, carbonation, and solution.

Hydration involves the absorption, or adhesion, of water molecules to the molecules of a mineral. It is not a permanent chemical combination and can be reversed with the application of heat. An important example is the conversion of anhydrite ($CaSO_4$) in the presence of water to gypsum ($CaSO_4 \cdot 2H_2O$). Hydration increases the mass and volume of a mineral and softens it. The alternating hydration and dehydration of the minerals in some rocks in response to changing temperature conditions produces shrink-swell pressures that can dissociate the rock constituents. In some rocks it produces the exfoliation of thin sheets, and in others it aids in granular disintegration (Figure 15.8).

Hydrolysis is a permanent chemical combination with water. Like hydration, it results in the expansion and weakening of the rock material. Hydrolysis is especially effective on igneous rocks because it particularly affects feldspars and micas, which these rocks contain in quantity. The hydrolysis of igneous rocks by groundwater, especially in the humid tropics, can occur at considerable depths.

Oxidation involves the chemical combination of minerals with oxygen that normally has first been dissolved in water. It is most effective on iron compounds, especially in warm climates, and primarily affects igneous rocks containing iron. The rusting of automobiles, iron railings, or other objects when exposed to moisture is a familiar example of oxidation. The presence of large quantities of iron oxides in the soils of humid tropical and subtropical locations is responsible for their distinctive red color (Figure 12.14).

Carbonation consists of chemical reactions of rock material with carbonic acid (H_2CO_3), which is produced when carbon dioxide is dissolved in water. The carbon dioxide involved generally is produced by plant decay, so that carbonation is most active in humid, well-vegetated environments. Carbonation results in the conversion of oxides to more soluble carbonates. It is highly effective on the feldspar minerals that comprise a large proportion of most igneous rocks. Carbonation is also the process by which limestone and dolomite are converted to calcium bicarbonate, a physically weak and highly soluble mineral that can be removed readily by groundwater flow.

Solution refers to the dissolving of solid rock material in water. The water often contains substances such as carbonic acid that enhance the solution process. In effect, the rock is

(a)

(b)

(c)

◆ **FIGURE 15.6** Chemical weathering results in the formation of rounded core stones when rectangular blocks of rock in a jointed rock mass are attacked by chemical solutions along their exposed edges and corners.

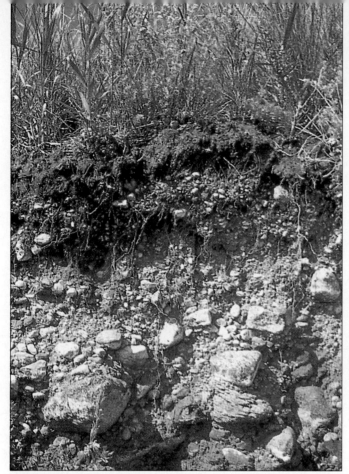

◆ **FIGURE 15.7** A surfaceward increase in chemical weathering intensity is evident in this view of a freshly exposed bank in North Dakota.

SOURCE: Bonnie L. Heidel/Visuals Unlimited

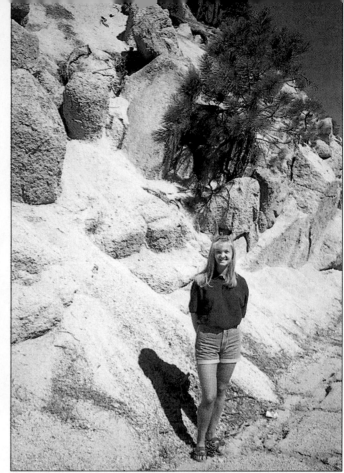

◆ **FIGURE 15.8** Granite exposed in this roadcut near Lake Tahoe in western Nevada is undergoing granular disintegration. The author's daughter Jill provides scale.

SOURCE: Ralph Scott

converted from a solid to a dispersed molecular state and therefore loses all resistance to erosion. Solution is commonly considered a chemical weathering activity, but it is not necessarily accompanied by permanent chemical change because the dissolved substances eventually may be precipitated back into their original chemical forms.

All substances are soluble in water to some degree. Solution is most effective by far, however, on highly soluble rock types. As noted above, solution also plays a role in the weathering of limestone and dolomite following their conversion to calcium bicarbonate. Limestone and dolomite are widely distributed. In areas where they are the dominant bedrock type, a combination of carbonation and solution, coupled with erosion, produces a distinctive set of landforms called karst. Karst landscapes will be discussed later in the chapter.

Human actions, which have resulted in the widespread acidification of both precipitation and surface water bodies, have substantially increased rates of solution in many areas. This has resulted in the rapid deterioration of many structures, especially buildings, statues, and tombstones constructed of soluble rock types such as limestone or marble.

Further discussion of this subject occurs in the Case Study beginning on page 263.

Global Distribution of Weathering Activities

Chemical and physical weathering vary considerably with regard to the climatic and lithologic conditions under which they operate most effectively. As a result, the relative importance of these two groups of processes differs significantly in different parts of the world.

For the world as a whole, chemical weathering is more active than physical weathering. This is because chemical weathering operates at a much faster rate than physical weathering under favorable conditions and produces a much more complete transformation of the rock material. The end product of physical weathering is merely smaller rock particles of the same composition as the original. The end products of most chemical weathering processes, on the other hand, are completely new substances. In fact, the

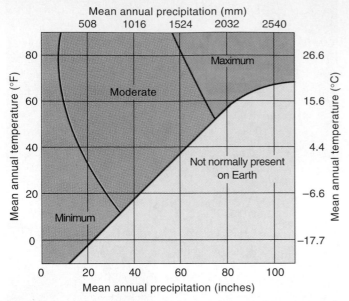

◆ **FIGURE 15.9** Relationship between chemical weathering rates and climate.

complete chemical weathering of rock-forming silicate minerals forms the clay minerals that are essential components of most soils.

Because most chemical weathering processes generally operate most efficiently under warm, wet conditions, the most rapid chemical weathering occurs in the humid tropics (Figure 15.9). Here, the deepest and most highly weathered soils exist and the fastest topographic modification takes place. Conversely, chemical weathering processes are least active in polar and arid regions, and landscapes in these environments evolve relatively slowly. Chemical weathering is also aided by the presence of fragmented, soluble, or chemically unstable rocks, so that geology as well as climate is important in controlling the rate of chemical weathering within a region.

Physical weathering occurs most rapidly in regions that favor the operation of its primary mechanism—frost wedging. It therefore is most vigorous in the subpolar regions and in high mountain areas in the low and middle latitudes (Figure 15.10). Physical weathering also attains relatively great significance in deserts, largely because the lack of water inhibits chemical weathering activities. Desert regions are frequently littered with fragmented rocks resulting from the effects of salt crystal growth and frost wedging. Physical weathering processes also are aided by certain lithologic characteristics. These include large surface exposures of brittle, fractured, thinly bedded, or physically soft rock. In most regions, physical and chemical weathering processes work together so intimately that it is difficult and rather arbitrary to separate them, and the existing landscape represents their collective contributions.

The zone of weathering varies in depth from the surface in some polar and arid environments to well over 30 meters (100 ft) in portions of the tropics. In most middle latitude areas, the weathering front is some 3 to 9 meters (10 to 30 ft) below the surface. The great majority of the rock materials that have been weathered and eroded at some time during the Earth's long history have been deposited and relithified and now exist in the form of sedimentary and metamorphic rock.

Regional variations in the rates, types, and products of weathering are largely responsible for differences in the geomorphic characteristics of regions with differing climates. For example, humid tropical environments dominated by chemical weathering processes typically display a deep soil cover and rounded topographic features. In contrast, many arid regions exhibit stark, angular features that result from the lack of a soil and vegetation cover and from the dominance of physical weathering activities (see Chapter 18). The correlation between climate and topography would be still greater were it not for the topographic variability imposed by differing rock types and especially by variations in the nature and degree of tectonic activity. These two topographic controls are independent of climate, so that a newly formed volcano, for example, looks much the same in Alaska as in Mexico.

In concluding this section, it should again be stressed that weathering is not a landscape producer in itself; it is merely the necessary preparatory step in the gradation process. Because weathering activities normally occur near the surface and involve little movement of materials, features produced by differential weathering alone are usually not apparent unless some erosional mechanism exposes them. Such exposure is generally accomplished by the selective

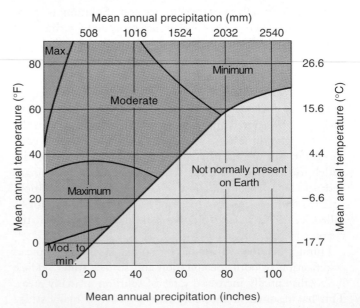

◆ **FIGURE 15.10** Relationships between physical weathering rates and climate.

relocation or removal of the more highly weathered materials by water, ice, wind, or gravity, leaving the less weathered materials behind. Differential weathering therefore facilitates differential erosion. In general, high-standing gradational landforms are composed of materials that are relatively resistant to weathering, and low-standing features are composed of easily weathered materials. For example, in the folded Appalachians, the ridges are composed of physically hard and chemically resistant sandstone, while the valleys are composed of soluble limestone or physically weak shale.

MASS WASTING

Once weathering processes have weakened surface materials, these materials become much more susceptible to the influence of outside forces, including gravity. Gravity is the direct motivating force in the process of gradation. Most of the work of gradation is accomplished by water, ice, and wind, which, impelled by gravity, transport surface materials downslope. A vital role in the gradational process, though, is performed by gravity acting alone. Downslope movements of surface material under the direct influence of gravity are termed **mass wasting** or **mass movements.** Although no medium such as water or ice serves as a transporting agent, one or both of these substances are present and act as destabilizing factors in most individual mass movements.

Mass wasting has a much greater impact on the Earth's surface than is often realized. It performs the vital role of transferring the products of weathering from their original sites, primarily on hillsides or other upland surfaces, to the valley or lowland sites where they are in position to be picked up for long-distance transport by the erosional tools of water and ice.

Mass movements exhibit a tremendous range of physical attributes. They vary in size from the motion of a single grain of sand to the massive displacement of many cubic kilometers of rock. The materials involved may consist of unweathered or weathered rock, soil, vegetation, water, ice, or, very commonly, a mixture of several of these ingredients. The speed at which mass movements travel varies from the imperceptible to velocities exceeding 160 kilometers per hour (100 mph). Distances range from fractions of a centimeter to many kilometers.

The larger and faster types of mass movements not only are awesome phenomena that are capable of sweeping away everything in their path, they also are likely to occur suddenly and unexpectedly. As a result, they can produce great property destruction and loss of life. In fact, mass movements comprise one of the major categories of natural hazards that have afflicted humanity for millennia. Even minor mass movements are quite capable of producing damage such as cracked road surfaces and building foundations, broken water mains, bent trees and telephone poles, and highway debris following rainstorms in mountainous terrain.

Factors Influencing Mass Wasting

The susceptibility of a slope to mass wasting is determined by the relative effectiveness of two opposing groups of forces: those that act to maintain slope stability, and the downslope forces that act to produce slope failure, or mass wasting. Three factors collectively aid in the maintenance of slope stability (Figure 15.11). The first is the **normal stress,**

◆ **FIGURE 15.11** Factors influencing slope stability. Stabilizing factors include normal stress, the cohesion of the slope materials, and internal friction along the potential slip plane. The destabilizing factor is the downslope force caused by the pull of gravity. The overall stability of the slope is determined by the relative magnitudes of these opposing sets of forces.

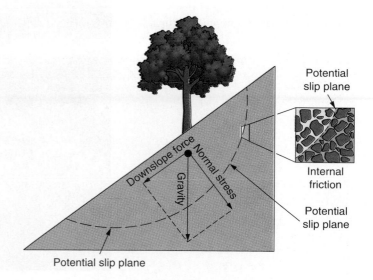

or the inward compression of the slope materials caused by their weight. The second is the **cohesion** of the slope materials, which is a measure of the extent to which the individual particles comprising the slope adhere to one another. The third is the **internal friction** along any potential failure surface, resulting largely from the roughness of that surface. Collectively, these attributes determine the **shear strength** of the hillslope materials, which is a measure of their ability to resist downslope movements.

Opposed to the shear strength is a single force—the downslope force caused by the component of gravity acting parallel to the slope. This force becomes greater as slope angles increase. The steepest angle that can be maintained on any given slope without failure of the slope materials is called the **angle of repose.** Its exact value depends upon the type, amount, and condition of the slope materials. The angle of repose represents a threshold at which the downhill pull of gravity is exactly counterbalanced by the shear strength of the materials. As long as the angle of repose is not exceeded, mass movements will not occur. Factors that foster mass movements cause slope angles to exceed the angle of repose either by increasing the steepness of the slope or by producing conditions that at least temporarily reduce the shear strength of the slope materials.

Oversteepening of Slopes

As a general rule, the steeper a slope, the less stable it is. Steep slopes therefore are likely to experience more frequent and more rapid mass movements than are gentle slopes. The angle of repose for slopes covered by unconsolidated materials normally ranges from 25° to 40°. At progressively steeper slope angles, there is a general decline in the thickness of the weathered mantle as mass wasting and other erosional processes become more efficient in transporting it downslope. Slopes steeper than 40° usually have bare rock surfaces.

A number of processes may increase slope steepnesses past the angle of repose. Probably the most important is undercutting by water, so that mass wasting of various types is common along stream banks. Likewise, wave undercutting, especially during storms, often produces mass movements along oceanic coasts or the shores of large lakes. The California coastline and portions of the Great Lakes shoreline are noted for such occurrences (Figure 15.12). Oversteepening leading to slope failure also may result from valley glaciation within mountainous regions such as the Alps.

Human activities can also produce oversteepened slopes. For example, the mass movements that sometimes block roads constructed through mountainous terrain are generally caused by road cuts that have been made at too steep an angle. Waste piles from mining activities may also become oversteepened, with disastrous consequences. On the morning of October 21, 1966, in Aberfan, South Wales, 150,000 cubic meters (5,300,000 ft³) of debris from coal mining waste that had become saturated with water from subsurface springs suddenly collapsed. The resulting mudflow (Figure 15.13) overwhelmed 18 houses and a school, taking 144 lives.

Weaknesses in Slope Material

Weak or loose slope materials are much more prone to mass wasting than is solid bedrock. Steep or even vertical cliffs can be maintained in a relatively stable condition in massive, solid rocks such as granite, limestone, or basalt. Fractures, steeply inclined bedding planes, and other zones of weakness, though, can greatly reduce the angle of repose. Weak rocks, such as shale and many sandstones, are also unable to

◆ **FIGURE 15.12** The Point Fermin landslide in San Pedro, California, was caused by the oversteepening of the coastline as a result of wave undercutting.
SOURCE: Bernard W. Pipkin, University of Southern California

◆ **FIGURE 15.13** Aerial view of the Aberfan mudflow in which 144 people died.

SOURCE: T. Spencer/Colorific

gradually more susceptible to mass movement. Most mass movements are actually triggered by an "active" factor that temporarily reduces the shear strength of the slope materials below the critical angle of repose. Although no clear-cut triggering mechanism exists for some mass movements, most are initiated either by strong vibrations or by the temporary presence of large amounts of water. Vibrations that trigger mass movements often result from earthquakes, which are relatively common in tectonically unstable mountainous regions prone to mass movements. If a slope is unstable enough, a lesser disturbance such as a loud clap of thunder may suffice to initiate slope failure. **Avalanches,** which are rapid movements of masses of snow and ice down steep mountain slopes, may be triggered by the sound of a gunshot or even a shout.

Large quantities of water that may produce slope failure usually are provided by heavy storms. The water adds weight to the slope and lubricates and buoys the individual slope particles so that their cohesion is reduced. Water is especially effective on slopes with a high clay content, since clay is not only capable of holding large quantities of water but also consists of platy minerals that readily slide over one another when wet. (This accounts for the slickness of mud, as opposed to the firmness of dry soil.) Several types of mass movements occur only when the water content of the slope materials is unusually high.

Types of Mass Wasting

Mass movements, like most other natural phenomena, exist in a continuum of types, and one type often grades into another with distance. Although this means that no mass movement classification system can adequately describe the existing spectrum of types, a number of systems have been

maintain steep slopes. In particular, the presence of a strong **caprock** (such as basalt) overlying a much weaker rock (such as shale) may lead to unstable slope conditions, since the more rapid erosion of the weak rock will likely lead to the undermining and subsequent collapse of the caprock. Many steep slopes in western North America are destabilized by this set of conditions.

The presence of a deep cover of soil and weathered rock also favors mass movements. Many mass movements, in fact, involve only the displacement of weathered materials; the underlying bedrock is largely or entirely unaffected. This factor brings into consideration the indirect influence of climate, which is crucial in determining the types and rates of weathering processes (Figure 15.14). The humid tropics are particularly susceptible to mass wasting of chemically weathered materials. Mass movements involving physical weathering products, conversely, achieve great geomorphic significance in alpine and high latitude environments. Mass wasting is so important in the periglacial environment that it is sometimes regarded as the leading geomorphic process operating in that region (see Chapter 17).

Triggering Mechanisms

The slope instability factors just described are considered "passive" factors, because over time they tend to make a slope

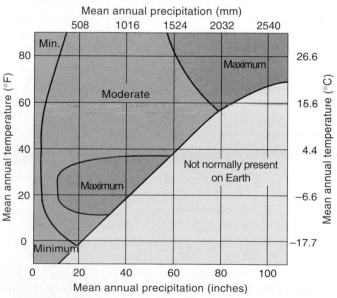

◆ **FIGURE 15.14** Relationship between climate and mass wasting potential.

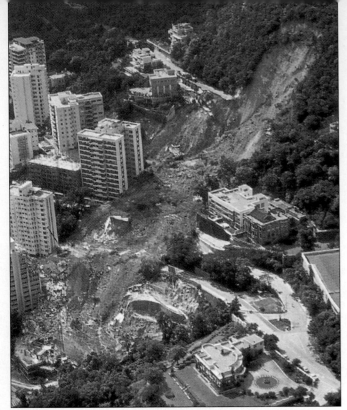

◆ **FIGURE 15.15** This debris avalanche, which occurred on June 18, 1972 in Hong Kong, killed 68 people when a portion of the steep hillside at the upper right suddenly failed, destroying a 4 story building and a 13-story apartment block.

SOURCE: Hong Kong Government, Geotechnical Control Office

◆ **FIGURE 15.16** The Saidmarreh rockslide, the world's largest known mass movement, was apparently triggered by an earthquake that dislodged a thick sequence of limestone and shale strata. These rocks had been destabilized by undercutting by the Karkheh River.

devised. Mass movements usually are classified according to whether their motion consists primarily of sliding, falling, or flowing; whether the material involved is predominantly rock, earth, or **debris** (a mixture of soil, rock, and vegetation); or whether their motion is relatively slow or rapid. We shall use the last of these criteria to divide mass movements into two groups and will briefly describe the major types of mass movements in each group.

Rapid Mass Movements

Rapid mass movements are those in which visually perceptible motion occurs (Figure 15.15). As a group, they are much briefer, more destructive, potentially more dangerous, but less geomorphologically significant than the slow mass movements. They tend to occur on relatively steep slopes and are likely to contain unweathered rock as a component of their material. Rapid mass movements in which motion occurs primarily by sliding are sometimes collectively referred to as **landslides.** This term is not employed here because of its lack of specificity.

ROCKFALLS. Rockfalls are among the smallest, most common, and most rapid mass movements. They simply involve the falling of individual rocks from nearly vertical cliffs. Rockfalls usually result from frost wedging in alpine envi-

ronments and are common in mountainous portions of western North America. Accumulations of angular rock fragments from innumerable individual rockfalls form the talus cones described earlier in the chapter (Figure 15.4).

ROCKSLIDES. Rockslides occur when masses of detached rock slide down moderate to steep slopes. They vary tremendously in size, ranging from individual rocks to entire mountainsides. Large rockslides are usually triggered by earthquakes. They may begin as coherent (solid) masses but quickly become fragmented by the violence of their motion. On long, steep slopes in high mountain areas such as the Andes or Himalayas, they can attain speeds well in excess of 160 kilometers per hour (100 mph).

The world's largest known single mass movement of any kind was the prehistoric Saidmarreh rockslide, which occurred in southwestern Iran (Figure 15.16). It consisted of approximately 21 cubic kilometers (5 mi³) of limestone and marl that became detached from an anticlinal ridge and slid into the adjacent valley and well up the ridge on the valley's opposite side. It left a scar some 14 kilometers (9 mi) long, and the resulting debris covered an area of 166 square kilometers (64 mi²).

DEBRIS SLIDES, DEBRIS AVALANCHES, AND DEBRIS FLOWS. These mass movements all involve the rapid transport down moderate to steep slopes of a mixture of soil and rock material in various stages of weathering, and vegetation. They differ from one another primarily in their water content and their consequent fluidity of motion. The debris slide is the driest of the three, while the debris flow is the most fluid. They are all similar in shape, tending to be elongated, narrow, and to follow hillside depressions or hollows (see Figure 15.15). When they reach the base of the slope, they typically spread out in fan-shaped deposits. All three normally are triggered by exceptionally heavy rainstorms and can inflict considerable destruction.

One of the major natural disasters of recent times resulted from a huge debris flow that began when an earthquake triggered an avalanche on Mt. Huascarán, Peru's highest peak, in 1970. Accelerating down the steep mountainside and picking up debris as it went, it may have attained speeds as great as 320 kilometers per hour (200 mph). It overwhelmed and destroyed several towns near the base of the mountain, taking an estimated 25,000 lives. The Case Study at the end of this chapter discusses the causes and consequences of the occurrence of a large group of **debris avalanches** in central Virginia as a result of torrential rainfalls associated with Hurricane Camille in 1969.

MUDFLOWS. Mudflows are the most fluid of the major types of mass movements. They consist of a perceptible and often rapid downhill flow of mud, usually mixed with rocks and other debris. Mudflows generally follow preexisting channels and are long and relatively narrow. Most occur in arid and semiarid regions, where they are triggered by heavy thunderstorms, but they may also form in alpine and ash-covered volcanic regions. In all three of these environments, the surface is covered by loose, fine-textured materials that are not stabilized by vegetation.

In interior portions of the western United States, mudflows caused by mountain thunderstorms may suddenly and unexpectedly debauch from canyons onto flat farmland in adjacent basins, burying large areas in mud. In southern California, slopes denuded of vegetation by summer brush fires often become the sources of destructive mudflows following the onset of winter rains. The Aberfan disaster (Figure 15.13) was a mudflow produced by human negligence.

Slow Mass Movements

Slow mass movements advance at a visually imperceptible rate and may take days, weeks, or even years to become evident. Although they are not nearly as spectacular or dangerous as the rapid mass movements, they are more geographically widespread and are responsible for the downhill transport of a greater total volume of weathered materials.

SLUMP. A **slump** is the intermittent movement of a mass of earth or rock along a curved slip plane. It is characterized by the backward rotation of the slump block so that its surface may eventually tilt in the opposite direction from the slope from which it became detached (see Figure 15.17). Slumps are most likely to occur on steep slopes with deep, clay-rich

◆ **FIGURE 15.17** A slump is caused by the slow sliding of one or more slabs of material along curved slip planes.

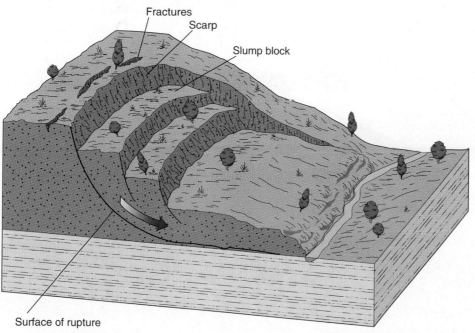

Fractures

Scarp

Slump block

Surface of rupture

◆ FIGURE 15.18 A hummocky surface resulting from multiple earthflows on a hillside near Healdsburg, California.

SOURCE: © Don Duckson

soils after saturation by heavy rains. The movement generally extends over many days or weeks and is nearly impossible to control or halt once it has begun.

Slumps are common along the California coast, where slopes frequently have been oversteepened by wave undercutting (Figure 15.12). They also occur along the side slopes of river gorges in various parts of the western United States. Small slumps capable of blocking traffic are frequent occurrences on steep road cuts following heavy rains.

EARTH FLOW. **Earth flows** involve the slow flow of nearly saturated soil down moderate to steep slopes. They are more fluid, shallower, and usually smaller in size than slumps. An earth flow typically has a depressed upper portion and a bulging "toe" at its base where the material has come to rest, perhaps beneath a mat of vegetation (Figure 15.18). Earth flows commonly occur in humid climates on grassy, soil-covered slopes following heavy rains. Large numbers may form on a single hillside under favorable conditions.

CREEP. **Creep** is the slowest and least noticeable, but the most widespread and geomorphologically important of all the categories of mass movements. It consists of the very slow downhill movement of soil or rock material over a period of years. Creep involves the entire hillside and probably occurs, to some extent, on any sloping, soil-covered surface. Because the near-surface layers creep fastest, it is commonly responsible for the downhill tilting of buried or partially buried objects such as trees, telephone poles, stone walls, and inclined rock strata (Figures 15.19 and 15.20).

Any disturbance to the surface mantle causes creep because such disturbances almost invariably result in a net downhill displacement of surface particles. Some types of disturbances contributing to creep include **frost heaving** (the loosening of soil by growing frost crystals), wetting and drying, heating and cooling, plant root growth, raindrop impact, and the trampling of livestock. Creep is especially active on slopes covered by deep soils with a high water content and is therefore much more significant in humid regions than in arid or semiarid regions. It is thought to be largely responsible for the generally rounded topographic features of humid landscapes. The subdued profiles of the Appalachian Mountains, for example, stand in stark contrast to the angular features of mountain ranges in arid portions of the West. This contrast is probably more attributable to variations in rates of chemical weathering and creep than it is to rock type or geologic age. Creep may be the chief mechanism by which weathered materials are transported down hillsides in soil-covered areas to the valleys of the streams that ultimately carry these materials to the sea.

◆ FIGURE 15.19 The effects of creep include the bending of near-surface rock strata and tree trunks, as well as the tilting of such features as telephone poles, fences, walls, and tombstones.

KARST TOPOGRAPHY

In some areas underlain by thick sequences of limestone and dolomite, the weathering processes of carbonation and solution can themselves be considered responsible for the major characteristics of the topography. Even here, a transporting agent (in the form of groundwater) is needed to remove the weathered materials from their original sites. Because these materials have first been converted to an individual molecular state, however, carbonation and solution often are considered the primary geomorphic mechanisms in the development of the resulting topography.

Cause and Global Distribution of Karst Topography

Groundwater in humid regions tends to become somewhat acidic from the solution of atmospheric carbon dioxide as well as from the acquisition of organic acids from the soil. It becomes, in effect, a weak carbonic acid solution. Carbonic acid reacts with carbonate rocks to produce calcium bicarbonate, which is highly soluble in water and is readily transported away by groundwater flow. An area with landform features produced largely by the differential solution and subsequent erosion of limestone and dolomite is called a **karst** area. The term is derived from the Kras region of Croatia near the Adriatic Coast, where some of the earliest research on this topographic assemblage was conducted.

A combination of lithologic, topographic, and climatic factors determines the favorability of a given site for the formation of karst topography. There are four prerequisites for optimum karst development. First, thick sequences of relatively pure limestone or dolomite must be present at or just beneath the surface. Second, the rock should be highly jointed to provide ingress for surface water, but should otherwise have low porosity and permeability. This is because karst features develop as a result of differential solution rates. A third and very crucial factor is a moderate to abundant supply of precipitation. This provides the water needed for the solution process. Plentiful precipitation also favors the growth of vegetation and the formation of a soil humus layer that lowers the pH of the downward-moving water, thereby greatly increasing its solution potential. The final prerequisite for maximum karst development is for the zone of groundwater saturation to be situated well below the surface. This zone marks the lower limit of downward percolation of the acidic water that accomplishes the most rapid solution. A near-surface water table inhibits the vertical development of solution features.

Karst topography is geographically widespread; approximately 15 percent of the Earth's land surface contains some karst features. Conditions generally are favorable for karst development in humid portions of the middle latitudes, and several major karst regions are located there, especially in the United States and Europe. It reaches optimum develop-

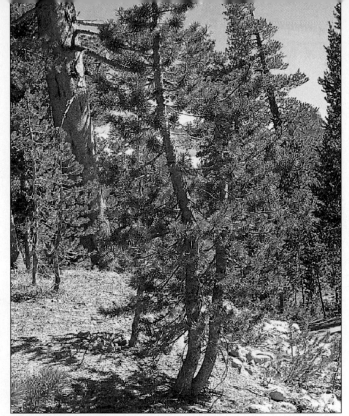

◆ FIGURE 15.20 The downhill bowing of tree trunks resulting from soil creep on the east slope of the Sierra Nevada, California.

SOURCE: Bernard W. Pipkin, University of Southern California

ment, however, in the humid tropics, where heavy precipitation and organic acids produced by the decomposition of the dense vegetation combine to favor the rapid solution of limestone and dolomite. Major areas are depicted in Figure 15.21. The United States probably contains more karst than does any other country. Important areas include Kentucky and Indiana, the central Appalachians, Florida, the Ozark Plateau, western Texas and eastern New Mexico, and the Black Hills of South Dakota. Karst is also well developed in southern Europe, especially in the southern Alps and in a number of regions bordering the Mediterranean Sea. Australia contains an extensive but little-publicized karst region in the Nullarbor Plain, located in the south-central part of the continent. The most important areas of tropical karst include South China and adjacent portions of Vietnam, Thailand, and Burma; parts of Indonesia and Malaysia; a large area in eastern Brazil; Mexico's Yucatán Peninsula; and the West Indian islands of Jamaica, Cuba, and Puerto Rico. Finally, a number of relict (no longer active) karst areas exist in currently arid areas such as the Sahara and Kalahari Deserts.

Karst Features

Karst regions vary greatly in their topographic characteristics (Figure 15.22). Some may be characterized as plains, some

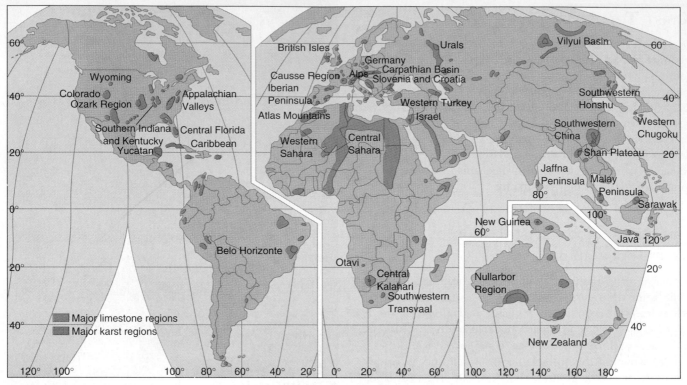

◆ FIGURE 15.21 The distribution of the major limestone and karst areas of the world.

Major limestone regions
Major karst regions

◆ FIGURE 15.22 Important karst features.

Solution valleys

Karst valley

Springs

Sinkholes

Disappearing streams

Deeply intrenched permanent stream

Cave

KARST-RELATED HAZARDS IN EXPANDING URBAN ENVIRONMENTS

United States Bureau of the Census data show that between 1985 and 1990 approximately 1.4 million Americans migrated to the southern region of the nation. As a result of rapid population growth, many areas in the South are experiencing urban sprawl. Tens of thousands of acres of forested and agricultural land are being developed and converted into commercial and residential properties, resulting in a rapid expansion of the human-built environment. In locations underlain by karst topography, planners, developers, and engineers face significant challenges in maintaining drinking water quality and preventing structural damage. Over 20 percent of the entire United States and 40 percent of the area east of the Mississippi River is underlain by soluble rock formations.

HIGHWAY RUNOFF MANAGEMENT

In many southern urbanized areas, the extension of transportation facilities to accommodate rising traffic volumes has become a top priority. Corridors for new highways are being delineated, and additional lanes are planned for existing roads. As highway systems are augmented, their accompanying sewage systems must be expanded to contain the additional runoff produced by the increase in impervious surfaces. Unfortunately, it is common practice in karst regions to directly link drainage structures (pipes, drains, culverts, etc.) with cave networks, thereby creating "natural sewers." As a result, toxic constitu-

ents in road surface runoff are granted clear passage into karst aquifers.

Because they are especially vulnerable to contamination, karst aquifer water supplies are designated by the U.S. Environmental Protection Agency (EPA) as "Class I–Special Groundwaters." Class I aquifers are those having specific hydrogeological characteristics such as flow velocity rates in excess of several meters per day, water tables at or near the land surface, and recharge zones in close proximity to pollution sources. An aquifer's irreplacability and ecological value are also factors considered for Class I denomination.

A recent U.S. Geological Survey study was conducted in Florida to examine retention ponds used to collect excess water discharged from interstate highways. Groundwater beneath several of the ponds was found to contain high levels of hazardous organic compounds. Contaminant metals including lead, zinc, iron, and copper have been discovered in highway runoff in other investigations. For automobile transportation corridors traversing vulnerable karst terrains, proper management of runoff is a significant engineering concern as the majority of stormwater discharge eventually becomes groundwater.

Many districts potentially affected by runoff-related aquifer contamination require that highway construction projects be conducted using "best management practices" (BMPs) to ensure proper design of drainage system elements such as detention basins, retention ponds, and swales. Hydrogeological studies indicate that BMP structures are effective in de-

creasing the concentrations of toxins in runoff waters.

WASTE FACILITY SITING

Increasing numbers of southern municipalities are in the process of siting and designing new solid waste management facilities to absorb growing accumulations of trash and garbage. In many cases, the need for these new facilities is critical, as existing landfills built without protective underlying clay or plastic liners are allowing contaminants to enter underlying karst aquifers. The landfills must therefore be closed before the turn of the century. In Florida, where over 90 percent of the population depends upon the karstic Floridan Aquifer for drinking water, hazardous waste landfills are banned. In nearly all cases, state and local governments will not grant permits for waste facility construction in areas underlain by karst and/or Class I aquifers.

In the extremely difficult efforts for siting much needed waste management facilities in limestone regions, karst features (sinkholes, sinking streams, dry valleys, etc.) serving as recharge zones must be avoided because infiltration from the surface may be almost immediate. In Alachua County, Florida, public officials are faced with finding a suitable location for a new solid waste management facility by 1998. Figure 15.23 is a map created using the

continued on next page

◆ **FIGURE 15.23** DRASTIC map of Alachua County, Florida. DRASTIC Indexes indicate relative vulnerability to hydrogeological contamination for various local terrains.

Alachua County, Florida

"DRASTIC" method illustrating various levels of hydrogeological vulnerability within this county. The "DRASTIC" method, an expert system created for the EPA by the National Water Well Association, assigns rankings (Indexes) to selected terrains by their susceptibility to groundwater contamination. DRASTIC Indexes range from 65 to 251, with higher scores indicating increased hazard. Note that the western portion of the county, where the water-bearing Ocala Group limestones exist at or near the surface, exhibits much higher Indexes than the eastern portion where the aquifer is overlain by the 30-meter (100 ft) thick clay sediments of the Hawthorn Formation.

SINKHOLE COLLAPSE DAMAGE PREVENTION

Research conducted by the Florida Sinkhole Research Institute has shown that the growth of impervious surfaces associated with urban expansion is a major contributor to increased incidences of sinkhole collapse in karst regions. Runoff that would normally be absorbed into the soil is instead diverted into drainage structures that raise the velocity and erosion capacity of stormwater flow. In karst formations, subterranean erosion, called "soil piping," creates large subsurface cavities with relatively weak soil ceilings. The weight of any overlying impervious structures (parking lots, roads, buildings, etc.) eventually exceeds the supporting capacity of these ceilings. The structures then collapse into the cavities, often causing expensive prop-

are mountainous, but most frequently they are hilly. The larger relief features tend to be very abrupt and steep-sided, giving the landscape a picturesque and sometimes spectacular appearance, but typically reducing accessibility and land-use potential.

The solution rates of limestone and dolomite can vary greatly over very small distances. As a result, even in areas of low relief, differential solution produces an extremely rough and intricately sculpted surface topography on exposed bedrock outcrops. Rock outcrops are common and frequently extensive in karst regions. These outcrops typically exhibit gullies, grooves, solution pits, and various other small-scale differential solution features in abundance. In addition, loose stones and boulders often litter the surface. The reddish, clay-rich soils of karst areas are fertile but discontinuous and often quite thin, sometimes existing only in patches within low-lying sites.

Sinkholes, also termed **sinks** or **dolines,** are the most widespread and undoubtedly the best known of the major karst features. They are rounded and often nearly circular depressions formed by subsurface solution. Most develop on bedrock joint intersections, are rather small, and are produced by the gradual subsidence of the weathered surface mantle into an enlarging solution hollow. Sinkholes can also

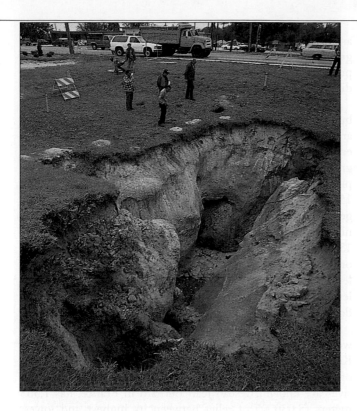

◆ **FIGURE 15.24** This collapse sinkhole in Chiefland, Florida, is about 9 meters (30 ft) deep. It appeared after north-central Florida received 25 centimeters (10 in) of rain in 24 hours.

SOURCE: John Moran

erty damage and sometimes personal injury (Figure 15.24). Rapid, catastrophic collapses are especially prevalent in karst regions following heavy rainfalls.

The geomorphologically slower phenomenon of sinkhole subsidence may also result in serious harm to human-built facilities. Increased pressure on groundwater resources produced by growing populations often leads to the depletion of karst aquifers. The resulting water table drawdown has been found to contribute signifi-cantly to sinkhole subsidence and collapse.

Remote sensing methods have been effectively employed to identify areas of potential hazard. Black and white aerial photography was used to locate subsurface karst cavities in central Florida. Low-altitude multispectral photography and thermal infrared imagery were used to locate areas of potential sinkhole collapse within a proposed interstate highway corridor in northern Alabama. Airborne multispectral scanner data were applied in west-central Florida to detect moisture-stressed vegetation commonly found adjacent to collapse-prone areas. The increasingly effective technology for translating images into GIS (geographic information system) format should greatly enhance future efforts to apply remote sensing tools to karst-related problems. ∎

This focus box was written by David A. Padgett, Assistant Professor, Environmental Geography Program, Austin Peay State University.

form rather suddenly and spectacularly by the collapse of the surface into an underground void (Figure 15.24). Such **collapse sinkholes** tend to be larger, deeper, and steeper than the more common **subsidence sinkholes**. Sinkholes commonly range up to 30 meters (100 ft) in diameter and 10 meters (30 ft) in depth, but within a given area, most are likely to be considerably smaller and shallower than this.

Although most individual sinkholes are highly rounded, where they are especially numerous, they may overlap to form irregularly shaped compound sinkholes. Sinkholes typically are dry if their bottoms are situated above the water table, but are partially filled with water if they extend below it. The rounded lakes of the Central Florida Lake District, for example, occupy large sinkholes in an area with a high water table. Sinkholes are most abundant and conspicuous in low-relief areas. **Karst plains,** such as those in portions of Indiana and Kentucky, are plains covered with sinks that may number a hundred or more per square kilometer (Figure 15.25).

Well-developed karst areas display an almost complete lack of organized surface drainage because the underlying carbonate rocks are honeycombed with solution channels. The surface is almost like a sieve, and runoff cannot travel far before it enters a sinkhole and disappears underground.

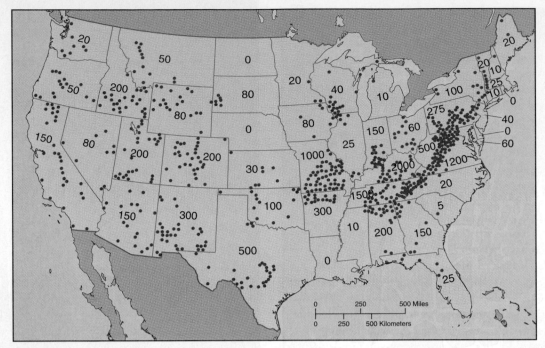

◆ **FIGURE 15.29**   Distribution of karst cave areas in the coterminous United States. Numbers indicate the estimated total number of caves in each state.

◆ **FIGURE 15.30**   Tower karst in the Li River Valley of southern China.
SOURCE: R. Krubner/H. Armstrong Roberts

▌SUMMARY

This chapter begins our examination of the gradational processes and their effect on the Earth's surface. Most gradational landforms are produced by erosion and deposition resulting from the movement of water, ice, and wind. Before these agents can operate effectively, however, the surface rock must first be acted upon by weathering processes that greatly reduce its resistance to erosion. Solar energy and gravity are the forces that power the gradational tools, and gravity also is capable of producing a variety of mass movements of the surface on its own. Another set of gradational features not normally considered a product of the "traditional" operation of the tools of gradation are the karst features of limestone regions.

In order to summarize more systematically the three chapter topics of weathering, mass wasting, and karst, we return to the Focus Questions:

1. *How do the weathering processes operate, and in what environmental settings are they most effective?*

Weathering involves the actions of physical and chemical processes that reduce the size of rock materials and often produce softer, more easily eroded substances. Physical weathering involves only a physical reduction in rock size and is most important in cold and dry environments. The chief physical weathering processes include the formation of joints and fractures caused by tectonic stresses, cooling contractions, and unloading; frost wedging caused by ice crystal growth; the growth of salt crystals in rock fractures; and the mining and construction activities of humans.

Chemical weathering has greater overall geomorphic importance than physical weathering. It operates most rapidly in warm, moist environments. Chemical weathering processes alter the chemical composition of the rock and produce new substances. The processes of hydration and hydrolysis involve chemical combinations with water, while oxidation and carbonation involve reactions with oxygen and carbon dioxide that generally have been first dissolved in water. Solution, another important chemical weathering process, causes the molecules of rock material to become detached from one another and to be suspended in water.

2. *What causes mass wasting, and what types of mass wasting are most widespread and important?*

Mass wasting, or mass movement, is the downward movement of surface materials under the direct influence of gravity. Because it is gravitationally induced, it is most likely to occur in areas that have steep slopes or physically weak surface materials, or in places where subsurface support has been removed. Factors resisting mass movements are the compression caused by the weight of the slope materials, as well as their cohesion and internal friction. Most mass movements are triggered by heavy rainfalls or earthquake vibrations.

Rapid mass movements such as rockfalls, rockslides, and debris avalanches are brief, spectacular, and often highly destructive. They are most common in geologically youthful mountain regions along or near lithospheric plate boundaries. Slow mass movements, especially creep, apparently have a greater total geomorphic impact because they are more widespread and collectively involve the downward transport of a greater volume of surface material.

3. *What are the causes and characteristics of karst topography, and how it is distributed around the world?*

Karst topography is created by the differential solution and erosion of soluble rocks, notably the carbonate rocks limestone and dolomite. Carbonate rocks are distributed widely around the world, but are especially susceptible to karst processes in areas of heavy precipitation and abundant vegetation. These conditions favor the production of large quantities of acidic surface water. Most important karst regions are located in humid portions of the tropics and middle latitudes.

A karst landscape is characterized by a variety of differential solution features. Small-scale surface features include grooves, solution pits, and loose rocks in rugged areas. The most important larger surface features are circular depressions known as sinkholes. Subsurface solution features also are very prominent. Most drainage takes place through underground routes; this frequently results in the formation of cave systems containing a variety of travertine features.

QUESTIONS FOR THOUGHT AND REVIEW

1. What is weathering? Why is it an essential prerequisite for gradation?
2. What is the difference between physical weathering and chemical weathering in terms of their effects on the rock? Give two examples of each weathering category.
3. How do the global distributions of physical and chemical weathering rates vary? Why do they do so? For the world as a whole, which of the two categories of weathering processes has the greater geomorphic significance? Why?
4. What is mass wasting? Discuss the importance of mass wasting in the gradational process.
5. Explain the significance of the angle of repose in mass wasting. Describe several ways by which human activities can cause slope angles to exceed the angle of repose.
6. Why can creep be considered the most important of the individual types of mass movements?
7. Describe the environmental conditions that favor the development of karst topography. Why is karst more widespread in the humid tropics and middle latitudes than in the subtropics and high latitudes?
8. Discuss the ways in which drainage in a karst area differs from the drainage elsewhere. What karst features develop as a result of these drainage characteristics?
9. How are karst caves believed to have formed? What distinctive features do they contain that make them more "scenic" than caves formed by other processes? How are these features produced?

KEY TERMS

physical weathering	debris avalanche
unloading	mudflow
exfoliation	slump
frost wedging	creep
chemical weathering	karst
mass wasting	sinkhole
shear strength	sinking creek
angle of repose	cave
rockslide	tower karst

Debris Avalanches Produced by Hurricane Camille

When conditions become highly favorable for their formation, mass movements sometimes occur in large numbers within a restricted area, resulting in considerable destruction and loss of life. A major example of such an occurrence was the formation of several hundred debris avalanches in western Virginia as a result of torrential rains associated with Hurricane Camille in 1969.

THE STORM

The storm that was to become Hurricane Camille first appeared on August 5, 1969, as a weak tropical disturbance in the eastern Atlantic off the West African coast. After drifting westward for several days, the disturbance began to intensify rapidly on August 14 in the western Caribbean Sea. Camille, as the storm was named, attained hurricane strength early on August 15, and later that day crossed the western tip of Cuba with 185 kilometer per hour (115 mph) winds. Emerging into the warm waters of the Gulf of Mexico, the storm continued its rapid intensification, with wind speeds approaching a phenomenal 320 kilometers per hour (200 mph) by the evening of August 16. Camille moved ashore in the vicinity of Gulfport, Mississippi, on the evening of August 17, accompanied by winds of at least 300 kilometers per hour (185 mph) and a 7-meter (23 ft) storm surge. The combination of winds and the storm

surge resulted in some 171 deaths in Mississippi and Louisiana, but timely warnings and a massive evacuation are credited with having saved as many as 100,000 lives.

By the evening of August 19, as the weakened remnants of the storm were crossing the central Appalachians, its rainfalls briefly intensified to unprecedented levels, producing widespread flooding and debris avalanches in several Appalachian counties of West Virginia and Virginia.

The heaviest rainfall was centered in Nelson County, Virginia (Figure 15.31). It is a rural county of about 10,000 inhabitants situated in the west-central part of the state on the eastern flanks of the Blue Ridge Mountains. Excessively heavy rain began falling in Nelson County about 9:30 P.M. and continued until about 3:30 A.M.—a period of six hours. The National Weather Service later conducted bucket surveys (field surveys of water levels in containers known to be empty preceding the storm) and at one point confirmed a storm rainfall total of 80 centimeters (31.5 in). Unofficial indications existed that in a few sites the storm total may have exceeded 100 centimeters (40 in). These figures are far larger for a six-hour time span than any officially recorded rainfall totals anywhere in the United States.

The cause of the sudden and rather localized increase in storm rainfall intensities is apparently related to a combination of three circumstances. First, the entire region was covered by an unusually moist maritime tropical air mass. Second, the center of the storm passed just south of Nelson County, producing an easterly wind flow. Because the county is on the easternmost flanks of the Appalachians, a significant orographic lifting factor was produced. Finally, an area of pre-frontal thunderstorms from an

approaching cold front reached the area at the same time as the remnants of Hurricane Camille.

THE DEBRIS AVALANCHES

Even before the onset of the storm, soils were very moist from one of the wettest summers on record. As the torrential rainfalls caused the saturated soils to lose coherency, small roadbank slumps began by 10 P.M. and the first debris avalanches by about 11 P.M. By 1 A.M., debris avalanches were occurring in large numbers, and new ones continued to form until about 4 A.M.

The sites on which the debris avalanches occurred were the mountainside hollows that serve as conduits for subsurface drainage. The highly porous soils on these slopes normally absorb water as rapidly as it is received and direct it as groundwater into the hollow systems. The water eventually surfaces at the foot of the slopes to form small streams.

The debris avalanches apparently were initiated by the coalescence of extraordinarily large amounts of subsurface water into the hollow systems. This water may have literally lifted the soil in each hollow from the underlying bedrock, allowing the entire detached mass to slide and flow down the hollow and into the stream valley below.

Witnesses reported that the debris avalanches occurred in groups. The noise and vibration from one avalanche were apparently sufficient to trigger several others nearby in a chain reaction sequence. In all, more than 275 occurred, including many with multiple branches totaling more than 1,000 meters (3,300 ft) in length. The debris avalanches left narrow, elongated scars on the hillsides, while the debris—a chaotic jumble of soil, rocks, giant boulders,

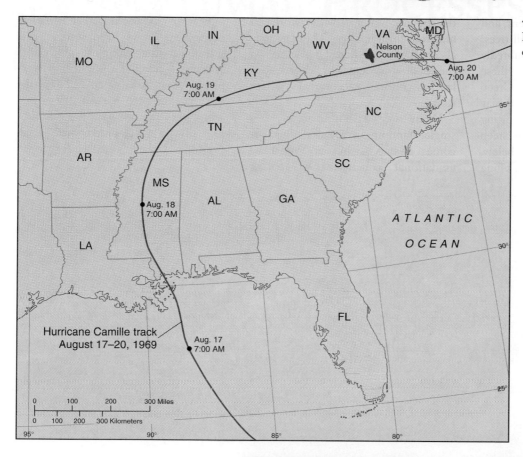

and trees—was deposited in the adjacent flat stream valleys (Figure 15.32). Many houses in the affected area were swept away by either the debris avalanches or the flooded streams, and approximately 100 people died in Nelson County alone.

In the years following the storm, the debris avalanches have been healing rapidly. The debris itself was quickly cleared, with government assistance, from the agriculturally productive valleys. The hillside scars remain, but are much less conspicuous than they once were because they are being recovered slowly by soil that has been washing and creeping onto them from upslope. Vegetation in this humid temperate setting returns

continued on next page

◆ **FIGURE 15.32** Debris from the three debris avalanche scars on the hillslope has been deposited in the foreground. This photo was taken in Nelson County, Virginia, a few months after the area was devastated by torrential rains from the remnants of Hurricane Camille.

SOURCE: Donald Poole

The Grand Canyon, featured in the chapter-opening photo, is considered one of the natural wonders of the world. The canyon is more than 1.6 kilometers (1 mi) deep in places, is 349 kilometers (217 mi) long from the Little Colorado River to Lake Mead, and ranges from 6 to 29 kilometers (4 to 18 mi) in width. Peering over the canyon's rim, in many places one can catch a glimpse of a narrow ribbon of water—the Colorado River—far below. It seems almost inconceivable that this stream, which is dwarfed by the canyon it occupies, could have been responsible for the excavation of the entire Grand Canyon, and yet this is precisely how the canyon was formed. In fact, the Earth's surface features are more the work of streams than of any other geomorphic agent. In Chapter 9, discussion centered on the hydrologic aspects of streams, especially on their widespread distribution and the role they play in the hydrologic cycle. In this chapter, we examine the physical characteristics of streams and stream systems, with particular emphasis on their role as shapers of landforms. The term **fluvial** (from the Latin *fluvius*, "river"), which commonly is used to identify stream-related processes and features, will appear frequently during the discussion.

GEOMORPHIC IMPORTANCE OF STREAMS

Stream systems, and the geomorphic work they accomplish, are powered jointly by insolation and gravity. Solar energy causes surface water to evaporate, chiefly from the oceans, and drives the global wind systems that transport the water vapor to upland regions, where it condenses and precipitates. Gravity then returns a portion of this water, as runoff, to the sea. Because of their rapid rates of flow, streams are a dynamic component of the physical environment and can undergo major changes over relatively short periods of time. As they flow, streams perform two vital natural functions. First, they remove excess water from the land. The water transported by streams to the ocean comprises about one-fifth of all the precipitation that falls on the Earth's land surface. By flowing in response to gravity, streams gain enough kinetic energy to perform their second function. This is the erosion, downhill transport, and eventual deposition of weathering products supplied to the streams from adjacent slopes largely through mass wasting processes. The removal and redistribution of these materials by streams produce fluvial landforms.

Streams have a profound impact on human activities. They have served vital functions as transportation routes and irrigation sources since the dawn of recorded history. They have formed the fertile agricultural floodplains on which we raise many of our food crops. From a broader perspective, fluvial processes acting over lengthy geologic time spans are responsible for most of the lowland surfaces on which the majority of our planet's human inhabitants reside. Two examples of human interactions with stream systems are provided in the Case Study at the end of the chapter and the Focus Box on Rio Grande boundary changes.

Satellite imagery indicates that fluvial processes are the chief shapers of the landforms on perhaps three-fourths of the Earth's land surface (Figure 16.1). This geomorphic dominance results from the nearly global distribution as well as the great gradational power of stream systems. Only the 10 percent of the land surface covered by continental glaciers does not contain at least intermittently active streams. Certain other nonglaciated regions contain some fluvial features, but are dominated by landforms produced by other geomorphic processes. Chief among these are the coastal margins (discussed in Chapter 19), some tectonically active areas (Chapter 14), and areas covered by shifting sands (Chapter 18).

The geomorphic effect of flowing water, though important nearly everywhere, displays significant spatial variations in its nature and intensity (Table 16.1). These variations are

◆ **TABLE 16.1**   **Rates of Regional Erosion in the United States***

| DRAINAGE REGION | DRAINAGE AREA (1,000 KM²)[†] | RUNOFF (1,000 M³/S) | LOAD (T/KM²/YR) | | | EROSION, (CM/1,000 YR) | AREA SAMPLED (%) |
			DISSOLVED	*SOLID*	*TOTAL*		
Colorado	629	0.6	23	417	440	17	56
Pacific Slopes, California	303	2.3	36	209	245	9	44
Western Gulf	829	1.6	41	101	142	5	9
Mississippi	3,238	17.5	39	94	133	5	99
South Atlantic and Eastern Gulf	736	9.2	61	48	109	4	19
North Atlantic	383	5.9	57	69	126	5	10
Columbia	679	9.8	57	44	101	4	39
Total/average	6,797	46.9	43	119	162	6	

*After Sheldon Judson and D. F. Ritter, "Rates of Regional Denudation in the United States" *Journal of Geophysical Research* 69 (1964): 3399. †Great Basin, Saint Lawrence, and Hudson Bay drainage not considered.

◆ **FIGURE 16.1** A fluvially-dominated landscape in subarctic eastern Russia, about 320 kilometers (200 mi) southeast of Yakutsk in the Yakutskya Republic. The rivers, appearing in white, were frozen when the photograph was taken by a Landsat satellite on February 1, 1989. The reddish brown areas are mostly unvegetated. The dark green areas are covered with northern coniferous forest.

SOURCE: EOSAT

related chiefly to climatic and vegetative patterns and are influenced strongly by local soil and bedrock characteristics. One might expect fluvial effects on the landscape to be least important in arid regions and to increase directly with precipitation averages, making the wettest regions the areas with the best-developed and fastest forming fluvial features.

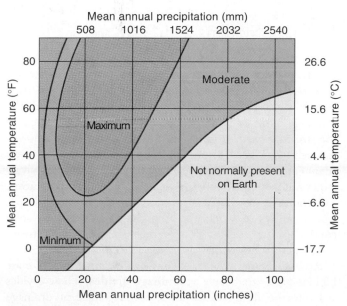

◆ **FIGURE 16.2** Relationship between climate and the effectiveness of fluvial geomorphic processes.

Figure 16.2 indicates that this is not the case. Flowing water actually is most effective as a landscape modifier in semiarid and subhumid regions. Very dry regions are minimally influenced because of a lack of water, but humid regions with abundant precipitation have only a moderate rate of development of fluvial features due in large part to the protective influence of vegetation, which generally increases in density with increased precipitation. The climatic zone of maximum fluvial effectiveness receives enough rainfall to produce considerable storm runoff, but is dry enough for the year as a whole to support only a limited cover of protective vegetation.

Although the topography of most of the Earth's land surface results primarily from fluvial processes, streams at any given time occupy and actively modify only a very small proportion of the total surface area. All stream action and energy are concentrated in a small area, producing features that stand in marked contrast to the remainder of the surface. The differential operation of fluvial processes therefore produces well-developed features, while their uniform operation would not. The majority of the surface in most areas consists of **interfluves,** or areas between streams.

The subsequent material is divided into two sections. We first analyze the causes and characteristics of stream flow—how and why streams develop, the mechanics of fluvial erosion and deposition, and the forms assumed by individual streams and the drainage systems of which they are a part. Second, we examine the major landforms produced by fluvial erosional and depositional processes.

CAUSES AND CHARACTERISTICS OF STREAM FLOW

Streams are formed by water that reaches the Earth's land surface and cannot be absorbed immediately. The surface in each locality has a maximum rate at which it can absorb water; this is known as its **infiltration capacity.** It is related primarily to the texture, degree of compaction, and existing water content of the surface materials. If precipitation exceeds the infiltration capacity, water begins to collect on the surface. Depressions are filled first, and then, assuming that some slope exists, the excess water begins to flow downhill. In the absence of existing channels, the initial flow of water is disorganized and spreads over the surface in a nonchannelized form known as **overland flow.** It may flow as a sheet of water on a smooth surface or, more commonly, as uneven threads around obstacles like plants and stones. Because this flow is not concentrated, it is very shallow and friction with the underlying surface is great. As a result, even on relatively steep slopes, its flow is slow, and its capacity for eroding and transporting sediments or debris of any kind is highly limited. Overland flow dominates near the crests of interfluves, producing zones of no erosion there.

Proceeding downhill, the flow increases in volume as the total upslope area being drained becomes progressively larger. **Channelization,** the beginning of a stream, occurs where the water depth and speed enable the flow to overcome the cohesion of the soil particles. On unprotected slopes, such as those cleared for agriculture, a nearly parallel series of small shoestring channels called **rills** forms first (Figure 16.3). Farther downhill, the rills may grow through coalescence into steep-sided **gullies** (Figure 16.4). The continued deepening of a gully may eventually enable it to

◆ **FIGURE 16.3** Newly developed rills leading into small gullies on agricultural land in Tennessee.
SOURCE: Tim McCabe/Soil Conservation Service

reach the water table and to begin tapping groundwater supplies. At this point the gully has been converted to a perennial stream channel and constitutes an extension of the stream system it joins.

Rills and small gullies normally do not form on undisturbed, well-vegetated slopes because the soil is highly absorbent and is held firmly in place by the vegetation mat. Under these conditions, water from even heavy storms typically is absorbed as fast as it is received. This water travels downslope as throughflow (Chapter 9), until it surfaces as a spring or seeps out along the side of a stream bank at a point where the surface intersects the water table. Deeper groundwater flow travels much more slowly than either surface water flow or throughflow. Thus, in a humid climate, a stream supplied by groundwater is likely to maintain a stable base flow even during dry periods.

Drainage Systems and Patterns

Streams are not separate, independent entities, but instead are organized into **drainage systems** consisting of numerous interconnected stream segments. These stream segments act collectively to remove excess water and sediments from the **drainage basin,** which consists of the total area drained by the stream system. In form and function, a drainage system is much like a tree. It normally consists of a master or trunk stream at its base that branches upstream into a progressively larger number of increasingly small segments, or **tributaries.** Tributaries extend into all portions of the drainage basin. Like branches of a tree, the branches of a drainage system conduct water and other materials (in this case, stream sediments) from one part of the system to another.

Around the peripheries of drainage basins of any size are highland rims that serve as **drainage divides.** These divides separate the flow and direct it toward adjacent drainage systems. In some areas, the divides are low and indistinct, while in others they form lofty mountain systems. The

◆ **FIGURE 16.4** Major gully development in the upper basin of the Rio Reventado in Costa Rica. Gullies result from the removal of the protective cover of dense natural vegetation in a region subject to heavy downpours.
SOURCE: H. H. Waldron/USGS

largest and most important drainage divide in the United States is the Continental Divide, which separates runoff flowing east into the Mississippi River system and ultimately the Gulf of Mexico from that flowing west through the Colorado and Columbia River systems to the Pacific (Figure 16.5).

◆ FIGURE 16.5 The North American Continental Divide separates water draining into the Pacific from that draining into the Atlantic, Arctic Ocean, and Gulf of Mexico.

◆ **FIGURE 16.6** The Horton-Strahler stream ordering system. A stream of a given numerical order is formed by the juncture of two streams of the next lower order.

Stream Orders

The size of a stream, a basic and important parameter from both a hydrologic and a geomorphic standpoint, can be described in several ways. Figures such as length, width, volume of water discharged, and areas drained all can be used as indices of stream size. A general idea of the regional significance of a stream, though, can be conveyed in comparatively simple fashion by the concept of **stream order.** The stream ordering system was first proposed by the American geomorphologist Robert Horton in the 1940s. In applying the system, streams are ranked from 1st order to approximately 12th order based on the number and nature of their tributaries—factors that correlate closely with the size and regional significance of the stream (Figure 16.6). By definition, a 1st order stream is a small stream with no tributaries. It is analogous to the outermost twigs on a tree. In humid climates, 1st order streams generally are fed directly by subsurface water outflow from adjacent interfluves. A 2nd order stream is formed when two 1st order streams join. Subsequent orders follow the same pattern, increasing by whole number units only when two streams of the next lower order join. Within a given drainage system, as the order becomes higher, the total number of stream segments of that order becomes smaller, but the physical dimensions of the streams increase. The numerical value of the highest order depends, of course, on the size of the stream system and on factors such as slope steepness, vegetation density, and rock type that influence surface erosional resistance and infiltration capacity. The Mississippi River, the largest in North America, is a 10th order stream, while the world's largest river, the Amazon, is a 12th order stream.

Horton and other researchers have discovered statistical relationships between stream order and a host of other stream and drainage basin characteristics, including mean stream length, width, discharge, depth, size of drainage basin, number of tributaries, and slope gradient (Figure 16.7). Of these, all except the slope gradient increase in value with an increase in stream order; within a given drainage system, all these factors can usually be related numerically to one another with a reasonable degree of accuracy. The relationships among these characteristics, and the differences they display between regions, provide another indication that drainage systems develop in response to natural controls.

Drainage Patterns

The drainage pattern is the surface pattern collectively formed by the streams in a drainage system. All the stream systems in a given region typically exhibit the same general pattern. The existing pattern is largely a response to differences in rock types and structures as well as to preexisting surface features. It therefore provides an indication of the region's geologic and geomorphic history as well as its present characteristics. A large number of drainage pattern types exist, and, as might be expected, some regions exhibit patterns that are poorly defined, are intermediate between two types, or are complex because of multiple influences. The types described here and illustrated in Figure 16.8 are among the most widespread.

The **dendritic pattern** is the most common drainage pattern type. It develops on gently sloping surfaces that are homogeneous in their resistance to erosion and therefore exert no significant structural control on stream location. It

◆ **FIGURE 16.7** Quantitative relationships between stream order and stream frequency, length, gradient, and size of drainage basin. The left graph indicates that the number of streams of progressively higher stream orders within three representative drainage basins decrease at relatively constant geometric rates. The right graph indicates that as stream order increases, average stream length and drainage basin area increase, but that mean stream gradient declines.

SOURCE: Stanley A. Schumm, "The Evolution of Drainage Systems and Slopes in Badlands at Perth Amboy, New Jersey," *Bulletin of the Geological Society of America*, vol. 67, 1956, pp. 603–604.

Ridges of resistant rock

a

b

c

d

◆ **FIGURE 16.8** Diagrammatic depictions of four major drainage pattern types: (a) dendritic pattern, (b) trellis pattern, (c) radial pattern, (d) deranged pattern.

is especially likely to occur in areas of horizontally bedded sedimentary rocks and glacial deposits. The dendritic pattern is treelike in form, with a large degree of variability in the orientation of tributaries, and with stream junctures normally forming angles well under 90° (see also Figure 16.1).

The **trellis pattern** typically develops in regions of parallel ridges and valleys where resistant and nonresistant sedimentary strata alternate. It consists of a parallel pattern of major streams that occupy valleys eroded into the weaker rocks, while much smaller and shorter tributaries flow down the steep ridge flanks to join the major streams at nearly right angles. In North America, trellis drainage patterns are well developed in the Ridge and Valley portion of the central Appalachians and in the Ouachita Mountains of Arkansas and Oklahoma.

The **rectangular pattern** usually is produced when the drainage pattern is controlled by intersecting fault or fracture systems in areas of crystalline rocks such as granite. Streams alternate between unusually straight segments, where they flow along one fault or fracture line, and sudden right-angled bends, where they are redirected onto another intersecting line. The rectangular pattern is similar to the trellis pattern in its angularity, but different in that it has no single predominant direction of flow. A rectangular pattern is well developed in portions of the Adirondack Mountains of upstate New York.

The **radial pattern** is perhaps the simplest of the basic drainage pattern types. It consists of the outward flow of streams in all directions from a central peak or upland. It is

especially well developed in the vicinity of large, isolated volcanic peaks such as those in the northern Cascades of Oregon and Washington. Conversely, a converging flow of streams toward a central basin produces a **centripetal pattern.** Centripetal patterns are typical of both karst regions and arid regions with interior drainage; consequently, they are widespread.

A more complex variety of the radial pattern, termed the **annular pattern,** develops when an uplifted region is surrounded by inclined bands of sedimentary rock strata that vary in their resistance to erosion. This structural pattern produces an alternation of divergent stream segments and of radial segments that follow the outcrops of the less resistant strata. An annular drainage pattern often develops around large igneous masses, such as the Black Hills of South Dakota and Wyoming (Figure 14.17), that have been intruded into sedimentary strata.

The **deranged pattern** displays a great degree of spatial disorganization. Streams are highly irregular in their directions of flow, sometimes exhibiting sharp, V-shaped bends. In addition, they do not flow smoothly downhill, but pass alternately through depressions occupied by lakes and swamps, and steep segments marked by rapids and waterfalls. The existence of a deranged pattern indicates that the area drained is of geologically recent origin and that an organized drainage pattern has not yet had time to develop. It is common in glacially scoured regions such as the

Canadian Shield and Scandinavia, where the previous drainage systems were erased by the ice.

Stream Erosion, Transportation, and Deposition

It is through the ability of streams to erode, transport, and deposit sediments that fluvial landforms develop. Because the characteristics of fluvial features are determined by the nature of the activities that produce them, we now examine the processes by which stream erosion, transportation, and deposition occur.

Erosional Processes

Stream erosion results from solution and from the friction and shear forces generated between the flowing water (including any sediments being transported by this flow) and the nonmoving materials with which it comes into contact. If the stresses brought to bear by this action exceed the cohesive strength of the materials, they will be eroded and transported downstream. Approximately 95 percent of the gravitational energy that causes stream flow is dissipated through friction between the water molecules of the stream and is converted to heat. The critical remaining 5 percent is used for sediment erosion and transport. The great erosional power of streams indicates that this is still a large amount of energy.

The erosive energy of a given stream is related to its velocity and volume of flow and to the amount of friction generated with its **banks** (sides) and **bed** (bottom). An increase in any of these factors will increase the erosive power of the stream. At the same time, a change in any of these three factors will affect the other two, since they are all interrelated. A fourth factor that must be taken into account when assessing a stream's ability to erode is the erosional resistance of the surface on which it flows. If the stream banks and bed consist of solid rock, erosion is normally very slow. If the stream flows within accumulations of unconsolidated materials, however, it can effect very rapid erosional modifications.

The speed at which the water flows and the amount of friction it encounters are both critical factors in determining the **turbulence** of the flow. Turbulence refers to the tumbling eddy motions of small parcels of water within the stream that are superimposed on the main downhill direction of the current. This phenomenon has already been encountered as it influences wind-flow characteristics (Chapter 5). Just as friction with the Earth's surface produces wind turbulence, which causes the wind to be gusty and to have the capacity to erode and transport sediments, so too the friction of a stream with its banks and bed causes the stream to flow in a turbulent manner and gives it the ability to erode and transport sediments with a higher specific gravity than water. Water flows at a much slower average

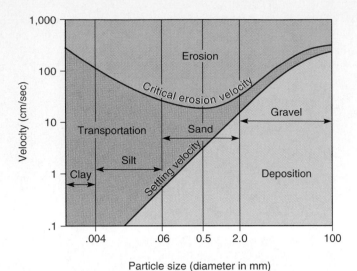

◆ **FIGURE 16.9** Stream sediment erosion, transportation, and deposition regimes as related to particle size and water flow velocity.

speed than the wind, but it is a much denser medium and has correspondingly greater erosional and transportational capabilities. An increase in either streambed roughness or speed of flow will increase turbulence and give the stream a greater erosional ability. The stream's speed of flow is a critical factor; it has been estimated that the erosive power of a stream increases as approximately the cube of its increase in velocity. This means that if the speed of flow of a stream doubles, its capacity to erode sediments increases by 2^3, or eight times.

In order for stream erosion to occur, the water must flow rapidly enough that the friction of the moving water, coupled with the lift produced by eddies within the flow, can dislodge the materials over which the water passes. The minimum stream velocity needed to erode a given streambed particle is called its **critical erosion velocity;** this is illustrated graphically in Figure 16.9. This velocity has been found to be lowest for sand-sized particles with diameters of about 0.5 millimeters (0.2 in). Smaller and lighter silt and clay particles have higher critical erosion velocities because of their cohesiveness and because they collectively produce such a smooth streambed that turbulence is reduced. Particles larger than 0.5 millimeters, on the other hand, require greater flow speeds for erosion because of their weight. As the settling velocity line in Figure 16.9 indicates, the coarser the texture of the stream-borne particles, the higher the flow speeds needed to continue to transport them. Note the large gap between the critical erosion velocity line and the settling velocity line on the left side of the graph. This indicates that small streambed particles are difficult to erode, but that, once eroded, they can continue to be transported at very low flow speeds. Conversely, high-velocity flow is required for both erosion and transport of coarse materials. Most streams do not maintain a high flow velocity for very long; as a result, most transported stream sediments consist of sand-sized or finer particles.

During times of high flow volumes, streams have the capacity to erode and transport a much greater sediment load than at times of normal flow. This occurs for several reasons: the quantity of water available to erode and transport sediments is greater; the "wetted perimeter," (the total surface area covered by the stream) increases, giving the stream access to more sediments; and, most important, the speed of flow increases. As a result, more stream erosion and fluvial landscape modification generally occur during floods than during the much longer periods of normal or low-flow conditions.

The actual process of sediment erosion by streams is accomplished in three different ways. **Hydraulic action** consists of the direct sweeping away of loose materials by the friction and turbulence generated by the moving water itself. Although very effective on unconsolidated materials, hydraulic action has a negligible influence on solid rock. **Abrasion** involves the scraping of particles carried in the water against the stationary materials of the banks and bed. This process can slowly erode even solid rock if the stream-borne particles are harder than the rock they are scraping. Abrasion also occurs within the flow itself as the transported sediments are repeatedly scraped or knocked together by the turbulence of the flow. As a result of this process, the sediments are gradually rounded and reduced in size. **Corrosion** or solution is the dissolving of rock materials so that they enter the flow in a molecular state. On the average, about 30 percent of the total sediment load of streams is transported in solution.

The volume of water flowing past a given point on a stream will typically display considerable variation on a daily, and sometimes even hourly, basis. As a result, the stream's ability to erode and transport sediment is constantly changing. This can produce rapid fluctuations in the size and shape of the channel, as the stream struggles to maintain an equilibrium among the interacting variables of discharge, sediment load, and channel width, depth, and slope.

Stream velocity, turbulence, and resulting erosional capacity can also vary over short distances. Erosion therefore does not occur simultaneously everywhere along the course of a stream, but is concentrated at certain points at any given time. These points are located where active currents are in contact with bed and bank materials, especially if these materials are weak or unconsolidated. Such short-term variations in the channel depth and location can create constant problems for agencies responsible for the maintenance of navigable waterways.

Over much longer time spans, most streams erode downward vertically, while simultaneously eroding laterally. The relative proportion of erosion that takes place in each of these directions is critical to the nature of the resulting topography, since a predominance of vertical erosion will produce a high relief surface containing deep, steep-sided valleys while a predominance of lateral erosion will result in the formation of broad, gently sloping valleys and interfluves. In general, vertical erosion is dominant where the

◆ **FIGURE 16.10** Methods of stream sediment transport. The arrows in the velocity profile at the right are proportional to flow velocity, indicating that water flows fastest just below the surface of a stream and most slowly along its bed.

stream gradient is steep, and lateral erosion dominates where it is gentle, although other factors such as volume of flow, channel form, and surface erosional resistance also are important. A gentle gradient reduces the stream's velocity, thereby lessening the rate of downcutting sufficiently that valley-widening processes such as sheet flow, tributary inflow, and mass wasting can keep up with deepening so as to maintain a gentle valley side-slope.

Transportation

The same water turbulence that lifts sediments from the streambed keeps them from settling back to the bed, allowing the current to transport them downstream. As illustrated in Figure 16.9, the critical erosion velocity is always greater than the settling velocity for a given particle. As a consequence, as long as the stream velocity and turbulence that eroded the sediments are maintained, stream sediments will continue to be transported downstream without being deposited.

Streams transport sediments in a number of ways, depending upon the sizes and physical state of the sediments as well as the speed and turbulence of the flow (Figure 16.10). A substantial proportion of the total sediment load is carried in solution. The **dissolved load** tends to be especially large in relation to the total load in areas of readily soluble rock such as limestone, in areas where the velocity of flow and resulting turbulence are low, and in humid areas where vegetation is abundant and the water is mildly acidic. Most dissolved materials are transported through the entire stream system.

More than half the total sediment load collectively transported by streams travels as **suspended load.** Suspended materials consist of generally fine-textured particles that are carried within the flow itself. They are kept from settling by the upward component of the turbulent eddies generated by friction. Streams flowing through areas in which unconsolidated materials are plentiful carry the largest volumes of suspended sediments. The greatest quantities of such materials are picked up and transported during floods (Figure 16.11).

◆ FIGURE 16.11 The relationship between stream discharge (volume of flow) and suspended sediment load in Brandywine Creek near Wilmington, Delaware. Each point represents a separate measurement. A modest increase in discharge produces a large increase in suspended load.

The coarsest particles that a stream is capable of moving are carried as **bed load.** As the term implies, these sediments are moved along the streambed because they are too heavy for turbulence to transport them for any great distance within the flow. Depending on their size and shape, they may travel by rolling, sliding, or bouncing (**saltation**) along the bed. The bed load, unlike the suspended and dissolved loads, moves intermittently. It typically is transported a short distance downstream, temporarily deposited, and later picked up and carried further. Many bed load particles are transported only during floods, when the carrying capacity of the stream is temporarily greatly augmented. This accounts for the often observed presence of giant flood-transported boulders on the bed of a small stream that seems utterly unable to move them. As it travels, the bed load is gradually rounded and reduced in size by abrasion.

The form of a stream channel is controlled largely by the quantity and coarseness of the sediment load it transports (Figure 16.12). Streams that carry a limited quantity of predominantly fine-textured sediments tend to be deep and narrow, to have a gentle downhill gradient, and to display a high degree of sinuosity (i.e., to have numerous curves or meanders). If the sediment load is large in quantity and coarse in texture, the stream will be broad and shallow, steeper in slope, less sinuous in form, and likely will develop a braided flow pattern (see Figure 16.18 on page 429).

These characteristics of channel form bring about an adjustment between the quantity and texture of the sediment load and the stream's ability to transport that load. The basic mechanics of the adjustment process are relatively simple. If the sediment-carrying capacity of a stream exceeds its load, the excess energy is used to erode the stream channel vertically. This, in turn, lowers the elevation of the stream and reduces its downhill gradient, reducing its speed of flow, turbulence, and sediment-carrying capacity. In the absence of any disrupting factors, downcutting will continue until the carrying capacity of the stream is reduced to the point that it becomes just sufficient to transport its sediment load. No further significant downcutting will then occur, because there is no longer a supply of excess stream energy.

By the same token, if the stream's sediment supply exceeds its carrying capacity, the excess sediment will be deposited. This will **aggrade** (build up) the channel, increasing the stream's downhill gradient, which in turn will increase its flow speed, turbulence, and sediment-carrying capacity. Deposition and aggradation will continue until the ability of the stream to transport sediment has been in-

◆ FIGURE 16.12
Diagrammatic depictions of the channels of three streams with the same cross-sectional area, but with markedly different shapes. The stream on the left is likely to have the steepest gradient and to carry the coarsest sediment load. The middle stream is likely to have the most gentle slope and the finest sediment load.

	Broad, shallow channel	Narrow, deep channel	Semicircular channel
Cross-sectional area	10 m²	10 m²	10 m²
Perimeter in contact with water	12 m	12 m	7.9 m

◆ **FIGURE 16.13** Measured stream longitudinal profiles. Both streams have concave profiles, although the Arkansas River is a much longer river with a gentler average gradient. The vertical to horizontal scale ratio is approximately 275:1.

creased sufficiently that it can carry all of its load. As we have seen, short-term variations in volume of flow and sediment load can cause a stream to rapidly switch back and forth between periods of erosion and aggradation at any given point along its course.

The extent to which a given stream has adjusted its downslope gradient to achieve an overall balance between its sediment-carrying capacity and sediment load can be estimated roughly by examining its **longitudinal profile.** This is a graphic depiction of the downhill gradient of the stream from its head to its mouth (Figure 16.13). A stream that has not achieved balance between flow and load typically has an irregular longitudinal profile (Figure 16.14a). Such a profile is likely to contain **knickpoints,** or points at which slope steepness increases abruptly. Most knickpoints occur where the stream passes over a resistant bedrock outcrop. They are typically the sites of rapids or of waterfalls.

On the other hand, a stream that has become **graded**—that is, has attained a balance between flow and load—ideally displays a smoothly concave longitudinal profile on a bed of uniform resistance (Figure 16.14b). The development of a graded profile means that the stream flows along a gradually declining gradient from head to mouth. Streams

(a)

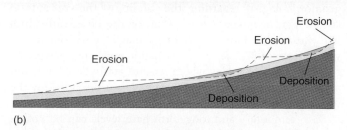

(b)

◆ **FIGURE 16.14** The ungraded stream depicted in diagram (a) has irregularities called knickpoints in its longitudinal profile. The graded stream profile shown in diagram (b) is smoothly concave.

are able to continue to transport their sediment loads at progressively reduced slope angles because their increased volumes of flow result in a gradual reduction in the amount of friction per unit volume of water. This often causes the stream's speed of flow to increase for some distance in a downstream direction, despite its gradual reduction in slope. In addition, although the volume of stream sediments progressively increases downstream, their size gradually decreases because of the effects of abrasion, making it easier for the stream to transport them. As an illustration of the influence of discharge on speed of flow, the lower Amazon River, with an extremely gentle gradient of only 4.75 centimeters per kilometer (3 in/mi), flows at nearly 2.4 meters per second (8 mph) during times of high water. This is faster than many mountain streams with much steeper gradients.

Deposition

Streams deposit sediments when a reduction in their speed, turbulence, and/or volume of flow makes them incapable of transporting all of their sediment load. As their carrying capacity declines, they first deposit the coarsest materials, which have the greatest settling velocities, and then deposit progressively finer sediments. The finest sediments of all, including the stream's dissolved load, generally are transported completely through the stream system to the water body into which the stream empties. This association between stream-transporting ability and the size of the particles being deposited causes fluvial deposits to be **sorted,** or separated by size. A high degree of sorting, in fact, is one of the most important and easily recognizable characteristics of fluvial deposits (Figure 16.15).

The accumulation of thick deposits of sediments is common in the lower reaches of stream systems, especially at the mouths of large rivers that empty into the sea. Deposition also occurs locally and temporarily at numerous points along the courses of nearly all streams. Sites favoring deposition include shallow stretches along the stream course such as the insides of bends, places where the channel widens, and around the bases of obstructions to flow such as boulders lying on the streambed. Human actions can greatly influence the pattern of stream deposition, with dams and reservoirs

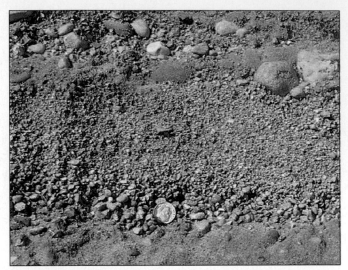

◆ **FIGURE 16.15** Close-up view of rounded and sorted fluvial deposits on a stream bank near Lafayette, Indiana. The dime provides scale.

SOURCE: Bruce F. Molnia, Terra Photographers

acting as sediment traps. The effect of channel modifications on the depositional pattern of the Mississippi River is discussed in the Case Study at the end of the chapter.

Accumulations of fine-textured stream-deposited sediments are called **alluvium.** The chief source of the alluvial deposits of most streams is soil eroded from areas upstream. Thick accumulations of alluvium cover the lower valleys of most large river systems, and their generally high content of soluble plant nutrients often makes them agriculturally productive.

FLUVIAL LANDFORMS

Streams are not the most powerful of the natural forces shaping the Earth's surface features. That distinction, among the major groups of forces, goes to the tectonic and glacial processes. The restricted size of the areas affected at any given time by tectonic activity and glaciation, however, reduces their global significance, so that streams are the leading geomorphic agents in terms of their overall influence on the Earth's surface topography. In many parts of the world, the three processes of weathering, mass wasting, and fluvial erosion act to erode the surface and produce most gradational landforms. Weathering acts first, producing unconsolidated surface materials; mass movements transport these materials to stream valleys; and streams then transport them, eventually, to the sea. During the last stage of the sequence, a variety of fluvial erosional and depositional features is likely to be formed.

In a direct sense, the dominant fluvial feature is the **valley.** A valley is a linear lowland produced by stream erosion. Most streams, of course, produce valleys, making them common and familiar surface features. The dimensions of a

valley are determined largely by the size (or, more precisely, the volume of flow) of the stream that produced it, the rate of stream erosion, and the length of time over which the valley has developed. Valleys contain a variety of both erosional and depositional features, as we will see shortly.

In a very general sense, fluvial landforms can be divided into two groups: those formed primarily by erosion and those formed primarily by deposition. Classifying many fluvial features as either erosional or depositional is actually rather arbitrary, however, because they are composite in nature. In other words, they are produced by both erosional and depositional processes operating in different places or at different times. Nonetheless, these two groups of processes are used here as a convenient way of organizing the discussion of fluvial landforms.

Fluvial Erosional Features

Streams erode when their volume, speed, and turbulence of flow give them the ability to remove and transport a greater quantity of sediment than they already contain. Excess erosional capacity is most common in the upper reaches of drainage systems. In these areas the downhill gradient is normally considerably steeper than it is farther downstream, the sediment load is usually much smaller in volume, and significant tectonic uplift is more likely to have occurred in the recent geologic past. The two major categories of features produced by fluvial erosion are valleys and erosional hills and mountains.

Erosional Valleys

Nearly all river valleys are produced largely or entirely by fluvial erosion, but the term **erosional valley** as used here refers to a valley still being actively deepened by a river that is downcutting toward base level. **Base level** is the level or elevation at any point along a stream's course that provides the stream with a gradient just sufficient for it to transport its sediment load past that point with no erosion or deposition. It is, in other words, the level of flow/load equilibrium for the stream. Any erosion or deposition can be viewed as a response by the stream to a changing or as yet unattained base level. If base level is achieved all along its course, the entire stream system is considered to be graded.

A great deal of controversy has occurred among fluvial geomorphologists regarding the validity of the concepts of stream grade and base level. One of the chief difficulties with these related ideas is that they may be envisioined over very different time scales. Two major opposing viewpoints regarding the process by which streams attain base level, and fluvial landscape evolution in general, are discussed in the Focus Box beginning on page 435.

Both temporary and long-term base levels can be postulated. A temporary base level may be achieved quickly through the reworking of stream sediments. During floods, as we have seen, a stream gains excess erosional capacity and rapidly erodes sediments lying within its channel. The base level in this case has been temporarily lowered by the stream's increased sediment-carrying capacity, and the stream therefore reduces its gradient in order to restore a balance between flow and load. When the flood subsides, the stream is located at too gentle a gradient to transport its sediment load with the now reduced volume of water it contains. It therefore deposits enough sediment to aggrade the channel back to its original level. Most streams are continuously eroding or depositing at various points along their courses in order to adjust to shifting temporary base levels.

On a long-term basis, an ungraded stream may have to erode through great thicknesses of rock over lengthy geologic time spans in order to reach a stable base level. A classic example is the Colorado River, which, over many millions of years, has incised itself deeply into the surface of the Colorado Plateau of northern Arizona to produce the Grand Canyon (see the chapter-opening photo). The river in this area apparently has not yet achieved its base level elevation, in part because tectonic uplift is still taking place, and active downcutting continues. Superimposed on this long-term erosional trend, though, the Colorado River locally erodes and deposits sediments at various points along its course in response to short-term variations in its volume of flow and sediment load.

If a stream is downcutting rapidly and the near-surface rock material is strong enough to maintain steep slopes, the erosional valley will be deep, steep walled, and relatively narrow (Figure 16.16). The width of such a valley is restricted because valley-widening processes depend largely upon the slow progress of weathering to weaken valley-side materials so that they may be removed by mass wasting. Valley deepening, on the other hand, can be effectively accomplished by stream particle abrasion on unweathered rock and by enhanced chemical weathering on the perennially moist valley floor materials. Even the deepest and most steep-walled stream valleys rarely have depths that exceed their widths. If it is exceptionally deep and narrow, a valley is referred to as a **gorge.** A steep-walled but broader and often flat-bottomed valley is called a **canyon.**

The side-slope characteristics of stream valleys and especially of rapidly expanding erosional valleys depend largely on the nature of the materials of which they are composed. If the materials are homogeneous, a smooth side slope is produced. If the materials consist of alternating layers of varying resistance to erosion, a highly irregular slope may be produced, as in the case of the Grand Canyon. This slope normally will consist of protruding ledges and cliffs, formed on the resistant strata, and inset slopes of intermediate steepness on the weaker strata.

Erosional valleys exist mostly in upland areas such as plateaus or in hilly or mountainous regions. Most large river systems have their headwaters in such regions and eventually enter plains farther downstream. Erosional valleys are especially steep, narrow, and spectacular in arid or semiarid

◆ **FIGURE 16.16** Devil's Canyon in Wyoming has been produced by active fluvial downcutting in an area of horizontally bedded sedimentary rocks.
Source: JLM Visuals

regions. Here they are less masked by vegetation, and much less moisture is available for chemical weathering to facilitate the valley-widening processes.

Erosional Hills and Mountains

Erosional hills and mountains are the highland areas that remain on the landscape following the fluvial dissection of a plateau, fault block, or other raised surface. We have in previous chapters given the credit for the widespread distribution of such features to the tectonic forces that produced the initial surface uplift, but perhaps equal credit should be given to the gradational forces that have carved these raised blocks into their present intricate patterns of hills or mountains and intervening valleys.

The pattern of hills or mountains within a region is largely determined by the type of drainage pattern that develops. The drainage pattern, in turn, is controlled by the nature of the uplift and by the rock types and structures within the area. If, for example, a plateau is formed by the broad uplift of a mass of homogeneous or horizontally bedded rock, a dendritic drainage pattern will develop. The randomness in the location of the stream segments in this pattern will cause the intervening hills or mountains to display a random pattern also. The Appalachian and Ozark Plateaus are both good examples of rugged but disorganized upland surfaces formed by the development of dendritic drainage patterns in regions of uplifted horizontal sedimentary rocks (Figure 16.17).

In places where folding or faulting has exposed strata that differ in their resistance to erosion, streams develop linear drainage patterns such as the trellis pattern as they preferentially carve valleys in the weaker rock layers. This action will produce a parallel system of resistant linear ridges and

◆ **FIGURE 16.17** Aerial view of maturely dissected topography displaying dendritic drainage on the Appalachian Plateau of central West Virginia.
SOURCE: USGS

alternating valleys such as that found in the Appalachian Ridge and Valley region and in portions of the Coast Ranges of California.

Fluvial Depositional Features

Deposition occurs on the largest scale in the lower portions of drainage systems, where slopes are normally gentle, the streams and their sediment supplies are large in volume, and a stable long-term base level is more likely to have been achieved than farther upstream.

The dominant landform resulting from fluvial deposition is the floodplain. Floodplains, in turn, contain a number of important secondary features such as meander deposits, natural levees, deltas, and alluvial terraces.

Floodplains are broad, flat-floored valleys covered by alluvium and subject to flooding at times of high water. Floodplains comprise the floors of valleys originally formed by vertical erosion and are normally in the process of being slowly widened by lateral erosional processes. In one sense, then, the floodplain, like any stream valley, can be considered more a product of erosion than of deposition. On the other hand, the stream occupying a floodplain typically has reached its long-term base level and has essentially ceased downcutting. Furthermore, nearly all floodplains are covered with stream-deposited alluvium, and the small-scale features

within the floodplain are alluvial and depositional in origin. A floodplain therefore is a depositional feature located within a larger erosional feature. Because the erosional processes that result in valley formation already have been discussed, the emphasis in this section is on the depositional aspects of floodplains.

One of the most important characteristics of a floodplain, from both a geomorphic and a human standpoint, is the fact that it is subject to occasional flooding. Streams flowing within floodplains are normally not incised into the surface, but instead flow close to the level of the surrounding land. During times when they carry large amounts of runoff, these streams can readily overtop their banks. As a result, overbank floods occur in a typical floodplain section about two years out of three. The relative flatness of the surface usually causes these floods to be extensive but rather shallow. Large quantities of sediments are transported by a stream during floods and are deposited in a layer over the surface, especially as the water recedes. Such deposits add valuable topsoil eroded from upstream portions of the drainage basin. The short-term economic hardships of floods for floodplain residents must therefore be weighed against the long-term benefits of naturally maintained soil fertility.

Extreme floods, which fortunately are quite rare on any given stream, may cover the floodplain so deeply and with such swiftly flowing water that erosion rather than deposition predominates. These floods help to maintain a long-term surface level equilibrium because they offset the slow buildup of sediments from "normal" floods by periodically flushing out the excess.

Many floodplains are covered by only a relatively thin veneer of alluvium, and streams retain the capability of temporarily exposing part or all of the underlying bedrock floor during extreme floods. Other floodplains, however, currently are covered so deeply by alluvium that bedrock exposure is impossible. In some cases, this has occurred because the floodplains are located in slowly subsiding tectonic basins. This is true, for instance, of the Mississippi, Amazon, and Ganges River Valleys. The slow subsidence of their valleys has caused these rivers to deposit huge quantities of sediment in order to maintain a graded state. For example, the Mississippi River Valley near Vicksburg, Mississippi, contains an average of more than 45 meters (150 ft) of sediment, and in places sediment is twice as thick. Pleistocene sea level changes and human activities that have accelerated erosion upstream also have increased deposition in the floodplains of many rivers.

Because a stream occupying a floodplain flows near the surface in unconsolidated alluvium, it can shift its course much more readily than can a stream occupying a narrow erosional valley. Over long periods of time, then, the stream is likely to flow over all portions of its floodplain in turn. On occasion, it will impinge on portions of the sidewalls of the floodplain, undercutting the bank and widening the floodplain (see Figure 16.23). Fluvial features therefore are not restricted to the immediate vicinity of the present channel, but are distributed widely over the floodplain surface. Most of the modest surface relief found within floodplains is associated with these features.

When a stream flows through a region of unconsolidated and easily eroded materials like sand, the weakness of the banks causes the stream to become wide and shallow. Under such conditions, an excess of sediment will likely enter the stream, choking the channel and dividing it into a shifting series of deeper and shallower channel segments. The deepest and most actively flowing segments will consist of a number of interweaving channels separated by elongated **bars** of sand or gravel that are deposited at places where the speed of flow diminishes. During times of low water, these bars may protrude well above the water. Because they are composed of loose sediments and are located within the stream channel, they are highly vulnerable to erosion and are subject to constant and rapid fluctuations in size, shape, and position. A stream that exhibits a pattern of multiple interwoven channels separated by intervening bars is said to have a **braided flow pattern** (Figure 16.18).

Meanders and Associated Features

The development and migration of broad looping bends, or **meanders,** in the courses of streams occupying floodplains are responsible for many floodplain features (Figure 16.19). Stream meanders display a distinctive cycle of formation, growth, and disappearance (Figure 16.21). A meander begins as a slight bend in the channel, perhaps initiated by differing degrees of cohesiveness of the materials on the opposing stream banks or by the inflow of water from a tributary. The diversion of flow results in the erosion of the unconsolidated material on the outside of the bend, allowing it to become more pronounced. As the bend increases in amplitude, centrifugal force becomes the primary factor furthering its growth by deflecting the main current toward the outer bank, where erosion occurs, producing an **undercut bank** (Figure 16.20a). As it grows outward, the meander slowly migrates down-valley.

The reduction in the speed and depth of flow on the inner side of the developing meander causes the deposition of sediments, which form a series of **point bars** (Figure 16.20b). Centrifugal force, acting on the current as it passes through the meander, causes the sides of the meander to converge gradually to form a narrowing **meander neck.** Eventually, the sides intersect to produce a **meander cutoff.** When this happens, the meander is abandoned by the active current as the river takes the shorter and steeper route through the cutoff (see Figure 1.2).

The abandoned meander eventually is separated from the stream by bank deposits along its new course so that it becomes an arcuate lake called an **oxbow lake** (Figure 16.21).[1] Lakes are relatively ephemeral features on the landscape, and an oxbow lake gradually fills with fine sediments and organic matter to form first a swamp and eventually dry land. The meander scar may long be visible on

◆ **FIGURE 16.18** A semiarid climate and plentiful sand supply have caused Nebraska's Platte River to develop a braided flow pattern.
SOURCE: John S. Shelton

the landscape, especially from the air, because its vegetation usually differs from that of its surroundings (Figure 16.22).

Over long periods of time, the development and migration of meanders allows the stream to flow over all parts of its floodplain in turn (Figure 16.23). Many floodplains have well-defined margins in the form of steep **bluffs** that rise abruptly above the flat valley floor.

The shifting flow patterns of meandering rivers occupying floodplains can have important political consequences, particularly for rivers whose courses have been designated as

---

[1]An oxbow, which gives this lake type its name, is the U-shaped part of an ox yoke that passes under and around the animal's neck and helps attach it to a cart or plow.

◆ **FIGURE 16.19** Aerial view of a meandering river flowing within a floodplain. A number of oxbow lakes can be seen.
SOURCE: John S. Shelton

◆ **FIGURE 16.20** (a) The fastest water flow speeds occur on the outsides of stream bends, while the slowest flow speeds occur on the insides of the bends. This results in the erosional steepening of the outer banks and the deposition of point bars on the inner banks. (b) Point bars occur on the insides of the meanders on this small stream.

SOURCE: Courtesy of J.S. Monroe

◆ **FIGURE 16.21** Stages in the development of a stream meander. As centrifugal force deflects water flow toward the outside of the meander and causes the meander to grow outward, its neck is gradually constricted (diagram b). This leads to the eventual cutoff of the meander (diagram c) and the formation of an oxbow lake (diagram d).

◆ **FIGURE 16.22** A variety of meander scars, as well as ridge and swale topography produced by the Tallahatchie River in Mississippi. Dark green areas are densely vegetated swamps occupying abandoned meanders.
SOURCE: USGS

river flowing within a broad floodplain attempts to take an entirely new course through another portion of the floodplain. The current attempt of the Mississippi River to change its course and the actions taken to maintain the river in its current channel are explored in the Case Study at the end of the chapter.

Natural Levees

When a river within a floodplain spills out of its channel during a flood, the water outside the main channel is subjected to a sudden increase in friction due to the shallowness of the water and the irregular surface provided by submerged vegetation; its speed of flow consequently is diminished greatly. This reduction in flow reduces the water's sediment-transporting capacity and results in the deposition of the excess sediment on the floodplain surface. The coarsest materials, which have the largest settling velocities, sometimes are deposited atop the stream banks where they accumulate to form natural levees.

Natural levees take the form of low, often broad ridges of fine sand and coarse silt that parallel each side of the stream channel (Figure 16.24). They are sometimes the highest natural features within the entire floodplain and, as such, increase channel stability by helping to keep the stream within its established channel. Natural levees along the lower Mississippi River, for example, reach heights of up to

international boundaries. Boundary disputes between the United States and Mexico that arose from changes in the course of the Rio Grande are explored in the Focus Box on page 432. Even more serious consequences can ensue if a

◆ **FIGURE 16.23** Diagrammatic depiction of major floodplain features discussed in the text.

INTERNATIONAL BOUNDARY CHANGES ALONG THE RIO GRANDE

Rivers have long served an important political function as boundaries between nations. The tendency of some rivers to shift their courses, however, can cause major diplomatic problems for the countries involved. A good example is provided by the Rio Grande, which forms the international boundary between the United States and Mexico over a distance of roughly 1,600 kilometers (1,000 mi).

Soon after the 1848 Treaty of Guadalupe Hidalgo established the middle course of the Rio Grande as the international boundary, it became evident that the river's course occasionally experienced major changes during floods. This raised a sticky political problem: Should these changes cause the international boundary to shift, or should the original boundary be maintained, leaving increasingly numerous enclaves of each country on the "wrong" side of the river?

Most changes in the Rio Grande's course during the nineteenth and early twentieth centuries were more a nuisance than a major problem, because the lands involved were barren and unpopulated. In 1905, Mexico and the United States signed the Banco Treaty in which they agreed to exchange approximately equal parcels of land from time to time in an attempt to offset shifts in the river's course as much as possible.

A more serious situation had developed, however, in the densely populated area where the Rio Grande separated El Paso, Texas, from the neighboring Mexican city of Ciudad Júarez. Southward shifts of the river occurred in 1852 and again in 1864, placing tracts of Mexican land on the American side of the river. Beginning in the late 1800s, Mexico repeatedly requested that the lands involved be returned. The U.S. government's position at the time was that, since the

shifts in the river's course resulted from natural processes of erosion and deposition, the lands involved now belonged to the United States.

In 1910, the United States finally agreed to settle the problem by arbitration, and a joint Mexican-American commission was appointed to decide the matter. The commission eventually decided that Mexican land lost by the 1852 shift should remain as part of the United States, but that the larger "El Camizal" tract lost in 1864 should be returned to Mexico. The United States refused to accept this finding for a long time, although it had originally agreed to abide by the commission's decision. The transfer of the land finally took place in 1963 as the result of an act signed by U.S. president John Kennedy and Mexican president Lopez Mateos. ■

5 meters (16 ft) above the main floodplain surface. If the river is aggrading its channel, the natural levees may enable the water surface to stand above the level of the surrounding floodplain. This is an unstable situation, because a breach in any portion of the levee will allow the river to abandon its channel and form a new channel at a lower level in another portion of the floodplain. Humans, who usually have a vested interest in keeping rivers within their existing channels, may reinforce natural levees in an attempt to avoid such breaches, which usually occur during floods (see the Case Study at the end of the chapter). In places the water level of China's Huang Ho River flows more than 10 meters (33 ft) above the level of its surrounding floodplain because it has long been kept from shifting its course by a combination of natural and artificial levees.

The tendency of rivers in floodplains to meander and to form natural levees, then, leads to the development of a variety

of relatively short-lived features on the floodplain surface. In general, these features are most recently formed and abundant near the current river courses. Their presence gives the relatively flat floodplain floor a complex microrelief that shows up well on topographic maps employing small contour intervals. They are also often clearly visible on aerial photographs because of the vegetative patterns they produce (Figure 16.22).

Deltas

Deltas are deposits of alluvium formed when streams enter standing water bodies. They exist on a wide variety of scales, from that of a small brook entering a pond to that of a giant river such as the Amazon or Mississippi entering the sea. Deposition takes place because the sudden reduction of flow that occurs as the stream enters the standing water body causes it to lose its sediment-carrying capacity. This causes

◆ FIGURE 16.24 Relationship of natural levees to a hypothetical stream occupying a floodplain. The coarsest sediments are deposited in the stream channel, where the stream's velocity is greatest. Deposits of intermediate texture (predominantly sand and silt) are deposited during floods on the levee surface as the stream overtops its banks. The finest sediments (consisting mostly of clays) are deposited on the floodplain surface well away from the stream.

all but dissolved sediments and clay-sized particles to settle to the bottom.

The outward spread of sediments that produces the fanlike shape of a typical delta is the result of active deposition and a lack of confining channel banks. Deposition causes the stream to aggrade itself repeatedly out of its channel in order to shift to a lower route. This process eventually causes a fan-shaped deposit to be built up. Frequently, the stream splits into several **distributary channels** that divide the river into a number of smaller channel segments.

In general, deltas are proportional in size to the sizes and sediment loads of the streams that produce them. All of the world's major deltas are produced by large rivers that transport heavy sediment loads to the sea. Some large rivers, though, carry little sediment and have not formed deltas. For example, Canada's St. Lawrence River cannot obtain the necessary sediment load for delta formation because the Great Lakes, which it drains, act as settling basins. In contrast, many of the large rivers of Asia carry heavy sediment loads and have developed extensive deltas. The Ganges delta, with a surface area of 78,000 square kilometers (30,000 sq mi), is the world's largest.

Delta formation is also greatly facilitated if the standing water body that the river enters is shallow and has weak wave and current action. If the water is very deep, great quantities of sediment must accumulate before the deposits reach the surface to form a delta. This is one reason why few major deltas exist along the west coasts of North and South America. Similarly, if wave and current action are strong, stream deposits will be carried away or spread out along the coastline rather than accumulating in one place to produce a delta. For this reason, most of the best-developed deltas are formed in bays, partially enclosed seas, or other protected waters.

The term delta was taken from the Greek letter delta (Δ), which originally was used to describe the shape of the "classic" delta of the Nile River (Figure 16.25b). Other deltas, like those of the Ganges or Niger Rivers, are more fanlike in shape (Figure 16.25c). A few deltas, most notably that of the Mississippi River, contain long fingerlike projec-

tions of land. These projections result from a combination of very weak currents that are unable to smooth off the outer perimeter of the delta and a slow subsidence of the surface that leaves only the crests of natural levees along the major delta channels above the water level. This form has been termed a **bird's-foot delta** (Figure 16.25a). Finally, a number of deltas are located in semi-enclosed water bodies (**estuaries**) either in drowned river valleys or in submerged tectonic troughs. Examples of **estuarine deltas** include the deltas of the Yangtze, Susquehanna, Colorado, and Seine Rivers (Figure 16.25d).

River deltas form seaward extensions of existing floodplains. Like floodplains, delta surfaces are low, flat, and often swampy and are composed of fine-textured sediments that normally contain a large proportion of organic matter. When adequately drained, they may form highly productive agricultural land and support dense populations. Unfortunately, coastal deltas may be rapidly eroded or inundated by the sea as a result of human activities that reduce their influx of alluvial materials. The current loss of land in the Mississippi delta is discussed in the Case Study at the end of the chapter.

Terraces

As noted earlier, the formation of a floodplain indicates that a stream has reached a stable base level and that downcutting is no longer significant. If the base level is lowered for any reason, a stream occupying a floodplain will renew downcutting and may eventually carve a new and narrower floodplain within the old floodplain surface (Figure 16.26). The now elevated remnants of the original floodplain form **terraces** that usually end abruptly in steep slopes well back from the current river channel. In most cases, a pair of terraces will be present, one on each side of the valley. The height of the terraces depends upon the extent to which base level has been lowered. The width of the new floodplain, and therefore the distance that separates the opposing terraces, is related to the amount of time that has elapsed since the river eroded to its new base level. If the new level is maintained long enough, lateral erosion eventually will widen the new

(a)

(b)

(c) Niger delta

(d) Seine delta

◆ **FIGURE 16.25** Maps of four deltas with differing forms. Delta forms are controlled by such factors as the rate of sediment deposition, the bottom topography of the water body in which the delta is being formed, and the patterns of coastal currents. The Nile and Niger Rivers have produced "classic" deltas, while the Mississippi has produced a bird's-foot delta and the Seine an estuarine delta.

◆ **FIGURE 16.26** Fluvial terraces have formed along the margins of the Inn River Valley, in Austria, as tectonic uplift periodically caused the river to downcut into the floor of its floodplain.

SOURCE: Jerry Scott

CHANGES IN FLUVIAL LANDSCAPES THROUGH TIME

Streams and their surrounding landscapes can undergo profound changes through time. For instance, the geological structure of the Earth's shield areas and of the Piedmont of the eastern United States indicates that these regions once contained lofty mountain ranges. Today, however, they exhibit a rather subdued topography and in places contain broad, gently sloping river valleys.

The idea that the topographic features of a region could be used to determine its geomorphic age was first suggested by the famous American geographer William Morris Davis in the late nineteenth century. Davis saw an analogy between the progressive changes in the fluvial landscape with time and the changing physical features of human beings with increasing age. He developed an elaborate fluvial "cycle of erosion" that classified streams as being in youth, maturity, or old age on the basis of their flow characteristics and the character of the surrounding landscape (Figure 16.27). The validity of this cycle has been hotly debated ever since its initial publication.

DAVIS'S INITIAL ASSUMPTION

In order to trace the fluvial cycle of erosion from the earliest possible point, Davis had to begin with a surface that was totally unaffected by stream action. He therefore envisioned an irregular seafloor being uplifted rapidly by tectonic forces to a considerable elevation. Erosion of this surface by flowing water, of course, begins with the first rainfall. Overland flow occurs initially, but very quickly a drainage network starts to form.

YOUTH

During youth, the surface still retains many characteristics of its previous seafloor setting. Early in the youthful period, streams are highly ungraded. Surface irregularities cause the stream courses to contain lakes or swamps, rapids, and even waterfalls. Stream gradients, although very irregular, are on the average quite steep. This gives the streams the ability to downcut at a relatively rapid rate.

By middle youth, the major streams in the drainage network have formed deep, narrow, and steep-sided gorges (Figure 16.26a). Lakes and waterfalls gradually disappear, and rapids become less numerous as the stream courses are gradually smoothed by erosion and deposition. The entire uplifted surface increases in ruggedness and relief as gorges are incised into it by the vigorously flowing streams. Late in the youthful period, the major streams approach their long-term base level of erosion, and the rate of downcutting tapers off considerably. This allows the walls of the gorges to be worn back to less steep angles by lateral erosional forces.

MATURITY

Davis considered a stream mature when it reaches long-term base level and becomes graded. This achievement marks a major change in the geomorphic activity of the stream. Until this time it has had excess energy and has been involved primarily in downward incision. From maturity onward, it remains essentially stable in elevation, deposits as well as erodes sediments, and is involved chiefly in valley widening rather than deepening.

With the stabilization of valley bottom levels, floodplain formation begins. At this point, the relief of the surrounding landscape is at a maximum. Nearly the entire surface is in slope (Figure 16.26b), and the last remnants of the original upland surface are being consumed by erosion. As the mature stage progresses, the river floodplain gradually widens, and various floodplain features such as meanders and natural levees begin to develop. The widening floodplain reduces the areal extent of the remaining upland surface as its elevation and ruggedness gradually diminish.

OLD AGE

By the time it has reached old age, the river flows sluggishly through a broad floodplain that may be covered by a considerable depth of alluvium. A variety of floodplain features develop and are erased in constantly changing patterns. The combination of continual weathering, fluvial erosion, and other gradational processes has converted the surrounding region to a **peneplain,** or a plain formed by the long-continued erosion of an upland surface (Figure 16.26c). A few isolated highland remnants, called **monadnocks,** may survive as the last vestiges of the former uplands, but in time even they

continued on next page

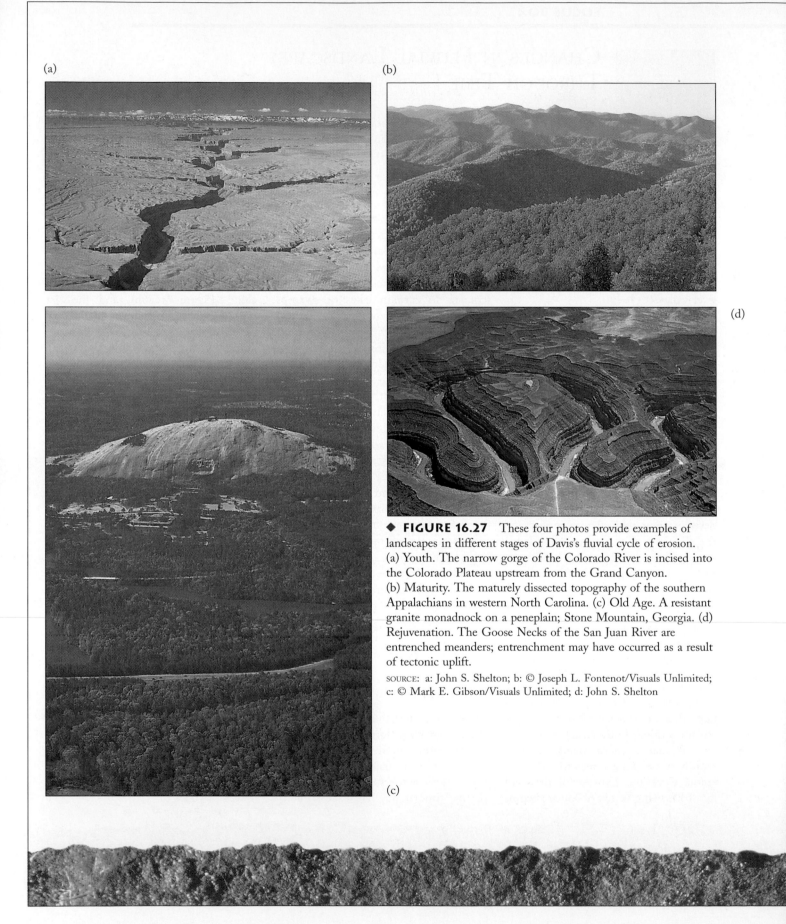

(a)

(b)

(d)

◆ **FIGURE 16.27** These four photos provide examples of landscapes in different stages of Davis's fluvial cycle of erosion. (a) Youth. The narrow gorge of the Colorado River is incised into the Colorado Plateau upstream from the Grand Canyon. (b) Maturity. The maturely dissected topography of the southern Appalachians in western North Carolina. (c) Old Age. A resistant granite monadnock on a peneplain; Stone Mountain, Georgia. (d) Rejuvenation. The Goose Necks of the San Juan River are entrenched meanders; entrenchment may have occurred as a result of tectonic uplift.

SOURCE: a: John S. Shelton; b: © Joseph L. Fontenot/Visuals Unlimited; c: © Mark E. Gibson/Visuals Unlimited; d: John S. Shelton

(c)

are destroyed by the inexorable progress of erosion. This leaves an essentially timeless landscape characterized by a low surface relief and slow-flowing, freely meandering streams.

REJUVENATION

Unlike a human being, a fluvial landscape can remain in old age indefinitely. On the other hand, old age or any of the earlier developmental stages can at any time be interrupted by surface uplift or a fall in sea level. Such an event will result in a reduction in base level and a **rejuvenation** of the stream system. The increased elevations will again initiate a period of rapid downcutting, resulting in the return of the stream system to a youthful condition. If the change in elevation is of limited extent, the chief topographic result will be the formation of terraces. If a major uplift occurs, the streams, assuming they maintain their original courses, will incise themselves well below their former surface level. If this occurs to a meandering stream, the increase in river flow speed will result in the development of **entrenched meanders** (Figure 16.26d). Rejuvenation will cause the

stream system and the surrounding landscape to pass through the life cycle a second time, and any later uplifts will produce additional rejuvenations of the system.

VALIDITY OF THE CYCLE OF EROSION

As already noted, the accuracy and even the basic validity of Davis's cycle of erosion have long been debated among geomorphologists. Probably the chief point in favor of the concept stems from the simple fact that streams transport sediments in only one direction—downhill. In the absence of uplift, therefore, the landscape *must* eventually be worn down. From what we know of fluvial erosion processes, Davis's basic erosional sequence of initial vertical incision to a base level elevation followed by the lateral expansion of the surface at this elevation seems logical. Landscapes apparently representing all the stages of Davis's cycle have been identified.

Although there have been many criticisms of the cycle of erosion concept, the most fundamental is the lengthy geological time span that would be necessary for the cycle to be completed. Most geomorphologists

believe that the Earth is far too active tectonically to allow the cycle to run its entire course. They therefore maintain that the concept is of only theoretical applicability. Many geomorphologists also contend that streams can quickly adjust their courses by means of changes in their gradient or flow pattern. This factor allows streams to remain in **dynamic equilibrium** with respect to changes in their environment.

It is likely that both the cyclical and the dynamic equilibrium concepts of fluvial landscape development possess validity, but over differing magnitudes of geological time. More research is needed to improve our understanding of the entire process of long-term landscape evolution. Regardless of the final verdict on the merits of Davis's fluvial cycle, it has played a key role in our understanding of geomorphic processes by stressing that streams and their surrounding landscapes do not display a random assortment of features. Rather, landforms are organized into a logical system that progresses toward an ultimate state of equilibrium with existing environmental conditions. ■

level sufficiently to destroy the terraces. In some cases, multiple terrace levels exist within a valley. This indicates that the base level is being sporadically lowered, with intervening periods of stability.

The primary cause of terrace formation is an increase in the gradient of the stream course. This can be accomplished by tectonic or isostatic uplift of the surface or by a decline in sea level. River terraces are common in areas undergoing tectonic uplift, including portions of California and the Alps. Another important cause of terrace formation is a reduction in the sediment load supplied to a stream. During the waning stages of the Pleistocene epoch, for example, streams draining the melting ice sheets of North America

and Eurasia received tremendous influxes of glacial sediments. Because they were unable to transport all of this load, they deposited much of it in their valleys to form thick alluvial fills. When the glaciers disappeared, the sediment supply was greatly reduced, and the rivers actively eroded into their alluvial deposits, leaving the remnants standing as terraces well above the present valley floors.

River terraces have become favored settlement sites because of their relatively flat surfaces, fertile alluvial soils, and proximity to water. In addition, their high-standing position makes them immune to the floods that periodically inundate the lower floodplain.

∎ SUMMARY

Streams perform two vital natural functions. The first is the removal of excess water from the land. The second is the removal of surface weathering products supplied to streams largely from adjacent slopes by mass wasting processes. Because of the gradational power and the nearly global distribution of streams, most of the Earth's land surface is dominated by fluvial landform features.

We now return to the Focus Questions:

1. *How and why do fluvial drainage systems develop?*

Streams develop in places where there is a sufficient downhill flow of surface water to erode a channel. Each stream channel conducts water from a drainage basin, and the coalescence of stream channels forms a drainage system. Drainage systems display spatial patterns of organization dictated largely by existing patterns of rock types and structures as well as surface topographic features.

2. *How do streams erode, and what are the most important landforms produced by stream erosion?*

Streams erode by the processes of solution, hydraulic action, and abrasion. All rock materials are soluble to some degree, and on average about 30 percent of the sediment load of streams is carried in solution. Hydraulic action resulting from stresses generated by the force of the flowing water itself is highly effective in eroding the unconsolidated materials that form the banks and beds of most streams. Abrasion, the scraping action of stream-borne particles, can slowly erode solid rock. Once sediments have been entrained into the flow of a stream, they continue to be transported as long as the turbulence generated by the flow keeps them from settling to the streambed.

The primary fluvial erosional feature is the valley. A stream with excess erosive energy will downcut rapidly, producing a deep, narrow valley. Downcutting, however, reduces the stream gradient and therefore the gravitational energy available to the stream. This eventually allows valley-widening processes to become dominant in tectonically inactive areas. The fluvial erosion of uplifted blocks of land also carves them into hills and mountains. The topographic pattern displayed by these upland features depends largely upon the drainage pattern that develops, while their relief is determined by the rate of downcutting.

3. *Why do streams deposit sediments, and what are the most important landforms produced by stream deposition?*

A stream deposits sediments when a reduction in its speed or volume of flow, or a change in the geometry of its bed, reduces the turbulence of the flowing water to the point that it can no longer transport all of its sediment load. When this occurs, the coarsest sediments are deposited first, followed by progressively finer sediments as the stream's carrying capacity diminishes. As a result, fluvial sediments are sorted, or separated by size. The basic fluvial depositional feature is the floodplain, which is a broad, flat-floored valley covered by alluvium and subject to occasional flooding. Floodplains exhibit a variety of depositional features formed largely by lateral stream migrations. These include bars, natural levees, and various meander features such as oxbow lakes and meander scars. When a stream deposits sediments in a standing water body, a delta often forms. Deltas vary greatly in size and shape, but most display a fanlike pattern caused by shifts in the stream's course through time as well as the formation of a branching pattern of distributary channels.

When a stream occupying a floodplain renews downcutting, it incises itself into the floor of the floodplain. The remnants of the original floodplain often are preserved along the sides of the valley as terraces. Some streams have formed multiple terraces at varying levels above the present valley floor.

∎ QUESTIONS FOR THOUGHT AND REVIEW

1. Why are fluvial processes more effective landform producers, for the Earth as a whole, than are glacial or eolian processes? What parts of the world are, and are not, dominated by fluvial landforms?
2. Briefly describe the likely general appearance of the surrounding landscapes of areas that have each of the following drainage pattern types: radial, trellis, deranged, dendritic, annular.
3. How do streams erode loose sediments? How do streams erode solid bedrock?
4. Describe each of the three methods by which streams transport sediments. Which moves the most sediments worldwide? Which moves the least?
5. What causes streams to deposit sediments? Describe the physical characteristics of stream deposits, and explain why these characteristics exist.
6. What is a graded stream? How can the stream's longitudinal profile indicate whether it is graded?
7. Define the concept of base level. What is the difference between short-term and long-term base level? Explain how the concepts of base level and grade are related.
8. Explain the correlation between the topographic relief of an area and the relative rates of vertical and lateral stream erosion.
9. What is a floodplain? How does it differ from an erosional valley? Describe several important floodplain features.
10. How do stream meanders form? Describe the meander development cycle that eventually results in the production of an oxbow lake.
11. Explain why deltas form. Why are bird's-foot deltas like that of the Mississippi River relatively rare?
12. What does the development of a series of alluvial terraces within a valley indicate about the recent geomorphic history of the valley? Explain.

13. Discuss the advantages and disadvantages of living in a river floodplain.
14. Discuss the differences between the "cycle of erosion" and the "dynamic equilibrium" theories of stream development.
15. Why does the lower Mississippi River currently have a tendency to change its course? What would be the chief adverse economic consequences of such a change?

▌Key Terms

fluvial	abrasion
interfluve	suspended load
infiltration capacity	bed load
drainage system	meanders
drainage basin	graded stream
stream order	base level
dendritic pattern	floodplain
trellis pattern	braided flow pattern
rectangular pattern	point bars
radial pattern	oxbow lake
deranged pattern	natural levee
critical erosion velocity	delta
hydraulic action	terrace

Will the Mississippi Change Its Course?

Humans have been modifying the flow of rivers for thousands of years in an effort to increase their usefulness and reduce the danger of flooding. Most early alterations involved small-scale diversions of flow for irrigation, but modern technology has greatly expanded the magnitude and variety of alterations that can be undertaken. A few examples of modern alterations of rivers include channel straightening and the construction of dams, reservoirs, locks, and levees.

Because all the components of the environment are complexly interrelated, purposeful human alterations occasionally produce unfavorable and unforeseen consequences. Our technological abilities sometimes seem to surpass our understanding of the more subtle interactions among the Earth's physical systems. The history of control efforts of the Mississippi River provides a good example of how desirable human adjustments to the flow of a major river have produced a highly unstable situation that offers the prospect of eventual disaster for much of southern Louisiana.

The Mississippi River system, one of the world's largest and most important, drains 41 percent of the United States as well as a portion of Canada. Its drainage basin is exceeded in size only by those of the Amazon, Nile, and Congo Rivers. The heavily farmed Mississippi Basin is the agricultural heartland of the nation, and the river system itself provides the means for transporting a tremendous volume of agricultural, industrial, and petrochemical products.

The task of developing and controlling the Mississippi River has been assigned largely to the U.S. Army Corps of Engineers. The two major responsibilities of the Corps have been to reduce the danger of destructive floods and to develop the system's navigation potential. On the whole, they have been very effective in accomplishing these goals. The Corps maintains the river throughout its length by means of a complex system of locks, dams, diversion channels, and artificial levees. Channels have been straightened and shortened in various places, and some channels have even been paved to increase their stability. The environmental consequences of all these activities have been modest in the upper portion of the drainage basin, but are far more profound in the Lower Valley, where deposition predominates.

The Lower Mississippi is an alluvial river. It transports 58 million kilograms (64,000 tons) of sediment per day. Some of this material is deposited within the river channel, but most is carried into the Gulf of Mexico. Accelerated erosion from upstream agricultural activities has augmented the sediment load so that it currently is twice as large as it was 60 years ago. In the past, the Mississippi repeatedly changed its course due to channel aggradation, and the deposits thus were spread over a large area to form an extensive delta. Because of the artificial stabilization of its channel, however, widespread sediment distribution can no longer occur.

The major problem resulting from channel stabilization is that the present channel of the Mississippi has aggraded itself well above the level of the surrounding floodplain. This has caused the river to display an increasing tendency to change its course to a more westerly route that is lower in elevation and shorter in its path to the Gulf. Continuing alluviation within the present channel, which is occurring at a substantial rate because of the large influx of sediment from upstream agricultural sources, is building the river ever higher. This continues to make the situation progressively more unstable and necessitates the constant heightening and strengthening of the levees along the river's current channel.

The site at which the Mississippi River would abandon its present course if it had its own way is known. It is located at the juncture of the Mississippi with the Red and Atchafalaya Rivers near the southwestern corner of the state of Mississippi about 80 kilometers (50 mi) northwest of Baton Rouge (Figure 16.28). The Atchafalaya is a distributary river that carries excess water from the Mississippi southward to the Gulf along a route located well to the west of the course of the Mississippi River itself. In 1839, a logjam on the upper Atchafalaya was cleared in order to open this river to navigation. The resulting increase in flow caused the Atchafalaya to become gradually deeper and wider and slowly to divert more and more water from the Mississippi. During this century, the process accelerated, and in 1953 the Corps of Engineers notified Congress that, unless stopped, the Mississippi's flow would be completely diverted into the Atchafalaya by 1990. The consequences of such a diversion would be enormous. Baton Rouge, New Orleans, and many other cities and towns along a more than 480-kilometer (300 mi) stretch of the Mississippi would be left high and dry, while towns along the Atchafalaya—

◆ **FIGURE 16.28** Map showing previous channels (in red) and the present channel of the Mississippi River. The location of the Old River Control Structure (ORCS) is indicated, as well as the possible new route of the river down the Atchafalaya.

notably Morgan City—would be flooded out of existence.

To prevent such an occurrence, Congress hurriedly authorized the Old River Control Project. This project had as its goal the permanent diversion of 30 percent of the flow of water from the Mississippi into the Atchafalaya through a system of 84 floodgates. The project was completed in 1963 and seemed at first to be highly successful. In 1973, however, a major flood caused the collapse of part of the structure and nearly destroyed it. At present, such a state of imbalance exists between the two channels that the undermining of the

Old River Control Structure (ORCS)by scour has become a very real danger. In 1986, a backup structure, the Old River Control Auxiliary Structure, was completed. It was sited at a sediment-rich bend of the Mississippi River and was designed to capture a larger percentage of the Mississippi's bed load sediments than did the ORCS. These two systems together, referred to as the Old River Control Complex, are monitored daily, and extensive channel surveys are taken three or four times a year to ascertain that no new scour channels are forming. The construction of the Auxiliary Structure, as well as the

completion of a hydroelectric power plant just upstream from the ORCS in 1990, appear to have considerably slowed the rate of increase in the elevational imbalance between the two rivers. Even if a future major flood does not destroy the Old River Control Complex and produce a permanent shift in the Mississippi's course, floodwaters would be diverted down the Atchafalaya in order to save Baton Rouge and New Orleans. This places the future of rapidly growing Morgan City in grave doubt.

Another major problem has been created by the tight control of the

continued on next page

◆ **FIGURE 16.29** The lower portion of the Louisiana delta is gradually disappearing due to subsidence and a lack of sedimentation in the various distributary channels of the Mississippi River. The sediment plumes produced in the Gulf of Mexico by the outflow of Mississippi River water are clearly evident.

SOURCE: Science VU/Visuals Unlimited

course of the Mississippi River. In the past, an extensive network of distributary rivers and swamps spread vast quantities of sediments over a broad area of southeastern Louisiana. These deposits offset the slow tectonic subsidence of the area and allowed for the maintenance and slow growth of the Mississippi delta. With the flow now tightly confined to a 1.6-kilometer (1 mi) wide channel, most sediments are currently shot out into the Gulf of Mexico. The deposition that does occur takes place only

within the existing channel where it is not wanted. The lack of sediment is causing the remainder of the delta to slowly subside into the Gulf (Figure 16.29). Each year, about 80 square kilometers (30 mi^2) of land are lost. The intrusion of salt water into other portions of the gradually submerging delta is killing vegetation and freshwater fish and is polluting domestic drinking water supplies.

Nature's price for channel stability thus has been the threat of a catastrophic shift in the course of the Mis-

sissippi River and the progressive loss of large amounts of valuable land from the lower Mississippi delta. These events provide eloquent testimony that environmental manipulations performed without adequate knowledge of possible long-term consequences may come back to haunt us.

17 GLACIAL AND PERIGLACIAL LANDFORMS

FOCUS QUESTIONS

1. What are glaciers, and why do they form?
2. What major categories of glaciers exist, and where are they found?
3. In what ways do glaciers modify the landscape?
4. What was the Pleistocene, and why did it occur?
5. Where do periglacial environments occur, and what are the chief characteristics of periglacial landscapes?

Only 15,000 years ago (a very brief time span in geological terms), the North American continent had a far different appearance than it does today. Nearly all of what is now Canada, as well as a large portion of the northern United States, was buried beneath a huge ice sheet several kilometers thick. The present sites of many North American cities, including Boston, Montreal, Toronto, Chicago, and Detroit, were covered by ice. Today, no trace of this ice sheet remains. During that geologic epoch, referred to as the **Pleistocene** (or the "Great Ice Age"), the global coverage of ice was as much as three times that of today. The effects of the ice on the landscape were profound, and the Pleistocene glaciation occurred so recently that, as yet, many landforms have been little modified. The importance of ice as a geomorphic agent therefore is much greater than its current limited global distribution would indicate, and its significance is increased by the possibility that some day the Pleistocene glaciers may form again. In this chapter, we examine the geomorphic influences of ice on the land and discuss the surface characteristics of landscapes dominated or strongly modified by glacial and ground ice.

GLACIER FORMATION AND DISTRIBUTION

A **glacier** is a mass of freshwater ice, formed on land, that is or has been in motion. Glaciers, like rivers, move largely in response to gravity and perform the basic hydrologic function of removing excess precipitation from the land. The chief source of glacial ice is compacted snow, and in order for a glacier to develop, the accumulation of snow must exceed losses by melting and sublimation over an extended period. These losses are collectively referred to as **ablation.** Therefore, not only cold temperatures, but also significant snowfalls are required for glacier formation. Because of a lack of snowfall, some very cold but dry areas, such as parts of Alaska and much of northeastern Asia, have apparently never been glaciated.

When snowfall accumulation exceeds ablation over a prolonged period, the accumulated excess is converted gradually to glacial ice. A fresh snowfall contains mostly air and has a specific gravity of approximately 0.10, a tenth that of water. After a short period, it changes form to a much more compact granular state called **firn** that has a specific gravity of 0.55 or more. Further compaction, melting, and refreezing result in the expulsion of the remaining air and a slow conversion to glacial ice, with a specific gravity of about 0.90.

Present Extent of Glaciation

Glaciers are divided into two general categories on the basis of location, size, and shape. **Continental glaciers,** as the name implies, cover extensive land areas and form in the high latitudes. **Alpine glaciers,** also called "mountain glaciers" or "valley glaciers," form at high elevations in mountainous regions. In total, glaciers cover approximately 10 percent of the Earth's land surface.

The great majority of the Earth's currently glaciated surface is covered by the two surviving continental glaciers, or ice sheets—the Antarctic and Greenland ice sheets (Figures 17.1 and 17.2). The Antarctic ice sheet is by far the larger; it covers 12.6 million square kilometers (4.9 million mi^2), an area larger than the United States and Mexico combined. It has a volume of approximately 24 million cubic kilometers, (5,750,000 mi^3), an average thickness of about 2,160 meters (7,090 ft), and a maximum thickness of 4,000 meters (13,000 ft or 2.5 mi).

The Greenland ice sheet is considerably smaller, with an area of 1.8 million square kilometers (696,000 mi^2), a volume of 2.3 million cubic kilometers (550,000 mi^3), and a maximum thickness of 3,350 meters (11,000 ft). These two ice sheets, if melted, would release enough water to supply all the world's rivers at their present discharge rates for nearly 900 years and, in doing so, would raise sea levels at least 45 meters (150 ft). They contain about 75 percent of the Earth's freshwater supply (Table 6.1).

In contrast to continental glaciers, alpine glaciers are of limited areal extent, with a total area of approximately 508,000 square kilometers (196,000 mi^2) and a combined volume of 80,000 cubic kilometers (19,200 mi^3). Several thousand are scattered throughout the higher mountain ranges of the world. The more significant of these glaciers exist in Alaska, the Canadian Rockies, the Canadian Arctic islands, the Andes, the European Alps, and several mountain ranges of Asia, including the Himalayas, Karakoram, Tien Shan, and Caucasus (Table 17.1). A few small glaciers are found in the central Rockies of the United States, the

◆ **TABLE 17.1 Present-Day Ice-Covered Areas (km²)**

Antarctic	12,588,000
Greenland	1,802,600
Northeastern Canada	153,200
Central Asian ranges	115,000
Spitsbergen group	58,000
Russian Arctic islands	55,700
Alaska	51,500
South American ranges	26,500
West Canadian ranges	24,900
Iceland	12,200
Scandinavia	3,800
Alps	3,600
Caucasus	1,800
New Zealand	1,000
USA (excluding Alaska)	500
Others	about 100

Total volume of present ice: 28 to 35 million km^3

Source: C. Embleton and C. A. King, *Glacial Geomorphology* (New York: Halsted Press, 1975), p. 25.

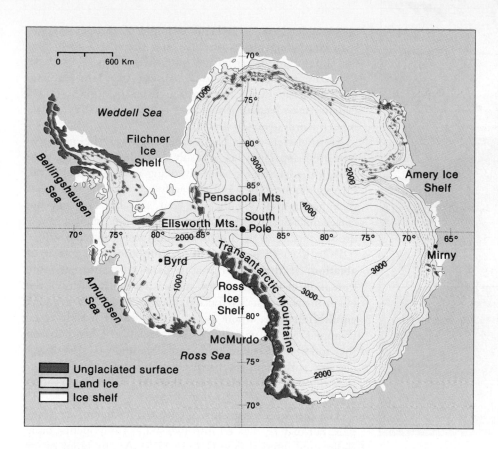

◆ **FIGURE 17.1** The Antarctic ice sheet. Antarctica is almost completely buried by an ice sheet that averages over 2,100 meters (7,000 ft) thick. It reaches a maximum thickness of 4,000 meters (13,000 ft) at a location about 480 kilometers (300 mi) inland from the former Soviet base of Mirny. Contour lines give surface elevations in meters.

◆ **FIGURE 17.2** The Greenland ice sheet covers the interior of the world's largest island and has a maximum thickness of approximately 3,350 meters (11,000 ft).

Cascades, and in East Africa. Australia is the only continent with no glaciers at all.

GLACIAL PROCESSES AND FEATURES

Glacier Motion

Glacier motion is caused by the force of gravity, which compacts, deforms, and may locally melt the ice as it is drawn constantly downward. Unlike streams, the motion of a glacier is not always downhill; rather it is down-ice—that is, in the direction that the ice surface slopes.

In discussing glacial motion, we must distinguish between absolute and relative motion. The **absolute motion** is the actual forward movement of the particles of ice in the glacier. This motion nearly always is imperceptible and usually totals less than 300 meters (1,000 ft) per year. Stagnant or "dead" glaciers, located in areas where rapid losses by ablation are occurring, do not move at all, but decay in place.

The absolute motion of a glacier may be far from steady. Most glaciers experience **surges,** or sudden increases in their rates of advance. The causes of these surges are not always apparent. Sometimes they are associated with one or more seasons of heavy snowfall in the ice accumulation zone. At

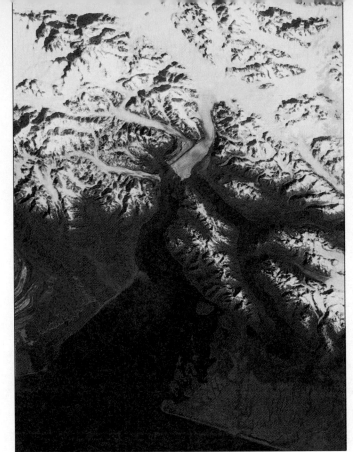

◆ FIGURE 17.3 A surge in the flow of Alaska's Hubbard Glacier, which occurred in early 1986, temporarily blocked Russell Fjord (the inland water body to the right of the lower portion of the glacier), trapping a large number of marine animals.

SOURCE: USGS

ablation. A glacier is a classic example of an open system, with continuing inputs of snow and outputs of meltwater and water vapor. A glacier will gain mass and is said to have a positive **mass balance,** if its inputs exceed its outputs. Such a glacier usually advances. Conversely, a glacier has a negative mass balance if its losses of ice (outputs) exceed its inputs; it will generally retreat.

Most glaciers have been retreating over the past few centuries, apparently as a result of global warming, even though the ice itself is still advancing in an absolute sense. Glacier motions are watched carefully, not only because of their geomorphic significance, but also because they are considered sensitive climatic indicators. This subject is addressed further in the Focus Box beginning on page 453.

Every glacier has a **zone of accumulation** and a **zone of ablation.** If the glacier maintains a constant mass, the formation of new ice in the former zone just balances the losses in the latter. Separating the two zones is a so-called **equilibrium line,** where gains and losses are exactly equal (Figure 17.4). In general, a size correlation exists between the zones of accumulation and ablation, so that larger glaciers, with more extensive zones of accumulation, typically extend to lower elevations than do smaller neighboring glaciers.

The precise mechanisms of glacier movement are complex, but may be divided into two major components: basal sliding and internal flow (Figure 17.5). At any point on its surface, the absolute motion of a glacier is the sum of the velocities of basal sliding and internal flow.

Basal sliding involves a true sliding of the ice on the bedrock surface underlying the glacier. Where the bedrock floor is uneven, a process termed **regelation** makes ice motion possible. To envision this process, imagine a securely

other times, they are initiated by the accumulations of subglacial meltwater, which helps to lubricate the zone of contact between the base of the glacier and the underlying bedrock. This factor was apparently responsible for a well-documented surge in Alaska's Hubbard Glacier in 1986. For most of the 90 years its movements have been recorded, the Hubbard Glacier has flowed into Yakutat Bay, near the northwest tip of the Alaskan Panhandle, at a relatively steady rate of about 150 meters (500 ft) per year. This motion has not been entirely counterbalanced by a loss of ice at the glacier's snout, so the glacier has been slowly advancing into the bay. During the winter of 1986, however, the Hubbard Glacier suddenly began to surge—moving as fast as 14 meters (46 ft) per day. A tributary glacier, the Valerie Glacier, also began to surge forward at speeds of up to 34 meters (112 ft) per day. This movement soon blocked off Russell Fjord at the head of the bay and turned it into a lake, temporarily trapping numerous seals and other sea animals (Figure 17.3).

The **relative motion** of a glacier refers to the advance or retreat of the ice front. It is determined by the relationship between the absolute motion of the glacial ice and its rate of

◆ FIGURE 17.4 Zones of ice accumulation and ablation on a glacier. Rock debris is buried beneath newly forming ice in the zone of accumulation, producing a clean surface, but eventually resurfaces in the zone of ablation due to the melting of the overlying ice. The equilibrium line typically coincides with the summer firn line.

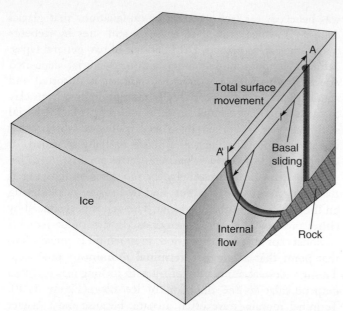

◆ **FIGURE 17.5** Mechanisms of glacier movement. The motion of warm ice glaciers is accomplished by both basal sliding and internal flow. Cold ice glacier motion takes place only by internal flow.

anchored rock outcrop, a few meters high, protruding from the bedrock floor into the ice. As the ice is pushed toward the outcrop, the pressure generated lowers the melting point of the ice, so that it melts and the water flows around the obstruction. As the water reaches the lee side, the decreased pressure raises the melting point, so that refreezing, or regelation, occurs. Basal sliding, which is normally considered to include the regelation process, occurs only with **warm ice glaciers,** which have an underlying film of meltwater acting as a lubricant to aid slippage. Most alpine glaciers are of this type.

The second component of glacier motion, internal flow, is believed to occur primarily as a smooth (laminar) flowage of horizontally oriented ice crystals that shear over those below. It takes place at depths greater than about 50 meters (160 ft) because the weight of the overlying ice causes the ice below this level to lose its rigidity and deform in a plastic fashion. Unlike basal sliding, internal flow has little or no effect on the underlying surface. It is the only mechanism of movement employed by **cold ice glaciers,** which are frozen fast to the underlying bedrock. Cold ice glaciers exist largely in the colder high latitude environments where continental glaciers form.

Because the nonmoving rock floor and walls of a glacier (if it is laterally confined) resist ice flow, the surface ice moves fastest. In addition, if the glacier takes the form of a stream of ice, as do most alpine glaciers, flow rates are greater near the center than near the margins. In these respects, the flow pattern of a glacier is very similar to that of a river. Major differences are the much slower flow velocities of glaciers and the predominance of laminar flow rather than turbulent flow.

Glacier Erosional Processes and Features

Ice is the most powerful of the tools of erosion. The great power of glaciers as eroders and transporters of regolith stems from several factors. One is the fact that, unlike water and wind, glaciers exist in the solid state. This produces the physical hardness and firmness needed to dislodge vast quantities of weathered rock material effectively. A second factor is the tremendous pressure that hundreds or even thousands of meters of ice exerts on the surface. Although water depths and pressures equal or exceed those of glaciers on most ocean floors and in some deep lakes, the bottom water usually is nearly stationary and has little or no erosional capacity. Indeed, the deep ocean basins are the Earth's major depositional sites. A third factor aiding glacial erosion is the great extent of ice coverage within a glaciated area. While only a tiny proportion of the surface in a fluvially dominated landscape is covered by flowing water at any one time, continental glaciers bury the surface, modifying the entire landscape. Even alpine glaciers, which are confined largely to mountainside depressions or valleys, may cover a significant proportion of the surface.

On the other hand, glaciers do have some erosional limitations. Probably the most important is the reduced rate of chemical weathering of the underlying bedrock, which is inhibited by cold temperatures and sometimes a lack of liquid water. The prior weathering of the bedrock is crucial in preparing material for glacial transport. Hard as ice is, most unweathered rock is harder yet and will not be affected significantly by direct glacial erosion. Frequently, however, pieces of bedrock become frozen into the base or sides of a moving glacier. They then scrape over the surface, abrading it in much the same manner as the hard grains of sand on a sheet of sandpaper scrape a piece of wood.

A second factor limiting the effectiveness of glacial erosion is the slow rate of ice movement over the underlying surface. Ice is by far the slowest moving of the tools of erosion. In addition, much of the ice motion occurs internally and therefore does not affect the underlying surface. Indeed, polar glaciers, frozen rigidly to their bases, perform little or no erosion.

Glacial erosion occurs by the processes of scraping and plucking. **Scraping** is accomplished primarily by rocks frozen into the base of the ice and dragged over the bedrock surface. Larger rocks may produce scratches called **striations,** which are sometimes enlarged to form grooves if rock fragments are dragged repeatedly over the same point (Figure 17.6). Striations are valuable indicators of the direction of past ice flow. The abrasion performed by silt-sized particles, on the other hand, smooths and even polishes some rock surfaces. Because these smaller particles are much more widespread than the larger rock fragments along the glacier base, the overall effect of glacial abrasion is to smooth and round rock surfaces.

Plucking involves the excavation of angular, often relatively large, rock fragments from the bedrock. The regelation

◆ **FIGURE 17.6** Glacial striations in basalt at Devil's Postpile National Monument, California.

process described earlier seems to be the key mechanism in making plucking effective. Ice melting on the up-glacier side of an obstacle such as a hill or boulder flows around the obstruction, freezes in rock fractures on the down-glacier side, and plucks out rock fragments as the ice moves on. The process is aided greatly by the presence of weathered or highly jointed bedrock. Some previously glaciated areas contain numerous bedrock features with their up-glacier sides smoothly rounded and their down-glacier sides irregularly plucked (Figure 17.7). These features are sometimes referred to as **roche moutonnées,** a French term meaning "rock sheep."

Glaciers are most effective, by far, in eroding unconsolidated materials. This most commonly occurs when a glacier advances over terrain that has not been glaciated in the recent past. The hundreds of meters of glacial ice involved readily bulldoze the soil and weathered rock or incorporate it into the ice itself, leaving the underlying surface stripped to unweathered bedrock.

Because the sediment-carrying ability of the ice is virtually unlimited, glacially transported material is carried in an unsorted state. Warm ice glaciers not bounded by rock walls carry most of their debris along or near their bases. Alpine glaciers contain much basal material but also transport large amounts of material, commonly dislodged by frost wedging, on their surfaces.

Glacier Depositional Processes and Features

Glaciers transport debris in a conveyor-belt fashion. Even if the forward edge of the ice is retreating, the absolute glacier motion is still down-ice, and this motion does the transporting. The general term for all deposits of glacial origin is glacial **drift.** This term is a holdover from the time when it

was believed, for lack of a better explanation, that glacial deposits had been rafted to their present sites by icebergs during the biblical Flood. Drift consists of two general types, defined by their mode of deposition. Material deposited directly by glacial ice is known as **till.** It is unsorted and angular, with constituent particles ranging in size from clay to occasional large boulders. In contrast, **glacio-fluvial sediments** are reworked by glacial meltwaters after their initial deposition by glaciers and are partially stratified, or sorted, because the largest particles settle out first.

Till is deposited primarily by the processes of dumping at the glacier terminus, or **snout,** and by lodging or plastering on the surface beneath the glacier. The deposits produced by either process are termed **moraines.** If the snout remains stationary for an extended period, so much till is conveyed to that point that a ridgelike **terminal moraine** is produced. This is a loose accumulation of till with its long axis oriented perpendicular to the direction of ice flow (Figure 17.8). Terminal moraines are often arcuate, because most glacier snouts are convex in shape. If the glacier front is retreating, a series of such moraines may be formed, with each moraine marking the site of a temporary halt in the glacier's retreat. In this case, only the outermost moraine is referred to as the terminal moraine, and the others are termed **recessional moraines.** Terminal and recessional moraines, where undisturbed, provide precise information on past glacial positions and are invaluable to researchers attempting to reconstruct the chronology of events in a previously glaciated area. Between these moraines, a thinner covering of **ground moraine** is deposited by the more rapidly retreating ice. It has been found that, although continental glaciers may transport some rock debris for hundreds of kilometers, most glacial deposits are derived from relatively nearby sources.

Glacio-fluvial materials normally are deposited at or beyond the margins of the ice, although in some cases subglacial deposits may occur. In general, the farther these deposits, called **outwash,** have been transported by water, the better sorted they are. Glacial meltwater streams usually are overloaded with sediment and frequently aggrade their courses, typically developing shifting braided channels as they do so. If multiple streams issue from the ice front, a broad outwash plain may be formed. If only one fairly

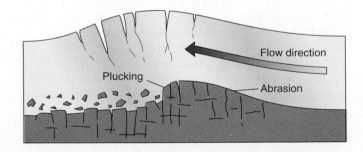

◆ **FIGURE 17.7** A roche moutonnée forms as ice passes over a jointed rock outcrop, smoothly rounding the up-glacier side, while irregularly plucking the down-glacier side.

(a) During glaciation

(b) After glaciation

◆ **FIGURE 17.8** Block diagrams depicting the formation of a terminal moraine, as well as the appearance and relative positions of terminal, recessional, and ground moraines after the melting of the glacier that formed them.

channelized stream emerges, a single long **valley train** of deposits is produced, perhaps extending for tens of kilometers from the ice front.

ALPINE GLACIERS AND ASSOCIATED LANDFORMS

Alpine glaciers form in mountainous regions at high elevations and flow downhill under the influence of gravity. Because global temperatures tend to increase equatorward, alpine glaciers become progressively less numerous with decreasing latitude and their zones of formation occur at progressively higher elevations. The lower limit of ice accumulation lies essentially at sea level within 10 degrees of the poles and increases to elevations of 4,600 to 5,000 meters (15,000 to 16,000 ft) in the Central Andes and the Ruwenzori Range of East Africa near the equator. The zones of ablation of well-fed glaciers can extend significantly below these limits, though, so that the lower limit of glaciation depends not only on temperature, but also on snowfall amounts and on the elevation and size of the glacial accumulation area.

Characteristics of Alpine Glaciers

Most alpine glaciers originate in relatively small mountainside hollows called **cirques.** If conditions are marginal for ice accumulation, the glacier may remain as a small, rounded **cirque glacier.** If conditions favor ice accumulation, however, the glacier will outgrow its cirque and flow down the mountainside as a long, narrow tongue-like mass of ice. Upon reaching the base of the mountain, continued glacier expansion usually will result in the ice advancing down a preexisting valley as a **valley glacier.** (Figure 17.9). Large valley glaciers, like those of the Alps and Alaska, typically are fed by several glacier tongues, so that a dendritic pattern of intersecting glaciers is developed.

The surface features and appearance of an alpine glacier can vary considerably from point to point. In the zone of accumulation, the ice usually is covered by snow and cannot be seen, while in summer, at least, the surface of the glacier in the zone of ablation is visible. Where the glacier passes over convexities on its bed, the stretching of the near-surface ice causes it to fracture, forming open fissures called **crevasses.** The deepest known crevasses extend downward approximately 50 meters (160 ft). Below this level, the weight of the overlying ice gives it enough plasticity to close any gaps that may form.

Lateral moraines frequently take the form of thick stripes of rock debris along the sides of alpine glaciers. This material accumulates both from glacial erosion and from weathering processes, especially frost shattering, which dislodge rocks from the slopes above. Where two glacial tongues merge, their innermost lateral moraines coalesce to form a single **medial moraine** within the glacier. A large alpine glacier often contains several medial moraines, the

◆ **FIGURE 17.10** St. Mary Lake occupies a glacial trough in Montana's Glacier National Park. A horn and a knife-edged arête further enhance this area's spectacular alpine scenery.
SOURCE: Ralph Scott

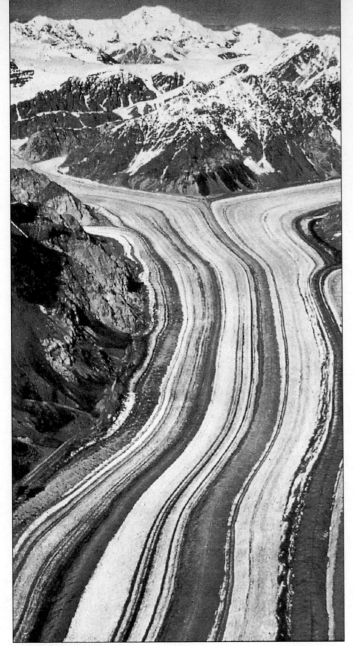

◆ **FIGURE 17.9** A large valley glacier in Alaska is banded by medial moraines that formed when glacial tongues merged.
SOURCE: Photo courtesy of Reed Wicander

total number generally indicating the number of individual valley glaciers that have combined to form the glacier.

Proceeding into the zone of ablation, the glacier takes on a progressively dirtier, more debris-laden appearance, because the relative proportion of morainic material gradually increases as the ice melts. Near the glacial snout, debris may completely cover the surface, making it impossible to accurately determine the most forward position of the ice.

Alpine Glacier Erosional Landforms

Erosion by alpine glaciers tends to increase the relief and ruggedness of the topography. It has been largely responsible for producing the spectacular alpine scenery that annually draws millions of tourists to highly glacially modified mountain systems such as the Alps and northern Rockies (Figure 17.10). A number of landform features are produced by alpine glacier erosion, but two of the most important are cirques and glacial troughs.

Cirques are bowl-shaped depressions eroded into the rock faces of mountainsides in glaciated areas (Figures 17.11 and 17.12). They vary greatly in degree of development and perfection of shape, but well developed cirques consist of three basic sections: a steep headwall, a smooth central basin, and a slightly raised lip or threshold (Figure 17.13). Some uncertainty exists as to the precise mechanism by which cirques develop, but they generally are believed to form where mountainside snow patches provide water for accelerated physical and chemical weathering. As the hollow enlarges, sufficient snow collects to prevent complete summer melting, and a cirque glacier forms. The cirque is subsequently enlarged and molded by frost shattering on the headwall and by erosion of the cirque floor by scraping and plucking. Many cirques that formed during the Pleistocene are now devoid of ice and instead contain small lakes called **tarns.**

As growing cirques erode headward from opposite sides of a linear mountain or upland area, they eventually may reduce the intervening mass to a narrow, serrated ridge called an **arête.** Headward cirque growth from three or more sides can produce a sharp-crested pyramidal peak termed a **horn** (Figure 17.10). The Matterhorn of Switzerland is the most famous example of this process, but most isolated peaks in glaciated mountain regions have formed in essentially the same manner. The growth of cirques therefore is a highly effective erosional mechanism and is largely responsible for the upland features in extensively glaciated mountains.

The valley features in glacially modified mountainous regions result largely from the development of glacial troughs. A **glacial trough** is a deep valley, usually of fluvial

(a)

(b)

◆ **FIGURE 17.11** The glacial modification of a mountainous landscape is represented in this sequence of diagrams. Part (a) shows a fluvially dissected landscape prior to the onset of glaciation. Note the smoothly rounded appearance of the slopes. Part (b) depicts the same landscape at the height of glaciation. In Part (c), the glaciers have retreated, leaving in their wake a rugged landscape containing numerous glacial erosional features.

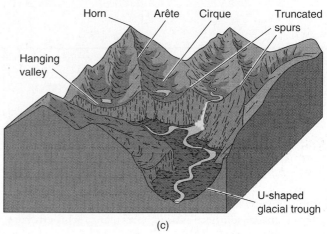

(c)

origin, that has been subsequently occupied and modified by a valley glacier (Figure 17.14). Glacial troughs are noted for their broad, rounded U-shapes, as opposed to typical fluvially eroded mountain valleys, which are more V-shaped. In contrast to river valleys, they also possess markedly ungraded longitudinal profiles and greater overall linearity.

The U-shape of a glacial trough results from the large size of a valley glacier as compared to a river draining the same area. This size is necessary because glacial ice flows much more slowly than river water. Indeed, a valley glacier usually fills its entire trough, often to a considerable depth. Thus, a glacial trough serves the same function for a glacier that a channel serves for a stream. The deepening and steepening of the trough by its glacier considerably increase the local relief because nearby mountain peaks have not been similarly reduced in elevation.

Smaller tributary glaciers entering a valley glacier also carve U-shaped valleys, but do not usually erode nearly as deeply. When the ice melts, their valleys intersect the main glacial trough high on its sides, forming **hanging valleys.** As streams issuing from these hanging valleys cascade into the major trough, spectacular waterfalls may result. Yosemite Falls, in California's Sierra Nevada, is a famous example.

The ungraded longitudinal profile of a glacial trough results from variations in the rate of vertical erosion by the ice in different parts of the valley. Erosion is aided by factors such as a narrowing of the valley (causing a thickening of the glacier), the influx of a tributary glacier, and the presence of relatively weak rock. Such areas are eroded to form localized basins within the trough; they usually are occupied by lakes after the ice melts.

At higher latitudes, glacial troughs may be eroded below sea level and occupied by ocean water when the ice melts. Such drowned troughs are called **fjords.** They are typically

long, winding, steep-walled, and relatively narrow (Figure 17.18). Fjords have formed because the great thickness of the ice enabled it to erode deeply even though it was partially buoyed by the ocean water into which it advanced.

◆ **FIGURE 17.12** Classic cirque development producing "biscuitboard topography" in the Uinta Range of Utah. These cirques are Pleistocene features, and no glaciers currently occupy them.

SOURCE: John S. Shelton

◆ **FIGURE 17.13** A topographic map depiction of a well-developed cirque in Norway. Note the precipitous headwall (A), the remnant cirque glacier (B), the central basin containing a tarn, or small lake (C), and the raised threshold (D). Elevations are in meters.

Sea level rises subsequent to the melting of the glaciers resulted in further flooding. As a consequence, some fjords in Norway and southern Chile attain depths of more than 1,300 meters (4,000 ft). They therefore provide excellent protected routes for coastal shipping and fishing interests. Fjords are best developed along the coasts of Norway, western Canada, southern Alaska, Greenland, southern New Zealand, and southern Chile.

Alpine Glacier Depositional Landforms

The depositional features of alpine glaciers typically are more poorly defined and less obvious than are the erosional features for several reasons. First, they have much less relief than the erosional features. In addition, they consist of unconsolidated materials and therefore do not have the permanency of the erosional features, which generally are

◆ **FIGURE 17.14** A U-shaped glacial trough in northwestern Montana.

SOURCE: Courtesy of J.S. Monroe

RECENT GLACIER VARIATIONS IN THE PATAGONIA ICEFIELD, SOUTH AMERICA

The Patagonia Icefield, located near the southern tip of South America, consists of two separate icefields, commonly referred to as the Northern Patagonia Icefields (NPI or Hielo Patagonico Norte) and the Southern Patagonia Icefield (SPI or Hielo Patagonico Sur) (Figure 17.15). Together they cover an area of 17,200 square kilometers, forming the third largest ice mass in the world, after the Antarctic and Greenland ice sheets. They also form the largest temperate ice mass in the southern hemisphere. The Patagonian glaciers are characterized by large accumulation and ablation rates. The area's mean annual precipitation is estimated at 7,000–10,000 millimeters, which is equivalent to a snow accumulation of 25–30 meters.

The behavior of individual glaciers within a large ice field is complex, because they are affected not only by climatic conditions such as precipitation and air temperature, but also by local topo-climatic effects and glacier dynamics. The general trend of glacier variations in a region, however, is believed to reflect the condition of the regional climatic trend. Glaciers in general, and particularly temperate glaciers, are regarded as good indicators of climatic change. Since the Patagonian glaciers have large rates of ice

continued on next page

◆ **FIGURE 17.15** Location map of the Southern Patagonia Icefield.

Southern Patagonia Icefield and Major Outlet Glaciers

formation and ablation, they should respond quickly to climatic changes and hence serve as sensitive climatic indicators.

Unfortunately, there are practically no inhabitants in the vicinity of the Patagonian ice sheets who can periodically observe the condition of the glaciers. Consequently, there are few reports on glacier variations. As a result, despite the generally inclement weather that prevails in this region, remotely sensed data are the only means by which we can obtain information on glacier conditions and variations. These data include aerial photographs taken in 1944/45 by the U.S. Air Force—the oldest record in Patagonia. Aerial photographs were also taken between the late 1960s and the early 1980s by the governments of Chile and Argentina for topographic mapping purposes. Also available, in limited numbers, are Landsat imagery (since 1972) and SPOT imagery (since 1986).

THE NORTHERN PATAGONIA ICEFIELD

The NPI is centered at approximately 47°S 73°30′W and is about 100 kilometers long and 30–40 kilometers wide (Figure 17.16). The icefield's area was about 4,400 square kilometers in 1944/45 and 4,200 square kilometers in 1976. Thus in 31 years, it had shrunk by about 5 percent. There are 22 major outlet glaciers around the periphery of the icefield. Among them, the San Rafael Glacier, with an area of about 760 square kilometers, is one of the largest. Located at 46°41′S, it is the tidewater glacier (i.e., one that reaches and enters the ocean) closest to the equator of any in the world. Most of the icefield's other glaciers currently terminate in proglacial (generally moraine dammed) lakes.

The San Rafael Glacier retreated rapidly between 1944/45 and 1990/91, with an average recession rate of 83 meters/year for a total loss of 3.8 kilo-

◆ **FIGURE 17.17** Landsat image of the Southern Patagonia Icefield. Ice-covered areas appear blue.

SOURCE: Landsat

the trend of rapid glacier recessions has ceased in the NPI.

THE SOUTHERN PATAGONIA ICEFIELD

The SPI extends for about 350 kilometers between 48°20′S and 51°30′S along an axis centered near 73°30′W (Figure 17.17). Its width is generally 30–40 kilometers, with its narrowest part only a few kilometers wide. The icefield's area in 1944/45 was 13,500 square kilometers, while it was 13,000 square kilometers in 1986. This glacier therefore was reduced by about 4 percent in areal extent in 41 years. In the SPI, only a few of the more than 25 major outlet glaciers have been studied for variations since 1944/45, and the results indicate that most glaciers have been retreating. However, there are a few glaciers whose behaviors have been contrary to this trend, a situation that raises intriguing questions as to the mechanisms controlling glacier variations.

For example, Bruggen Glacier, located on the west (wet) side of the icefield, has advanced since 1944/45. It has blocked a fjord, and its snout has split into northern and southern tongues. As a result, the head of the fjord was impounded and has become a freshwater lake. Since there are no remote sensing data or reports between 1945 and 1976, we do not know exactly when this glacier started advancing. However, its apparent advance in

meters. However, its retreat speeds have varied; an aerial survey in 1993 showed that the most recent period of recession had stopped, and, on the glacier's right half, a slight advance had actually occurred. Other glaciers have continued to recede, although at slower rates than before. It seems that

31 years was about 9 kilometers (290 meters/year). A comparison of Landsat images taken in 1976 and 1986 revealed that while the northern tongue had continued to advance by 1,200 meters (or 120 meters/year), the southern tongue had retreated up to 600 meters (60 meters/year). Only 60 kilometers to the northeast of Bruggen Glacier and across the icefield, O'Higgens Glacier is located on the east (dry) side of the Andes. This glacier had retreated during the same period about 13.4 kilometers (327 meters/year), which is by far the fastest rate of recession in Patagonia. The accumulation fields of these two glaciers lie at similar latitudes.

About 70 kilometers south of O'Higgens Glacier is Viedma Glacier. This glacier was very stable between 1968 and 1981, but after 1981 it retreated slightly. Sharing the divide in the accumulation area with Viedma Glacier is Upsala Glacier, one of the largest in South America. Upsala Glacier retreated about 3 kilometers between the early 1960s and 1990 (110 meters/year). However, other remote sensing data revealed that the middle

part of the glacier advanced up to 400 meters between 1968 and 1978. The surface elevation near the snout was lowered by more than 100 meters between 1968 and 1990 and has further lowered by 30–40 meters during 1990–93. This lowering rate is one of the fastest in the world.

About 60 kilometers farther south, Ameghino and Moreno Glaciers are juxtaposed north-south on Lake Argentino. They share the same accumulation area. Moreno Glacier in this century has periodically blocked the lake channel by reaching the opposite bank. At the end of the last century, it started advancing, closing the lake for the first time in 1917. Since then it has closed the lake channel many times. The snout has oscillated by up to 500 meters since 1944, suggesting that the glacier has been stable. A recent survey of the glacier surface near the snout has confirmed this stability, with almost no change observed in surface elevations. On the other hand, Ameghino Glacier, which had reached a maximum length as recently as 1870–1880, has been retreating rapidly since 1970.

DO GLACIER VARIATIONS IN PATAGONIA RESULT FROM THE RECENT GLOBAL WARMING TREND?

Most Patagonian glaciers have been retreating since 1944, and it may be that this trend is the result of recent global warming. However, the rates of retreat differ greatly for individual glaciers, and some rates are accelerating while others are decelerating. Unfortunately, there are no meterological stations near the icefields with records sufficiently long to allow a long-term analysis of these variations. Caution must be taken before concluding that the ice field is in overall retreat due to global warming because of the analomous behavior of some of the icefield's outlet glaciers. It seems certain that in addition to the effect of the recent global warming trend, more subtle local climatic changes and/or local glacier dynamics are influencing the behavior of individual glaciers within the Patagonia Icefield. ■

This box was written by Masamu Aniya, Institute of Geoscience, University of Tsukuba, Japan.

composed of solid bedrock. Finally, because most glacial deposits are located in valleys, the rivers that reoccupy the valleys after the departure of the ice rework and often largely remove these materials.

The major till features are moraines of various types. Fresh moraines consist of ridges of loose, unsorted rock debris. Lateral moraines form terracelike deposits along valley margins. Those produced by large valley glaciers are sometimes more than 100 meters (330 ft) in height and often become progressively larger and better defined down-valley. Medial moraines, which are more likely to have been removed by fluvial processes, may form ridges well within the glacial trough. Terminal and recessional moraines form arcuate ridges, bulging down-valley, that are oriented perpendicular to the lateral and medial moraines. Many for-

merly extensive alpine glaciers have disappeared completely, so that morainic deposits occur from their lower valleys all the way up to their cirque floors.

Glacio-fluvial deposits already have been discussed. Of those produced by alpine glaciers, the most significant are the valley trains deposited by debris-laden glacial meltwater streams.

CONTINENTAL GLACIERS AND THE PLEISTOCENE

Continental glaciers differ from alpine glaciers in several major respects. The most important difference is in their size and thickness. While alpine glaciers extend, at most, a few

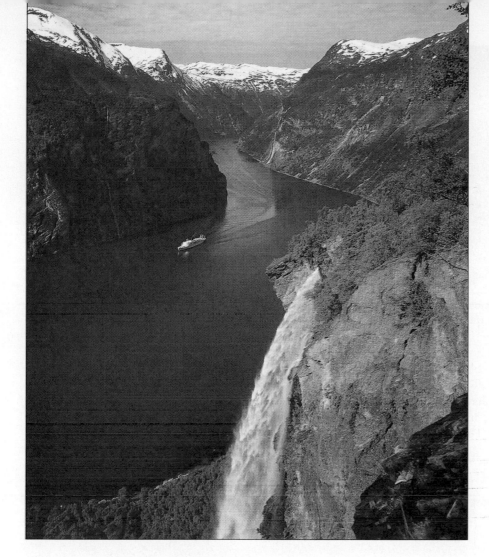

◆ **FIGURE 17.18** Geirangerfjorden, in Norway, is a fjord with numerous waterfalls issuing from hanging valleys on its sides.

SOURCE: Tony Stone Images

tens of kilometers and frequently have widths of only a kilometer or two, continental glaciers have horizontal dimensions of hundreds or even thousands of kilometers. Alpine glaciers lie cupped in mountainside hollows or in deep valleys, while continental glaciers usually cover the entire surface deeply.

The underlying surface exerts much less control on the movement of continental glaciers than it does for alpine glaciers. The movement of the entire ice sheet is not unidirectional, but rather is radial, with the ice spreading outward in all directions from one or more centers of accumulation. The cause of this radial spread is not so much the force of gravity as the pressure exerted by the great weight of the ice itself. Radial expansion gives the glacier a rounded, domelike overall appearance. Examined in more detail, however, the ice front advances as a series of lobes whose positions are dictated by the locations of the centers of ice accumulation and by the topography and geology of the underlying surface.

The geographic settings of the zones of ice accumulation of alpine and continental glaciers also differ in one major respect. While high elevations are responsible for the cold climates that support alpine glacier growth, high latitudes are responsible for the low temperatures that support the growth of continental glaciers. Since larger areas of the Earth are at high latitudes than at high elevations, continental glaciers are much more extensive than are alpine glaciers.

The Pleistocene Epoch

The geologic record indicates that for most of its existence the Earth has had no continental glaciers. There may always have been alpine glaciers in the highest mountains, but continental glaciers seem to be a rare phenomenon. Scientists have evidence of only five relatively brief periods of continental glaciation in the past billion years. The first four occurred hundreds of millions of years ago. The fifth—the **Pleistocene**—began less than two million years ago and is still continuing.[1]

---

[1] Many scientists consider the Pleistocene to have ended about 10,000 years ago, and that we are currently in a mild epoch termed the **Holocene.** There is increasing evidence, however, that the cyclical pattern of continental glacier formation and retreat will continue in millennia to come. Even if this does not occur, the continuing existence of massive glaciers in Antarctica and Greenland indicates that the present glacial stage is not completed.

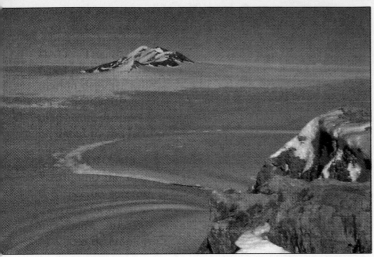

◆ **FIGURE 17.19** Nunataks, or isolated mountain peaks, protrude above a sea of ice on the margin of the Antarctic Plateau.

SOURCE: John Craddock

Over the past one million years, there have been approximately 10 major glacial periods, or stages, and as many as 40 more minor periods of glacial advance. The **Wisconsinan Stage** (or **Würm Stage,** as it is known in Europe) is the

most significant because of its relative recency. The Wisconsinan Stage began some 50,000 to 100,000 years ago and generally is considered to have ended about 10,000 years ago, when the retreating ice left the north-central United States. The last ice may not have melted in central Canada, however, until about 4,000 years ago. The recency of this period is such that only a superficial modification of the landscape has occurred in many of the glaciated areas; as a result, these areas tend to be dominated by relict glacial features.

Between the various glacial stages were lengthy interglacial stages. Evidence suggests that during at least one interglacial stage the ice may have disappeared largely or entirely from Greenland and Antarctica.

Geographic Extent of Pleistocene Glaciation

During the Pleistocene glacial stages, ice covered as much as 30 percent of the Earth's present land area. Glacial ice now covers some 14 million square kilometers (5.5 million mi^2), but during the Pleistocene it covered up to 44 million square kilometers (17 million mi^2). Then, as now, the Antarctic ice sheet was the largest of all. (Figure 17.19). During the Pleistocene, it enlarged and spread farther out over the adjacent sea than at present. The Greenland ice sheet was considerably larger than it is today, extending to the coast

◆ **FIGURE 17.20** Centers of ice formation and maximum extent of Pleistocene glaciation in North America. Areas of both glacial ice and sea ice are shaded in blue.

♦ FIGURE 17.21 Areas of continental glacier coverage and directions of ice flow during the Pleistocene in Europe. Both the maximum extent of the ice sheets and the limits of ice advance during the last (Würm) glacial stage are shown.

and probably forming ice shelves in protected coastal waters. The major addition of ice, however, consisted of two giant ice sheets that formed over the northern half of North America and over northern Europe and adjacent northwestern Asia.

The North American or **Laurentide ice sheet** spread outward from two or more centers in interior Canada to cover virtually all of Canada, including Hudson Bay and most of the Canadian Arctic islands (Figure 17.20). The United States was covered as far south as a line extending westward from Long Island through northern Pennsylvania, then along the courses of the Ohio and Missouri Rivers, and finally westward near the international border to northern Washington. Surprisingly, much of Alaska remained unglaciated, apparently because of a lack of snowfall and the presence of orographic barriers. It should be stressed that the line just described is a composite maximum line of glacial advance. During each major glacial stage, the ice advanced to a different boundary. The total area covered by the Laurentide ice sheet was approximately 10.2 million square kilometers (4 million mi²), about 80 percent as large as the Antarctic ice sheet is today. Its thickness is unknown but was probably comparable to that of the Antarctic ice sheet. It was at least thick enough to cover completely the 1,918-meter (6,288 ft) crest of Mt. Washington, New Hampshire, which was located near the presumably thinner margin of the ice.

The Eurasian or **Fennoscandian ice sheet** radiated from a center in eastern Sweden to cover a maximum of 5.1 million square kilometers (2 million mi²). Its southernmost boundary extended through southern England, eastward through the Netherlands, then east-southeastward across central Europe to southern Ukraine, and finally east-northeastward to a point in north-central Siberia (Figure 17.21). The northern terminus of the ice extended some distance past the present continental margin into the Arctic Ocean. The North Sea and the Baltic Sea also were under glacial ice, which reached far enough westward to cover Ireland.

The Pleistocene glacial stages were also periods of formation or expansion for smaller glaciers in many other parts of the world. A subcontinental glacier may have formed over eastern Siberia, which was excessively cold during the Pleistocene but was too dry to support the growth of a more massive glacier. Alpine glaciers in many areas also expanded greatly, and many currently unglaciated mountains supported extensive Pleistocene glaciers. For example, the Canadian Rockies developed a system of coalescent alpine glaciers, sometimes termed the **Cordilleran ice sheet,** that merged with the Laurentide ice sheet east of the continental divide. The Alps developed an ice cap above which only the higher peaks protruded. Glaciers formed in the mountains of southeastern Australia where none exist today. The southern Rockies, which contain only a few cirque glaciers at present, were extensively glaciated. Glaciers may have descended to 2,000 meters (6,500 ft) in the Andes at the equator, where today the glacial limit is at least 4,300 meters (14,000 ft).

Causes of the Pleistocene

The determination of the cause of the Pleistocene has produced one of the greatest controversies in geomorphology since nineteenth-century Earth scientists first confirmed the past existence of an ice age. There are actually two facets to the problem, and their respective solutions may or may not be related.

The first question is why the Pleistocene occurred at all. More specifically, why, after many tens of millions of years of nearly global warmth, did the Earth become cold enough within the past two million years to permit the growth of massive continental glaciers? The answer to this question is still a mystery, although a number of hypotheses have been developed. One is that the Sun's energy output experiences long-term variations. A second hypothesis is that major volcanic eruptions may periodically inject huge quantities of insolation-blocking dust into the atmosphere, resulting in global cooling. A third hypothesis is that the Gulf Stream, which transports tremendous amounts of heat to the high latitudes of the northern hemisphere, may have crossed the Atlantic at a much lower latitude during the colder stages of the Pleistocene. Yet another hypothesis currently arousing considerable interest is that temperatures may be controlled by fluctuations in atmospheric carbon dioxide levels (Figure 17.22). Recent analyses of air bubbles trapped in ice cores from Greenland and Antarctica do indeed seem to indicate

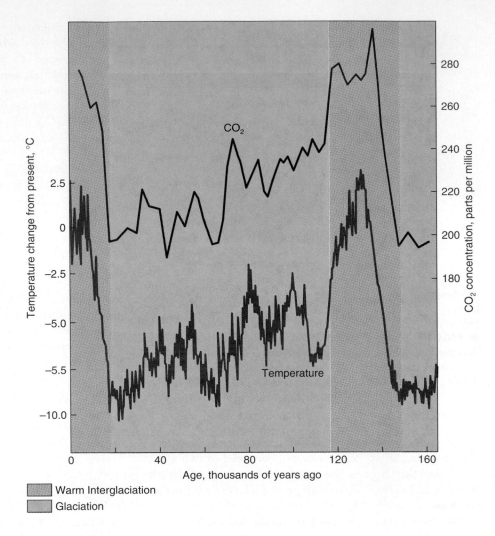

◆ **FIGURE 17.22** Changes in atmospheric CO_2 and air temperature for the last 160,000 years, determined from air bubbles trapped in the Vostok, Antarctica, ice core.

Warm Interglaciation
Glaciation

that colder temperatures are strongly correlated with times when CO_2 levels are reduced.

Whatever the answer, it appears likely that the present position of the North Pole in the restricted waters of the Arctic Ocean is of great significance. Continuing plate tectonic movements have positioned the land and water bodies of the high latitudes of the northern hemisphere in such a way that a small ocean—the Arctic Ocean—is nearly surrounded by two massive continents that provide potential sites for continental glacier growth. In addition, the Arctic Ocean has a sufficiently restricted pattern of water circulation with the adjacent warmer Atlantic and Pacific that its surface water can freeze.

The second facet of the Pleistocene problem is of greater potential import, because its answer relates strongly to the likelihood of future glacial epochs. It is this: What is the cause of the alternation of glacial and interglacial stages within the Pleistocene? Scientific investigation over the past few decades has shown that the answer apparently is is at least partially related to variations in the Earth's orbital geometry. Specifically, these variations include (1) the precession of the equinoxes, which determines the time of year

that the Earth is at aphelion and perihelion, and (2) long-term variations in the tilt of the Earth's axis. The influence of these factors is discussed in more detail in the Focus Box on pages 462–463. An important implication of these findings is that we are probably now in an interglacial period and that the ice sheets of North America and Eurasia may well re-form in the near geologic future.

Landforms Produced by Continental Glaciers

Continental glaciers produce many of the same types of landforms as do alpine glaciers. The sizes, locations, and relative abundance of most of these features exhibit important differences, however. As in the case of alpine glacial features, those produced by continental glaciation can be categorized as erosional or depositional in origin (although some are produced by both processes). The areas that undergo net erosion by a continental glacier generally are located beneath the central portion of the ice sheet, in the zone of ice accumulation. Surrounding this, in the zone of ablation, are the major areas of deposition.

Continental Glacier Erosional Landforms

A basic difference exists between the erosive actions of alpine and continental glaciers. Alpine glaciers, as we have seen, sharpen the topography and usually increase local relief. Continental glaciers have just the opposite effect. Because they cover everything, they subject the entire surface, not just lowland areas, to erosion. Surface features such as hills or mountains are especially susceptible to erosion because they protrude well into the faster moving zones of the ice. Lowland areas, on the other hand, are somewhat protected and may even be sites of local deposition within an area dominated by erosion. The surface as a whole therefore is smoothed and rounded by the ice.

The zone of erosion of the Laurentide ice sheet included the Canadian Shield area, the Canadian Maritime Provinces, and northern New England. The Fennoscandian ice sheet zone of erosion included Scandinavia and adjacent areas currently beneath the shallow waters of the North Sea and Gulf of Bothnia.

For the most part, the surfaces in the zones of erosion of both the Laurentide and Fennoscandian ice sheets were shield areas of ancient crystalline rocks, with predominantly low surface relief. These areas were further smoothed by the ice, so that today **ice-scoured plains** are the most extensive features of continental glacier erosion. These areas now consist of vast expanses of irregularly rolling surface, largely stripped of soil and covered by myriad lakes and swamps (Figure 17.23). The glaciers originally removed virtually all soil and weathered bedrock from these areas. Most areas today, though, have a thin covering of soil produced either by deposition as the ice melted or by postglacial weathering.

Among the most important features of ice-scoured plains are their numerous lakes and swamps. These features have formed because erosion and deposition by the glaciers obliterated the preglacial drainage network. In addition, areas of deeply weathered rock were scooped out by the ice, forming countless rock basins that subsequently filled with water to form lakes. Wandering through these plains in a haphazard fashion from lake to lake are stream systems displaying markedly ungraded longitudinal profiles and deranged drainage patterns. The presence of glacial lakes has earned Minnesota the nickname "The Land of 10,000 Lakes." The glacially derived lakes of Canada are so large and numerous that they reputedly contain more than half the world's fresh surface liquid water. The Great Lakes were either formed or greatly enlarged when the ice excavated areas of weathered rock.

The extensive glacial erosion of areas such as the Canadian Shield has greatly reduced their economic potential. The soil that once covered the surface was removed, and most weathered rock material was transported southward, where it now enhances the agricultural productivity of the midwestern United States. In addition, much land is agriculturally unusable because of glacially induced poor drainage. As a consequence, most of interior eastern Canada

◆ **FIGURE 17.23** View of the rocky, lake-strewn surface of an ice-scoured plain in the Northwest Territories of Canada.
SOURCE: Alan Kesselheim/Mary Pat Ziter/JLM Visuals

today is an almost uninhabited rocky, lake-strewn wilderness.

Continental Glacier Depositional Landforms

Continental glaciers produce a variety of depositional features (Figure 17.25). Some are similar in origin to those produced by alpine glaciers, but occur on a much larger scale; others are unique to continental glaciation. Most recognizable depositional features in North America were produced within the past 20,000 years and are the product of the Wisconsinan ice sheets.

The most extensive depositional feature is the **till plain,** a thick accumulation of ground moraine covering the preglacial surface over an extensive area (Figure 17.26). Depths of till often exceed 30 meters (100 ft), with thicker accumulations in originally lower areas such as valleys. In central Ohio, for example, till thicknesses over preglacial uplands average about 15 meters (50 ft), increasing to about 60 meters (200 ft) over lowland areas. Surface relief therefore is reduced by glacial deposition, although the preglacial topography is still generally reflected in a subdued fashion on the surface of the till plain.

Till plains cover much of the north-central portions of both the United States and Europe. The general flatness of the terrain, the high mineral content of the still lightly leached till, and the favorable present-day climate have combined to produce fertile soils. The American Midwest and the North European Plain are renowned for their agricultural productivity and currently support substantial human populations.

Most of the till forming a till plain is fine in texture and consists of sand, silt, and clay. Occasional rocks or boulders are mixed in, however, and in some areas they are quite numerous. The entire mixture is unsorted, since it was deposited directly by the melting ice. In some localities, the

Earth's Orbital Geometry and Climatic Change

Ever since nineteenth-century scientists discovered that the Earth has been experiencing a recent series of glacial and interglacial periods, a search has been underway for the mechanisms causing these changes. During the 1920s and 1930s, the Serbian mathematician Milutin Milankovitch argued that the glacial and interglacial stages resulted from long-term variations in the Earth's orbit. As late as the 1970s, most experts believed that changes in insolation resulting from these orbital variations would be insufficient to produce the necessary climatic changes. In the past two decades, however, knowledge of the chronology of the Pleistocene glacial and interglacial stages has been refined to the point that it has been possible to compare the timing of these stages with those of terrestrial orbital variations. The correlation has been so close that it is now almost certain that long-term variations in the Earth's orbit are a major cause of the glacial and interglacial stages.

Three orbital variations are involved (Figure 17.24). First are changes in the tilt of the Earth's axis with respect to the plane of the ecliptic. The tilt currently is approximately 23.5°, but it varies between 22.1° and 24.5° over a 41,000-year cycle. An increase in tilt concentrates insolation toward the poles and decreases insolation received at the equator. A decrease in tilt, which is currently taking place, has the opposite effect.

A second orbital variation is the precession of the equinoxes. The Earth's axis has been found to gyrate or "wobble" like that of a slowly spinning top. In addition, the elliptical orbit of the Earth is slowly rotating. The net result of these two independent movements is that the time of year that the Earth is at different points in its orbit—at aphelion and at perihelion, for example—gradually changes, completing a full cycle each 22,000 years.

A final orbital variation is the change in the eccentricity of the Earth's orbit over cycles of 100,000 and 400,000 years. When the eccentricity of the orbit increases, the Earth takes a less circular orbit around the Sun, and consequently there is a greater annual variation in distance between the Earth and Sun.

surface has been depressed by the weight of huge blocks of ice that have separated from nonmoving or "dead" glaciers during the melting process. This has produced pitted till plains; the pits, called **kettles,** are often currently occupied by small lakes.

Systems of terminal and recessional moraines form the major relief features on till plains in many areas of both the United States and Europe. These moraines take the form of long belts of low, gently sloping ridges or hills. They are arcuate in shape, bulging southward; this reflects the fact that the ice sheets, along whose melting edges the moraines were deposited, were lobate in form. Well-developed recessional moraine systems are especially numerous in the vicinity of the Great Lakes and in the area to their south and west in Ohio, Indiana, Illinois, Minnesota, and the Dakotas. In Eurasia, a series of moraines extends some 4,800 kilometers (3,000 mi) from northern Germany through Poland and western Russia to north-central Siberia. Both Long Island, New York, and the Danish Peninsula consist largely of morainic deposits.

Another, smaller feature frequently encountered on till plains is the **drumlin.** Drumlins are smooth, elongated hills composed of till. When well formed, they are oval in shape, resembling an egg that has been cut in half lengthwise and placed with its flat surface down. They generally occur in groups and are aligned with their tapered ends pointing down-ice (Figure 17.27). Some question exists as to their exact method of formation, but many apparently were produced by the plastering of till around subglacial obstructions lying in the path of the ice in places where the ice was overloaded with debris. A typical drumlin is perhaps 30 meters (100 ft) high and 600 meters (2,000 ft) long. Especially large groups or "swarms" are located in the Lake Ontario Plain of upstate New York, as well as in southeastern Wisconsin, northern Lower Michigan, the Niagara Peninsula of Ontario, and Ireland and Scotland.

Till plains, moraines, and drumlins are all features produced by direct glacial deposition. **Glacio-fluvial features** are also of great importance in areas of Pleistocene continental glaciation. Some of these features formed above,

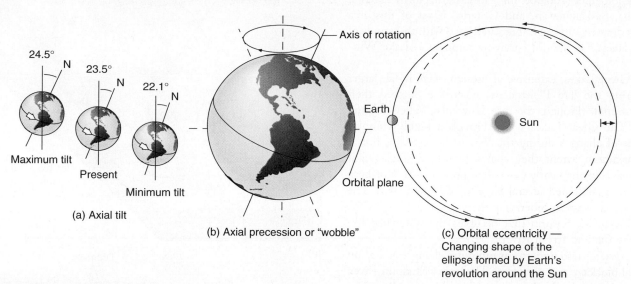

(a) Axial tilt

24.5° N — Maximum tilt

23.5° N — Present

22.1° N — Minimum tilt

Axis of rotation

Earth

Orbital plane

(b) Axial precession or "wobble"

Sun

**(c) Orbital eccentricity —
Changing shape of the
ellipse formed by Earth's
revolution around the Sun**

◆ **FIGURE 17.24** The three Earth orbital variations believed to be collectively responsible for the sequence of Pleistocene glacial and interglacial stages.

It has been found that, of these motions, the precession of the equinoxes and the changes in axial tilt have the greatest influence on temperature changes and on continental glacier formation and retreat. The influence of the Earth's orbital eccentricity is of secondary importance and mainly influences the amplitude of the precession cycle. Although human activites may well enter into the picture from this point onward, the net effect of these natural orbital variations over the next 10,000 or so years would seem to be the gradual ending of the present relatively warm interglacial stage and the beginning of a new period of continental glaciation. ■

within, or beneath the glacier, but most formed beyond the melting edge of the ice.

Foremost among the glacio-fluvial features in importance and areal extent are outwash plains and lacustrine plains. **Outwash plains** may be regarded as the glacio-fluvial counterparts of till plains. They consist of extensive areas covered by stratified drift deposited by meltwaters flowing outward from glaciers. Not only are the deposits vertically stratified, they also tend to be horizontally graded, with coarser sediments deposited near the ice front and progressively finer sediments away from the ice. Most outwash plains have rather smooth surfaces, but pitted outwash plains, containing numerous water-filled kettles, have formed where large blocks of ice became detached from the glacier and settled onto the outwash before melting.

Outwash plains are scattered throughout the areas of Pleistocene glacial deposition, primarily in the Great Lakes region and Upper Midwest of the United States and in the North European Plain. Areas of outwash plains are interspersed with moraine systems, till plains, and lacustrine

plains in complex patterns because of the numerous changes in position of the ice front. Where the ice front remained stationary for an extended period, however, an orderly sequence of deposits resulted, with till plain deposits beneath the ice, a terminal moraine along the forward edge of the ice, and outwash plain deposits beyond the ice margins (Figure 17.25).

Outwash plains formed where the surface sloped away from the glacier front; this allowed sediments to be transported and deposited by braided meltwater streams. In places where the surface sloped toward the ice front, however, drainage was impeded and ice-dammed lakes frequently formed. Very fine sediments accumulated on the floors of these lakes, and their eventual drainage produced **glacial lacustrine plains.** These areas now have exceptionally smooth, level surfaces and, like the outwash plains, have been converted largely to productive agricultural land. Many of the Pleistocene lakes were very large, and the lacustrine plains they created currently cover extensive areas. In North America, the largest lake of all was Lake Agassiz. Late in the

Wisconsinan glacial stage, it covered more than 260,000 square kilometers (100,000 mi²) in eastern North Dakota, Manitoba, Saskatchewan, and Ontario. Most of this area has now drained, although it is still dotted with thousands of smaller lakes. The largest surviving remnant is Lake Winnipeg.

The Great Lakes, extensive at present, were much larger still during the late Pleistocene, when ice blocked their current outlet through the St. Lawrence Valley (Figure 17.28). The Great Lakes went through a highly complex sequence of changes during the Wisconsinan Stage, but at their maximum extent they coalesced into a single giant inland sea, covering nearly twice their present area. Northern Eurasia also contained several large ice-dammed lakes. The largest was in western Siberia on the present-day site of the huge Vasyugansk Swamp that occupies the watershed between the Ob and Irtysh Rivers.

River systems also underwent profound changes as a result of glacial blockage and rerouting. The upper Missouri River, for example, originally flowed north to Hudson Bay. The growth of the Laurentide ice sheet blocked this course and redirected the river southward, adding it to the Mississippi River drainage system. The present courses of both the Ohio and Missouri Rivers developed roughly parallel to the line of maximum glacial advance and served as lateral drainage routes for glacial meltwaters. During the late Wisconsinan Stage, when the St. Lawrence River was blocked by ice, the Great Lakes drained southward through several routes into the Mississippi River system, as well as into the Atlantic through the Mohawk and Hudson River valleys and the Susquehanna River basin.

The rapid melting of great volumes of ice during the Pleistocene summers resulted in the production of enormous quantities of debris-laden glacial meltwater. Where valleys

(a)

(b)

◆ **FIGURE 17.25** Block diagrams of landscapes during and after glaciation, showing the origin of various continental glacier depositional features.

◆ **FIGURE 17.26** The checkerboard pattern of productive farmland attests to the fertility of the soil on this flat till plain southeast of Aberdeen, South Dakota.

SOURCE: John S. Shelton

were present to channelize the flow, large meltwater rivers were formed. These rivers, in most cases overloaded with glacial debris, formed large glacial spillways floored by extensive valley train deposits. The deposits often extended hundreds of kilometers from the ice front. The Mississippi River Valley contains deep accumulations of glacially derived alluvium and can be regarded as a largely buried valley train of giant proportions. Sporadic flows of glacial meltwater on a scale even larger than that of the Mississippi River periodically flooded a large portion of Washington State. The story of these flows, which radically altered the area's topography, appears in the Case Study at the end of the chapter.

Eskers are among the more unusual glacio-fluvial features associated with continental glaciation. They are long, meandering ridges of stratified drift, usually located on till plains (Figure 17.29). Most apparently were deposited by subglacial streams flowing through tunnels in the ice near the margin of the glacier. Because the surrounding ice locked the stream in place, it was unable to shift its channel as it deposited sediment and aggraded its bed. This produced an inverted stream channel deposit—higher, rather than lower, than its surroundings. Eskers are especially numerous in New England, near Lake Michigan, and throughout the Canadian Shield and Scandinavia. Most formed during the final melting of the Wisconsinan ice sheets, some 6,000 to 10,000 years ago.

Another feature of glacio-fluvial origin is the **kame,** a mound of slightly sorted water-deposited debris. The exact origin of kames can vary, but they sometimes form when glacial meltwater, flowing on the ice surface, enters a crevasse and descends to the base of the glacier. When the water reaches the bedrock floor of the glacier its velocity is reduced, and the debris it carries is deposited to form the kame.

Indirect Pleistocene Effects

The effects of the Pleistocene were not limited solely to those areas covered by glaciers or glacially derived meltwaters; probably every part of the Earth experienced changes in

◆ **FIGURE 17.27** These elongated hills in Antrim County, Michigan, are drumlins. Agricultural fields, which generally follow the contour of the land, typically run parallel to the orientation of the drumlins.

SOURCE: Photo courtesy of B.M.C. Pape

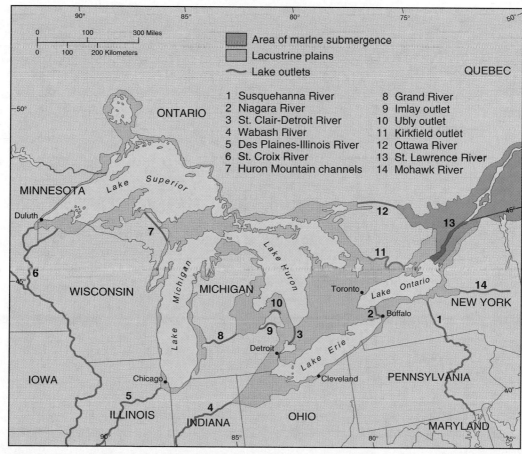

◆ **FIGURE 17.28** This map shows the expansion in areal extent of the Great Lakes during the late Pleistocene, when their present drainage route through the St. Lawrence River was blocked by ice. Major Pleistocene drainage outlets are numbered.

◆ **FIGURE 17.29** An area of pitted ground moraine and an esker in Coteau des Prairies, South Dakota.

SOURCE: John S. Shelton

◆ **FIGURE 17.30** Present-day lakes of the western United States occupy only a small proportion of the area covered by lakes during the cooler and wetter Pleistocene epoch.

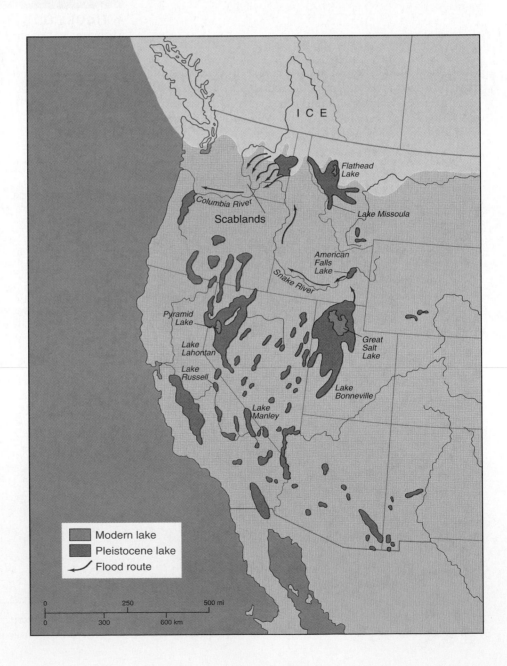

climate during this period. Pleistocene temperatures over most of the Earth are estimated to have averaged some 3 to 6 C° (5–10 F°) lower than at present. Pleistocene temperature fluctuations were responsible for major redistributions of plants and animals, including the extinction of numerous species. The cooler world climate during the Pleistocene glacial stages also influenced the locations of the global air pressure and wind belts. These belts, with their associated climatic conditions, generally shifted equatorward during cold periods, giving many areas markedly different precipitation patterns than they have at present.

The cooler, wetter Pleistocene climate resulted in the formation of large lakes in a number of currently dry areas with interior drainage. The currently arid Great Basin of the western United States, for instance, contained a large number of extensive Pleistocene lakes that have now either dried completely or have shrunk to small, often highly saline remnants (Figure 17.30). The largest lake of all was Lake Bonneville, which at one point occupied more than 51,000 square kilometers (20,000 mi^2) in Utah, Nevada, and Idaho. Its chief remnant is the Great Salt Lake, with an area of 2,400 square kilometers (940 mi^2). Lake Bonneville fluctuated considerably in extent, and a series of elevated terrace-like abandoned shorelines are readily visible at sites as much as 300 meters (1,000 ft) above the present surface of the Great Salt Lake. Numerous other large lakes existed in western Nevada, southern Oregon, and southeastern California, as well as in other parts of the world.

The drying of the climate in some areas apparently has continued into recent times. Large portions of the northern Sahara, for example, are covered by giant drainage networks that rarely, if ever, carry water at present. Extensive areas in extreme North Africa were forested only a few thousand years ago. The disappearance of these forests has been attributed primarily to human activities, but the present climate apparently is too dry to permit their return.

Another Pleistocene effect with worldwide implications was changing sea levels. The alternate formation and melting of the ice sheets involved the withdrawal and addition of vast quantities of ocean water. During the height of the glacial stages, enough water was locked in the glacial ice to lower global sea levels some 90 to 150 meters (300 to 500 ft). As a result, the Earth's appearance was altered; the surface was covered by less water and more land, and the shapes and locations of coastlines differed greatly from those of the present. For example, the British Isles were connected to mainland Europe, and North America and Eurasia were joined across what is now the Bering Strait. It is believed that the Bering Strait land bridge allowed Asiatic nomads who were the forebears of Native Americans to migrate from eastern Asia to the Americas.

Conversely, during the interglacial stages, sea levels were substantially higher than at present. A series of seven Pleistocene marine terraces, each representing a past position of the coastline, has been recognized along the Atlantic and Gulf coastal plain of the Untied States. The continued slow melting of the world's remaining glaciers is partially responsible for the gradual rise in ocean levels observed on most present coastal margins.

A related Pleistocene influence is the **isostatic rebound** of previously glaciated areas. The great weight of the continental glaciers depressed the underlying surface much as the pressure of a person's thumb might depress the surface of a rubber ball. When the pressure is removed, the ball springs back to its original shape. In similar fashion, the recent melting of the Laurentide and Fennoscandian ice sheets has allowed the areas formerly covered by the ice to rebound slowly to their original elevations. This process has been continuing for thousands of years and is now perhaps two-thirds completed. The areas that are rebounding are increasing in elevation with respect to sea level, causing their coastlines to advance seaward. As a result, countries such as Canada and Finland gain land each year. The displaced ocean water, which is relocated southward, has contributed to the current slow rise in sea level over much of the world. The resulting widespread problems of coastal flooding and beach erosion are discussed in the Case Study at the end of Chapter 19.

THE PERIGLACIAL ENVIRONMENT

The term **periglacial** literally means "near glacial." It is the climate of periglacial areas, however, rather than their location, that is near to glacial in its characteristics. Although conditions are warm enough, or dry enough, to prevent a permanent ground cover of ice and snow from forming, the climate is still severely cold. Landscape development in the periglacial environment is associated largely with the alternate freezing and thawing of the near-surface layers.

Location and General Appearance

At present, periglacial processes are most active in tundra climate areas. The tundra zone, as described earlier, includes areas with a warmest monthly mean temperature of 0°C to 10°C (32°F to 50°F). The absence of periglacial conditions in areas averaging below 0°C (32°F) in all months of the year results from the permanently frozen and generally snow- and ice-covered nature of the surface. Conversely, areas with temperatures averaging above 10°C (50°F) in summer generally are forested. Trees reduce the effectiveness of the freeze-thaw cycles and hold the soil firmly in place. The periglacial realm therefore is restricted at present largely to the far northern portions of North America and Eurasia, to some unglaciated fringe areas of Antarctica, and to scattered high mountain and plateau areas including, notably, much of the Tibetan Plateau.

Periglacial conditions dominated much larger areas during the Pleistocene. As the Pleistocene glaciers advanced and retreated, periglacial environments occupied shifting zones generally adjacent to the ice margins. These included, at different times, most of Canada and Alaska, the northern United States, much of northern and western Europe, and much of northern Asia, as well as extensive upland areas in other locations. In some of these areas, relict periglacial features still are discernible.

The periglacial landscape, especially in colder locations, often consists of a lightly vegetated surface covered with a thin, stony soil or regolith. Physical weathering processes are important, and in more rugged areas, the surface is often littered with a jumble of angular rock fragments of various sizes. These rocks generally have been freed by frost shattering from adjacent hillsides. In flatter areas, which are extensive, the surface material is mostly fine in texture, and a thin soil cover has formed. Drainage tends to be very poor, and irregularly shaped lakes and swamps, generally of glacial origin, are common in many lower and flatter sites. The entire area is frozen and largely snow-covered for 8 to 10 months of the year. Vegetation, which is more abundant toward the warmer margins, is of the tundra type and consists of a mixture of grasses, small flowering plants, heaths, mosses, and lichens (Figure 11.22).

Major Periglacial Features

Periglacial areas display a variety of distinctive surface features. The most prominent of these are patterned ground features and mass movement features. Both owe their existence largely to the alternate freezing and thawing of the surface.

Permafrost, or perennially frozen ground, underlies extensive land areas of the higher latitudes, including not only glaciated Antarctica and Greenland, but also much of northern North America and Eurasia. It extends well beyond the periglacial margins and into much of the northern conifer-

◆ **FIGURE 17.31** Northern hemisphere areas of continuous and discontinuous permafrost.

ous forest zones of these continents (Figure 17.31). Permafrost penetrates to an average depth of about 300 meters (1,000 ft), but is as much as 600 meters (2,000 ft) deep in parts of northeastern Siberia.

Frozen ground serves as a barrier to the downward movement of water. Although the upper meter or so of soil generally thaws in summer in unglaciated areas, the underlying permafrost traps water in the surface layer, resulting in nearly saturated conditions in the summer melt season. Poor drainage in many areas is reinforced by the recent Pleistocene disruption of the surface drainage network, as well as by cool summer temperatures that reduce evapotranspiration rates.

Repeated freezing and thawing in the so-called **active layer** near the surface, with its attendant expansion and contraction, produces a churning action in the upper soil and rock material, often with rather striking results. The surface material tends to develop polygonal patterns of a symmetrical nature, often covering vast areas (Figure 17.32). These so-called **patterned ground** features may be sorted or unsorted. Sorted patterns display a gradation from the finest material in the center of each polygon to the coarsest material, often in the form of large rocks, filling the cracks between the polygons. The sorting mechanism apparently is volume changes associated with the repeated freezing of near-surface water. This pushes rocks first to the surface and then laterally into the cracks. Nonsorted patterns show no significant size gradation of particles, but are evidenced by the slightly raised or lowered edges of the polygons or by

◆ **FIGURE 17.32** Patterned ground is caused by the contraction and cracking of the near-surface soil layers as they freeze. The cracks are occupied by vertical ice wedges that push the cracks farther apart as they expand.

SOURCE: Steve McCutcheon/Visuals Unlimited

variations in vegetation. Water may also accumulate and freeze in these cracks, eventually forming large vertically oriented **ice wedges** that taper downward and usually are covered with a thin layer of soil or regolith.

Mass movements also play a crucial role in periglacial landscape evolution. The chief factors favoring mass movements are the existence of saturated soil conditions in summer, a scarcity of stabilizing vegetation, and, most important, the action of ground ice. The volumetric expansion associated with the freezing of water loosens the regolith. On flatter surfaces this produces patterned ground features, which are actually small-scale mass movement phenomena. On sloping ground it results in a net downhill motion, generally consisting of a form of creep. In mountainous regions, rocks released from the adjacent slopes by frost shattering may accumulate to form **rock glaciers** that slowly advance downhill.

During the summer melt season, the upper surface of the permafrost on many slopes serves as a sliding plane for the overlying saturated masses of soil. The result is a type of earth flow termed **solifluction,** which may cause an entire hillslope to be covered by hundreds of solifluction lobes bulging downhill (Figure 17.33). The combined actions of creep and solifluction typically produce a steady transport of weathered material down slopes that in most other environments would be too gently inclined to experience mass movements of any significance. Other, more rapid mass movements also occur on steeper slopes under periglacial conditions, but are less widely distributed. Taken as a whole, mass movements exceed fluvial processes in importance in the periglacial environment, so that landscape evolution is accomplished more by slope retreat through mass wasting than by fluvial erosion.

▌SUMMARY

This chapter has examined the influence of ice on the Earth's land surface. Ice in the form of glaciers is the most powerful of the three tools of gradation. It is also an effective landscape modifier in periglacial environments, where it exists in a less concentrated form as ground ice. Despite its power as a geomorphic agent, ice is geographically restricted to low-temperature environments at high latitudes and elevations. In the geologically recent past, however, colder worldwide temperatures prevailed, and ice was of much greater geomorphic importance than at present. Many relict glacial and periglacial features still exist in areas that formerly served as sites of ice accumulation.

We now return to the Focus Questions posed at the beginning of the chapter:

1. *What are glaciers, and why do they form?*

Glaciers are large accumulations of freshwater ice derived primarily from the compaction of snow. They form on land areas where snowfall receipts exceed melting rates over an extended time span.

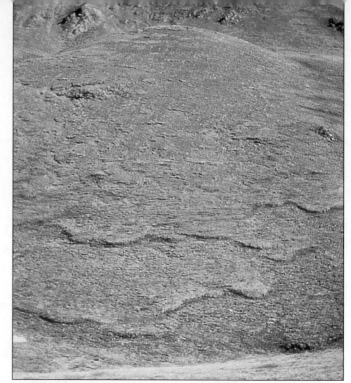

◆ **FIGURE 17.33** Solifluction flows near Suslositna Creek, Alaska, display the typical lobate topography characteristic of solifluction conditions.
SOURCE: NOAA

2. *What major categories of glaciers exist, and where are they found?*

Glaciers are divided into two basic types—alpine glaciers and continental glaciers. Alpine glaciers are much smaller than continental glaciers but are more widely distributed. They generally assume the form of long, relatively narrow tongues of ice occupying depressions in high mountain regions. Continental glaciers, which form over extensive high latitude land areas, are domelike bodies that move outward in all directions from their centers of accumulation. At present, continental glaciers occupy the continent of Antarctica and the interior of Greenland.

3. *In what ways do glaciers modify the landscape?*

Glaciers modify the landscape by eroding and depositing surface materials. Erosion is accomplished by the entrainment or "bulldozing" of weathered surface materials, and, to a lesser extent, by the scraping of rocks frozen into the base of the ice against the surface. The chief erosional landforms of alpine glaciers are cirques and glacial troughs; the ice-scoured plain is the chief erosional landform of continental glaciers.

Most glacial deposition occurs after the ice melts sufficiently to become overloaded with sediment. Sediment is transported in a conveyor-belt fashion from the glacial erosional area to the glacial depositional area. The chief depositional features of alpine glaciers are lateral and terminal moraines. Major depositional landforms produced by continental glaciers are terminal and recessional moraines and till plains. In addition, glacial meltwaters can carry large amounts of glacially derived materials away from the ice front to produce glacio-fluvial features such as lacustrine plains, outwash plains, and valley trains.

4. *What was the Pleistocene, and why did it occur?*

The Pleistocene was a geologically recent period of colder temperatures and widespread glaciation. There have been about 10 periods or "stages" of extensive glaciation over the past two million years, separated by relatively warm interglacial stages. During the glacial stages, continental glaciers formed over northern North America and northwestern Eurasia, leaving a major imprint on those areas today. Alpine glaciers also increased greatly in number and extent. The cause of the Pleistocene is not known, but the present location of the North Pole in the restricted waters of the Arctic Ocean is probably an important factor. The cyclic alternation of glacial and interglacial stages within the Pleistocene apparently results at least partially from long-term variations in the Earth's orbit around the Sun. There is growing evidence that the Pleistocene has not yet ended, and that the Earth is currently in an interglacial period.

5. *Where do periglacial environments occur, and what are the chief characteristics of periglacial landscapes?*

Periglacial environments develop in severely cold areas, usually those with tundra climates. Although the surface is not glaciated, landscape development is associated with the alternate freezing and thawing of the near-surface layers, which are usually underlain by a zone of permanently frozen ground, or permafrost. Most periglacial areas today exist in Alaska, northern Canada, and northeastern Eurasia. The permanently frozen subsurface layers act as a barrier to the downward movement of water, producing nearly saturated conditions during the summer, particularly in low-lying areas. This can lead to the occurrence of lobate earth flow movements, termed solifluction, over even gentle slopes. The repeated expansion and contraction of the surface layer as it freezes and thaws also act to produce polygonal patterned ground features of various types.

▌ QUESTIONS FOR THOUGHT AND REVIEW

1. What is a glacier? What conditions are necessary for a glacier to form?
2. What areas currently are covered by continental glaciers? By alpine glaciers?
3. How do glaciers transport and deposit regolith? How do the characteristics of direct glacial deposits differ from those of fluvial deposits? What are glacio-fluvial deposits?

4. Describe the overall impact of extensive alpine glaciation on a mountainous region such as the Alps. What are some of the more important erosional and depositional landforms?
5. How do the physical characteristics of alpine and continental glaciers differ?
6. What was the Pleistocene? Briefly describe three theories regarding its origin, including the Milankovitch theory.
7. Describe the maximum geographical extent of Pleistocene glaciation in North America and Eurasia. Why do you think glaciation was more extensive in North America than in Eurasia, even though Eurasia is a larger, colder landmass?
8. Describe the present surface characteristics of the areas that experienced a predominance of continental glacial erosion during the Pleistocene. Where are these areas located?
9. What are till plains? What are the present surface characteristics, geographic locations, and economic significance of the major till plains produced by the Pleistocene glaciers of North America and Eurasia?
10. What are the two most extensive types of glacio-fluvial features produced by continental glaciers? What are their surface characteristics? Where are they found in relation to till plains?
11. In what major ways would the physical geography of North America be different if the Pleistocene had never occurred?
12. What changes occurred during the Pleistocene in areas of the low and middle latitudes far removed from the ice sheets?
13. Under what circumstances do periglacial conditions exist? What parts of the world currently have periglacial conditions?
14. What is permafrost? What environmental problems are caused by the presence of permafrost?
15. What were the Spokane Floods? Explain the apparent cause of these floods.

▌ KEY TERMS

glacier	Pleistocene
continental glacier	Laurentide ice sheet
alpine glacier	ice-scoured plain
till	till plain
moraine	drumlin
crevasse	glacio-fluvial features
cirque	outwash plain
horn	esker
glacial trough	periglacial
fjord	patterned ground

A Glacial Catastrophe: The Spokane Floods

Geomorphologists and other Earth scientists are on the whole committed to the principle of **uniformitarianism,** which is often described by the expression, "the present is the key to the past." The idea is that natural Earth phenomena of the geologic past were produced by the same physical processes that operate at present. On occasion, however, landforms are found that differ so greatly in appearance or magnitude from those with which we are familiar that at first they seem to defy the uniformitarian concept. An area containing such landforms exists in the Columbia Basin of southeastern Washington. After more than 50 years of study, Earth scientists have concluded that many of its features were produced by the largest verified floods in the history of the Earth.

The Columbia Basin is a low-elevation portion of the Columbia Plateau. The climate is arid, and the topography varies from flat to hilly. Despite its present aridity, however, a large portion of this area shows strong evidence that short-lived floods of gigantic proportions occurred during the Pleistocene. An area of approximately 28,500 square kilometers (11,000 mi²) is covered by a variety of fluvial erosional and depositional features, some on a scale unknown anywhere else in the world. Among these features are numerous dry canyons (called **coulees**) containing stream-lined mesas and buttes, rock-rimmed basins, low basaltic ridges apparently eroded into segments by rushing floodwaters that overtopped them, dry waterfalls, giant gravel bars, and, perhaps most amazingly, ripple marks 15 meters (50 ft) in height and spaced 150 meters (500 ft) apart (Figure 17.34).

Debate on the origins of this area, called the Channelled Scablands because of the abandoned river canyons scarring its surface, continued for many years following its first description in a 1923 study by J. H. Bretz.[1] Many geomorphologists originally believed that it resulted from a long-continued flow of glacial meltwater. It gradually was realized, though, that features such as the overtopped drainage divides and giant ripple marks could be explained only as the result of one or more gigantic, but very brief, floods. The problem was then to determine how and where such floods could have originated.

In 1942 an answer finally was found. It had been known for some time that, during the Wisconsinan glacial stage, much of what is now western Montana was covered by a glacially dammed lake that was given the name Lake Missoula. J. T. Pardee, a geomorphologist employed by the U. S. Geological Survey, hypothesized that the lake was blocked by a lobe of ice that extended down the Purcell Trench in northern Idaho.[2] The melting or breaching of this ice lobe allowed the estimated 1,250 cubic kilometers (300 mi³) of water in Lake Missoula to be released within a period of a few hours or days, creating the catastrophic flood that carved the Channelled Scablands (Figure 17.35). This origin now is generally accepted.

Recent fieldwork in the area indicates that Lake Missoula drained suddenly and catastrophically on numerous occasions during the Late Pleistocene. Geologist R. B. Waite has presented evidence of 40 catastrophic floods during the Late Wisconsinan period (between about

continued on next page

◆ **FIGURE 17.34** Giant ripple marks produced by the Spokane Flood.
SOURCE: John S. Shelton

This chapter examines the geomorphic aspects of two related topics. We begin with a discussion of the surface effects of the wind. The wind is the weakest of the tools of gradation; it attains importance as a landscape modifier only in areas devoid of a protective vegetation cover. Most such areas have arid or semiarid climates, although sandy beaches and other localized environments can also be strongly influenced by the effects of the wind. Geomorphic processes performed by the wind are called **eolian** processes after Aeolus, the Greek god of the winds.

The second section departs from the single geomorphic process approach of the last several chapters and examines the landscape characteristics of desert regions. Desert landscapes are not dominated by one geomorphic process; instead, they are polygenetic in origin, with tectonic, fluvial, and eolian processes being most important. The specific actions of each of these processes differ from place to place, causing deserts to vary greatly in appearance. As a group, though, deserts display surface characteristics that differ in several basic respects from those of humid landscapes.

THE WIND AS A GEOMORPHIC AGENT

The ability of windstorms to erode and transport enormous quantities of loose sand and dust has long been demonstrated to travelers in arid lands, and until fairly recently, the wind was believed to be the dominant geomorphic agent within desert regions. This assessment of the wind's surface influence has been downgraded considerably, however, because the erosive power of the wind on solid rock has been found to be minor. Its weakness results largely from the gaseous condition of the atmosphere and the air's very low density ratio of about 1:2,000 as compared to rock. (Water and ice, by contrast, have density ratios of about 1:3 when compared to rock.) As a result, the wind is restricted primarily to the redistribution of loose, fine-textured surface materials. Where such materials are plentiful and are not immobilized by moisture or vegetation, wind can perform substantial erosional and depositional work and produce a variety of distinctive surface features.

Wind Sediment Transport Mechanisms

The wind transports sediments in much the same manner as does flowing water, although some important differences exist. Both wind and water are fluids whose flow over a surface generates turbulence. The upward component of the turbulent eddies allows both fluids to pick up loose sediments and to carry them in suspension. Although the wind typically flows much faster than water, its ability to carry suspended particles is reduced by its lower density. As a result, normally only silt and clay particles are carried in suspension by the wind. These fine particles, however, can be lifted high into the air and quickly transported for great distances before being deposited.

Where winds are exceptionally strong and large quantities of loose soil are available, **dust storms** are sometimes

◆ **FIGURE 18.1** The awesome front of a dust storm approaches Clayton, New Mexico, on the morning of May 29, 1937. Storms such as this were common in the southern Great Plains during the "Dust Bowl" years of the 1930s.
SOURCE: AP/Wide World Photos

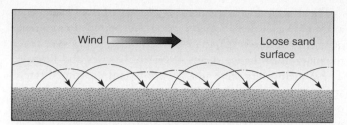

◆ **FIGURE 18.2** Most sand is moved near the surface by saltation. Sand grains are picked up by the wind and carried a short distance before falling back to the ground where they usually hit other grains, causing them to bounce and move in the direction of the wind.

produced. These storms can reduce surface visibilities to only a few meters and can last for hours. Tremendous volumes of valuable topsoil can be lost from agricultural regions, as was discussed in the Case Study at the end of Chapter 12. The best-known dust storms in North America occurred during the 1930s in the "Dust Bowl" region of the southern Great Plains of the United States (Figure 18.1). Although improved agricultural practices and moister climatic conditions have greatly reduced the frequency of such storms within this region, severe dust storms are still common in portions of Africa, Asia, and Australia.

If sediments are too coarse textured to be carried in suspension, wind—like water—can transport them by surface creep or by saltation over the surface. **Surface creep** refers to the slow shifting of surface materials by rolling or sliding. It chiefly transports particles ranging from 2 to 4 millimeters (0.08–0.16 in) in diameter and accounts for approximately one-fifth of all eolian surface sediment transport. **Saltation** is an asymmetrical bouncing motion of surface grains (Figure 18.2). It is caused both by the aerodynamic lift resulting from the increase in wind speed with height above the ground and by the impact of other saltating grains. Saltation moves materials rapidly and accounts for about four-fifths of all eolian surface sediment transport. It is through saltation that sand grains are aggregated into shifting sand dunes. Because saltation rarely lifts sand grains more than a short distance above the surface, true **sandstorms** consist of a stinging mass of wind-whipped saltating sand grains extending only a meter or two above ground level.

To initiate the eolian transport of particles of any given diameter, a critical wind speed threshold is needed to produce the necessary frictional pull or **drag velocity** (Figure 18.3). Once particles begin saltating over the surface, though, their impacts with other particles can sustain the motion at somewhat lower wind speeds. As is true of sediments entrained by flowing water, sand-sized particles are the first particles in an undisturbed area to be eroded, chiefly because of the compaction and adhesion of finer particles.

Despite the general similarities between the transport of sediments by water and by wind, several important differ-ences exist between the two. Wind transport is much more rapid but is restricted to finer-textured sediments than is water transport. In addition, the wind does not carry a dissolved sediment load. Finally, the wind is much less restricted than water in the directions it can travel and in the sites in which it can erode or deposit sediments. Water always transports sediments downhill, while the wind can transport sediments either uphill or downhill (although downhill movement predominates). Fluvial processes are confined largely to the valley or lowland sites where water flow is concentrated. In contrast, the wind flows over the entire surface and is capable of eroding or depositing materials almost anywhere within a suitable region.

Eolian Environments

In order for eolian processes to be effective, surface materials must be fine textured, dry, and loose. In addition, the affected area must experience at least occasional strong winds. A vegetation cover greatly diminishes the effectiveness of the wind by producing a large amount of near-surface friction that reduces wind velocities. It also covers the soil and holds the soil particles in place. For these reasons, areas subject to significant wind erosion must be at least partially devoid of vegetation.

◆ **FIGURE 18.3** The relationship between wind drag velocity and particle diameter. The fluid threshold is the minimum wind speed needed to initiate grain movement by the force of the wind alone. The impact threshold is the minimum wind speed needed to sustain the movement of a grain that has first been set in motion by the impact of another particle.

The most extensive geomorphic environments meeting these criteria are deserts. Every continent contains large deserts in which the wind plays an important role in landscape development. In most deserts, wind erosion predominates over deposition. The surface in such areas is covered largely by stones of various sizes because the finer sand, silt, and clay particles have been winnowed out by wind action. Approximately one-fourth of the world's deserts, though, have been the sites of net deposition and are covered partially or completely by sand.

Another well-known but much less areally extensive category of eolian environment consists of the coastlines of large water bodies, including the ocean, seas, and large lakes. Along coastlines, waves and currents supply weathering products susceptible to wind action, and vegetation is discouraged by storms and tides. The coastal environment is described in detail in Chapter 19.

Three types of more localized environments may also contain large quantities of loose, fine-textured sediments capable of removal or redistribution by the wind. One includes areas underlain by weakly cemented sandstone. In such areas, weathering activities detach the individual grains to form loose sand.

The valleys of streams draining arid and semiarid regions may also be important sites for eolian processes. Because their volumes of flow fluctuate greatly, these streams support relatively little stream bank vegetation. During periods of low water, extensive strips of sandy alluvium are often exposed on the streambed. Strong winds are able to remove the silt and clay particles and to collect the sand grains into dunes. The floodplains of several large rivers of the American Great Plains, including the Missouri, Platte, and Arkansas, are important sites for these eolian activities.

The third local environment is a legacy of the Pleistocene, a time when glacial meltwaters transported tremendous quantities of fine-textured glacial sediments southward from the margins of the ice sheets of North America and Eurasia. These meltwaters were not always organized into discrete streams, but often spread over large areas in braided or overland flow patterns. They frequently were overloaded with sediments, which they deposited over extensive areas. The wind removed great volumes of predominantly silt-sized materials from these deposits, and their downwind deposition produced extensive windblown sediment or **loess** (from the German *loss,* or "loose") accumulations. Most glacial outwash deposits currently are stabilized by vegetation.

Over the last few centuries, eolian processes unfortunately have been augmented considerably by human activities. Especially affected are semiarid and seasonally dry regions in Africa and Asia that are used for agricultural or pastoral pursuits. The activities of the rapidly growing human populations within these regions have led to the extensive removal of the already marginal vegetation cover. The loss of protective vegetation, in turn, has allowed the wind to erode tremendous quantities of irreplaceable soil. The human element in eolian sediment transport is explored further in the Case Study at the end of this chapter.

Wind Erosional Processes and Features

The wind erodes materials from the surface in two different ways. **Abrasion** occurs from the impact of saltating sand grains and often is described as a "sandblasting" effect. **Deflation,** on the other hand, occurs when the wind removes loose surface materials directly.

The effect of wind abrasion on solid rock is relatively minor and generally superficial. Wind normally does not produce major bedrock erosional features; it merely modifies their surfaces by pitting, etching, smoothing, or polishing. These activities are most effective on weak sedimentary rocks such as soft shales and poorly consolidated sandstones. On such rocks, the wind tends to etch out micropatterns that reflect differences in rock hardness. **Ventifacts** are desert rocks with prominent wind-polished, grooved, or faceted surfaces. Sand grains large enough to perform abrasion are transported only a small distance above the surface. As a consequence, features standing more than a few meters above the ground are not significantly affected by this process (Figure 18.4).

Deflation is a much more important mechanism of wind erosion than abrasion, but it requires the presence of unconsolidated surface materials such as sand or dry soil. In many instances the entire surface is rather evenly lowered by deflation; as a result, no striking surface features are produced, and it is difficult to estimate the quantity of materials removed. On occasion, though, remnants of the original surface, perhaps anchored by the roots of a tree or shrub, are left to provide an indication of the depth of surface removal.

◆ **FIGURE 18.4** The effect of wind abrasion on solid rock is generally considered to be relatively minor. Repeated sandstorms, though, are sometimes capable of producing "mushroom rock" formations in soft rock outcrops because erosion is most vigorous just above the ground. The example here is in southeastern Iran.
SOURCE: Rodman Snead/JLM Visuals

◆ **FIGURE 18.5** A deflation hollow in Death Valley, California.
SOURCE: Martin G. Miller/Visuals Unlimited

One widespread category of features resulting from the differential erosion of surface materials is **deflation hollows** or "blowouts" (Figure 18.5). These are shallow, often circular depressions that can vary greatly in size. The largest are more than 1.6 kilometers (1 mi) in diameter and up to 15 meters (50 ft) deep, although most are much smaller. Many thousands of such depressions dot portions of the southern Great Plains of the United States. These features, which may have been formed in part by solution and past bison wallowing, become the sites of shallow ponds after heavy rainfalls.

Deflation is the only natural mechanism that can remove sediments from enclosed desert basins. Large desert basins, such as the Qattara Depression, which covers 18,000 square kilometers (7,000 mi²) of northwestern Egypt, may owe their origin primarily to deflation; in effect, they may be gigantic blowouts.

In many arid regions the eolian removal of sand–sized and finer sediments has left behind deposits of pebbles, stones, and boulders that are too large and heavy to be blown away. In time, through the effects of gravity, salt crystal growth, and ephemeral floodwaters, these deposits can accumulate and be arranged into a **desert pavement** of close-fitting pebbles or stones that protect the underlying surface from erosion (Figure 18.6). A stony desert from which the fine particles have been removed by the wind is called a **reg** in Arabic, and this term has been applied to similar deserts in other parts of the world. Regs cover large areas in the southwestern United States, the Sahara, and interior Australia.

Wind Depositional Processes and Features

Wind, like water, deposits sediments when its speed of flow and resulting turbulence become insufficient for it to continue to transport these materials. The transport of sand requires higher wind speeds and greater turbulence than does the transport of finer sediments, and this factor leads to the sorting of eolian deposits on the basis of mass. Where both sand and silt are derived from the same source region, the coarser sand particles usually are deposited relatively near the source of supply, while the windblown silt is deposited as loess over a larger region farther downwind. Not only are these two eolian deposits geographically separated from one another, but each has distinctive physical characteristics, landform assemblages, and human implications. They therefore are treated separately in the following discussion.

Topography of Sand-Covered Surfaces

Sand consists of loose grains of rock ranging from 0.0625 to 2.0 millimeters (0.0025 to 0.08 in) in diameter. Most eolian sand deposits are composed chiefly of the stable mineral

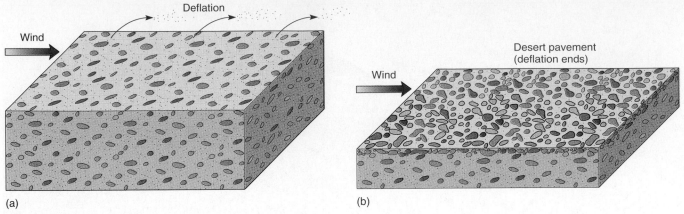

◆ **FIGURE 18.6** (a) Desert pavement forms when deflation removes fine-grained material from the surface, leaving larger-sized particles. (b) As deflation continues, the larger particles are concentrated and form a desert pavement, which protects the underlying material from additional deflation.

quartz and have average grain diameters of about 1 millimeter (0.04 in). Such grains, when dry and loose, begin to move over the surface when wind speeds exceed 4.5 meters per second (10 mph).

Because sand normally does not travel far from its region of origin and can be moved by the wind only when it is dry and loose, it accumulates primarily in deserts. The impression that many people have of most deserts being blanketed by sand, however, is incorrect. Sand does not cover the majority of any large desert. Rather, it is accumulated within certain areas by the action of the wind, and, to a lesser extent, of ephemeral streams. Sandy deserts, sometimes referred to as **ergs** (another Arabic term), comprise at most one-fourth of the world's deserts. They contain not only local sand, but also sand derived from surrounding nonsandy desert areas. The quantities of sand within such an accumulation area vary from scattered patches in some locations

(Figure 18.12) to a massive cover in others (Figure 18.7).

The major sandy deserts of the world are depicted in Figure 18.8. The largest ergs are located in the central Sahara, eastern Saudi Arabia, west-central Australia, southeastern Iran, northwestern India and adjacent Pakistan, and western China. Additional smaller sandy deserts are scattered throughout other arid lands, including the southwestern United States. A well-known sand accumulation in the United States is located within White Sands National Monument near Alamogordo in southern New Mexico. It is unusual in that it contains snow-white gypsum sand derived from deposits on a nearby dry lake bed. Great Sand Dunes National Monument in south-central Colorado is another sand accumulation site that has been preserved as a tourist attraction.

The most characteristic and important surface features associated with sand-covered regions are dunes. A **dune** is a mound or ridge of wind-deposited sand. Most sand-covered regions exhibit dunal topography. Sand dunes vary greatly in size and shape; small dunes, which are most common, usually are 3 to 15 meters (10–50 ft) in height. Where sand is plentiful and winds are strong, though, dunes can attain heights of 180 meters (600 ft) and extend for many kilometers. The dimensions and forms of sand dunes are determined chiefly by the quantity of available sand, the strength and direction of the wind, and the presence and extent of stabilizing vegetation. Areas with winds that are relatively constant in direction generally have the most organized dune patterns.

Dunes form and grow largely because of the frictional resistance they offer to the wind. They are self-generating features because, as they grow, they become more effective barriers to the wind, reducing its speed and causing it to deposit any sand it may be transporting. Because other surface features also offer frictional resistance to the wind, dunes are best developed on flat, unvegetated surfaces.

The sides of most dunes are asymmetrical, with a gentler windward-facing **back slope** up which the sand grains are

◆ **FIGURE 18.7** Giant sand dunes in the Sahara (Hammada du Guir, Morocco).

SOURCE: R. Krubner/H. Armstrong Roberts

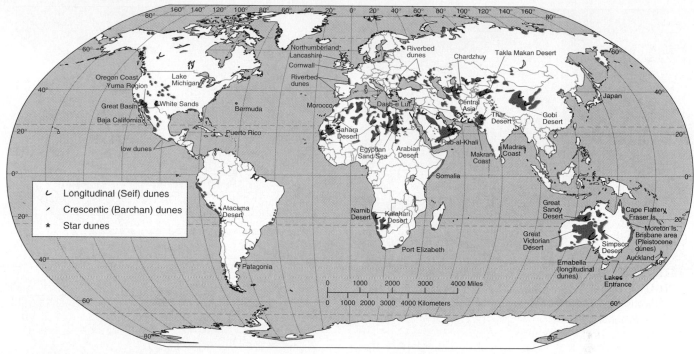

◆ **FIGURE 18.8** Global distribution of sand-covered regions and their characteristic dune forms.

blown, and a steeper leeward-facing **slip face** down which the grains tumble until they land within a niche in which they can lodge (Figure 18.9). The slip face normally rests at the angle of repose of dry sand, which is approximately 34° from the horizontal. Through the mechanism of individual sand grains saltating up the back slope and masses of sand sliding down the slip face, the dune gradually changes its location and shape.

Dunes are described as "active" if they are not covered by vegetation and are currently moving and changing shape. Conversely, they are "inactive" if stabilized by a partial or complete cover of vegetation. Stabilized dunes often are relict features of a past drier climate. For example, the Sand Hills of central Nebraska is an area of sandy Pleistocene outwash deposits, reworked by the wind, that is currently covered by sod and grasses. The Palouse region of Washington is another important area of stabilized sand dunes (Figure 18.10). In addition, large areas of stabilized dunes exist along the equatorward margins of the Sahara and Kalahari Deserts. They provide evidence that the subtropical deserts, at least in Africa, have shifted poleward in post-Pleistocene times.

Local wind systems frequently organize sand dunes into complexes that give a distinctive pattern to the topography. Dunes tend to be similar within a given area because the controlling factors usually are fairly constant. A basic distinction is normally made between longitudinal dunes, which are elongated parallel to the prevailing winds, and transverse dunes, which are elongated perpendicular to the prevailing winds. The following paragraphs describe, and Figure 18.11 illustrates, some of the major categories of sand dunes on the basis of form.

Probably the most common and familiar of the dune forms are the **barchans.** Relatively small and crescent shaped, they are oriented with their long axes perpendicular to the prevailing winds and their tapered ends or "horns" pointing downwind (Figures 18.11a and 18.12). Barchans,

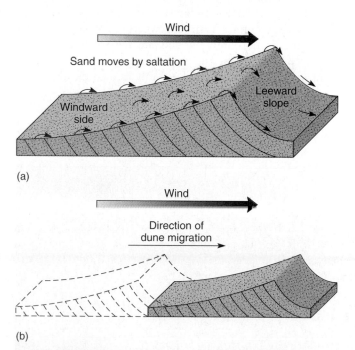

◆ **FIGURE 18.9** (a) Profile view of a sand dune. (b) Dunes migrate as sand is blown up the windward side and slides down the leeward slope. This movement of the sand grains produces a series of inclined beds that slope in the direction of wind movement.

◆ **FIGURE 18.10** A sea of stabilized Pleistocene sand dunes in the Palouse region of Washington State.
SOURCE: H. Armstrong Roberts

which are found in portions of all large deserts, develop in areas of relatively constant wind direction and restricted sand supply. They are the most mobile of the major dune types.

Longitudinal dunes are long ridges of sand generally aligned parallel to the direction of the prevailing winds (Figure 18.11b). They may become very large, attaining heights of more than 100 meters (300 ft) and lengths of many kilometers. Longitudinal dunes are especially well developed in central Australia, where they cover nearly one-fourth of the continent and are remarkable for their even spacing. Most of the Australian dunes are currently stabilized by vegetation and apparently formed during a past

◆ **FIGURE 18.11** (a) Barchan dunes are crescent-shaped dunes that have their tapered edges or "horns" pointing downwind. They exist on the hard, flat floors of deserts where there is constant wind and a limited sand supply. (b) Longitudinal dunes are linear dunes oriented parallel to the wind. They exist in deserts with a limited sand supply and strong winds from one prevailing direction. (c) Transverse dunes form wavelike sand ridges oriented perpendicular to the wind direction. They exist in areas with abundant sand and little vegetation.

drier period. Currently active longitudinal dunes cover extensive desert areas in Saudi Arabia, Egypt, and Iran.

Transverse dunes are elongated perpendicular to the prevailing winds (Figure 18.11c). Common in regions of abundant sand, they typically assume a wavelike pattern with undulating crest lines and have been likened in appearance to a picture of a storm-tossed sea. They are much less symmetrical and continuous than longitudinal dunes. Some transverse dunes develop a clearly distinguishable linked barchan form and may separate into individual barchan dunes if they extend into an area containing less sand. Transverse dunes frequently form on sandy beaches that have a prevailing onshore wind flow.

Sand dunes do not always develop the symmetrical forms just described. Complex dunes of almost any shape may form, especially in areas of variable winds. These include the so-called **star dunes** of Egypt and the Arabian peninsula,

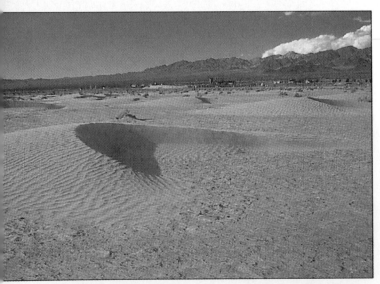

◆ **FIGURE 18.12** Closeup view of a field of barchan dunes.
SOURCE: Dick Hilton

which are among the highest in the world. They consist of pyramidal mountains of sand with sharp crest lines on the various radiating ridges that form the "points" of the star. At the other extreme, some deserts are covered by flat **sand sheets** that display little or no tendency toward dunal formation. One of the best examples of a sand sheet covers a large portion of the Libyan desert.

Most sand dunes, unless stabilized by vegetation, are free-moving forms that gradually change their sizes, shapes, and locations. In contrast, fixed dunes, sometimes called **sand shadows,** frequently form in the lee of an obstruction to the wind such as a building or a rock outcrop (Figure 18.13).

Superimposed on all sand deposits are small **sand ripples** that contain only 2 to 5 centimeters (1–2 in) of surface relief. They are highly similar in cause and appearance to the small ripples on water that are superimposed on the larger wave forms on a windy day. Sand ripples are erased and re-formed during each windy period and develop an undulating pattern elongated perpendicular to the direction of the wind.

Topography of Loess-Covered Surfaces

Loess consists predominantly of silt-sized eolian deposits. It is typically buff-colored and is composed of a well-sorted mixture of quartz and calcite ($CaCO_3$) particles. Loess is derived from the same general areas as sand. The three most important loess sources are deserts, Pleistocene outwash deposits, and the floodplains of rivers in semiarid regions.

Despite its general similarity in origin to sand, loess differs markedly in most other geographic and geomorphic characteristics. Because it must be stabilized by moisture and a cover of vegetation in order to accumulate, it is not associated with desert environments. Most loess deposits are currently overlain by both vegetation and a surface soil layer; consequently, they are not as easily recognizable as sand deposits, which often are devoid of both soil and vegetation. Soils derived from loess, in contrast to those developed from sand, often are highly fertile and agriculturally productive.

Uneroded loess deposits usually do not form distinctive topographic features because the wind spreads the loess as a smooth blanket over the surface. As a result, some of the most level terrain on Earth occurs in undissected loess regions. These regions become much more topographically distinctive, however, when subjected to stream erosion. Loess particles are angular and tend to interlock and develop a columnar structure. At the same time, the unconsolidated nature of loess readily permits streams to erode into the deposits and form nearly vertical banks or cliffs. As a result, eroded loess areas typically are dominated by deep, steep-walled valleys subject to rapid lateral and headward expansion (Figure 18.14).

Loess, predominantly of Pleistocene origin, currently covers approximately 10 percent of the Earth's land surface and 30 percent of the United States. Depths range from as much

◆ **FIGURE 18.13** Sand shadows form in the lee of an obstruction because of the localized reduction in wind speeds.
SOURCE: Frank Hanna

as 90 meters (300 ft) immediately downwind from a major source to only a few centimeters. In general, the deposits taper in a downwind (usually easterly) direction. The major loess-covered regions of the world are shown in Figure 18.15. Among the most important are portions of the North European Plain from Belgium to Ukraine, parts of central Asia, northern China, the Pampas of Argentina, and the Great Plains, Midwest, and Mississippi River Valley in the United States. It is notable that most large deposits exist within the semiarid and subhumid areas of the middle latitudes that collectively comprise most of the world's major grain-producing regions.

◆ **FIGURE 18.14** Loess particles are angular and tend to interlock, allowing them to form steep cliffs when subjected to stream erosion. These nearly vertical loess cliffs are located along the Yukon River in northwestern Canada.
SOURCE: Steve McCutcheon/Visuals Unlimited

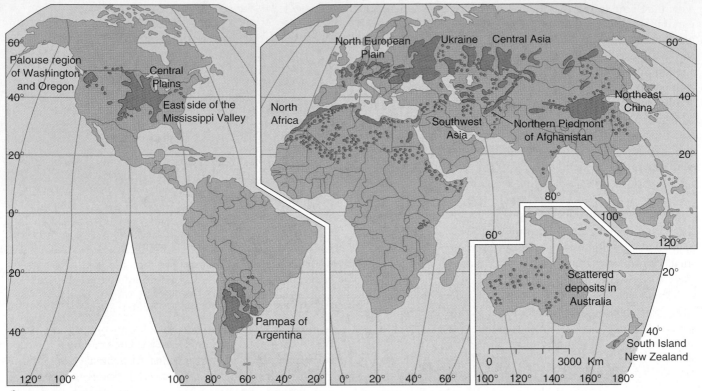

◆ **FIGURE 18.15** Global distribution of major loessal regions.

Most loess deposits within North America were derived from braided river courses such as that of the Missouri River, dry lake beds, and glacial outwash plains at times when these surfaces were largely devoid of vegetation. Maximum thicknesses of 15 to 30 meters (50–100 ft) occur on the eastern bluffs of the Lower Mississippi River and just east of the major Pleistocene outwash plains in Nebraska, Iowa, and northern Kansas. A smaller area of loess occurs in the Palouse region of southeastern Washington. The loess deposits of Europe and central Asia also had a glacial eolian origin, but those of northern China and Argentina apparently were derived largely from deserts to their west.

◆ **FIGURE 18.16** Tilted fault-block mountains and salt-encrusted playas are common geomorphic features in the Great Basin of the western United States. This photo shows Racetrack Playa, in Death Valley, California, with the Inyo Mountains in the background.

SOURCE: John S. Shelton

DESERT LANDSCAPES

A **desert** is a large area with little or no vegetation. This broad definition can be taken to include the so-called polar deserts, where plant life is absent because of a combination of cold temperatures and a surface cover of ice and snow. In this section, however, the concept is restricted to deserts resulting from aridity.

The causes, climatic characteristics, and global distributions of arid regions were discussed in Chapter 8. It should be recalled that arid regions are those in which potential evapotranspiration rates exceed precipitation receipts on a long-term basis. Most are located in the subtropics and owe their dryness to the subsidence of air within the subtropical high pressure belts. Additional deserts exist in the middle latitudes. Middle latitude deserts sometimes are described as topographic or rainshadow deserts because their aridity results largely from the moisture-blocking influences of

◆ **FIGURE 18.17** The angular nature of many desert landscapes is well illustrated in this view of Monument Basin in Canyonlands National Park, Utah. Downcutting by the Colorado River and its tributaries has produced these steep cliffs and isolated sandstone "monuments."

SOURCE: Photo from Van Diver, B. B. 1988: *Imprints of Time: The Art of Geology*, Mountain Press Publishing Co., Missoula, MT

mountain barriers. In all, some 19 percent of the Earth's land surface consists of arid deserts.

Surface Characteristics of Deserts

The first-time visitor to a desert is struck immediately by the great contrast between the desert landscape and those of more humid environments. Perhaps the most fundamental impression is of the starkness of the land (Figure 18.16). The surface is not covered by a mantle of soil and vegetation, but instead is likely to consist largely of a loose jumble of rocks of various sizes. The dominant colors are the buffs, grays, and browns of rock in various states of weathering rather than the soft greens of live vegetation. The topography often appears highly angular, with a prevalence of horizontal and vertical surfaces and a lack of the smoothly rounded slopes typical of many humid landscapes (Figure 18.17). Even the clear weather and unlimited visibilities that prevail in most desert regions add to the impression that the universe consists predominantly of rock and sky.

The underlying factor controlling all the differences between desert and nondesert regions is the reduced availability of surface moisture. The low precipitation totals received in deserts not only discourage the growth of vegetation, but also profoundly affect the processes of landscape modification. Chemical weathering processes, which are highly dependent upon the presence of water, are especially affected. Although still important, they progress much more slowly than in humid climates and generally do not reach the clay mineral production stage. Physical weathering activities are less affected, because more of them require either no water or else just the short-term presence of water. The major physical weathering processes in deserts are salt crystal growth, frost wedging in colder areas, and perhaps thermal expansion and contraction related to daily heating and cooling. Figures 15.9 and 15.10 summarize the relationship between climatic conditions and the relative importance of physical and chemical weathering processes.

The reduced rates of weathering and limited vegetation cover cause bedrock to be exposed much more extensively in deserts than in most humid environments. Based on the nature of their surface materials, about one-fourth of the world's deserts may be classified as ergs, or sandy deserts. Over half are **regs,** or stony deserts. A third category of desert contains a solid bedrock surface, sometimes covered by a scattering of boulders. Known as **hamada,** they are less extensive than the other two categories. True desert regions generally contain only relict soils (left from a previous more humid period) or, locally, soils developed on alluvial deposits.

The angular topography that characterizes many desert regions has been attributed largely to the virtual absence of creep as a downhill transporter of weathered materials. Instead, most downhill movement of weathering products occurs by surface water transport during heavy rainstorms or by rapid mass movements such as rockslides and mudflows.

Despite its restricted availability within desert regions, water is still the most effective erosional agent for deserts as a whole, with wind relegated to a distant second place. Because less water is available for weathering and erosional activities, the overall rate of landscape evolution is much slower in arid than in humid environments. On the other hand, weathered surface materials within deserts are highly mobile because they are not stabilized by vegetation. This allows flowing water and strong winds to be effective transporting agents and results in the continual rearrangement of surface materials.

The long-term importance of fluvial processes in current desert regions is increased by the relatively frequent shifts in position of the global climate belts. Most deserts apparently have not been arid for long periods of geologic time; many therefore contain relict fluvial landscape features that remain

◆ **FIGURE 18.18** Dense urban development is located in close proximity to barren desert in the Nile Valley, as here in Aswan, Egypt.
SOURCE: Jon Gray/Tony Stone Images

well preserved because of the slow operation of current gradational processes.

A basic fluvial characteristic of desert regions is the small quantity of water normally carried by stream systems. Streams serve the hydrologic purpose of removing excess precipitation from the land. Within deserts such an excess exists only rarely and locally. Water tables in desert regions generally are well below the surface. As a result, sites of groundwater outflow are rare, and most streams are fed only by surface runoff. This causes most desert streams to be ephemeral, carrying water for only short periods at infrequent intervals.

Major exceptions to the ephemeral nature of most desert streams are the **exotic rivers** that cross some deserts. These rivers originate in humid regions and pass through deserts on their way to the ocean. Although they lose large volumes of water to seepage and evaporation as they traverse the desert, the larger rivers have sufficient flow to survive the passage. A large proportion of the human inhabitants of desert regions reside along the banks of exotic rivers (Figure 18.18). Some examples of important exotic rivers are the Colorado and Rio Grande in North America, the Nile and Niger in Africa, and the Indus, Tigris, and Euphrates in Asia.

Except in regions adjacent to exotic rivers or coasts, surface runoff in deserts does not reach the ocean, but instead flows into enclosed basins, where it eventually evaporates. Such desert regions are characterized by **interior drainage.** Runoff usually travels only a short distance before either sinking into the ground or entering an enclosed basin. Because weathering products are not transported out of the area, depositional features of various kinds—dunes, mass movement debris, and stream deposits—are especially abundant.

Desert regions are notoriously susceptible to destructive flash floods, which play a vital role in landscape development (Figure 18.19). Flash floods are common in many areas for a variety of reasons. First, despite the overall paucity of precipitation, desert thunderstorms can produce localized torrential downpours. Flooding also is promoted by the impermeable nature of most desert surfaces. These surfaces are likely to contain a water-resistant **duricrust layer** formed by the near-surface precipitation of salts, or to consist of a solid bedrock surface beneath a thin cover of sand or stones. No deep soil layer exists to absorb and hold moisture. Due to the lack of vegetation and soil organisms, the weathered surface materials experience little working and loosening, so they are typically highly compacted. Major drainage channels capable of carrying off large volumes of water are few

◆ **FIGURE 18.19** Approaching thunderstorm over the desert north of Las Cruces, New Mexico.
SOURCE: © LINK/Visuals Unlimited

and far between. This lack of surface drainage capacity causes runoff to flow in sheets or shallow rivulets over much of the surface during a heavy storm. Because runoff can flow with little friction or absorption over the surface, it can travel very rapidly, especially on moderately to steeply sloping terrain. Dry stream channels therefore can be quickly transformed into raging torrents. Desert storms typically are highly localized and short in duration, though, so floods can subside almost as quickly as they arrive.

Although drainage routes in deserts usually are dry and large streams are rare, small stream channels often are closely spaced and highly visible. On a very localized scale, the surface in many areas is covered by an interweaving network of miniature drainage lines that may have widths of only 0.3 to 0.6 meters (1–2 ft) and vertical dimensions of 2.5 centimeters (1 in) or less. Somewhat larger and slightly deeper channels are termed **washes** in the western United States. On a still larger scale, ephemeral stream channels with typical widths of several meters and depths of half a meter (2 ft) or more are called **wadis** in parts of the Mideast and **arroyos** in the American West. They generally are steep-sided and flat-bottomed features (Figure 18.20). At the largest scale of magnitude are the gorges and canyons of larger, often exotic rivers like the Colorado.

Desert Landform Assemblages

There is no predominant desert landscape. Deserts vary greatly in their landform assemblages because in different regions they have been subject to differing combinations of geomorphic controls. Factors such as temperature, precipitation amounts and intensities, and wind speeds and directions are all significant. Probably the most critical controls, however, are lithologic factors and recent climatic and tectonic events. Lithologic features such as differing rock strata, bedding planes, and fractures often are strikingly exposed because of the general absence of soil and vegetation. Recent tectonic processes are especially important in deserts because the gradational processes that modify and eventually remove tectonic features have been slowed greatly by a lack of water. Because tectonic features are so important, the nature and recency of their actions strongly influence the three categories of desert landform assemblages discussed in this section. These are desert plains, deserts containing plateaus, and mountainous deserts.

Desert Plains

The Australian, Arabian, Turkestan, and Kalahari Deserts, as well as most of the Sahara, have developed in stable continental interiors where little tectonic activity has occurred for long periods of geologic time. These areas have low to moderate elevations and limited surface relief, so that they are classified topographically as plains. Desert plains contain some of the most uniformly flat surfaces on Earth, chiefly because vertical erosion is restricted by interior

◆ **FIGURE 18.20** A dry arroyo in Arizona.
SOURCE: © JLM Visuals

drainage. Because no fluvially transported materials can leave the area, local base levels prevail. The erosion of higher surfaces generates sediments that are deposited in low-lying sites, eventually filling them so that an almost perfectly level surface is produced.

If some slope is available, the surface is likely to be covered by a braided pattern of rills or shallow washes. The finer-textured, damper soils within the channels encourage the growth of vegetation, frequently causing them to become covered or paralleled by growths of small shrubs. These plants accentuate the drainage pattern so that it is readily visible from the air.

Surface materials vary considerably in different desert plains. A reg surface of closely fitting stones is most common, but extensive desert plains are covered by shifting sands. Sandy regions typically display highly irregular although usually rather small-scale surface features in the form of dunes. Also in contrast to other desert plains, sand-covered surfaces usually are free of visible drainage channels because of the high porosity and shifting nature of the sand. Still other desert plains are bedrock-surfaced hamada or contain stony residual soils.

Deserts Containing Plateaus

Tectonic forces have raised some arid and semiarid regions well above sea level with little deformation so that they now form plateaus. The Colorado Plateau of northern Arizona and southern Utah is an excellent example of such a region (see the chapter-opening photo for Chapter 16). Other desert plateaus are located in northern Mexico, southern Argentina, the Middle East, western Arabia, southwestern Africa, and western China.

Typically, the majority of a desert plateau surface is flat, making it topographically indistinguishable from a desert plain. Higher elevations and some orographic influence are

◆ **FIGURE 18.21** Monument Valley, in northeastern Arizona, contains a large number of residual sandstone buttes and mesas. Seen here is one of the Mittens.
SOURCE: E. D. McKee/USGS

likely to make plateaus cooler and less arid than their low-elevation counterparts, however, so they usually are better vegetated. Most flat upland surfaces are developed on sedimentary rocks such as sandstone or limestone, which are resistant to weathering and erosion in dry climates. They may even be resistant enough to form a **cap rock** layer that protects the weaker underlying strata from erosion.

Among the most notable features of desert plateaus are their steep, clifflike escarpments. These escarpments usually exist either where a river has incised a deep gorge or canyon into the surface or at the eroded margins of the plateau. In either case, the likely presence of a resistant cap rock stratum and the reduced importance of chemical weathering and creep help to maintain slope steepness. Along escarpment faces, differing rates of erosion of resistant and nonresistant strata often are clearly evidenced by the development of a steplike series of alternating cliffs and slopes.

Escarpments in all climates typically retreat in an irregular fashion, so that they have highly uneven margins. Uneroded segments of a retreating scarp frequently become isolated by the erosion of the surrounding rock. These outliers are termed **mesas** if they still retain a flat top and **buttes** if erosion has reduced them to narrow spires with no flat upper surface. Isolated mesas or buttes also may exist as the last remnants of plateaus that have otherwise been completely consumed by erosion (Figure 18.21).

Mountainous Deserts

Mountainous deserts contain the greatest variety and complexity of surface features. The major mountainous deserts of the world include the Great Basin of the western United States and adjacent northern Mexico; South America's Atacama Desert; large portions of southwestern Asia, including much of Iran, Afghanistan, and Pakistan; limited areas of the central Sahara and central Australia that contain isolated mountain masses; and portions of western China and Mongolia.

Structural features in mountainous deserts typically are exposed clearly. Rock strata, folds, fault lines, differential weathering features, and many other phenomena are easily seen and often sharply delineated, making such regions ideal geomorphological laboratories.

Two important gradational activities that are confined largely to arid regions and are of great importance in shaping the landscapes of mountainous deserts are back-wasting and interior drainage. Due apparently to the reduced chemical weathering and the virtual absence of creep, the erosional reduction of highland masses in arid regions takes place by **back-wasting** (Figure 18.22). The significance of back-wasting is that mountain masses, while being reduced in size and areal extent, maintain their steepness and ruggedness.

Interior drainage is critically important in mountainous deserts because it allows no weathering products to leave the region (except in limited quantities by wind transport); consequently there is no significant reduction in mean surface elevations. Instead, as highland areas are gradually worn back, the adjacent lowlands that receive the products of erosion are aggraded. This results in a gradual reduction in surface relief unless tectonic activity produces further uplift.

The Great Basin of the western United States, which lies between the Rocky Mountains and the Sierra Nevada, has been studied more intensively than any other mountainous desert region. It consists largely of linear, north/south–oriented, block-faulted mountain ranges separated by broad flat basins. The discussion of mountainous desert features that follows stresses the landforms of this region. Other regions in different lithologic, tectonic, and climatic settings may contain features that vary to some degree in appearance

Humid climate   Arid climate

◆ **FIGURE 18.22** Down-wasting processes (left), which predominate in humid regions, result in a gradual reduction of slope angles through time. Back-wasting processes (right), which are characteristic of arid regions, cause the slopes to maintain their steepness as they are eroded. Back-wasting apparently results from the reduced action of chemical weathering and creep in arid climates. This allows rock masses to retain their steep slopes while being eroded by flowing water at their bases.

Pediment

Basin fill

Fault

Barchan dune

Playa

◆ **FIGURE 18.23** The erosion of plateaus or mountain masses in an arid region produces planed-off surfaces termed pediments. Interior drainage exists, so that sediments are transported only into the adjacent basins (bolsons). The photograph shows a pediment surface north of Mesquite, Nevada.

SOURCE: Photo courtesy of Reed Wicander

or extent of development, but all mountainous deserts display a basic similarity of landform types.

Mountains in most desert regions rise abruptly from a level or gently sloping surface. In many localities, however, the bases of mountain fronts have been buried by such features as talus cones or alluvial fans, which are formed from weathered materials transported down the mountain flanks by mass wasting or storm runoff. The mountain masses themselves typically are steep, rugged, and rocky. If composed of sedimentary strata, the rock may remain largely intact, with the surface supported by a physically strong limestone or sandstone cap rock layer. If it is composed of an intrusive igneous rock such as granite, though, weathering processes operating along fractures frequently have reduced the mountain, at least near the surface, to a mass of blocks or boulders.

The surface at the foot of the mountain front often is covered by deposits eroded from the adjacent uplands. If

these materials have been transported away from the mountain front, the mountains are typically surrounded by a planed rock surface called a **pediment** (Figure 18.23). The origin of pediments has been the subject of much controversy, but they apparently are produced by the combined erosional activities of back-wasting, sheet floods following heavy rains, and the lateral shifting of ephemeral streams emanating from the adjacent mountains. The lack of vegetation to trap and hold soil also aids in the formation of a bare rock surface. Pediments increase in size with time as the mountain fronts are worn back at a constant angle. The pediments serve as surfaces of transportation because their slope angles of 4° or less are just sufficient for floods to carry weathered materials from the adjacent mountains across them. Some pediments have their solid rock surfaces exposed to view, but more frequently they are hidden beneath a veneer of loose rock debris that has been temporarily deposited while being transported across the pediment. The

◆ **FIGURE 18.25** Aerial view of a "classic" alluvial fan in California's Death Valley. A road winds around the outer margin of the fan.
SOURCE: Martin G. Miller/Visuals Unlimited

◆ **FIGURE 18.26** View of "The Olgas," a conglomerate inselberg in Australia.
SOURCE: E. D. McKee/USGS

and the intervening mountain masses are consumed to produce a **pediplain.** This erosion surface is produced largely by back-wasting and sheet floods rather than by the lateral shifting of rivers. A pediplain is likely to lie at a considerable elevation because it has developed in response to local base levels. These base levels have actually been raised as the adjacent bolsons have been progressively filled with debris because of the lack of exterior drainage.

Standing conspicuously above some pediplain surfaces in arid, semiarid, and savanna landscapes are isolated steep-sided mountain remnants called **inselbergs** (German for "island mountains"). They may be considered the arid land equivalent of monadnocks. Like monadnocks, they have survived for a longer period than the surrounding uplands largely because of their high resistance to weathering. Inselbergs are especially common in the Kalahari, parts of Algeria, northwestern Nigeria, and western Australia (Figure 18.26).

SUMMARY

This chapter deals with a geomorphic process, the wind, and an important surface environment, the desert, that are associated with areas largely devoid of vegetation. In vegetation-free settings, weathering products frequently lie loose on the surface and are eroded easily by even low-energy gradational forces.

In order to review these subjects more systematically, we return to the Focus Questions:

1. *In what geomorphic environments are eolian processes most important?*

Eolian processes are important in regions that contain loose, dry surface materials of sand size or smaller. The frictional drag and turbulence of the wind enable it to move these materials through the processes of surface creep, saltation, and aerial transport. The most geographically extensive of the eolian environments are the deserts. Areas that experienced desert conditions in the recent geological past or those located downwind from deserts also may contain wind-produced landforms. Coastlines with sandy beaches also are important eolian settings. Other areas in which wind-produced features are important include those underlain by weakly cemented sandstone, the floodplains of streams draining arid and semiarid regions, and regions containing fine-textured glacial deposits. Human activities, especially those associated with agriculture, also have allowed the wind to erode and redistribute vast quantities of soil.

2. *What are the primary erosional and depositional features produced by the wind?*

Wind erosion may act rather evenly to produce a general surface lowering, leaving little in the way of distinctive features. More commonly, though, it selectively removes finer or weaker materials while leaving those of greater resistance. Deflation hollows of various sizes exist in many areas as a result of wind erosion, and many desert regions contain reg or desert pavement surfaces from which the finer materials have been selectively winnowed.

Wind deposits of sand-sized particles differ considerably from those of silt-sized particles. Sand typically is deposited in dunes within collection areas known as ergs. Several major dune forms exist, including barchans, longitudinal dunes and transverse dunes. Dunes not stabilized by vegetation migrate slowly over the surface, mostly within desert regions, in response to prevailing local wind patterns. Silt-sized eolian deposits, commonly termed loess, typically are situated downwind of sand deposits. Loess deposits were derived chiefly from Pleistocene glacial outwash. At present most deposits are stabilized by vegetation, and their surfaces often have weathered to form agriculturally productive soils. Loessal surfaces are predominantly flat and nearly feature-less, but the angular nature of the loess particles has allowed streams in some areas to erode steep-walled gullies.

3. *What are the major landform features of deserts, and how do they differ from the landforms of more humid regions?*

The most apparent difference between deserts and most humid climate regions is the absence of a vegetation cover. The general lack of surface water also results in a greatly reduced rate of chemical weathering and of soil development. These factors cause bedrock features generally to be exposed at the surface and the topography to be angular. Streams in desert regions normally carry water only infrequently, but fluvial geomorphic processes dominate in most arid regions because of the lack of surface vegetative protection. Desert rainstorms, when they do come, can generate copious precipitation that can produce spectacular and geomorphically effective floods. Many surfaces are covered by shallow drainage channels that may be visually enhanced by vegetation patterns. Approximately one-fourth of the world's deserts are ergs, covered by shifting sand, but most are stony regs or bouldery hamada.

Despite the slow rate of landform evolution in arid regions, many deserts are situated in tectonically inactive regions and therefore consist of rather monotonous plains. Flat upland surfaces likewise may have arid or semiarid climates and often are bordered by steep-faced escarpments. Mountainous deserts like those in parts of the southwestern United States contain the greatest diversity of features. They are characterized by rocky and rugged mountain masses that tend to retain their steepness as they are worn back. The mountain masses typically are surrounded by gently sloping pediment surfaces, sometimes partially overlain by alluvial fan deposits. These features merge downhill with sediment-filled basins of interior drainage called playas.

QUESTIONS FOR THOUGHT AND REVIEW

1. How does the wind erode and transport materials that are much denser than air? How does the eolian transport of dust differ from that of sand?
2. In what environmental settings are eolian processes most important? What basic similarities of all these settings enable the wind to be an effective geomorphic agent?
3. What is the difference between abrasion and deflation? Which is more important as a gradational mechanism? Why? Name and briefly describe the major surface features associated with each process.
4. Are most desert regions covered with sand? Why or why not? Where are some of the most extensive ergs located?
5. What is a sand dune? Why and how do dunes form? What determines whether they are "active"?
6. What is loess? Why are loess deposits not normally found in the same places as sand deposits? Why does loess weather into better agricultural soils than sand?
7. Describe the basic differences between the surface appearance of a desert landscape and a humid landscape. Explain why these differences exist.

8. Explain why water, which is scarce in desert regions, still is normally the most effective desert erosional agent.

9. What is interior drainage? Why is it prevalent only in arid regions?

10. What is the difference between back-wasting and down-wasting? Which is more common in deserts? Why?

11. What is a pediment, and what function does it serve in the mountainous desert erosional cycle? How does a bolson differ from a pediment?

12. Explain the similarities and differences in the forms and functions of alluvial fans and deltas.

13. What human activities are most responsible for the process of desertification?

▌KEY TERMS

eolian	transverse dune
surface creep	desert
saltation	exotic river
sandstorm	interior drainage
loess	back-wasting
abrasion	pediment
deflation	bolson
deflation hollow	playa
desert pavement	alluvial fan
erg	pediplain
dune	inselberg
barchan	desertification
longitudinal dune	

The Expanding Deserts

The deserts, which already occupy approximately one-fourth of the Earth's land surface, have been growing at an alarming pace in recent decades. The degradation of the surface, leading to the development of desertlike conditions in areas where they did not exist previously, is called **desertification**.

The spread of deserts into nondesert lands has caused a great deal of human suffering, including economic hardship, malnutrition, starvation, and mass migration. Desertification has been occurring for a long time, but was brought forcibly to public attention by the Sahelian drought of 1968–1973. (The Sahel is the name given to the vast semiarid region of Africa lying just south of the Sahara.) This drought resulted in the deaths of up to 250,000 people, the loss of millions of livestock, and the southward displacement of hundreds of thousands of people. More recent major droughts in Ethiopia, Somalia, and other countries in East Africa have taken hundreds of thousands of additional human lives.

Most regions undergoing desertification lie on the fringes of existing deserts (Figure 18.27). Included are the areas to the north and south of the Sahara, the "horn" of East Africa, regions to the northwest and southeast of the Kalahari, and large portions of the Middle East and southwestern Asia, as well as smaller areas in Australia, eastern Brazil, southern Argentina, the central Andes, northern Mexico, and the western United States.

The expansion of desertlike surfaces does not take place in a smooth, continuous manner, but instead occurs in a highly irregular, piecemeal fashion (Figure 18.29). It may be halted during a period of wet years, only to progress rapidly during a succeeding dry period. Often the less dry areas degrade the fastest, because they are used more intensively than are drier, less productive areas. Each year some 60,000 square kilometers (23,000 mi^2) of land are converted to desert. This is an area the size of West Virginia! Additional vast areas annually suffer from extensive surface degradation and loss of agricultural supporting capacity. It has been estimated that the total area threatened

continued on next page

◆ **FIGURE 18.27** Global distribution of existing desert areas and of areas where desertification is taking place.

by this process covers 37,600,000 square kilometers (14,500,000 mi^2). This is one-fourth of the land surface of the Earth, or an area four times the size of the United States.

CAUSES OF DESERTIFICATION

Desertification is accomplished primarily through the loss of stabilizing natural vegetation and the subsequent accelerated erosion of the soil by wind and water. In some cases, the loose soil is blown completely away, leaving a reg or hamada surface. In other cases, the finer particles are removed, while the sand-sized particles accumulate to form mobile dune systems.

Even in areas that retain a soil cover, the reduction of vegetation generally results in the loss of the soil's ability to absorb substantial quantities of water. The impact of raindrops on the loose soil transports fine clay particles into the soil pore spaces, sealing them and producing a largely impermeable surface. Water absorption is greatly reduced, and as a consequence, runoff is increased, re-sulting in accelerated erosion rates. The gradual desiccation of the soil caused by its diminished ability to absorb water results in the further loss of vegetation, so that a cycle of progressive surface deterioration is established.

In some regions, these processes are occurring largely as the result of a trend toward a drier climate. The continued slow global warming following the end of the last glacial stage of the Pleistocene may have much to do with this process. As the Earth warms, the subtropical highs slowly expand in size and shift poleward. This has produced a well-documented increase in aridity for areas such as North Africa and the southwestern United States over the past few thousand years. (The process may well accelerate in subsequent decades if an intensified greenhouse effect resulting from human-produced air pollution and forest destruction occurs.)

There is little doubt, however, that surface degradation in most areas results primarily from current human activities. The semiarid lands bordering the deserts exist in a delicate ecological balance and have limited potential to adjust to increased environmental pressures. Most of these areas are located in less-developed countries that currently are experiencing rapid population increases of 2.5 to 3.5 percent each year. The expanding populations are subjecting the land to increasing pressures to provide them with food and fuel. The inequitable distribution of property in many of these countries compounds the problem, since the large landholdings of the wealthy often are effectively off-limits to agricultural development, thereby increasing pressures on the remaining land.

In wet periods, the land may be able to respond to these stresses. During the short-term droughts that are common and expectable phenomena along the desert margins, though, the pressure on the land often is far in excess of its diminished supporting capacity, and desertification results.

Four specific activities have been identified as the major contributors to the desertification process: overcultivation, overgrazing, firewood gathering, and overirrigation. The cultivation of crops has expanded into progressively drier regions as population densities have increased. These regions are especially drought-prone, so that crop failures are common. Because crop production generally necessitates the prior removal of the natural vegetation, crop failures leave extensive tracts of land devoid of a plant cover and susceptible to wind and water erosion.

The raising of livestock is a major economic activity in semiarid lands, where grasses generally are the dominant natural vegetation. This is especially true in parts of Africa, where the number of cattle a family owns are a measure of its wealth and status. Livestock numbers have been increasing at about the same rate as the human population, and in many regions they now far exceed the carrying capacity of the land (Figure 18.28). The consequences of an excessive number of livestock are the reduction of the vegetation cover and the trampling and pulverization of

the soil. This is usually followed by the drying of the soil and accelerated erosion by wind and water.

Firewood is the chief fuel used for cooking and heating in many developing countries. The increased pressure of expanding populations has led to the removal of woody plants on such a large scale that many cities and towns, especially in the Sahel, are surrounded by expanses of land completely denuded of trees and shrubs. Journeys to obtain firewood in these areas may take two or three days. The reduction in vegetation also decreases local evapotranspiration rates, resulting in the addition of less water vapor to the atmosphere. As plants are removed, the underlying lighter-colored soil is increasingly exposed, thereby increasing surface albedos. This may result in lowered temperatures, which in turn increase the stability of the air and decrease precipitation.

The final major human cause of surface degradation, soil salinization resulting from overirrigation, was the

subject of a Focus Box study in Chapter 12. Soil salinization has been especially severe in the Middle East, southwestern Asia, and North Africa and is a growing problem in the western United States.

The extreme seriousness of the desertification problem results from the vast areas of land and the tremendous numbers of people affected as well as the great difficulty of reversing or even slowing the process. Once the soil has been removed by erosion, only the passage of centuries or millennia will enable new soil to form. In areas where considerable soil still remains, though, a rigorously enforced program of land protection and cover-crop planting may make it possible to reverse the present deterioration of the surface. If the process of desertification is to be halted, the basic rule that must be followed is that the intensity of use of marginal lands must not exceed their carrying capacity in even the *poorest* years.

19 COASTAL LANDFORMS AND SEAFLOOR TOPOGRAPHY

FOCUS QUESTIONS

1. What geomorphic processes are chiefly responsible for the production of coastal features?
2. What are the most important types of erosional and depositional coastal landforms?
3. What are the major causes of coastal retreat and advance, and what categories of coastlines have resulted from these coastal changes?
4. What are the major topographic divisions of the ocean floor?

Water is the most geo-morphically effective of the gradational tools. This chapter examines two important groups of geomorphic features in which water plays a dominating role: coastal landforms and the seafloor. Coastal landforms actually are polygenetic in origin, since terrestrial, marine, and atmospheric processes all contribute to their development. Because some knowledge of each of these processes is necessary for an understanding of coasts, this subject has been placed in the final chapter of the book.

Coastal processes are often powerful and can perform relatively rapid geomorphic work, but their areal extent is highly restricted at any given time. Unlike fluvial, glacial, and eolian processes, they affect the land only along its margins. The coastlines of the world, where coastal processes perform all their work, are several hundred thousand kilometers long, but are generally quite narrow. Their extent is somewhat enlarged if the shorelines of large lakes, such as the Great Lakes, are included in the inventory. Lakeshores are subject to many of the same geomorphic processes as oceanic coasts, but the processes operate on a more limited scale.

The coastal zone, despite its restricted extent, is especially critical for human beings. It contains some of the world's highest population densities and is the site of some of the most intensive human activities. Many of the world's largest cities, including Tokyo, New York, Shanghai, Rio de Janeiro, Los Angeles, and St. Petersburg, have been built on the coast as seaports. The coastal zone also is one of the most dynamic and environmentally sensitive of all natural habitats, so the potential exists for major coastal alterations as a result of human actions. Although a thorough discussion of the multiple influences of humans on coastal processes and features is beyond the scope of this text, some aspects of this topic are examined briefly in the Case Study at the end of the chapter.

The final section of the chapter briefly investigates the major features of the seafloor. The ocean covers 71 percent of the Earth's surface, making the seafloor by far the most extensive of all geomorphic environments. Yet, it is still poorly known, and detailed mapping of its features is in its early stages. Studies already have revealed, however, that seafloor features are both diverse and often strikingly different from those on land.

◆ **FIGURE 19.1** Large wave breaking on the California coast near San Francisco.

SOURCE: Barbara J. O'Donnell/Biological Picture Service

available and the sediment supply is restricted, erosion is dominant; where sediment is abundant and the available energy is restricted, deposition predominates. The physical characteristics of a given coast, then, control the relationships between energy and materials.

Most coastal sediments undergo multiple episodes of erosion, transportation, and deposition in response to the shifting balance of forces that influence the coastal environment. In the absence of tectonic uplift or a decline in sea level, though, the ultimate tendency of coastal processes is toward erosion because of the inexorable downward force of gravity. A continual net seaward transport of sediments takes place on a global scale, and the deep seafloor is the ultimate repository for sediments derived from the land. Terrestrial sediments therefore must be transported out to sea through the coastal zone. Geomorphic processes continually strive to produce, as an equilibrium condition, a continuous, gentle graded transportation slope that facilitates this seaward transport. As a consequence, steep coasts tend to be eroded back, while low-lying coasts typically experience net deposition over the long term.

Waves, currents, and tides, whose origins and characteristics were discussed in Chapter 9, are the most active agents shaping coastlines. Of these three oceanic movements, waves generate the fastest water motions and are by far the most powerful and effective coastal geomorphic agent. Their strength results largely from the fact that all the energy they contain is released suddenly as they break, often in contact with the coastline (Figure 19.1). Waves and currents are set in motion by the wind. Through its influence on the ocean water, the wind, whose relative weakness as a direct geomorphic agent on the Earth's land surface was discussed in the last chapter, attains considerable additional importance as a shaper of landforms.

The combined motions of waves, currents, and tides are responsible for a variety of coastal gradational processes.

COASTAL PROCESSES

Coastal processes, like the gradational processes already discussed, can locally erode, transport, or deposit materials, depending upon the balance between available energy and the sediment supply. Where a large amount of energy is

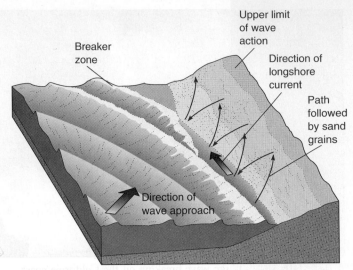

Breaker zone

Upper limit of wave action

Direction of longshore current

Path followed by sand grains

Direction of wave approach

◆ **FIGURE 19.2** Beach drifting occurs as wave swash carries sand grains obliquely up the beach. The return backwash then carries them back toward the ocean directly down the slope of the beach, resulting in a net lateral transport.

Because they are produced by water movement, these processes are very similar to those performed by streams. Most important is **abrasion,** caused by the scraping or impact of materials carried in the water against those on the shore. Breaking waves sometimes are capable of lifting even heavy rocks from the seafloor, and large storm waves have been known to hurl rocks for considerable distances. A related erosional process is **attrition**—the wearing down of loose rock materials by impact or rubbing action. Attrition is especially effective shoreward of the zone of wave-break because of the ceaseless alternation between the forward motion of wave **swash** and the return flow of wave **backwash.**

Solution is locally important on coasts containing soluble rock types, although it affects all rock types to some extent. Finally, **hydraulic action,** or the direct impact of waves on

the coast, can at times be a highly effective erosional mechanism. Pressures generated by the impact of large waves can be enormous. Water and/or compressed air can be forced repeatedly into rock fractures, eventually splitting the rocks apart. The greatest effect of hydraulic action, though, is upon unconsolidated shoreline materials. These materials may be removed rapidly during periods when large waves strike the coast.

One of the most obvious characteristics of a sandy beach is the smoothness of the surface in the zone of wave action. As waves alternately push fine-grained materials up the beach and pull them back, they produce a smooth, gently inclined surface. If waves approach the shoreline at an angle, sand grains are transported down the beach. This movement, termed **beach drifting,** occurs because the wave swash carries the sand grains up the beach at the oblique angle at which the waves intersect the shore, while the backwash carries them directly downhill in response to gravity (Figure 19.2). The resulting zigzag movement of the particles carries them laterally along the beach and in time may move large quantities of materials for substantial distances.

Localized **longshore currents** may also be produced by the lateral movement of water associated with breaking oblique waves. These currents consist of a steady flow of water paralleling the shore in the down-beach direction of oblique wave approach. They tend to augment the sediment-transporting effects of beach drifting by carrying fine particles somewhat farther offshore.

Although waves normally approach the coastline from a specific direction at any given time, coastal irregularities bend or refract the wave fronts as they approach the shore. **Wave refraction** occurs because shallow water with a depth of less than half the wavelength exerts a frictional drag on the waves that reduces their rate of landward propagation. If a portion of an incoming wave is thus slowed, but the remainder of the wave is not, the axis of the wave is rotated so that it is directed more toward the shallow water (Figure 19.3). Along an irregular coastline, this process concentrates

◆ **FIGURE 19.3** Wave refraction is caused by the reduced speed of landward propagation of the waves as their bases rub against the seafloor in shallow areas. This process leads to the concentration of wave energy on headlands and the dissipation of wave energy in bays. The arrows indicate the direction of movement of the wave fronts.

Headland   Embayment

Erosional cliff

Beach

Wave crests

High-energy zone

Wave troughs

Wave orthogonal

E = energy
$E_1 = E_2 = E_3$

E_1 E_2 E_3

Low-energy zone

wave energy on seaward-extending promontories or **headlands,** while wave energy in coastal indentations such as bays or inlets is reduced. The net geomorphic effect of this process is to straighten the coastline by encouraging the erosion of headlands and facilitating the deposition of sediment in indentations.

The motion of waves is always toward the coast, and this generates a net shoreward flow of water near the ocean surface. This flow is counterbalanced by a seaward movement of subsurface water, called **undertow.** Undertow serves as the major transporter of fine sediments into deep water. When large waves produce a vigorous landward flow of water, the return seaward flow may be concentrated in **rip currents.** These currents consist of narrow zones of rapidly seaward-moving water that extend all the way to the surface. Rip currents sometimes carry unwary swimmers well away from the shore.

The effects of eolian and fluvial processes on the coast are also important. The wind forms dunes on many sandy beaches and can erode or redistribute fine-textured, dry coastal sediments that are not stabilized by vegetation. Streams erode valleys that reach the sea to form coastal indentations. If sea levels rise to flood the lower portions of these valleys, extensive **estuaries** can be produced (see Figure 19.13). Streams that enter the ocean are vital to coastal processes because they supply most of the sediments that are distributed along the coast by waves and currents to form beach deposits. Low-lying coasts frequently are major sites of fluvial deposition and delta formation.

The rates at which coastlines are modified by erosion and deposition vary greatly. Modification rates depend primarily on wave size, the strength of the coastal materials, the angle of seaward slope of the coastal surface (determining how far offshore the waves will break), and the shape of the coastline (which may locally focus or diminish wave energy by refraction). Just as rivers in flood produce the most rapid fluvial landform alterations, large storm waves produce the most rapid coastal modification. Depending upon local conditions, inland floods can cause either rapid erosion or deposition; similarly, coastal storm waves can either transport sediment onshore, resulting in deposition, or carry it out to sea, resulting in erosion. At any given site, the erosional or depositional energy of the sea is concentrated in the zone of wave impact. Over a 12- or 24-hour period, this zone shifts between high and low tide levels. Areas with large tidal ranges therefore have coastal processes spread over larger vertical and generally larger horizontal distances than do areas with small tidal ranges.

If erosion occurs, it will do so most rapidly in unconsolidated coastal materials such as glacial or alluvial deposits or beach sand. Individual severe coastal storms, particularly hurricanes and intense winter frontal cyclones, can produce major erosion and even change the configuration of a coastline by eroding and redistributing sediments. Hurricanes crossing North Carolina's Outer Banks, for example, have altered the pattern of tidal channels approximately 30 times in the last 400 years by cutting new channels and filling others. Erosion can also be relatively rapid on a long-term basis along steep coastlines composed of weak sedimentary rock such as chalk or some sandstones. The famed Cliffs of Dover on the southeastern coast of England are being eroded at rates of up to 1.6 kilometers (1 mi) per thousand years. Conversely, coasts composed of massive resistant rock such as granite can withstand even intense wave action for long periods with little erosion.

COASTAL LANDFORMS

In this section, features produced directly by coastal marine processes are stressed. It should be noted, however, that although many coastal features have been modified by coastal processes, they were originally produced by tectonic activity or by the actions of streams, ice, wind, and gravity. A common method of categorizing coastal features, and the one employed in this section, is according to whether they are predominantly erosional or depositional in origin.

Erosional Features

Coastal erosion occurs when the near-shore water has the capacity to transport a greater quantity of sediment than is locally available. If coastal erosion has long been active in an area of consolidated materials, the coast generally is steep and rugged (Figure 19.4). The existence of such a coast indicates that terrestrial gradational processes such as streams, which erode the surface from the top, have been unable to keep pace with the undercutting action of marine erosion. A steep coast allows large waves to break at or near

◆ **FIGURE 19.4** A cliffed erosional coast in Oregon with a pocket beach.
SOURCE: David Butler

the shore, concentrating their energy on the coastline rather than dissipating it farther out to sea.

The most characteristic landform developed along an erosional coast is a **sea cliff** (Figure 19.5). Sea cliffs are produced by wave undercutting in areas of strong rock. The undermining of sea cliffs by wave hydraulic action, abrasion, and solution causes periodic slope failure as a portion of the cliff falls or slides into the ocean, where it is consumed by the waves. Along the California coast, where sea cliffs are well developed, most failures occur as rotational slumps (see Figure 15.12). In other places, they may take the form of rockslides or rockfalls. These mass movements temporarily stabilize the sea cliff, but the process of undermining immediately begins anew.

Cliffed coasts frequently are highly irregular in shape. This sometimes reflects differences in the erosional resistance of the rocks comprising the cliff. In other cases it results from differences in the elevation of the land along whose margin the sea cliff has developed. For instance, a ridge cut into by a sea cliff would contain much more rock for the sea to erode than would a valley. For this reason, the ridge would tend to be eroded back more slowly and would form a coastal promontory or **headland,** while the valley would become the site of a coastal indentation. Such indentations, depending upon their sizes and shapes, may be referred to as **bays, coves,** or **inlets.**

The process of wave refraction concentrates wave attack on headlands; consequently, the cliffs at their margins are likely to be especially bold and steep. A narrow headland attacked from opposite sides by the sea eventually may be cut through to form a **sea arch.** The collapse of the arch will produce a steep-sided island remnant called a **stack** (Figure 19.6). These features are common along portions of the west coast of the United States.

Conversely, coastal indentations between headlands are subject to much less vigorous wave attack. The quieter environment encourages the local deposition of sediments,

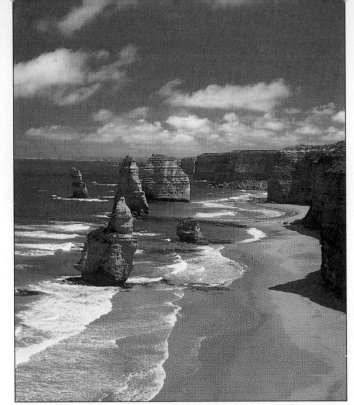

◆ **FIGURE 19.6** The "Twelve Apostles" sea stacks on the south coast of Australia near Apollo Bay.
SOURCE: Suzanne and Nick Geary/Tony Stone Images

often derived from the headlands, so that crescent-shaped **pocket beaches** may develop even along a coastline that, as a whole, is being worn back by erosion (Figure 19.4).

Coastal erosion of solid rock surfaces is effective only down to the lowest level of wave attack at low tide. As a sea cliff retreats, then, it leaves behind a planed-off rock platform at a level representing the base of erosionally effective wave action. This surface is termed a **wave-cut platform.** A wave-cut platform serves as a surface of transportation between an erosional site—the sea cliff—and a depositional site—the deeper water farther offshore. The platform may be bare, or it may be covered by a veneer of rock fragments derived from the sea cliff and in the process of being transported seaward (Figure 19.7). In form and function, then, a wave-cut platform is analogous to the pediment surface that develops at the foot of a desert mountain range.

Sediment frequently is deposited at the normally steeper outer edge of the wave-cut platform. In time, depending upon the steepness of the coast and the amount of available wave energy, these sediments may accumulate to form a **wave-built platform** consisting of mud, sand, gravel, or cobbles. The wave-built platform therefore extends the gradual slope of the continental margin farther seaward.

Depositional Features

Deposition occurs when the coastal waters are supplied with a greater quantity of sediment than they can transport and

◆ **FIGURE 19.5** Features produced by coastal erosion.

remove from the coastal zone. These materials are derived from various sources. Some are picked up well offshore and carried to the coastline by large storm waves. Others are eroded from sea cliffs or other coastal features. In most areas, though, the majority of sediments are transported to the coast by streams from inland locations. Waves and longshore currents then sort and spread these sediments laterally along the coastline. Conditions favor deposition where sediment supplies are plentiful and both onshore and offshore slopes are gentle. Gentle slopes cause large, erosionally effective waves to expend their energy well away from the shoreline.

The most characteristic feature of a depositional coast, and undoubtedly the best-known coastal landform of all, is the beach. Beaches annually attract tens of millions of tourists to coastal resorts within North America alone and add billions of dollars to coastal economies (Figure 19.8). A **beach,** by definition, is the gently sloping shore of a body of water that is washed by waves or tides and is normally covered by loose deposits of sand or gravel. Beach sediments often accumulate on a wave-cut platform during periods when conditions favor deposition. The depth of accumulation is highly variable, but is usually quite shallow.

The most familiar beaches are composed of whitish or buff-colored sand-sized quartz grains, but beach deposits can vary considerably in both composition and size. Some portions of the Hawaiian Islands, for example, have beaches of black basaltic sand. In some areas, shell fragments are an important component of beach sand, and on many tropical islands, especially in the Pacific, beaches are formed from crushed coral. Where coastal slopes are fairly steep and wave action is too vigorous for sand-sized deposits to be retained, beaches typically are covered by gravel or cobbles. In some locations, the type of beach materials may change from one time of year to another because of seasonal variations in climate. Some California beaches, for example, are covered with sand during the tranquil summer months, but winter storms transport the sand seaward, exposing an underlying cobble beach. With the return of summer, the sand once again is carried onshore where it re-forms the higher summer beach.

As the California example illustrates, even where beach deposits cover the coast for long periods, they are highly vulnerable to erosion and may be removed quickly under extreme conditions. Beach materials are transient because they are normally fine textured, are located in a high-energy environment subject to changeable conditions, and are usually not stabilized by vegetation. Material continuously arrives and departs and the beach remains relatively unchanged only if erosional and depositional processes offset one another.

Because beaches are highly sensitive to coastal processes, they normally develop in discontinuous fashion along an irregular coastline. Smooth, regular coastlines, conversely, are more likely to be continuously fronted by beaches. The Pacific coast of North America from California to Alaska is predominantly high, rugged, and irregular. As a result,

◆ **FIGURE 19.7** A wave-cut platform exposed at low tide along the California coast.
Source: John S. Shelton

beaches have developed only in protected sites, especially within bays or coves. Conversely, the Atlantic and Gulf Coast from Long Island to Texas is predominantly low-lying and gently sloping, and beaches have developed along virtually its entire length.

The gentle seaward slope and large sediment supply of a coast along which deposition is occurring allow waves, and

◆ **FIGURE 19.8** Sandy beaches annually draw millions of tourists to North American beach resorts. Shown here is the Grand Strand of South Carolina at Myrtle Beach.
source: Michael Slear

◆ **FIGURE 19.9** Cross-sectional profile of a typical beach.

especially large storm waves, to redistribute the bottom sediments to a considerable distance offshore. Just seaward of the **shoreline,** where land and water meet, there typically exists a trough of relatively deeper water excavated by moderate-sized waves breaking near the shore (Figure 19.9). Farther seaward, however, the water often becomes shallower again as the bottom rises to form a **sand bar** paralleling the coastline. A sand bar is a largely submerged ridge, composed chiefly of sand, formed in the near-shore waters by currents and larger waves. The top of the bar may be exposed at the surface at low tide, but it is submerged at high tide. It is located between the zones where large storm waves and moderate-sized "normal" waves break. While the beach contains an accumulation of sediments thrown forward by normal waves, the sand bar consists largely of an accumulation of sediments thrown forward by storm waves. Bars also serve as temporary repositories of sediments eroded from the beach during periods of larger waves, such as those occurring in winter in portions of the middle latitudes.

Continued deposition by storm waves, coupled with a reduction in sea levels, may convert sand bars to **barrier islands** that lie continuously above the ocean surface even at times of high tide (Figure 19.10). Like the bars from which they are formed, barrier islands are long and narrow and

◆ **FIGURE 19.10** A barrier island fronted by sandy beaches and backed by mudflats and a lagoon near Cape Lookout, North Carolina.

SOURCE: P. Godfrey

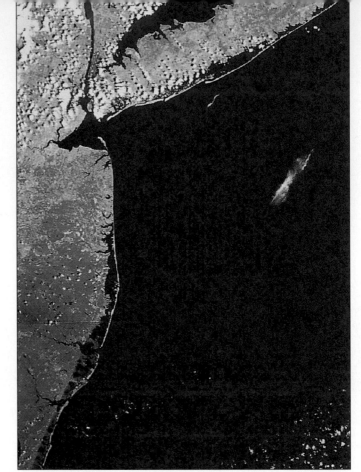

◆ **FIGURE 19.11** Satellite photo of the barrier island system off Long Island and New Jersey.

SOURCE: NOAA

parallel the coastline. They often form in chains along low-lying sandy coasts and are separated from one another by **tidal inlets** through which the sea flows. These inlets enable water levels seaward and landward of the barrier island chain to equalize during storms and as the tides change. The water between the barrier island chain and the mainland, referred to as a **lagoon,** is generally rather tranquil because it is protected from storms and large waves. The protected waters of coastal marshes and lagoons generally receive a continuous flow of fine sediments and organic materials from rivers entering them from the landward side. As a result, they typically contain a great quantity and variety of aquatic plants and animals and comprise one of the world's most naturally productive and sensitive ecosystems.

Barrier islands vary greatly in length, width, and distance from the coast. Some are situated only a short distance offshore, but others are 30 kilometers (20 mi) or more from the mainland, as along the North Carolina coast. Small barrier islands may be only 100 meters (300 ft) or so wide, but larger ones may exceed 2 kilometers (1.2 mi) in width. These ribbons of land may extend for many kilometers (Figure 19.11). Padre Island off the coast of Texas is one of the world's longest, with a length of about 160 kilometers (100 mi). Many barrier islands, as discussed in the Case

Study, have been developed extensively as coastal resorts and contain large human populations. In total, barrier islands front nearly 10 percent of the world's coastlines. The most extensive barrier island chain extends along the east coast of North America from New York southward to eastern Mexico.

Like beaches, bars and barrier islands are highly unstable features and are subject to rapid changes in size, shape, and position. These features tend to grow in the direction taken by the longshore drift of sediments. If the coastline is irregular, with headlands and embayments, they generally are attached at one end to a headland. In this case, the strip of land is a peninsula, rather than an island, and is more appropriately termed a **barrier beach.** Along less rugged, straighter coasts, they usually are entirely separated from the land. A barrier beach attached to the land at one end may extend partially across a bay to form a hook-shaped **spit** ending in open water. Cape Cod, Massachusetts, is an example of a giant coastal spit (Figure 19.12).

The lagoons on the landward sides of barrier beaches generally are tranquil enough that fine silt and clay particles can settle to the bottom. Plants such as marsh grasses that grow in the shallow water near the shore aid in trapping the sediment. Lagoons therefore tend to fill gradually with organic matter and fine-textured river sediments. This process may eventually convert them to **mudflats,** which typically are submerged only at times of high tide. Coastal mudflats may be diked, flushed of salts, and drained for settlement and agriculture; this has occurred on a large scale on the "polderlands" of the Netherlands. When coastal mudflats are colonized by marsh grasses, they become **salt marshes.** A decline in sea level will convert such marshes to meadows and eventually to dry land.

TYPES OF COASTS

A number of coastal classification systems have been devised, but no one system has proven completely satisfactory. The basic problem is the great variety of processes that affect coasts and the resulting complexity of forms exhibited by the coastlines of the world.

One very basic division of coasts, which is used as a starting point in some classification systems, is based on whether coasts are retreating (losing land) or advancing (gaining land). This system, on a very general level, is employed in this section.

Retreating Coasts

Coastal retreat caused by the erosion of the land and the deposition of the eroded materials on the seafloor is a slow process. In theory, in the absence of tectonic or other uplift, this process could continue until all land areas were completely removed and a global ocean prevailed. There is

another basic process of coastal retreat, though, that has much less potential for continental inundation, but can operate much more rapidly in the short term. This is coastal submergence caused by a rise in sea level. Submergence currently is causing a relatively rapid retreat of most coastlines because of the post-Pleistocene rise in sea level. The sea level rise results largely from the isostatic rebound of previously glaciated regions in the high latitudes, which displaces ocean water equatorward and increases its depth in the low and middle latitudes. Other related causes of the rise in sea level are the continuing melting of glaciers and the thermally induced expansion of the ocean water.

Rates of coastal submergence have been accelerating in recent decades. Along the east coast of the United States, for instance, sea level currently is rising at a rate of up to 0.3 meters (1 ft) or more per century. Depending upon the steepness of the coast, this generally translates to a horizontal shoreline retreat of 30 to 300 meters (100–1,000 ft) per century. The effects, of course, are most evident along low-lying coasts. It has been speculated that the recent acceleration in the rate of sea level rise is an early indication that the widely predicted global warming trend resulting from the increase of atmospheric greenhouse gases has begun (see Chapter 4).

Although erosion and submergence both result in coastal retreat, they produce coastlines that differ in basic characteristics. Each type of coast will now be examined briefly.

◆ **FIGURE 19.12**   Cape Cod, Massachusetts, is a classic example of a large coastal spit. It contains a number of smaller spits.

SOURCE: GEOPIC™, copyright Earth Satellite Corporation

◆ **FIGURE 19.13** The Chesapeake Bay is a large estuary. It formed when the lower portions of the valleys of rivers draining the east coast of the United States were flooded by rising sea levels at the end of the Pleistocene epoch.

SOURCE: GEOPIC™, copyright Earth Satellite Corporation

Erosional Coasts

Erosional coasts, as noted earlier, are generally steep, irregular, and rugged. They often contain bold cliffed headlands and intervening indentations, sometimes occupied by pocket beaches. The sea cliffs frequently exhibit a variety of secondary erosional features such as sea caves, arches, and stacks, and they are commonly fronted by wave-cut and wave-built platforms. Beyond the outer platform margins, water depths increase rapidly, so that the coastal zone, both seaward and landward of the shoreline, is narrow. The majority of erosional coasts, such as those existing along the west coast of North America, are tectonically active and are associated with the leading edges of lithospheric plates.

Submergent Coasts

Submergence brings the sea into contact with a land surface that has been shaped by terrestrial processes such as rivers or glaciers. Unless this inherited surface is exceptionally flat, the result is usually a highly irregular coastline.

The development of a coastline on a surface that has been dominated by fluvial processes produces a **ria coast** (from the Spanish *rio,* or "river"). Submergence results in the drowning of the lower portions of the river valleys to produce an extremely irregular coastline dominated by numerous bays or estuaries. The higher interfluves, conversely, stand out as broad headlands separating the coastal indentations. Where drowning is more extensive, the remnants of headlands may exist only as strings of aligned islands. With time, erosion of the headlands and deposition within embayments straighten the coast. This process has not proceeded far since the Pleistocene, though, because submergence has been taking place for only about 15,000 years and is an ongoing process.

The east coast of the United States from New York south to central North Carolina is a prime example of a ria coast. Numerous large embayments, including Delaware Bay, Chesapeake Bay, the Potomac and James River estuaries, and Albemarle and Pamlico Sounds, were produced by the drowning of the lower portions of large river valleys (Figure 19.13). The shores of these embayments, which are at least partially protected from oceanic storms, are the sites of several of the major port cities of the eastern United States, including Baltimore, Maryland, and Norfolk, Virginia.

In high latitude regions where Pleistocene valley glaciers once reached the sea, glacial erosion and coastal submergence have resulted in the development of **fjord coasts.** Their characteristics and global distributions are discussed in Chapter 17. Like ria coasts, fjord coasts contain deep

◆ **FIGURE 19.14** A view of the emergent coastline of central California. Such coasts are typically fronted by sea cliffs and backed by hills or mountains.
SOURCE: Jerry Westby

indentations. The fjords, however, tend to be much deeper, steeper walled, and more consistent in width than the drowned river valleys of ria coasts. Both ria and fjord coasts contain protected natural harbors that are well suited for commercial maritime activities.

Advancing Coasts

Because of recent rises in worldwide ocean levels, advancing coasts are much less common than retreating coasts. They occur only where surface uplift or coastal addition operates quickly enough to more than offset the loss of land from erosion and sea level rises.

Advancing coasts can be divided into two categories. First are coasts emerging from the sea because of tectonic uplift or isostatic rebound. A decline in water levels would also produce an emergent coast, and although this is occurring along the shores of many lakes, it is not currently taking place along oceanic coasts. The second category of advancing coasts consists of outbuilding coasts. These coasts are being built outward by geological or biological processes at a sufficiently rapid rate to allow the position of the coastline to advance seaward.

Emergent Coasts

Most actively emerging coastlines are located in the high latitudes of the northern hemisphere and around portions of the Pacific rim. Notable high latitude locations include the Canadian Shield region and Scandinavia. These two areas are still rebounding from the released weight of the Pleis-

tocene ice sheets. The coastal margins of the Pacific are rising as a result of tectonic uplift caused by the overriding of the Pacific Plate by lighter plates containing continental (sialic) crustal rock (Figure 19.14).

Emergent coastlines associated with tectonic uplift, such as the west coast of South America, are relatively straight and regular compared to submergent coastlines because the seafloor surface being exposed is generally quite smooth. In contrast, surfaces emerging as a result of isostatic rebound are highly irregular because glacial features such as fjords (as in Norway and Labrador) and drumlins (as in Finland) are being exposed. If an emerging coastline is steeply sloping, as is most commonly the case, no offshore bars or barrier islands develop. If it is gently sloping, and particularly if it is supplied with large quantities of sediments from rivers or other sources, bars and barrier islands do tend to develop.

The most characteristic features of emergent coasts in general are **marine terraces.** These elevated flat surfaces are the remnants of old wave-cut platforms, wave-built platforms, or beaches that are now located inland well above sea level. Coastal emergence seems in most cases to be sporadic rather than steady, and each terrace represents the position of the shoreline during a previous period of coastal stability. Along portions of the southern California coast, multiple marine terraces extending to elevations as great as 425 meters (1,400 ft) give a steplike appearance to the landscape (Figure 19.15).

Elevated terraces are also common along the shores of lakes that formerly contained more water. Well-developed lake terraces are found adjacent to the northern shores of the Great Lakes because of isostatic rebound. Lake terraces are

◆ **FIGURE 19.15** Multiple marine terraces resulting from sporadic tectonic uplift are evident on the west side of San Clemente Island, California.

SOURCE: John S. Shelton

DEPOSITIONAL COASTS. Depositional coasts are experiencing the active deposition of loose sediments. The sediments are transported to the coast largely by rivers and commonly are redistributed by waves and currents. The chief factors favoring the development of this type of coast are an abundant sediment supply and a gently sloping coastal margin. Such a situation occurs, for example, along the outer margins of growing deltas. Where wave and current action is sufficiently vigorous to spread these materials along the coastline, depositional coasts are characterized by beaches and by offshore bars and barrier islands.

VOLCANIC COASTS. Volcanic coasts are found where volcanic eruptions have raised new land above the sea. This may occur along the margins of already existing landmasses, such as those around the rim of the Pacific, but most frequently, it results in the formation of new volcanic islands. Volcanic coasts usually are very steep, but typically are fairly regular in form. A recently formed volcanic island is likely to be nearly circular, and a larger volcanic landmass along which several volcanoes have erupted typically has a coastal margin consisting of multiple arcs. After the volcanoes become dormant or extinct, their steep flanks are eroded rapidly and made more irregular by wave action.

CORAL COASTS. Coral coasts are composed of limestone deposited by coral polyps and other reef-building marine organisms. The coastline itself is fronted by a **reef,** which is a ridgelike accumulation of the exoskeletons of large colonies of these small animals. Reefs grow upward through time, since the living organisms construct their limestone structures only on the upper and outer portions of the reef. Reef-building organisms require clear, shallow, warm water to survive. As a result, coral coasts occur in the tropics between approximately 30°N and 25°S. In addition, they favor islands rather than mainland coasts containing rivers that discharge large quantities of sediments.

also associated with the western United States, particularly within the Great Basin, because the climate of this region has become warmer and drier since the end of the Pleistocene and numerous lakes have either disappeared or diminished greatly in size. Portions of the former margins of Lake Bonneville, in particular, are marked by multiple sets of terraces that extend through much of western Utah.

Outbuilding Coasts

Outbuilding coasts are advancing because new material is being deposited on the coastline at a rate rapid enough to counteract processes of erosion and submergence. The types of materials and the modes of deposition vary, leading to the formation of several distinct types of outbuilding coasts.

◆ **FIGURE 19.16** The three types of reefs surrounding a volcanic island tend to develop sequentially. At left, a fringing reef forms, but as the central volcanic island erodes and sinks, the reef becomes a barrier reef separated from the island by a lagoon. As the island disappears beneath the sea, the barrier reef continues to grow upward, forming an atoll. If the atoll itself is submerged, a seamount is formed.

Depending upon their relationship to land, coral reefs can be divided into three types (Figure 19.16). **Fringing reefs** are attached to the coast and indicate that local land levels have been rising relative to the sea. **Barrier reefs** are separated from the coast by a lagoon that is too deep for coral growth. In form, they are very similar to barrier beaches and, like barrier beaches, they protect the coast from storms. The largest and best-known barrier reef is the Great Barrier Reef off the east coast of Australia. Many volcanic islands, especially in the western Pacific, are surrounded by barrier reefs (Figure 19.17).

Atolls, the final type of reef, differ from the other two in that they are not associated with other landmasses. They typically consist of circular reef growths surrounding a central lagoon (Figure 19.18). Small low-lying islands of sand derived from crushed coral may surmount high points on the reef to give the complex a "pearl necklace" appearance from the air. It is believed that most or all atolls originated as fringing or barrier reefs surrounding volcanic islands. They later evolved into atolls as the central islands eroded and subsided beneath the ocean while the continued upward growth of the reefs allowed them to remain at or near the ocean surface. Most atolls are located in the western Pacific.

Compound Coasts

Compound coasts add complexity to the variety of coastal types by displaying evidence of both submergence and emergence. They are common around the world because of the numerous Pleistocene and Holocene (post-Pleistocene) fluctuations of sea level. A good example of a compound coast is the Mid-Atlantic Coast of the United States, where recent drowning of a generally emergent coastline has been taking place. As a consequence, the coast contains both large estuaries, formed by the submergence of the lower portions of river valleys, and a well-developed system of barrier islands. Something of a reverse situation exists along the west coast of North America. Here, erosional features such as sea cliffs and, farther north, fjords dominate a coast that is subject to sporadic tectonic uplift and emergence.

SEAFLOOR TOPOGRAPHY

The ocean covers some 71 percent of the Earth's surface, making the seafloor nearly three times as extensive as all other topographic surfaces combined. Our knowledge of the seafloor, although still far from complete, is increasing rapidly. Like our knowledge of the Earth's interior, it has been obtained largely through indirect means because environmental conditions make direct observations extremely difficult.

As data continue to accumulate, it becomes increasingly evident that the topography of the ocean floor is highly diverse and is at least as rugged and varied as that of the

◆ **FIGURE 19.17** A volcanic island surrounded by a barrier reef in the Society Islands of French Polynesia.
SOURCE: Marcel Isy-Schwart/The Image Bank

land. The fact that major differences exist between the topography of the land and the seafloor should not be surprising, since the forces that produce them differ greatly. Of the three gradational tools—wind, ice, and water—that largely sculpt the land surface, two—wind and ice—are not present, and the action of water is much modified and reduced by its general lack of movement along the seafloor. The dominant geomorphic processes therefore are largely tectonic and gravitational in nature. The weakness of the gradational forces allows the larger tectonic features to be maintained until they are buried in sediment or subducted into the Earth's interior.

While erosion exceeds deposition in quantity on the continents, deposition greatly exceeds erosion on the ocean floors. Sediments transported from the land drift to the

◆ **FIGURE 19.18** View of an atoll in the Pacific Ocean.
SOURCE: © Douglas Faulkner/Science Source/Photo Researchers

WATER BODY	CONTINENTAL MARGIN[1]		OCEAN BASIN			PERCENTAGE OF TOTAL OCEAN
	Continental Shelves and Slopes	Continental Rise and Partially Filled Sedimentary Basins	Deep Seafloor	Oceanic Ridges	Other Areas	
Pacific and adjacent seas	13.1%	2.7%	43.0%	35.9%	6.3%	50.1%
Atlantic and adjacent seas	17.1	8.0	39.3	32.3	2.7	26.0
Indian and adjacent seas	9.1	5.7	49.2	30.2	5.8	20.5
Arctic and adjacent seas	68.2	20.8	0	4.2	6.8	3.4
Total ocean	15.3	5.3	41.8	32.7	4.9	100.0

[1]The continental margin has a total area of about 74.5 million square kilometers (28.2 million mi²).

SOURCE: David A. Ross, *Introduction to Oceanography*, 3d ed., (Englewood Cliffs, N.J.: Prentice-Hall, 1982), p. 106. Ross obtained the data from Menard and Smith, 1966.

seafloors, especially near the margins of the continents, and eventually can accumulate to great depths. The deep seafloors are the ultimate sediment traps; because of the density of the underlying rock, they are unlikely ever to be uplifted sufficiently to become land. The oceanic rocks and their overlying sediments do not last forever, though, because they are recycled by plate tectonic movements (see Chapter 14).

The ocean basins occupy a geologically different portion of the Earth's crust than do the continents. Unlike the continents, they do not contain a thick upper zone of relatively low density granitic sial rock; instead, they consist of a much thinner layer of dense basaltic simatic rock that is covered by ocean water because it floats at a lower level on the underlying mantle.

If the ocean is considered to "belong" over the sima-surfaced portions of the crust, however, it is overfull of water. The ocean actually extends well beyond the margins of the basins and onto the geological rims of the continents. This allows the floor of the ocean to be divided into two segments—the continental margins and the ocean basins. Both segments have several major subdivisions and important accompanying topographic features.

The Continental Margins

The continental margins are composed geologically of the tapering edges of the sialic continental platforms. The crust in these areas is neither quite light enough nor thick enough to extend above sea level and therefore is overlain by shallow ocean water. In total, the continental margins comprise slightly over one-fifth of the ocean area (Table 19.1). This zone may be subdivided into three sections: the continental shelf, the continental slope, and the continental rise (Figure 19.19).

The **continental shelves** consist of the shallow portions of the seafloor that adjoin and surround most parts of the continents. They are quite shelflike in nature, as they are relatively smooth and level, and form the outer edges of the continental platforms.

Continental shelves are extensively developed along most continental margins and, in total, cover nearly one-sixth of the seafloor (Figure 19.20). They range up to 1,300 kilometers (800 mi) in width and average about 75 kilometers (47 mi) wide. They tend to slope seaward, but at a nearly imperceptible mean inclination of only 0°07' (seven-sixtieths of one degree). Their average depth below sea level is 60 meters (200 ft), and at their sharply defined outer edges, they average about 180 meters (600 ft) in depth.

The shelves apparently originate largely from the combined effects of erosion and deposition. Lower sea levels during the Pleistocene caused most shelf areas to be exposed as land during that period, and they were subjected to surface gradational processes. As sea levels rose and the shelves were flooded, they began receiving large quantities of sediments from the adjacent continents. These sediments buried most surface irregularities, producing a relatively smooth offshore platform.

The most impressive and controversial features of the otherwise relatively smooth continental shelves are the deep **submarine canyons** that are entrenched into their surfaces in places. These V-shaped canyons generally are cut entirely through the thick accumulations of shelf sediments and extend well into the underlying bedrock. Some of the largest rival or exceed in size the deepest river canyons on land. Like river canyons, they are winding and contain tributary systems of smaller branching canyons. Most submarine canyons are located immediately seaward of large rivers on the continents. They seem to exist on all continental shelves, and about 20 are located off both the east and west coasts of the coterminous United States.

The headward portions of most submarine canyons apparently formed first as river valleys on land during periods of lowered sea levels. They subsequently were enlarged and extended seaward into deep water by **turbidity currents**,

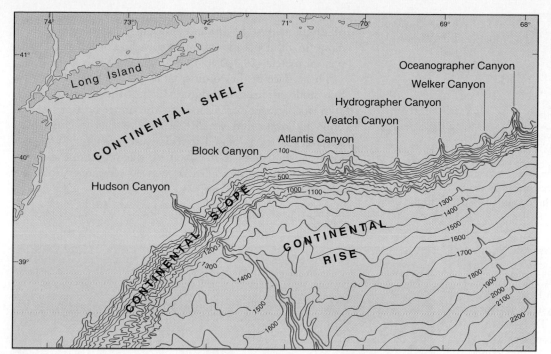

◆ **FIGURE 19.19** The continental shelf, continental slope, continental rise, and a submarine canyon system are all evident in this bathymetric map of the Atlantic off the east coast of the United States. Water depths are in fathoms. (One fathom equals 6 feet, or 1.83 meters.)

which are rapid flows of dense, sediment-laden water down the steep continental slopes at the outer edges of the shelves (Figure 19.21). Fluvially derived sediments apparently accumulate on the shelves until they attain unstable slopes. They are then carried by turbidity currents down the continental slope and onto the deep seafloor. If this idea is correct, turbidity currents are the chief mechanism responsible for completing the gradational cycle by transporting terrestrial sediments to their ultimate repositories on the deep seafloor. The submarine canyons serve as chutes to facilitate these movements.

The **continental slope** is the long slope that separates the continental shelf from the deep seafloor. It is the surface manifestation of the two major divisions of our planet's

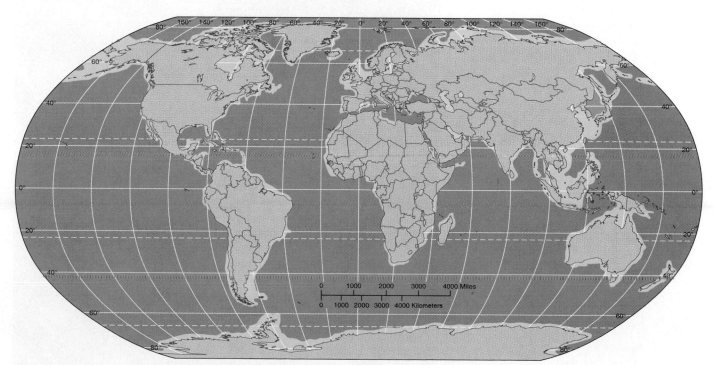

◆ **FIGURE 19.20** Continental shelf areas, shown here in light blue, comprise nearly a sixth of the total ocean floor.

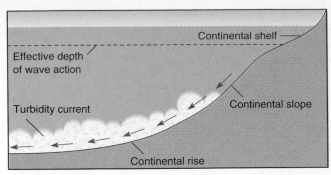

◆ **FIGURE 19.21** Turbidity currents serve to transport vast quantities of sediments from the continental shelves to the deep ocean floor.

◆ **FIGURE 19.22** The physiography of the ocean floors. The linear oceanic ridge systems can be seen extending through the mid-Atlantic and southeastern Pacific and Indian Oceans. These ridge systems are cut by innumerable transform faults. The western Pacific has a highly irregular topography because it contains numerous volcanic islands and seamounts. The shallow platforms of the continental shelves are also evident in places, such as along the east coast of North America.

SOURCE: Copyright © 1980 Marie Tharp

crust—the sial and the sima. The slope has an average width of 20 kilometers (12 mi) and an average inclination of 4° or more. In all, the seafloor drops an average of 3,000 to 3,650 vertical meters (10,000 to 12,000 ft) across this slope.

The surface of the continental slope is covered largely by fine muds of terrestrial origin. Much of this material apparently was transported directly to the upper portions of the slope by rivers and winds during the Pleistocene, when lowered sea levels caused it to lie just offshore. At regular intervals, the slope is trenched by the submarine canyons down which turbidity currents convey sediments from the adjacent continental shelf.

The **continental rise,** which occupies about 5 percent of the seafloor, is a broad, very gentle depositional ramp located at the base of the continental slope. It consists of a wedgelike apron of sediments that have spread out from the lower ends of the submarine canyons. The continental rise represents the final depositional site for most continental sediments that are transported to the sea. In some areas, sediments have accumulated on the continental rise to depths as great as 10 kilometers (6 mi). Sediment depths are especially great seaward of large rivers such as the Mississippi, Ganges, and Indus. Slope angles on the continental rise generally are less than half a degree (0.5°). Widths are highly variable, ranging from 100 to 1,000 kilometers (60–600 mi).

The Ocean Basins

The ocean basins begin at the seaward margins of the continental rise and collectively comprise nearly four-fifths of the ocean floor (Figure 19.22). In most places the ocean basin topography is rough and irregular, with various troughs, smaller basins, ridges, and plateaus, as well as hills and mountains existing either individually or in groups or ranges (Figure 19.23). One basic reason for the generally

◆ FIGURE 19.23 Comparative profile views of transects of the United States and of the Atlantic seafloor illustrate that the ocean bottom topography is at least as rugged as is the land.

rugged surface of the deep seafloor is that insufficient sediment is available to cover surface irregularities produced by tectonic movements. Only limited quantities of the finest terrestrial sediments are deposited so far from land, and marine sediments derived from sea plants and animal remains or direct chemical precipitation are in most places quite limited in volume. The major topographic features of the ocean basins are abyssal plains, trenches, the oceanic ridge complex, and seamounts.

Abyssal plains are deep, smooth-surfaced basins covered with a mixture of terrestrial and marine sediments and located at mean depths of 4.0 to 5.6 kilometers (2.5 to 3.5 mi) below sea level. Most occur near the western margins of the ocean basins, adjacent to continental east coasts that have large sediment-laden rivers entering the sea. The smoothness of these plains seems to result largely from the fact that enough terrestrial sediments reach them to have buried most surface irregularities. Deep sea cores taken from abyssal plains show that they contain interlayered deposits of unsorted sandy terrestrial sediments and well-sorted marine clays.

The systems of oceanic ridges and deep sea trenches were discussed in Chapter 14. They are opposite types of features both tectonically and topographically. The trenches are narrow slits in the ocean floor where old, deeply sediment-covered oceanic crust is subducted into the asthenosphere and destroyed. They are the sites of the world's greatest ocean depths. Conversely, the oceanic ridges are broad mountainous zones of newly formed crustal material with little or no sediment cover. They are produced by the extrusion of vast quantities of fluid basaltic magmas along the axes of seafloor spreading centers. The oceanic ridge

complex forms a system of imposing undersea mountains that have a combined length of some 64,000 kilometers (40,000 mi), rise from 1,500 to 3,000 meters (1–2 mi) or more above the deep seafloor, and cover nearly a third of the total ocean bottom (Figure 19.22). Together, the oceanic trenches and ridges divide the remainder of the ocean floor into a number of separate compartments. It has also recently been discovered that hydrothermal activity is very extensive along the volcanically active oceanic ridge systems and that geysers of superheated seawater are precipitating large quantities of valuable metallic minerals onto the seafloor. This subject is explored in the Focus Box on the next page.

Seamounts, which were produced by undersea volcanic eruptions, are individual circular or elliptical submarine mountains rising more than 900 meters (3,000 ft) from the deep seafloor. Most apparently form near mantle plumes or oceanic ridges and are then shifted laterally by seafloor spreading motions. As they move away from the zone of volcanism, they become extinct, cool and contract, and gradually subside. Although most seamounts are located in the Pacific, they occur in all oceans. More than 2,500 are known to exist, and undoubtedly thousands of others await discovery. For example, during a 1993 expedition, scientists mapping the floor of the South Pacific 1,000 kilometers (600 mi) northwest of Easter Island discovered a cluster of 1,133 seamounts in an area about the size of the state of New York. A few of the highest and most recent seamounts break the ocean surface to form volcanic islands, but the tops of most are situated well below sea level. As illustrated in Figure 14.6, the Hawaiian Islands lie at the southeastern end of a chain of seamounts that extends more than 6,500 kilometers (4,000 mi) to Russia's Kamchatka Peninsula.

SEAFLOOR METAL FACTORIES

Recent marine investigations have led to the discovery that rich deposits of metallic minerals currently are being formed on the ocean floors. The minerals are being deposited by "geysers" of superheated water rich in dissolved metallic solutions. The geysers apparently are widespread in volcanically active areas, especially in the vicinity of seafloor spreading centers. They are known to exist on the East Pacific Rise off the west coast of Mexico and in the Red Sea, and also were discovered recently on a portion of the Mid-Atlantic Ridge. They probably exist elsewhere as well, such as in the vicinity of volcanic "hot spots" like the Hawaiian Islands and the Azores, and in active subduction zones, such as those near Japan, Indonesia, and the Lesser Antilles.

The geysers are convectionally driven hydrothermal features powered by heat from the mantle. They exist because the rocks of the ocean floor, especially in volcanically active areas, are "leaky," allowing seawater to circulate freely through them. Cold, dense seawater enters bedrock fissures on the seafloor and descends until it encounters hot igneous rock. There, it is heated, expands, and rises back through the fissured rocks, dissolving large quantities of minerals as it does so. When it is discharged into the ocean water in geysers, the superheated water is almost immediately cooled, greatly reducing its ability to hold dissolved minerals in solution. These minerals consequently are precipitated onto the seafloor in the vicinity of the geyser vents.

The most potentially valuable mineral deposits are associated with "black smokers," the hottest type of geyser, with water temperatures higher than 315°C (600°F) (Figure 19.24). Their name is derived from their surrounding clouds of precipitating polymetallic sulfide crystals. These crystals build tubular vents around the geysers and eventually accumulate to form substantial ore bodies on the adjacent seafloor. Some of the geyser water minerals apparently precipitate at a later time, helping to form the manganese nodules that cover vast portions of the oceanic abyssal plains.

continued on next page

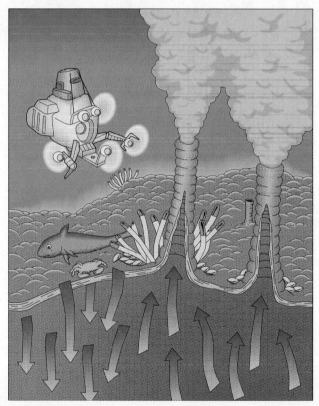

◆ **FIGURE 19.24** The submersible *Alvin* sheds light on hydrothermal vents at the Galapagos Rift, a branch of the East Pacific Rise. Seawater seeps down through the oceanic crust, becomes heated, and then rises and builds chimneys composed of anhydrite ($CaSO_4$) and sulfides of iron, copper, and zinc. The plume of "black smoke" is simply heated water saturated with dissolved minerals. Communities of organisms, including tubeworms, giant clams, crabs, and several types of fish, live near the vents.

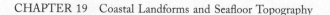

Seafloor geysers are also important contributors to the total dissolved mineral content of the ocean waters. Prior to their discovery in the 1970s and 1980s, most oceanic minerals were believed to have been washed into the oceans by rivers. It is currently thought that the geysers constitute at least as important a source of these minerals.

The geysers also seem to have been a key mechanism in the production of mineral deposits on land. These deposits, which often are similar in appearance to those forming on the sea-floor, are major commercial sources of gold, silver, copper, zinc, and other metals. The metallic ore bodies of eastern Quebec, for example, although found in rocks 2.6 billion years old, resemble polymetallic sulfide deposits currently forming on the seafloor around black smokers. It is likely that these deposits were precipitated onto the floors of ancient oceans that were at some later time lifted above sea level by tectonic forces.

Valuable as seafloor mineral deposits are, their commercial exploitation is not likely in the near future. Most of the deposits occur in deep water, far from land. No technology currently exists for breaking, lifting, and sorting these hard-rock deposits. Indeed, we are at present only in the early stages of discovering the existence and plotting the distribution of these minerals. Marine polymetallic sulfide (MPS) deposits, however, may constitute an economically important source of metals in the twenty-first century. ∎

This chain was formed by the northwestward movement of the Pacific Plate over the mantle plume that is currently situated beneath the big island of Hawaii. Many seamounts are a flat-topped variety termed a **guyot.** Guyots apparently once extended above the water surface, where they were planed off by subaerial erosional processes. In some cases, coral colonies became established around these islands and subsequently were able to grow upward rapidly enough to offset the subsidence of the seamounts and remain near the surface. They are now the sites of coral atolls (Figure 19.18).

▮ SUMMARY

This chapter has dealt with two categories of geomorphic features that differ greatly in their physical characteristics and human impact, but share the common bond of being dominated by marine processes. The coastline, located at the interface of air, land, and sea, is influenced by a complex combination of atmospheric, terrestrial, and marine processes. Despite its limited extent, this highly dynamic physical environment is crucial to human beings. Conversely, the ocean floor occupies more than 70 percent of the Earth's surface, but has so far been little used, or even explored, by humans. Water here plays a much less active role in the production of topographic features. In fact, because of its lack of movement, water on the ocean floor serves to protect features produced by tectonic and gravitational processeswhile permitting the deposition of vast quantities of both terrestrial and marine sediments.

We now return to the Focus Questions posed at the beginning of the chapter:

1. *What geomorphic processes are chiefly responsible for the production of coastal features?*

Atmospheric, terrestrial, and marine processes all contribute to the production of coastal features. The most powerful and effective geomorphic agent for most coastlines, however, is wave action. Waves erode shoreline materials through the processes of abrasion, attrition, solution, and hydraulic action. Waves also deposit sediments derived both from the land and from the ocean floor along the coast. The action of currents is important in transporting fine-textured sediments along the coastline.

2. *What are the most important types of erosional and depositional coastal landforms?*

Coasts dominated by long-term erosion typically are steep and rugged. The undercutting action of waves along an elevated coastline characteristically produces sea cliffs, which gradually retreat as they are undermined and collapse. The process of wave refraction causes erosion to be most active around headlands, while coastal indentations may be the sites of localized deposition. Headland erosion leaves wave-cut platforms that can serve as future sites for the accumulation of beach sediments.

The dominant coastal depositional feature is the beach. Beaches may front mainland shorelines, but many are developed on offshore barrier islands separated from the mainland by coastal lagoons.

3. *What are the major causes of coastal retreat and advance, and what categories of coastlines have resulted from these coastal changes?*

Coastal retreat can be caused by either erosion or submergence. Erosion typically produces a rugged coast characterized by slowly retreating headlands. Submergent coastlines are geographically widespread because of the nearly global post-Pleistocene rise in sea levels. The nature of the previously dominant geomorphic process largely determines the present surface characteristics of a given submergent coastline. Ria coasts, probably the most distinctive of the submergent coastal types, are formed by the partial drowning of river valleys. They are characterized by numerous bays or estuaries.

Coastlines advance seaward either because of tectonic or isostatic uplift, or by outbuilding resulting from the addition of material by geological or biological processes. Large areas in the high latitudes of the northern hemisphere are advancing because of isostatic rebound. Emergent coastlines around the Pacific are rising because of crustal compression caused by the subduction of the Pacific Plate. Outbuilding coasts include those coasts subject to rapid fluvial or marine deposition, volcanic eruptions, and the growth of reef-building organisms.

4. *What are the major topographic divisions of the ocean floor?*

The ocean floor consists of two basic segments—the continental margins and the ocean basins. The continental margins have formed on the tapering edges of the continental plates and are major depositional sites for sediments derived from the land. Immediately bordering the continents are the ramplike continental shelves. They generally are broad, covered by shallow water, and slope gently seaward. In places they are trenched deeply by submarine canyons. The continental shelves are bordered on their seaward sides by long and often steep continental slopes, which merge at their bases with the gently sloping and deeply sediment-covered continental rises.

The ocean basins are compartmentalized by their extensive systems of oceanic ridges and trenches. These systems separate deep and generally flat-floored abyssal plains. Scattered through the abyssal plains are individual rounded submarine mountains, termed seamounts, that are products of undersea volcanic eruptions.

▌Questions for Thought and Review

1. What is the justification for making coastal landforms the last landform group to be covered in this text?

2. By what processes do waves and currents erode coasts? Briefly describe each process.

3. How are sea cliffs formed? How is their steepness maintained? What features may be produced as a sea cliff is eroded back?

4. What coastal environmental conditions favor coastal erosion? What conditions favor coastal deposition?

5. What is the most characteristic feature of a depositional shoreline? Explain how it forms. From what sediment sources are its materials derived?

6. What similarities and differences exist between bars and barrier islands? Where and why do they form? What effects do they have on the characteristics of the mainland coast?

7. What are the two chief mechanisms by which coastlines retreat? Why are more shorelines around the world currently retreating than advancing?

8. What are the two basic causes of present-day coastal advance? In what geographical areas of the world is each occurring?

9. What are compound coasts? Why are they common at the present time?

10. Discuss the major differences in the physical characteristics of the coasts of the eastern and western United States. Explain why these differences exist.

11. What are the two basic divisions of the ocean floor? In what physical and geological respects do they differ?

12. How are submarine canyons produced and what is their geomorphic function? How do they differ in origin from deep sea trenches?

13. What are seamounts and how are they formed? What is their relationship to coral atolls?

14. Why do barrier beaches migrate under natural conditions? What human activities can accelerate the process of barrier beach migration?

15. Explain the mechanism by which large quantities of metallic minerals are currently being deposited on the ocean floor. Where is this process occurring? What problems exist with regard to human use of this resource?

▌Key Terms

beach drifting	ria coast
wave refraction	marine terrace
sea cliff	reef
wave-cut platform	continental shelf
wave-built platform	submarine canyon
beach	turbidity current
sand bar	continental slope
barrier island	continental rise
lagoon	abyssal plain
spit	seamount

Barrier Beach Migration and Coastal Development

Increases in affluence and leisure time, coupled with the development of efficient modern transportation networks, have resulted in a great increase in tourism in North America during the past few decades. The chief focus of the multibillion dollar tourist industry is the string of beach resorts that have been developed on the more than 300 barrier islands along the Atlantic and Gulf Coasts of the United States. Such cities as Atlantic City, New Jersey; Ocean City, Maryland; Virginia Beach, Virginia; and Daytona Beach, West Palm Beach, Fort Lauderdale, and Miami Beach in Florida, are only some of the largest and best-known of scores of tourist-oriented cities and towns along the East Coast.

Unfortunately, the barrier beaches upon which these cities and towns have been built are unstable and, geologically speaking, short-lived features. Furthermore, the human alteration of these strips of sand has in many places upset the delicate balance of nature and further destabilized conditions. A combination of ongoing natural processes and human influences has reduced the size of barrier beaches along the Atlantic and Gulf Coasts and has caused them to migrate slowly landward. This raises serious concerns for the future of the countless billions of dollars worth of nonmovable buildings and other property on these beaches and for the livelihoods of their hundreds of thousands of human inhabitants.

CAUSES OF BARRIER BEACH MIGRATION

The basic cause of barrier beach migration is the nearly global rise in ocean levels. Barrier beaches occupy a fixed position relative to the coast just shoreward of the zone of impact of storm waves. When sea levels rise—as they are doing now all along the U.S. Atlantic and Gulf Coasts—this zone shifts landward, resulting in the shoreward migration of the barrier beaches. Because slopes along barrier beach coasts tend to be quite gentle, a small increase in sea level produces a large horizontal migration of both the shoreline and the barrier beaches. For example, the approximate 0.3 meter (1 ft) per century rise in sea level experienced recently along much of the East Coast has produced a landward migration of about 60 meters (200 ft) in the shoreline and the offshore barrier beaches.

Barrier beach migration does not occur as a steady movement, but takes place sporadically during major storms. Large storm waves pound at the seaward sides of the barrier beaches, eroding the sand ridges behind the shoreline. If these ridges are eroded through, or if waves become high enough to breach them, waves can wash completely over the barrier beach from one side to the other. This **overwash** process causes sand to be removed rapidly from the seaward side of the barrier beach and deposited on the landward side (Figure 19.25a). Unfortunately, fixed structures are unable to move with the land, so a building originally located well away from the shoreline behind a protective beach ridge may have the shoreline migrate toward it until it is finally undercut by waves and destroyed (Figures 19.25b and c).

A number of human activities further destabilize barrier beaches, hastening their erosion and landward migration. One is the destruction of dune grasses that help to hold the sand in place and assist in dune formation. These grasses may be removed purposefully for "aesthetic" reasons or be destroyed by trampling or by the use of off-road vehicles. The sand dunes themselves, which form the backbone of barrier beaches, may be flattened to improve the view of the ocean for residents of homes situated behind them. Of more widespread significance is the upstream damming of rivers that supply sand to the coast. The reservoirs behind the dams serve as sediment traps, greatly reducing the quantity of sand available for barrier beach construction. This factor is believed to play a major role in the rapid erosional losses currently being experienced on both barrier islands and mainland beaches in many areas. If the widely anticipated global warming of the greenhouse effect becomes a reality, increased glacial melting and thermally induced seawater expansion will produce even more rapid rises in ocean levels than already experienced, further compounding the problem.

SOLUTIONS

The problem of barrier beach erosion and migration is so serious and complex that many differing types of solutions have been tried. One group involves the construction of structures designed either to protect the shoreline from wave attack or to encourage the local deposition of sand. These include **seawalls,** which are rock or wooden barriers built parallel to the coast to absorb and dissipate wave energy and protect the land behind

(a) (b) (c) (d)

◆ **FIGURE 19.25** (a) Ribbons of fresh sand have been transported by overwash nearly across the northern tip of Assateague Island just south of Ocean City, Maryland. (b) Housing development perilously perched on an eroding sea cliff in San Diego County, California. The cliff is receding most rapidly where soft sands overlie hard sandstone. Shoring is needed to keep the homes from falling. (c) Drastic undercutting has brought the Pacific Ocean to this home's back door. This house is the one on the right in figure (b). (d) These groins at Cape May, New Jersey, interrupt the flow of longshore currents causing sand to be trapped on their upcurrent side. On the downcurrent side of the groin, sand is eroded because of continuing longshore drift.

SOURCE: a: Bernard W. Pipkin, University of Southern California; b–c: Allan B. Mayo; d: John S. Shelton

them. Unfortunately, seawalls do not protect the beach in front of them and often deflect wave energy downward so that they eventually are undermined and destroyed.

Another type of structure is the **groin,** a narrow seaward-extending barrier, generally built of rock, that is designed to interfere with beach drifting and longshore currents and to encourage the deposition of sand. Series of groins have been constructed in numerous locations along the East Coast of the United States (Figure 19.25d). Although they may locally solve the problem of beach erosion, at least temporarily, their interception of sand often accelerates the erosion of beaches immediately down the coast. In addition, both seawalls and groins are expensive to build and negatively affect the appearance of the beach and shoreline they are designed to protect. They are at best only temporary solutions to the problem of beach erosion because they do not address the underlying factor of rising sea levels.

Another very temporary but aesthetically more pleasing solution is to

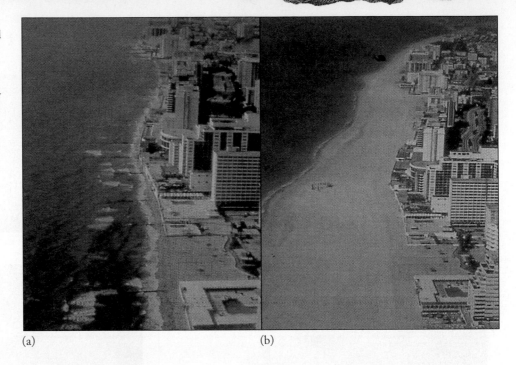

◆ **FIGURE 19.26** The beach at Miami Beach, Florida (a) before and (b) after the U.S. Army Corps of Engineers' beach nourishment project.

SOURCE: Official photographs, U.S. Army Corps of Engineers, Jacksonville District

(a)   (b)

haul in replacement sand from elsewhere or to pump sand onto the beach from further offshore. A number of East Coast cities, including Ocean City (Maryland) and Miami Beach have recently employed this method of beach restoration (Figure 19.26). In only a decade or two, however, or perhaps much sooner if these sites are visited by major storms, this sand will be eroded away. This is especially true where pumping has deepened the water just offshore, allowing larger waves to reach the beach.

A different type of solution, which is favored by various government agencies and many conservation groups, is not to fight the sea, but rather to limit development on barrier beaches and to let nature take its course. In recent years, many barrier beaches along the Atlantic and Gulf Coasts have been set aside as National Seashores, National Wildlife Refuges, and state parks. Coastal changes within these areas are expected and are adjusted to by those responsible for their care. This policy is undoubtedly wisest in areas where no major development has yet occurred, but it is too late for many other areas where enormous investments of money, energy, and materials already have been made.

The future of the developed portion of our barrier beaches is very uncertain. In most cases, development has occurred recently enough that beach migration has progressed only to the point of being an expensive annoyance rather than a major disaster. Technological developments and the expenditure of large sums of money may postpone the day of reckoning indefinitely for some areas, but future coastal storms are certain to cause great destruction in others. In the meantime, increased understanding of coastal processes and greater restraint in coastal development are necessary.

APPENDIX A

METRIC AND IMPERIAL MEASUREMENT EQUIVALENCIES

Throughout this book, dual systems of numerical values have been used because both our traditional English (or Imperial) system and the metric system are widely employed. In most cases, metric values are given first, and Imperial equivalents are provided in parentheses. Although the metric system is undoubtedly a superior system and is used in most parts of the world, the conversion to metric within the United States is proving to be slower and more difficult than earlier supposed, and it appears that both systems will be in widespread use for some time to come. For convenience, conversion formulas for units of measurement especially relevant to the subject of physical geography are provided here.

Length

- 1 kilometer = 1000 meters = 0.6214 mile = 3281 feet
- 1 mile = 5280 feet = 1.609 kilometers = 1609.3 meters
- 1 centimeter = 0.3937 inch
- 1 inch = 2.540 centimeters
- 1 micrometer (μm) = 10^{-6} meter = 10^{-4} centimeter = 3.937×10^{-5} inch

Area

- 1 square kilometer = 0.3861 square mile = 247.1 acres
- 1 square mile = 2.590 square kilometers = 640 acres

Volume

- 1 cubic kilometer = 0.2399 cubic mile
- 1 cubic mile = 4.1684 cubic kilometers
- 1 cubic meter = 1.308 cubic yards = 61,024 cubic inches
- 1 cubic yard = 0.7646 cubic meters = 46,656 cubic inches
- 1 liter = 1.0567 quarts
- 1 quart = 0.946 liters

Mass

- 1 metric ton = 1000 kilograms = 2205 pounds
- 1 short ton = 2000 pounds = 907.2 kilograms
- 1 kilogram = 1000 grams = 2.2046 pounds
- 1 pound = 0.4536 kilograms

Speed

- 1 meter per second = 3.281 feet per second = 3.6 kilometers per hour = 2.237 miles per hour
- 1 mile per hour = 1.467 feet per second = 1.6093 kilometers per hour = 0.8684 knots
- 1 kilometer per hour = 0.6214 miles per hour
- 1 knot = 1.152 miles per hour = 1.853 kilometers per hour

Temperature Scale Conversion Formulas

- $°C = \frac{5}{9}(°F-32°)$ or $°C = \frac{°F - 32}{1.8} -$
- $°F = \frac{9}{5}°C + 32°$ or $°F = 1.8\,°C + 32$
- $K = °C + 273.15$

Air Pressure

- Mean sea level air pressure = 1013.2 millibars = 29.92 inches of mercury = 101.3 kilopascals = 1 atmosphere

WEATHER MAP SYMBOLS

Surface weather maps are prepared daily by the National Weather Service in Washington, D.C., and it is possible to purchase a subscription to a weekly booklet entitled "Daily Weather Maps" from the superintendent of documents. Each map in these booklets depicts the surface weather conditions for the coterminous United States as well as adjacent portions of Canada and Mexico at 7:00 A.M., Eastern Standard Time. An example of the maps is shown in Figure 7.15. Also included are smaller maps depicting upper air windflow patterns, 24-hour high and low temperatures, and precipitation totals for a large number of reporting stations.

The weather information contained on the Daily Weather Maps is conveyed by the use of standardized symbols. The two major types of map features depicted are isobars, which are drawn at four-millibar intervals, and weather fronts, the symbols for which are shown in Table B.6. The positions of high and low pressure centers are also shown.

The locations of a large number of individual reporting stations are displayed on the maps. The weather information these stations provide serves as the basis for positioning the isobars and fronts depicted on the maps. The sample report in Table B.1 for the station shown below illustrates the format used for displaying reporting station data. A translation of the information appearing in the sample plotted report is indicated in the table. Additional weather symbol information is provided in Tables B.3 through B.6.

SAMPLE PLOTTED REPORT

◆ TABLE B.1 Sample Station Information

DATA	MEANING (SAMPLE STATION CONDITIONS IN PARENTHESES)
●	Circle represents station's location. Proportion blackened in represents amount of cloud cover (overcast).
(wind shaft symbol)	Shaft extending from station circle indicates direction from which wind is blowing. Lines on shaft indicate wind speed (northwest wind at 18–22 knots).
32	Current temperature in degrees Fahrenheit (32° F).
30	Current dew point in degrees Fahrenheit (30° F).
3/4	Visibility in miles, if under 10 miles (3/4 mile).
✳ ✳	Present weather of significance (continuous light snow).
247	Last three digits of air pressure reading to tenths of a millibar, with decimal point omitted (1024.7 mb).
+ 28/	Air pressure change in 3 hours preceding observation, including pressure tendency over the period (pressure has risen steadily a total of 2.8 millibars).
.4	Time precipitation last began or ended (precipitation began as rain between 3 and 4 hours ago).
.45	Amount of precipitation during past 6 hours in inches. If frozen, water equivalent is given (0.45").
(high cloud symbol)	Type of high cloud, if any (dense cirrus).
(middle cloud symbol)	Type of middle cloud, if any (altocumulus).
- - -	Type of low cloud, if any (fractostratus).
6	Fraction of sky covered by low- or middle-level clouds (seven-tenths to eight-tenths of sky covered by fractostratus clouds).
2	Height of cloud base or ceiling (ceiling 300–599 feet).

◆ TABLE B.2 Total Sky Cover

Symbol	Description
◯	No clouds
◓	Less than one-tenth or one-tenth
◔	Two-tenths or three-tenths
◑	Four-tenths
◑	Five-tenths
◕	Six-tenths
◕	Seven-tenths or eight-tenths
◕	Nine-tenths or overcast with openings
●	Completely overcast
⊗	Sky obscured

◆ TABLE B.3 Common Weather Symbols

• •	Light rain	⦙	Rain shower
•.•	Moderate rain	⋇	Snow shower
•.•.	Heavy rain	⧊	Showers of hail
* *	Light snow	⇄	Drifting or blowing snow
⋇	Moderate snow	⟳	Dust storm
⋇	Heavy snow	=	Fog
,,	Light drizzle	⦾⦾	Haze
△	Ice pellets (sleet)	⩫	Smoke
⦾⦾	Freezing rain	⧑	Thunderstorm
⦾⦾	Freezing drizzle	⑂	Hurricane

◆ TABLE B.4 Pressure Tendency

Symbol	Description	
╱	Rising, then falling	
⟋	Rising, then steady; or rising, then rising more slowly	
╱	Rising steadily or unsteadily	Barometer now higher than 3 hours ago
⟍	Falling or steady, then rising; or rising, then rising more quickly	
—	Steady, same as 3 hours ago	
⟍	Falling, then rising, same or lower than 3 hours ago	
⟍	Falling, then steady; or falling, then falling more slowly	
╲	Falling steadily, or unsteadily	Barometer now lower than 3 hours ago
╲	Steady or rising, then falling; or falling, then falling more quickly	

◆ TABLE B.5 Wind Information

Symbol	MILES (STATUTE) PER HOUR	KNOTS	KILOMETERS PER HOUR
◎	Calm	Calm	Calm
—	1–2	1–2	1–3
⌐	3–8	3–7	4–13
⟍	9–14	8–12	14–23
⟍	15–20	13–17	24–32
⟍	21–25	18–22	33–40
⟍	26–31	23–27	41–50
⟍	32–37	28–32	51–60
⟍	38–43	33–37	61–69
⟍	44–49	38–42	70–79
⟍	50–54	43–47	80–87
⌐	55–60	48–52	88–96
⌐	61–66	53–57	97–106
⌐	67–71	58–62	107–114
⌐	72–77	63–67	115–124
⌐	78–83	68-72	125–134
⌐	84–89	73–77	135–143
⌐	119–123	103-107	144–198

◆ TABLE B.6 Weather Front Symbols

Symbol	Description
▲▲▲▲	Cold front
●●●●	Warm front
▲●▲●	Occluded front
▼▲▼▲	Stationary front

APPENDIX C

ANSWERS TO PROBLEMS

Chapter 2

1. 15°S
2. 170°E
3. 5 P.M., Monday
4. 4 A.M., Friday
5. a. The Sun's noon altitude is higher in Florida.
   b. Maine has the longer daylight period.
   c. Alaska experiences the greatest number of hours of twilight because it is located at the highest latitude.
6. a. 1:506,880
   b. 1 centimeter would represent 5.07 kilometers
7. a. 1 inch = 15.78 miles
   b. 1 centimeter = 10 kilometers
8. a. 800 square miles
   b. 2,070 square kilometers
9. a. 4 times as much paper
   b. a sheet measuring 40 inches × 20 inches
10. a. 26 contour lines (27 if sea level is counted as a natural contour line)
    b. 650 feet

Chapter 4

1. a. 10° C
   b. 283K
2. −459.4° F
3. 89.6° F
4. 39° F
5. 47° F
6. −7.2° C

Chapter 5

1. ENE, or 70°
2. SW, or 210°
3. The high will be centered southeast of you (in the northern hemisphere).

Chapter 6

1. 14,750 calories
2. 26,800 calories
3. a. relative humidity is 61%
   b. temperature is 27° C or 80° F
   c. dew point is 14° C or 59° F
4. about 69%
5. unsaturated air is stable; saturated air is unstable

APPENDIX D

WORLD PHYSICAL-POLITICAL MAP

Maps of the Eastern and Western Hemispheres.

The following countries were part of the former Soviet Union
Armenia
Azerbaijan
Belarus
Estonia
Georgia
Kazakstan
Kyrgyzstan
Latvia
Lithuania
Moldova
Russia
Tajikistan
Turkmenistan
Ukraine
Uzbekistan

GLOSSARY

A horizon The soil horizon located between the O and E horizons. It contains a mixture of organic and inorganic matter, and is often referred to as the topsoil.

ablation The loss of ice from a glacier as a result of melting and sublimation.

abrasion Erosion accomplished by the scraping of water-borne or wind-borne particles against other materials.

absolute glacier motion A measure of the actual forward motion of a given particle of ice in a glacier.

abyssal plain A deep, smooth-surfaced submarine basin that is covered with a mixture of terrestrial and marine sediments.

acid rain Rainfall of more than usual acidity. Acid rain is believed to result largely from the human release of sulfur dioxide (SO_2) gas, especially from coal-fired electric generating plants, and has proven harmful to plant and animal life as well as to a variety of material substances.

acidic magma A magma that is relatively rich in silicon dioxide. Acidic magmas are typically viscous, and are produced by violent volcanic eruptions associated with subduction zones.

acidic soil A soil with a pH value of less than 7. Most acidic soils are associated with forest or tundra vegetation in relatively humid climate regions.

adiabatic temperature changes Temperature changes produced within a parcel of air because of the air's expansion or contraction. Adiabatic cooling normally occurs as a parcel of air rises and expands; adiabatic warming normally occurs as air descends and contracts. No energy is actually gained or lost by the air during the process.

advection The process of energy transfer through the horizontal movement of a fluid from one place to another. A type of convection.

advection fog A fog type that forms when moist air flows over a colder surface that cools it below the dew point.

aggrade To build up. For example, streams can aggrade the floodplains they occupy by depositing sediment.

air mass A large body of air with relatively uniform horizontal temperature and moisture characteristics.

air mass thunderstorm A thunderstorm that has developed within a single unstable air mass. Air mass thunderstorms typically show a lack of linear organization, and over land areas form most frequently during the afternoon hours.

air pressure The force exerted by the air on its surroundings because of the pull of gravity. Air pressure at sea level averages 1.0 kilograms per square centimeter (14.7 lbs/in^2).

albedo The short wave energy reflection rate of an object or substance, normally stated in percent.

alfisol A soil order containing fine textured soils with a moderately high cation exchange capacity that are low in humus and have a yellowish-brown A horizon. They are common in forest-grassland transitional environments in the low and middle latitudes.

alkaline soil A soil with a pH value higher than 7. Most alkaline soils are associated with grassland or desert vegetation in relatively dry climate regions.

alluvial fan A gently sloping fan-shaped deposit of alluvium that is formed when a stream flowing within a steep, narrow channel suddenly enters an open lowland and loses its capacity to transport its sediment load.

alluvial soil A soil formed from alluvium.

alpine glacier A relatively small glacier or glacier system that forms at high elevations in a mountainous region and moves downhill.

alpine meadow The name given to areas covered by tundra vegetation in high mountain (alpine) environments.

altitude Distance above sea level; generally used for a location above the Earth's surface.

altitude (solar) The angular distance in degrees of the Sun above the nearest horizon.

altocumulus A middle level cloud type typically consisting of small, circular cloud masses arranged in a linear pattern.

altostratus A middle level cloud type that typically appears as a smooth, featureless veil of light gray cloud covering the entire sky.

andisol A soil order consisting of soils formed from volcanic ash. Andisols are often fertile, but are geographically restricted to small areas that are volcanically active.

anemometer An instrument consisting of several rotating cups that measures wind speeds.

aneroid barometer A barometer that measures air pressure by means of the force exerted on a collapsible box, called a sylphon cell, from which some of the air has been removed.

angle of repose The steepest angle that can be maintained on a given slope without slope failure (mass movement) occurring.

annual A plant that has a lifespan of one year or less, usually because it is killed by cold or dry weather at the end of the growing season.

annual temperature range The number of degrees between the warmest and the coldest monthly mean temperatures for the year.

annual variability of precipitation The average annual percentage deviation of a site's annual precipitation total from the long-term mean total.

annular drainage pattern A stream drainage pattern characterized by an alternation of divergent stream segments and radial segments.

Antarctic Circle The most equatorward line of latitude in the southern hemisphere (66¹/₂°S) at which the Sun remains constantly above the horizon on the date of the southern hemisphere summer solstice, and constantly below the horizon on the date of the southern hemisphere winter solstice.

anticline An upward fold in one or more rock strata. A ridge consisting of upfolded strata is termed an *anticlinal ridge*; a valley consisting of upfolded strata is termed an *anticlinal valley*.

anticyclone A high pressure system, characterized by an outward spiraling windflow pattern.

antipode The location on exactly the opposite side of the Earth from a given point.

aphelion The point of the Earth's orbit that is farthest from the Sun. It occurs about July 4th, when the two bodies are approximately 152.5 million km (94.5 million mi) apart.

aquiclude A relatively impermeable rock or sediment layer that acts as a barrier to the penetration of water from an adjacent layer.

aquifer A porous and permeable rock or sediment layer that contains large quantities of ground water and permits its ready movement.

Arctic Circle The most equatorward line of latitude in the northern hemisphere (66¹/₂°N) at which the Sun remains constantly above the horizon on the date of the summer solstice, and constantly below the horizon on the date of the winter solstice.

arête A narrow, serrated mountain ridge in a region of present or past glaciation. It is formed by the headward growth, and eventual meeting, of cirques from opposite sides of the ridge.

argon An inert gas comprising almost 1% of the atmosphere.

area symbols Map symbols, such as colors, used to indicate the areal extent of a mapped subject.

arid climate A climate with a potential evaporation rate that greatly exceeds its mean precipitation receipts, so that desert conditions prevail.

aridisol A soil order containing predominantly shallow, stony, and mineralogically immature soils that have developed in arid environments.

aspect (of a slope) The compass direction in which the slope faces.

association See *vegetative association*.

asthenosphere The plastic upper portion of the mantle. Convection currents in the asthenosphere are believed to produce lithospheric plate motions.

atmosphere The gaseous envelope surrounding the Earth.

atoll An often ring-shaped reef that forms an island or chain of islands. Atolls are believed to form around oceanic volcanic islands. They persist because of the continued growth of the reef-building organisms after the volcanic island has disappeared due to erosion or submergence.

attrition The wearing down of loose rock materials (usually being transported by water or wind) by impact or rubbing action.

aurora australis Luminous bands or streamers in the night sky of the southern hemisphere apparently caused by the disruption of the stream of ionized solar particles by the Earth's magnetic field.

aurora borealis Luminous bands or streamers in the night sky of the northern hemisphere apparently caused by the disruption of the stream of ionized solar particles by the Earth's magnetic field.

autumnal equinox The day that the Earth's axial inclination is exactly perpendicular with respect to the Sun, following a period in which one's hemisphere was inclined toward the Sun. Marks the official beginning of autumn.

avalanche The rapid movement of a detached mass of snow and ice down a steep mountain slope.

average environmental lapse rate The average rate at which the temperature declines with altitude in the troposphere. This value is approximately 6.4C°/1,000 meters, or 3.5F°/1,000 feet.

axis The line, connecting the North and South Poles, about which the Earth rotates.

azimuth system A directional system that states directions in degrees clockwise from due north. Values range from 0° to 360°.

azimuthal map A map that retains true directions outward from one central point.

B horizon the soil horizon located between the E and C horizons. It consists chiefly of weathered rock material, and is often referred to as the subsoil.

bacteria Microscopic plants that play an essential role in the breakdown of organic and inorganic matter to aid in soil production.

backwash The gravity-induced return flow of wave swash to the ocean.

back-wasting Processes of weathering and erosion that result in a mountain mass maintaining its steepness and ruggedness while being reduced in size and areal extent. Back-wasting is characteristic of arid regions and contrasts with the down-wasting processes typical of humid regions.

bajada See *piedmont alluvial slope.*

bar A linear deposit of sand or gravel often found within a stream channel at a place where the speed of flow is diminished.

barchan A relatively small, crescentic sand dune that is oriented with its tapered ends pointing downwind.

barometer An instrument designed to measure air pressure.

barrier An unfavorable area, such as an ocean or mountain range, that serves to limit the spread of plant or animal species.

barrier island An elongated sandy island paralleling the coastline and formed by the continued deposition of sand by storm waves. This same feature, if connected to the mainland to form a peninsula, is termed a **barrier beach.**

basalt A dense, black colored extrusive igneous rock containing substantial iron and magnesium.

base level The elevation at any point along a stream's course that provides the stream with a gradient just sufficient for it to transport its sediment load past that point with no erosion or deposition. If a stream is at base level throughout its course it is considered a graded stream.

basic magma A magma with a relatively low silicon dioxide content. Basic magmas are hot, low in viscosity, and are generally associated with non-explosive volcanic eruptions or lava flows in areas of plate divergence or above mantle plumes.

basin An enclosed lowland.

basin and range topography A section of the Earth's crust that has been divided into a number of steep-sided mountain ranges and intervening lowlands by a system of parallel faults. The classic example is the Great Basin region of the western United States.

batholith A large mass of intrusive igneous rock, usually granitic in composition. Because of its resistance to erosion, it may form a mountain mass if exposed at the surface.

beach The gently sloping shore of a body of water that is subject to wave or tide action. It is normally covered by loose deposits of sand or gravel.

beach drifting The gradual movement of sediments along a beach in a zig-zag motion. This occurs as the sediments are repeatedly pushed up the beach by swash from oblique waves and then carried directly downslope by the wave backwash.

bed load Sediments that are moved along the bed of a stream by sliding, rolling, or bouncing (saltation).

bedding plane The boundary between adjacent sedimentary rock strata.

bedrock Solid surface rock. It is typically unweathered and undetached.

Bergeron process A process of precipitation formation, based on the fact that the dew point is slightly lower for water droplets than for ice crystals. This encourages the evaporation of the water droplets in a mixed cloud, and the subsequent sublimation of the moisture onto ice crystals.

big bang theory At present, the most widely accepted theory of origin of the universe. It holds that all matter in the universe was initially produced, and hurled outward, by the explosion of an inconceivably tiny and compact entity sometimes referred to as the "cosmic egg."

biogeochemical cycle The cycling of a nutrient (such as phosphorus or nitrogen) between an ecosystem and the Earth's atmosphere or surface.

biogeography The branch of geography concerned with the spatial characteristics of plants and animals.

biological community An assemblage of plants and animals whose members occupy an environmental setting of restricted size.

biomass A measure of vegetation density, defined as the mass of vegetation per unit of area.

biome A global-scale community of plants and animals that exists in relative equilibrium with its environment.

biosphere The sphere of living organisms inhabiting the surface and near-surface portions of the Earth.

biota The population of lifeforms that occupy a region or ecosystem.

bird's-foot delta A delta that contains long finger-like deltaic projections. It develops when a stream deposits sediments in a quiet, partially enclosed water body that lacks the wave and current action to spread out the deltaic materials.

bluffs Steep slopes; often those that form the margins of a river floodplain.

bolson An enclosed desert basin into which ephemeral streams transport water and sediments. Following rainfalls, they become the sites of temporary lakes termed playas.

boulder field A surface area littered with rocks, often as a result of active frost wedging in a mountainous region.

braided flow A stream flow pattern that contains multiple interwoven channels separated by bars. This pattern tends to develop when the stream is supplied with an excessive quantity of coarse sediments.

broadleaf Usually used to describe trees with leaves of considerable width, such as oak, maple, or elm, as opposed to needleleaf trees.

bush A woody plant with several permanent stems beginning near or below the ground surface.

butte A small, isolated, and steep-sided remnant of a former more elevated topographic surface. Buttes are often left as erosional outliers of retreating escarpments in arid regions.

C horizon A layer of weathered rock material that is just beginning to undergo the pedogenic processes that will convert it to soil.

calcification A pedogenic process that results in the concentration of calcium carbonate within the subsoil.

caldera A roughly circular volcanic depression caused either by the explosion or collapse of a volcano.

calorie The amount of heat needed to raise the temperature of one gram of liquid water by one Celsius degree (1.8F°).

canyon A deep, steep-walled valley of considerable width, usually carved by a stream that is rapidly downcutting.

cap rock A resistant surface rock layer that serves as a cap to protect the underlying less resistant strata from erosion. Most cap rock layers exist in arid regions.

capillary tension The electromagnetic attraction of a thin molecular film of water to solid surfaces, such as soil grains and plant root linings.

carbon dioxide An atmospheric gas essential to plant photosynthetic processes and largely responsible for the long wave radiation absorbing capacity of the atmosphere. It has the chemical formula CO_2.

carbonate mineral A mineral that contains a predominance of carbon and oxygen. It often has an organic origin.

carbonation A chemical weathering process involving the reaction of rock material with carbonic acid (H_2CO_3) produced when carbon dioxide is dissolved in water.

carnivore An animal whose diet consists chiefly of meat.

cartographer A professional map maker.

cartography The art and science of map making.

catastrophism The widely held view of a few centuries ago that the Earth's major surface features were produced in the recent past by cataclysmic events.

cation exchange capacity (CEC) The quantitative measure of the ability of a given soil to exchange cations with plants.

cave A natural cavity within or beneath the surface. Caves can be formed by a number of processes, but are especially numerous and important in karst areas, where they are produced by subsurface solution.

celsius The metric system temperature scale. It is based on the boiling point of water (under standard temperature and pressure conditions) at 100°C, and the freezing point of water at 0°C.

centripetal drainage pattern A drainage pattern consisting of a network of streams that converge into a central basin.

Channelled Scablands A region, located in Washington state, where the surface topography has been produced largely by repeated Pleistocene floods of massive proportions. These floods resulted from the breaching of an ice dam that blocked the outflow of glacial Lake Missoula.

chaparral The name used in California for a dense cover of woody shrubs and small trees ranging up to 6 meters (20 ft) in height. It is a variety of the Mediterranean Woodland and Scrub biome.

chemical sedimentary rock A sedimentary rock formed by the direct precipitation of chemicals in water.

chemical weathering Weathering processes involving chemical changes in the rock material that form new substances.

chinook A warm, dry wind that is adiabatically heated as it descends the eastern slopes of the Rocky Mountains in winter.

chlorofluorocarbon gases (CFC's) A group of synthetic gases that are apparently depleting the ozone layer, causing an increase in the quantity of short wave radiation reaching the Earth's surface.

chlorophyll A green plant pigment with the ability to convert water and atmospheric gases into food in the presence of light.

chronometer A highly precise clock. Its use enabled mariners to accurately determine their longitudinal position when at sea.

cinder cone A conical volcanic hill or low mountain constructed of loose pyroclastic material thrown into the air from a central vent.

cirques Bowl-shaped mountainside hollows sculpted by glacial scour and frost action, and in which alpine glaciers typically have their heads.

cirrus From the Latin word meaning "curl of hair," cirrus clouds are the highest of the ten basic cloud types, are composed of ice crystals, and have a wispy appearance.

cirrocumulus A high cloud type consisting of a large number of small, rounded, and sometimes iridescent cloud masses arranged in groups or lines.

cirrostratus A high cloud type that typically covers most of the sky in a smooth, thin layer through which the Sun and Moon can readily shine.

clastic sedimentary rock A sedimentary rock composed of particles of pre-existing rocks. Based on the sizes of the constituent rock particles, the four major categories are conglomerate, sandstone, siltstone, and shale.

clay-humus complex A complex combination of tightly bound clay and humus particles that is capable of attracting and exchanging cations needed by vegetation.

climagraph A graph of monthly mean temperature, precipitation, and sometimes other climatic data for a weather reporting station.

climate The long-term or overall condition of the atmosphere with respect to the weather elements and weather systems.

climatology The branch of physical science that deals with the study of climate.

climax community The stable biological community that eventually occupies an ecosystem at the end of a successional sequence. It is theoretically the best suited to survive under existing environmental conditions of all possible communities available to form the ecosystem.

closed system A system with a limited energy supply confined within it.

cloud A visible accumulation of minute water droplets or ice crystals, suspended in the air, that have condensed or sublimed around dust particles because the air has cooled below the dew point.

coal An organic sedimentary rock composed primarily of carbon, and formed from the burial, compaction, and lithification of plant remains in a poorly drained environment.

coalesence A process of precipitation formation in clouds, operating through the merging of droplets that come in contact with one another.

cohesion A measure of the degree to which the individual particles of a substance adhere to each other.

cold front A frontal boundary along which an advancing mass of cold air is underrunning relatively lighter warm air, forcing it to rise.

cold ice glacier A dry-base glacier that is frozen fast to the underlying bedrock surface. Glacial movement must therefore take place solely by internal deformation.

colloids Very fine clay particles that are capable of remaining suspended indefinitely in water and are soft and gelatinous in nature.

composite volcano A volcano constructed of alternating layers of pyroclastic materials and lava. Most large volcanoes are of this type.

computer mapping The use of computers to organize and display on maps a variety of geographical information about an area.

condensation The phase change of a substance from a vapor to a liquid.

condensation nuclei Atmospheric particulates that serve as the collection centers for water molecules, enabling clouds and ultimately precipitation to form.

conduction The transfer of thermal energy by physical contact between two objects or substances having different temperatures.

cone of depression The cone-shaped zone of lowered water table around a well.

conformal map A map that has the property of showing the correct shapes of the areas it depicts.

conglomerate A clastic sedimentary rock comprised largely of rounded gravel or larger sized clasts.

conifer A tree that produces its seeds within cones.

consumer The animals in an ecosystem, which consume carbohydrates produced by plants. A *primary consumer*, or herbivore, feeds directly on plants. A *secondary consumer*, or carnivore, consumes herbivores. A *tertiary consumer*, or "top carnivore," consumes other carnivores.

continental arctic air mass A very cold, dry air mass that has formed over a snow and ice covered surface.

continental climate Climatic characteristics typical of a large land mass, especially including relatively large daily and annual temperature ranges and low amounts of precipitation, fog, cloudiness, and humidity.

continental crust The relatively light, discontinuous layer of granitic rock that comprises that uppermost portion of the crust and forms the continental masses.

continental drift The somewhat outdated term for the movements of the lithospheric plates as a result of convection currents in the asthenosphere.

continental glacier A glacier that forms at high latitudes and covers an extensive land area. Continental glaciers currently cover Antarctica and interior Greenland, and existed during the Pleistocene in northern North America and northwestern Eurasia.

continental polar air mass A relatively cold, dry air mass that has formed over a high latitude land mass.

continental rise The broad, gently inclined submarine depositional ramp located at the base of the continental slope.

continental shelf The shallow portion of the ocean floor that adjoins and surrounds most parts of the continents.

continental slope The long slope separating the continental shelf from the deep sea floor.

continental tropical air mass A relatively hot, dry air mass that has formed over a low latitude land mass.

contour interval The vertical distance between consecutive contour lines on a contour map.

contour lines Lines of constant elevation on a topographic map that are generally numbered from mean sea level.

convection The process of energy transfer through the physical movement of a fluid from one place to another. On the Earth, the chief fluids involved are air and water.

convectional lifting The rising of air because it is warmer, and therefore lighter, than the surrounding air.

convergent lifting The inward and upward spiral of air into a cyclonic storm.

Cordilleran ice sheet The name given to the ice sheet that developed from the coalescence of alpine glaciers in the Northern Rockies during the Pleistocene.

core The innermost layer of the Earth, believed to be primarily metallic in composition. The core consists of two segments—a solid inner core and a liquid outer core.

Coriolis effect The apparent deflection of moving fluids, especially winds and ocean currents, on the Earth's surface, occurring as a result of the Earth's rotation. The direction of deflection is to the right in the northern hemisphere and to the left in the southern hemisphere.

corrosion The solution of rock or mineral material in water (usually within a stream).

cosmic radiation Charged atomic particles that enter the Earth's atmosphere at nearly the speed of light and are capable of destroying or mutating plant and animal cells.

coulee The name given to a dry canyon carved by former flash floods in the Channelled Scablands of the northwestern United States.

creep The very slow downhill movement of soil or rock material over a period of years. Creep is most important in humid climate regions, and involves the entire mass of near-surface weathered material on most slopes.

crest (of a flood) The peak flow of a flood, measured in terms of either discharge or water level.

crevasse A large crack or fissure in the near-surface portion of a glacier, caused by the brittle fracturing of the ice as it passes over a convexity in its bed.

critical erosion velocity The minimum velocity of flow that will enable a stream to erode streambed particles of a given size.

crust The relatively thin "skin" of rigid low density rock that forms the outermost lithospheric layer of the Earth. The crust varies in thickness from 5 to 70 kilometers (3–40 mi) and consists of two segments, the sima and the sial.

cumulonimbus The cloud type responsible for thunderstorms, hail, and tornadoes; probably the world's leading precipitation producing cloud.

cumulus From the Latin word meaning "pile" or "heap," cumulus clouds are the most common of the ten basic cloud types. They have a puffy appearance, with flat bases and uneven sides and tops.

current A flow of wind or water that may be capable of transporting loose materials.

cyclone A low pressure system or storm center, characterized by an inward spiraling windflow pattern.

cyclonic lifting Atmospheric lifting associated with the convergence of air into a low pressure system.

daily mean temperature The average temperature of the day, calculated by adding the daily high and low temperatures and dividing by two.

daily temperature range The number of degrees between the high and the low temperatures for the calender day.

Davis, William Morris An American geomorphologist best known for developing the concept of the fluvial landscape "cycle of erosion," patterned after the three human life cycle stages of youth, maturity, and old age.

daylight saving time A variation of standard time in which the time is set ahead by one hour in order to achieve greater correlation between the normal human activity period and the daylight period.

debris avalanche The rapid movement of a mixture of soil, rock, and vegetation down a pre-existing mountainside hollow. Movement is nearly always triggered by heavy rainfall; it typically begins as a slide and becomes a flow.

deciduous A leafy plant that loses all its foliage during a given season, usually as a mechanism for coping with cold or dry conditions.

decomposer An organism within an ecosystem that consumes organic wastes. Examples include bacteria, worms, and fungi.

deep ocean current An ocean current flowing at considerable depths that is produced by density differences resulting from variations in temperature and salinity from the surrounding water. It often displays significant vertical as well as horizontal components of motion.

deflation The erosion of loose surficial material directly by the force of the wind. It occurs when the wind speed exceeds the drag velocity of the materials.

deflation hollow A shallow, often roughly circular depression of variable dimensions that has been formed by deflation.

delta An often fan-shaped deposit of alluvium formed when a stream enters a standing body of water and loses its ability to transport its sediment load.

dendritic stream pattern A treelike drainage pattern that develops on homogeneous surfaces that exert no significant structural control on stream location. It is the most widespread of the drainage pattern types.

depression contour line A special type of contour line, drawn with inward-facing line segments (hachures), that surrounds a closed depression.

deranged stream pattern A drainage pattern containing stream segments that display a large degree of spatial disorganization such as sudden turns and changes in slope steepness.

desert A region of substantial size that is largely or entirely devoid of vegetation, usually as a result of aridity.

desert pavement A desert surface of close-fitting pebbles or stones that protects the underlying surface from erosion. It results largely from the eolian removal of fine textured particles.

desertification The degradation of the surface to produce desertlike conditions in areas where they did not exist previously.

developable shape A geometrical shape, such as that of a cone or cylinder, that can be flattened without distortion.

dew Water droplets that have condensed on surface objects because the overlying air has cooled below the dew point.

dew point The temperature at which a given parcel of air will become saturated with water vapor.

diastrophism Solid-state movements of the Earth's crust, including folding and faulting, that result from tectonic forces.

differential erosion Erosion occurring at differing rates within a given area because of the uneven operation of gradational forces.

dike A sheet of basalt that was injected as magma into a preexisting rock fracture that is discordant with (cuts through) the host rock structure.

dike ridge A wall-like basaltic ridge formed by the erosional exposure of a dike injected into weaker sedimentary strata.

dip The angle at which a rock stratum or any planar feature is inclined from the horizontal.

discharge The volume of stream flow past a given point during a specified timespan, normally measured in cubic feet per second (cfs) or in cubic meters per second (cms).

dissolved load The portion of the sediment load that is carried in solution within a water body.

distributary channels Channels that separate from a stream, thereby dividing it into multiple segments. They occur most frequently near the mouth of a stream, particularly within a delta.

doldrums Another name for the equatorial belt of variable winds and calms.

doline See *sinkhole*.

dome A rounded anticlinal structure. Smaller domes are usually formed by the surfaceward movement of material such as magma or salt.

Doppler radar A type of radar that holds great promise for tornado identification and prediction. It can remotely measure and display the speed of winds moving toward or away from the radar site, allowing for the detection of developing rotary motions in thunderstorms.

drag velocity The critical wind velocity needed to overcome frictional resistance and initiate the eolian transport of surface materials of a given diameter.

drainage basin The total surface area drained by a given stream system.

drainage divide A linear highland zone, such as a mountain range, that separates two drainage basins.

drainage system An organized system of interconnected streams.

drift See *glacial drift*.

drizzle Liquid precipitation droplets with diameters less than 0.5 millimeters (0.02 in).

drumlin A smooth, elongated hill composed of glacial till. They generally occur in groups or "swarms," and are oriented so that their tapered ends point in the direction the ice was moving.

dry adiabatic rate The rate at which unsaturated air cools as it rises (or warms as it descends) through the troposphere. This rate is 5.5F°/1000 feet, or 10C°/1000 meters.

dune A mound or ridge of wind-deposited sand. Dunes assume a variety of forms and, unless stabilized by vegetation, tend to migrate slowly over the surface because of wind action.

duricrust A hard, nearly impermeable surface or near-surface layer, typically formed by the precipitation of salts in an enclosed basin in an arid climate.

dust The collective term for all liquid and solid particulates suspended in the atmosphere.

dynamic equilibrium A system that undergoes constant change so as to remain in equilibrium with changing environmental conditions. This term is often applied to stream systems and to soils.

E horizon An eluvial soil horizon; a mineral horizon generally located between the A and B horizons in a humid climate.

earth flow The slow flowage of a mass of nearly saturated soil down a moderate to steep slope.

Earth system The Earth and its atmosphere, treated as a unit.

earthquake A sudden movement of the Earth's surface produced by the differential motion of the opposing sides of a fault.

easterly wave A north-south oriented low pressure trough in the trade wind belt that drifts westward and may have the potential to develop into a hurricane.

ecosystem A functioning biological entity consisting of all the organisms in a biological community, as well as the environment in which and with which they interact.

ecotone A transition area between two adjacent vegetative communities, such as forest and grassland, and therefore often exhibiting a complex ecology.

electromagnetic spectrum The division of electromagnetic radiation into categories based on wavelength.

elevation The distance above sea level of a point on the Earth's surface.

El Niño The anomalous warm ocean current that sometimes flows eastward across the southern Pacific to reach the coast of South America. It recently has been discovered to exert a major impact on global weather patterns.

eluviation The removal of solid or dissolved materials from a soil horizon by downward percolating water.

energy The capacity to perform work on some form of matter.

entisol A soil order containing poorly developed soils, mostly of recent origin, that display little or no profile development.

entrenched meander A stream meander that has been incised deeply into the surface, perhaps as the result of the renewal of stream downcutting following a period of uplift.

environmental lapse rate See *average environmental lapse rate*.

eolian The term used for geomorphic processes performed or products produced by the wind.

ephemerals Small, short-lived desert plants that pass through their life cycles quickly following a substantial rainfall.

ephemeral stream A stream that flows only during wet periods when it is supplied by surface or soil water.

epicenter The point on the Earth's surface directly above the focus of an earthquake.

epiphyte An "air plant" that grows on other plants, but derives its moisture and nutrients from the air and is therefore not parasitic. Epiphytes are especially diverse and abundant in the tropical rainforests.

equal area map A map that has the property of showing the correct proportional sizes of the areas it depicts.

equator The line of fastest rotation on the Earth. It is a full circumference, located midway between the North and South Poles, and is the starting (0°) line of latitude.

equatorial belt of variable winds and calms The zone of light and variable winds associated with the Intertropical Convergence Zone.

equatorial lows Weak low pressure centers that develop along the ITCZ and drift westward, attended by extensive cloudiness and rainfall.

equilibrium line The line on a glacier separating the zones of accumulation and ablation. No net gain or loss of ice occurs along this line.

equinox A time when the axial tilt of the Earth is precisely perpendicular to the direction of the Sun's vertical rays. This occurs only twice each year—on the first days of spring and of autumn.

esker A long, meandering ridge of stratified drift, apparently produced by the deposition of sediments by a subglacial stream.

estuary A semi-enclosed water body located either in a drowned river valley or a submerged tectonic trough.

evaporation The phase change of a substance from a liquid to a vapor.

evapotranspiration The total water evaporated from both the Earth's surface and from vegetation.

evergreen A leafy plant that retains its foliage throughout the year.

exfoliation The separation of rock into sheets paralleling the surface as a result of unloading and chemical weathering.

exfoliation dome A rounded dome-shaped mountain whose surface form is controlled largely by exfoliation.

exotic stream A large perennial stream, flowing through an arid region, that has its headwaters in a humid region from which it has derived most of its water.

extrusive igneous rock Igneous rock that was extruded onto the Earth's surface in a liquid state (as lava). Subsequent rapid cooling allowed little or no time for mineral crystals to form.

extrusive volcanism Processes by which molten rock reaches, and is extruded upon, the Earth's surface.

Fahrenheit The temperature scale most commonly used by the general public in the United States. It is based on the boiling point of water (at standard temperature and pressure) at 212°F, and the freezing point of water at 32°F.

fault escarpment (or fault scarp) Ideally, a linear, smooth-sided slope consisting of the exposed surface of the fault plane on the higher side of a fault.

fault A zone of separation between two rock masses in which there has been displacement of the sides relative to one another.

fault creep The gradual offsetting of two sides of a fault that are not locked together.

fault line The trace of a fault on the Earth's surface.

fault plane The two dimensional surface of the fault. It primarily exists below the surface, but a portion may be exposed along a fault escarpment.

Fennoscandian ice sheet The continental ice sheet that covered most of northwestern Eurasia during the glacial stages of the Pleistocene.

fertility The productivity of the soil with respect to vegetation. The fertility of the soil is determined largely by its mineral content.

fetch The distance the wind blows over water. It is generally of concern when calculating expected wave heights.

field capacity The maximum quantity of water the soil is capable of retaining against the pull of gravity.

firn A compact granular form of snow with a specific gravity of 0.55 or more.

fissure eruption The eruption of volcanic material from a linear fracture or fault rather than from a single vent.

fjord A glacial trough that has been partially occupied by sea water to form a deep oceanic coastal inlet.

flood The inundation of the land along the banks of a stream that occurs when the stream's discharge exceeds its carrying capacity.

flood basalt A huge outpouring of fluid lava from a fissure that can cover an extensive region to form a lava plain or plateau.

floodplain A broad belt of low, flat land adjacent to a stream, formed by fluvial erosional and depositional processes, that is inundated periodically during floods.

food chain The passage of food energy through a sequence of organisms within an ecosystem. If the energy can take a number of different routes in its passage, it is referred to as a *food web*.

fluvial An adjective used to identify stream-related processes and features.

foehn A warm wind in the European Alps that is heated by its descent into valleys or surrounding lowlands.

fog A visible accumulation of minute water droplets or ice crystals suspended in air immediately overlying the surface. Fog normally forms as the air is cooled below the dew point, forcing the condensation or sublimation of water vapor on dust particles.

folding The bending of rock layers subjected to tectonic stresses.

foot wall The side of a fault plane that extends beneath the hanging wall side.

forest An area in which trees are spaced sufficiently close together that their foliage overlaps to form a more or less continuous canopy.

fossil water Ground water supplies, generally in drier regions, that cannot be replaced under existing climatic conditions.

freezing rain Rain that freezes on contact with objects at the surface.

front See *weather front*.

frontal cyclone A low pressure system that has developed on a weather front. Most frontal cyclones have an associated warm front and cold front. Frontal cyclones are major precipitation producers for the middle and high latitudes.

frontal lifting The lifting of relatively light warm air by denser cold air along a weather front.

frontal thunderstorm A thunderstorm that has developed in association with a weather front. Frontal thunderstorms typically develop in a linear pattern, and are most frequently associated with cold fronts.

frontal wave A poleward bend in a weather front, normally associated with an area of weak low pressure. Frontal waves often develop into frontal cyclones.

frost Ice crystals that have sublimed on surface objects because the overlying air has cooled below the dew point.

frost heaving The loosening and lifting of the soil by growing frost crystals.

frost wedging The process of rock breakup resulting from the growth of ice crystals within rock fractures or hollows.

Fujita scale A tornado severity rating scale that ranks tornadoes from a minimum of F0 (weak) to a maximum of F5 (devastating).

funnel cloud A tornado funnel that remains aloft.

galeria Strip growths of forest in the moist soils along riverbanks in savanna grassland regions.

geographic information system A computer system that allows individuals to superimpose several categories of mapped data about an area in order to identify spatial patterns of surface phenomena.

geography The study of the distributions and interrelationships of Earth phenomena.

geographical approach The study of topics by analyzing their spatial (locational) characteristics.

geomorphology The systematic study of landforms, including their origin, characteristics, and distribution.

geothermal heat Heat from the Earth's interior, believed to result largely from the radioactive decay of uranium and thorium. It serves as the energy source for plate tectonic processes.

glacial drift A general term for all deposits of glacial origin.

glacial lacustrine plain A plain covered by fine textured lake sediments. These sediments were deposited at a time when the surface was covered by a lake whose drainage outlet was blocked by a glacier.

glacial trough A deep and generally "U" shaped valley, usually of fluvial origin, that was subsequently occupied and modified by a valley glacier.

glaciation The freezing of supercooled water droplets within a cloud to form ice crystals. *Also* the covering of a surface area by the advance of a glacier.

glacier A mass of fresh water ice, formed on land, that is or has been in motion as a result of gravity and/or internal pressure.

glacio-fluvial material Rock material that was partially sorted by meltwater after being deposited by a glacier. It therefore displays some stratification.

global radiation balance The long-term global balance between incoming short-wave insolation and outgoing long-wave terrestrial radiation.

gorge A narrow, deep, and steep-walled valley that is usually carved by a stream that is rapidly downcutting.

graben A block of land bounded by two normal faults that has been downdropped relative to the blocks on either side.

gradational forces Solar powered forces that, in conjunction with the force of gravity, act to lower the Earth's surface. The three major "tools" of gradation are water, ice, and wind.

graded stream A stream that has adjusted its flow parameters to the quantity and texture of its sediment load, so that it is neither eroding nor depositing materials (at least on a long-term basis).

granite A low density intrusive igneous rock consisting primarily of quartz, feldspar, and mica, that comprises much of the continental crust.

graphic map scale A map scale stated by means of a graph, usually taking the form of a line or bar.

grass A large and biologically diverse family of plants with bladelike, opposing leaves connected to a jointed stem by a sheathlike attachment. The cereal grains that form the basic foodstuffs of humans and livestock are grasses.

great circle A circle, drawn on a sphere, that bisects the sphere and is a full circumference of the sphere.

greenhouse effect The heat-trapping ability of the atmosphere, especially of the gases carbon dioxide and water vapor. The process is analogous to the heat-trapping ability of the glass in a greenhouse.

Greenwich meridian The starting line of longitude; also known as the prime meridian.

groin A narrow seaward-extending barrier, usually constructed of rock, that is designed to interfere with nearshore currents and thereby encourage the deposition of sand.

groundwater Gravitational water that completely fills the pore spaces or other voids in the soil or rock below the water table.

growing season The period of time during the year when weather conditions continuously favor the growth of vegetation. In North America, it is generally considered as the period between the last killing frost of spring and the first killing frost of fall.

gully An erosion channel so deep that is cannot be crossed by a wheeled vehicle or eliminated by plowing.

guyot A flat-topped seamount that apparently once extended above the ocean surface, where it was planed off by subaerial erosion.

gyre The geometrical pattern formed by surface ocean current systems, consisting of a giant elliptical loop.

habitat A specific site that has environmental conditions favoring its habitation by a given species or group of species.

hail A precipitation form consisting of relatively large, rounded particles of ice that typically exhibit internal layering. Hail forms as strong updrafts and downdrafts in a thunderstorm carry ice pellets alternately above and below the freezing level in a thunderstorm.

halophytes Salt-tolerant plants that normally occupy lowland basins of interior drainage in arid regions.

hamada A desert area with a solid bedrock surface, sometimes covered with a scattering of boulders.

hanging valley A small valley that enters a larger valley high up on its flank, and therefore "hangs" above the larger valley. This markedly ungraded situation generally results from either glaciation or fault displacement.

hanging wall The side of a fault that extends above the foot wall side.

haystack hill A rounded hill, somewhat resembling a haystack, that remains on a karst surface as an erosional remnant of a former higher topographic level.

headland A seaward projecting promontory, usually of significant steepness and elevation.

heat of condensation The quantity of heat energy, totaling approximately 585 calories per gram of H_2O, that is released when water vapor condenses to liquid water. It is the major direct energy source for many atmospheric processes.

heat of fusion The quantity of heat energy, totaling approximately 80 calories per gram of H_2O, that is released when liquid water freezes.

herbivore An animal whose diet consists chiefly of plants.

heterosphere The compositionally layered upper atmosphere, located above an altitude of 80 kilometers (50 mi).

high See *high pressure system.*

high latitudes The polar and subpolar regions; often considered as those regions poleward of about 60° N and S.

high pressure system An area of higher barometric pressure than the surrounding areas, characterized by an anticyclonic windflow pattern, and typically associated with fair weather.

histosol A soil order consisting of dark colored organic soils of poorly drained areas. Most histosols are highly acidic and low in fertility.

homoclinal The inclination of a group of tilted strata of the same dip. If they extend above the adjacent strata, they will form a *homoclinal ridge;* if they are lower than the adjacent strata, they will form a *homoclinal valley.*

homosphere The compositionally homogeneous lowest 80 kilometers (50 mi) of the atmosphere.

horn A sharp-crested, pyramidal peak, representing the remnant rock mass between three or more coalescing cirques.

horst A raised block of land bounded by two normal faults.

human geography The branch of geography concerned with the spatial aspects of Earth phenomena that are produced or are primarily controlled by humans.

humidity A term referring to the water vapor content of the air. Several types of humidity measurements exist.

humus Finely divided, decomposed organic matter with a brownish-black color that is a major supplier of plant nutrients.

hurricane A tropical cyclonic storm that forms over water and has sustained winds of at least 33.5 meters per second (75 mph). Hurricanes are the world's most destructive storm type.

hydration A chemical weathering process involving the absorption, or adhesion, of water molecules to the molecules of a mineral.

hydraulic action The direct removal of loose materials by the friction and turbulence generated by flowing water in a stream or by oceanic wave action.

hydrograph A graph of stream discharge at a given gaging station over a period of time.

hydrologic cycle The continuous cycle of phase change and movement of water between the Earth's surface and the atmosphere.

hydrolysis A chemical weathering process consisting of a permanent chemical combination of minerals with water.

hydrosphere The aqueous envelope of the Earth, including water in the atmosphere, at the Earth's surface, and in the near-surface layers of the ground.

hygrophyte A plant especially adapted to grow in a very moist or waterlogged surface environment.

hygroscopic Water attracting. Usually refers to dust particles that serve as condensation nuclei.

ice-scoured plain An extensive, lake-strewn, low relief surface that was subjected to glacial scour by a continental ice sheet. Ice-scoured plains presently occupy much of interior eastern Canada and Scandinavia.

iceberg An islandlike mass of fresh water ice produced by the breakup of a glacier as it enters the ocean or other large standing water body.

igneous rock A rock that has formed directly from the solidification of magma or lava resulting from tectonic activity.

illuviation The deposition, usually in a lower soil horizon, of solid or dissolved materials that were removed (eluviated) from another horizon.

inceptisol A soil order containing chemically immature soils with weak profile development.

infiltration capacity The maximum rate at which a given surface can absorb water. It is determined primarily by the texture, degree of compaction, and existing water content of the surface materials.

inselberg An isolated, steep-sided remnant of a previous mountain mass or other upland surface in an arid or seasonally dry region. Inselbergs are sometimes considered the arid climate counterparts of monadnocks.

insolation A shortening of the term "incoming solar radiation," referring to the total quantity of solar radiation reaching the Earth system.

interception The process by which vegetation intercepts precipitation before it reaches the ground. This process becomes ineffective once the vegetation has been fully wetted.

interfluve The generally well drained higher ground between streams.

interior drainage Drainage that does not reach the ocean, but instead enters an enclosed basin where it is absorbed into the ground or evaporates. Interior drainage is characteristic of most arid regions.

International Date Line This line is based on the 180th meridian and is located in the central Pacific. Time varies by a full 24 hours on either side, with the west side being a day later than the east side.

Intertropical Convergence Zone (ITCZ) The zone of low barometric pressure and wind convergence centered near the equator, caused primarily by the prevailing high temperatures.

intrusive igneous rock Igneous rock that has formed inside the Earth. The slowness of the solidification process usually allows visible mineral crystals to form.

intrusive volcanism The subsurface transfer of magma from one place to another. The major resulting features are batholiths, laccoliths, dikes, and sills.

invertebrate An animal without a backbone or spinal column.

isarithmic map A map that employs lines to connect places of equal numerical value.

island arc An arc-shaped group of islands paralleling, at a distance of several hundred kilometers, a submarine trench.

It is formed when magma from the melting subducted plate that produced the trench has been erupted in sufficient mass to protrude above sea level.

isobars Lines connecting places having the same barometric pressure.

isostatic rebound The buoyant uplift of a portion of continental crust that had formerly been depressed by the weight of a continental glacier.

isostatic equilibrium The floating of an object in a fluid to just the depth needed to support its mass.

isostatic uplift The broad, gentle upwarping of a region that has become buoyant because of its reduction in mass as a result of erosion or deglaciation.

isotherms Lines connecting places having the same temperature.

jet streams Relatively narrow bands of strong westerly winds centered in the upper troposphere above the middle latitudes and the subtropics of both hemispheres. They act as steering currents to direct the movements of surface weather systems.

jungle A nearly impenetrable growth of low growing plants that is common in warm, moist areas where sunlight can reach the near-surface vegetation layers.

kame A mound of slightly sorted glacio-fluvial debris, often originating as a crevasse filling.

karst Topography developed largely by the differential solution and subsequent erosion of highly soluble rocks, especially limestone and dolomite.

karst plain A limestone plain covered with large numbers of sinkholes and sometimes other solution features.

Kelvin temperature scale A temperature scale with its starting temperature (0K) at absolute zero—the temperature at which all molecular motion ceases. Like the Celsius scale, it employs 100 units between the freezing and boiling points of water (273K and 373K, respectively).

kettle A depression within a till plain that may be occupied by a small pond. Kettles are formed by the weight of detached blocks of ice from a stagnant glacier.

knickpoint A point where the downhill gradient of a stream abruptly increases.

knot A speed of one nautical mile per hour. It equals a speed of 1.15 statute miles per hour.

Köppen climate classification A system of classifying global climates according to their basic temperature and precipitation characteristics. (Details of the system, developed by Wladimir Köppen about 1918, appear in Chapter 8).

laccolith A dome-shaped igneous intrusion of relatively acidic magma that is injected into a sequence of near-surface sedimentary strata.

lagoon A shallow, protected body of salt water located between a barrier island chain and the mainland. It may consist of an extensive mudflat at the time of low tide.

lake A landlocked body of water with a horizontal surface level.

lake breeze A diurnally reversing windflow along the shore of a large lake, caused by air temperature differences between the water and land. An onshore breeze occurs during the day, and an offshore breeze develops at night.

land and sea breeze A diurnally reversing coastal windflow pattern established in some areas by land and water temperature differences. Sea breezes are generated during the day, and land breezes at night.

landforms Individual Earth surface features.

landslide General term for a rapid mass movement in which motion occurs primarily by sliding.

lapse rate See *average environmental lapse rate.*

latent heat Potential heat energy. Latent heat is stored in the molecules of water vapor and of liquid water, and will be released if these substances undergo a phase change to a lower energy state.

lateral moraine A linear accumulation of rock debris deposited along the present or former margin of an alpine glacier.

laterite A hard, iron-rich subsurface layer that sometimes develops in humid tropical regions following the removal of vegetation.

latitude Angular distance in degrees north or south of the equator.

Laurentide ice sheet The continental ice sheet that covered most of Canada and the northern United States during the glacial stages of the Pleistocene.

lava Molten rock material formed by the expulsion of magma onto the Earth's surface.

lava plain (or lava plateau) An extensive, gently rolling plain or plateau surface formed by consecutive fissure outpourings of basaltic lava.

lava flow The flow of molten rock material from a volcano or fissure.

leaching The removal of dissolved materials and fine textured particles from the upper soil by downward percolating soil water. These materials may be deposited deeper within the soil, or they may be removed completely from the soil.

leap year A year containing 366 days (including February 29).

leeward coast A coast with a prevailing land to water windflow.

legumes A group of plants that host bacteria in their root nodules that extract nitrogen from the atmosphere and convert it to a form available to vegetation.

liana Any of a variety of climbing plants that root in the ground. Large woody lianas are characteristic of the tropical rainforest.

limestone An organic sedimentary rock composed of calcium carbonate ($CaCO_3$). Most limestone formed in quiet tropical sea floor environments either by direct chemical precipitation or through the accumulation of the remains of marine organisms.

line symbols Map symbols taking the form of lines; usually used to denote linear features such as roads and power lines.

lithosphere The solid portion of the Earth, composed largely of rock and metal. More recently, this term has also been used to refer to the rigid uppermost portion of the mantle and the overlying crust that forms the lithospheric plates.

lithospheric plate A discrete segment of the Earth's crust and uppermost mantle that moves as an entity when transported by convection currents in the mantle.

loam A soil of intermediate texture that generally contains substantial proportions of sand, silt, and clay.

local time The time of day based on the east-west position of the Sun at a given locality.

loess Silt-sized eolian deposits that are typically buff-colored and are composed of a well sorted mixture of angular quartz and calcite ($CaCO_3$) particles.

longwave radiation Electromagnetic radiation of relatively long wavelength. Usually used to describe terrestrial radiation (as opposed to shortwave solar radiation).

longitude Angular distance in degrees east or west of the prime meridian, stated to values of 180° E or W.

longitudinal dune A long ridge of sand that is typically aligned parallel to the direction of the prevailing winds in an arid region.

longitudinal stream profile A graphic depiction of downhill gradient of a stream from its head to its mouth.

longshore current A current paralleling the shoreline, often generated by the momentum of waves approaching the shoreline obliquely.

low See *low pressure system.*

low latitudes The tropics. Usually refers to the regions equatorward of 30° N and S, or between the Tropics of Cancer and Capricorn.

low pressure system An area of lower barometric pressure than the surrounding areas, characterized by a cyclonic windflow pattern, and typically associated with cloudiness and precipitation.

magma Molten rock material inside the Earth. if it reaches the surface in a molten state, it is termed lava.

magnetic North Pole The northern locus of the Earth's magnetic field, and the place to which a compass needle points. It is located in the Canadian Arctic.

magnetic South pole The southern locus of the Earth's magnetic field, located near the coast of Antarctica.

mantle The most extensive of the internal layers of the Earth, occupying an intermediate position between the core and the crust. The mantle is approximately 2,900 kilometers (1,800 mi) thick, and is apparently composed primarily of peroditite rock.

map A pictoral representation of the geographic location of selected surface features at a reduced scale.

map distortion Inconsistencies in the sizes and/or shapes of areas depicted on maps. It results form the impossibility of accurately depicting the spherical Earth on a flat map surface.

map scale The ratio between the separation of places on a map and on the Earth's surface.

Maquis The name used in the Mediterranean region of Europe for an often dense vegetative cover of woody shrubs and small trees.

marine terrace An elevated surface of low relief that represents the remains of an old wave-cut or wave-built platform. These features occur in areas where the land has been uplifted or sea level has dropped.

Mariner's Compass A directional system that states directions by employing combinations of the words north, south, east, and west (or letters N, S, E, or W).

maritime climate Climatic characteristics typical of an area over or near a large water body, especially including relatively small daily and annual temperature ranges and high quantities of atmospheric moisture.

maritime equatorial air mass A very warm, moist air mass that has formed over a water body near the equator.

maritime polar air mass A relatively cold, moist air mass that has formed over a high latitude water body.

maritime tropical air mass A relatively warm, moist air mass that has formed over a water body in the low latitudes.

mass wasting The downslope movement of surficial material under the direct influence of gravity.

meander A broad looping bend in the course of a stream occupying a floodplain. Meanders typically display a distinctive cycle of formation, growth, and disappearance.

medial moraine A linear accumulation of rock debris, deposited or still being transported by a glacier, that extends, or extended, down the middle of the glacier.

mercurial barometer A barometer that measures air pressure by the height of the mercury column in a glass tube from which the air has been removed.

meridian Another term for a line of longitude.

mesa An isolated, steep-sided and flat-topped remnant of a previous more elevated topographic surface. Mesas are often left as erosional outliers of retreating escarpments in arid regions.

mesocyclone A cyclonic windflow several miles in diameter. The development of a mesocyclonic windflow in a severe thunderstorm may precede the development of a funnel cloud or tornado.

mesopause The boundary between the mesosphere and the thermosphere. At this level, temperatures stop falling with increasing altitude.

mesophyte A plant adapted to an environment with a moderate water supply that is available throughout the year. It is intermediate in characteristics between a xerophyte and a hygrophyte.

mesosphere The thermal layer of the atmosphere between the stratosphere and the thermosphere, in which temperatures fall with increasing altitude.

mesquite A drought-resistant, spiny, deciduous tree or shrub with an extensive root system that has recently proliferated in the American Southwest.

metamorphic rock Rock subjected to solid-state alteration from its original form by heat, pressure, and/or chemical activity. Each igneous and sedimentary rock type has one or more metamorphic rock equivalents.

meteorite A naturally occurring mass of rock or metal that has fallen to the Earth's surface from outer space.

meteorology The branch of physical science that systematically studies the weather.

middle latitudes The latitudinal zone located between the high and low latitudes. Most of the United States as well as southern Canada is located in this zone, which extends from approximately 30° N and S to 60° N and S.

Milky Way galaxy The organized aggregation of stars in which our solar system is located.

mineral A naturally occurring, solid, inorganic compound that has a specific chemical composition and usually exhibits a crystalline structure that results from a distinctive atomic arrangement of its constituent elements.

minute An angular distance of 1/60th of a degree; used for accurately stating latitude and longitude.

Moho (or Mohorovičić discontinuity) The relatively sharp boundary between Earth's mantle and crust.

mollisol A soil order containing soils with a thick, dark colored, humus-rich A horizon and a high cation exchange capacity. Most mollisols are highly fertile and agriculturally productive and are associated with major grain producing regions.

monadnock An isolated hill or mountain that stands as an erosional remnant of a previous higher and more mountainous topographic surface.

monsoon winds A seasonally reversing regional-scale wind-flow pattern caused by land and water temperature differences. They are especially well developed in southern and eastern Asia and in parts of North Africa.

monthly mean temperature The average temperature for the month, calculated by averaging the daily mean temperature values for each day of the month.

moraine A glacial depositional feature composed of till. Several types of moraines exist, depending on their sites of deposition. These include lateral, medial, terminal, recessional and ground moraines.

mountain and valley breezes A diurnally reversing wind-flow pattern established in some mountain areas as a result of surface heating and cooling.

mudflat A muddy, low-lying strip of land near the shore that is usually submerged at high tide.

mudflow The perceptible and often rapid downhill flow of mud, usually mixed with rocks and other debris.

muskeg A swampy area within the Northern Coniferous Forest that is covered with mosses, swamp grasses, shrubs, and scattered hygrophytic trees such as larches.

mutation A sudden and inheritable alteration in the characteristics of a plant or animal caused by a molecular rearrangement of its genetic material.

nappe A thick sheet of rock that has been thrust by extreme compressionary forces over other strata.

natural levees Low and often broad ridges of stream deposits that parallel each side of a stream channel within a floodplain. They are formed by deposition resulting from the sudden increase in friction as a flooding stream overtops its banks.

neap tides Tides with exceptionally small vertical ranges, occurring approximately twice a month, caused by the perpendicular alignment of the Moon and Sun with respect to the Earth.

needleleaf Used to describe trees with needle-shaped leaves, such as pine, spruce, or fir, as opposed to broadleaf trees.

net primary productivity The total amount of organic matter generated by a biological community.

niche A set of environmental conditions that is necessary for an organism to maintain a viable population.

nimbostratus A dark, textureless cloud associated with the large-scale uplift of stable air. It is the major winter precipitation producer for regions in the middle and high latitudes.

nimbus A Latin term meaning "rain cloud", used in conjunction with the two cloud types responsible for most of the world's precipitation—nimbostratus and cumulonimbus.

nitrogen A relatively inactive gas that comprises approximately 78% of the lower atmosphere.

nitrogen fixation The process by which atmospheric nitrogen is extracted by legumes and converted to a form available to vegetation.

non-variable gas An atmospheric gas that does not vary appreciably in relative abundance over a short timespan.

normal fault A fault in which the vertical displacement exceeds the horizontal and an expansionary component is present.

North Pole The northern end of the Earth's axis, located at 90° N.

northern hemisphere The half of the Earth north of the equator.

nuée ardente A dense cloud of superheated gases and fine particles that rushes down the flanks of an erupting volcano, charring or setting fire to anything combustible and burning or suffocating all living organisms in its path.

nutrient A chemical element needed by the lifeforms of an ecosystem.

oasis A desert locality where the water table intersects the surface, resulting in the growth of significant vegetation.

oblate ellipsoid A three-dimensional ellipse that is flattened at the poles. The Earth assumes this shape because the centrifugal force of rotation bulges out the equatorial regions.

obsidian A generally black-colored, glassy volcanic rock that has a relatively high silicon dioxide content.

occluded front A frontal boundary separating, at the surface, two relatively cold air masses, with a warmer air mass aloft. It is formed by the overtaking of a warm front by a cold front.

ocean The interconnected body of salt water that dominates the Earth's surface. It is commonly divided into segments named the Pacific, Atlantic, Indian, and Arctic Oceans.

ocean current The continuous movement of an ocean water stream of similar temperature and density.

oceanic crust The thin, continuous layer of dense basaltic rock that comprises the lowermost portion of the crust and forms the oceanic basins.

omnivore An animal whose diet contains significant amounts of both plants and animals.

open system A system that has access to an unlimited supply of energy from an external source.

organic sedimentary rock Sedimentary rock produced from the remains of plants and animals. Most are composed of carbonate minerals, especially limestone ($CaCO_3$).

orographic lifting The lifting of air over a topographic barrier.

outwash Glacio-fluvial materials that have been transported and deposited beyond the edge of a glacier by meltwaters.

outwash plain A relatively extensive and flat glacial outwash surface deposited by multiple meltwater streams issuing from the margin of a glacier.

overland flow A non-channelized, sheetlike flow of water that is common on agricultural fields and near drainage divides.

overwash The transport of sand from the seaward side of a barrier beach or island to its landward side by waves that wash completely over the surface.

oxbow lake An arcuate lake that occupies an abandoned stream meander.

oxidation A chemical weathering process involving the chemical combination of minerals with oxygen that has often first been dissolved in water.

oxisol The deepest and most highly chemically weathered of the soil types. They are quite limited in natural fertility and have developed on ancient surfaces within the humid tropics.

oxygen The second most abundant gas in the atmosphere, necessary to animal respiratory processes, and derived from carbon dioxide by means of plant photosynthesis.

ozone A tri-atomic form of oxygen (O_3), occurring in very small quatities in the atmosphere, that helps to protect the Earth's surface from cosmic ray bombardment.

ozone layer The layer of the upper atmosphere, extending from about 15 to 65 kilometers (10–40 mi) altitude, in which ozone is most abundant.

P waves Primary earthquake shock waves that result from compression.

Pangaea The hypothesized former supercontinent that fragmented to produce the present continents.

parallelism The idea that the Earth's axis is constantly oriented in the same direction with respect to the plane of the ecliptic.

parallels Another term for lines of latitude.

parent material The solid inorganic and organic material from which the soil develops. The inorganic parent material may be either residual (originally formed in that location) or transported (derived from elsewhere and deposited in that location).

patterned ground Polygonal patterns that develop on the surface in many periglacial regions. They are caused by the alternate freezing and thawing of the ground, producing a churning action that cracks the surface and shifts surface materials in a selective fashion.

ped A soil structural aggregate formed by the tendency of individual soil particles to adhere to one another. Different soil types tend to develop peds of differing sizes and shapes.

pediment A gently sloping bedrock erosional surface commonly found at the foot of a mountain mass in an arid region.

pediplain A desert erosional plain formed by the coalescence of pediments as the intervening mountain masses are worn away.

pedologist A soil scientist.

peneplain A land surface worn down by long-continued erosion to a nearly flat or gently rolling plain.

perennial A plant that has an expectable lifespan of at least several years. It is therefore not normally killed by adverse weather conditions at the end of the growing season, although it may experience a period of dormancy.

perennial stream A stream that flows continuously because it has a dependable source of ground water.

periglacial A geomorphic environment that develops in a severely cold climate. Although the surface is not glaciated, landscape development is associated with the alternate freezing and thawing of the near-surface layers.

perihelion The closest approach between Earth and Sun. It occurs about January 3rd, when the two bodies are approximately 147.5 million kilometers (91.5 million miles) apart.

permafrost The condition of perennially frozen subsoil and bedrock that underlies large areas in arctic, subarctic, and alpine environments.

permeability The ability of the soil or rock to transmit the flow of water.

photogrammetry The making of accurate measurements from photographs.

photosynthesis The formation of carbohydrates in the chlorophyll-containing tissues of plants exposed to light.

physical geography The branch of geography concerned with the spatial aspects of natural Earth phenomena, particularly with weather and climate, plants and animals, soils, and landforms.

physical weathering Weathering processes that reduce the size of rock masses without altering their chemical composition.

piedmont alluvial slope A ramplike alluvial apron, typically found along the base of a desert mountain range, that is formed by the coalescence of numerous alluvial fans. Also called a bajada.

place A specific location on the Earth's surface that contains a unique assemblage of spatial attributes.

plain An area of restricted local relief that has a predominantly flat to gently rolling surface.

plane of the ecliptic The angular plane on which the Earth's orbit around the Sun is located.

planimetric map A map that provides systematic information about only the horizontal characteristics of the areas it depicts.

plant An organism containing chlorophyll, which has the ability to convert water and atmospheric gases into food in the presence of light.

plant association A global-scale community of plants that exists in relative equilibrium with its environment.

plant community A plant assemblage whose members occupy a specific environmental setting of restricted size.

plate See *lithospheric plate*.

plateau A relatively level upland surface that is frequently bounded on at least one side by an abrupt descent to lower elevations. Also termed a tableland.

plate tectonics Encompasses the processes involved in the formation, movement, and destruction of lithospheric plates.

playa A shallow desert lake, often of considerable areal extent, that temporarily occupies the center of a bolson following a rainfall.

Pleistocene The geologic epoch, extending from about 2,000,000 B.P. to 10,000 B.P., during which continental and alpine glaciers alternately advanced and receded over large portions of the world, especially in North America and Eurasia.

plucking The process of erosion, by glaciers and streams, in which blocks are removed from bedrock along joints and stratification surfaces.

plunging fold A fold whose constituent rock strata were subjected to prior tilting so that the fold "plunges" into the ground in one direction.

pluton A body of intrusive igneous rock that has generally solidified well beneath the surface.

pocket beach A small, crescent-shaped beach located on a coastline that is, as a whole, being worn back by vigorous wave attack. These beaches, common along the U.S. West Coast, often develop in areas of relatively weak rock.

point bar A streambed bar of sand or gravel that develops on the inside of a meander. It is caused by the diminished stream flow at this point.

point symbols Non-linear map symbols, such as dots or squares, used to indicate the locations of specific mapped features.

polar easterlies The belts of easterly winds blowing from the polar high pressure centers of each hemisphere toward the subpolar lows.

polar front The major weather front of the middle latitudes, separating air masses of subtropical and subpolar origins. It is a zone along which frontal cyclones frequently form.

polar front zone The zone of highly variable winds associated with the subpolar low pressure belts in each hemisphere.

polar highs High pressure caps centered near the North and South Poles, resulting largely from cold surface temperatures.

polar projection A map projection centered on the North or South Pole.

polar zones of variable winds and calms The zones of light and variable winds associated with the centers of the polar highs.

pore spaces The voids between the solid soil particles that contain soil liquids and gases.

porosity The volume of soil pore spaces divided by the total volume of the soil.

potential evapotranspiration The total quantity of water that would evaporate and transpire from a given site if the available moisture supply were unrestricted.

prairie Tall grassland generally associated with subhumid portions of the middle latitudes. Most prairie grasslands have been destroyed by human agricultural activities.

precipitation Water in liquid or solid form falling through the atmosphere toward the Earth's surface. The major forms are rain, snow, sleet, hail, and drizzle.

pressure gradient The rate of change in air pressure per unit of distance. When uncompensated for by other factors, it causes the wind to blow.

primary production rate The rate at which a given plant, or the plants within a given area, produce energy by photosynthesis.

primary succession Vegetative succession that begins on a "new" surface completely devoid of vegetation, such as a lava flow or a previously glaciated surface.

prime meridian The starting line of longitude. A separation line between the eastern and western hemispheres.

producer A plant, which produces carbohydrate energy from water and carbon dioxide in the presence of light.

protozoa Microscopic, single-celled animals that feed on organic matter and hasten its decomposition.

pumice A vesicular, glassy lava, generally containing a high proportion of silicon dioxide.

pyroclastic Fragmental material erupted from a volcano.

radar An instrument used to determine the approximate size, shape, and intensity of areas of precipitation. It emits pulses of microwave energy that are reflected from the precipitation and displayed on a screen.

radial stream pattern A stream pattern characterized by the outward flow of streams in all directions from a central peak or upland.

radiation The process of energy transmission by electro-magnetic waves.

radiation fog A fog type that commonly forms over land areas on calm, clear nights because surface radiation chills the overlying air below the dew point.

radiosonde A weather instrument package that is sent aloft by means of a hydrogren or helium filled balloon and automatically radios back upper atmosphere data on temperature, air pressure, and humidity.

rain Liquid precipitation drops with diameters exceeding 0.5 millimeters (0.02 in).

rainforest A dense forest associated with perennially damp conditions. Most rainforests exist in the tropics and along windward coasts in the middle latitudes.

rain gage A funnel-shaped, tubular instrument used for the collection and measurement of precipitation.

rainshadow effect The reduction of precipitation by the blocking effect of a topographic barrier.

rectangular stream pattern A drainage pattern alternating between relatively straight segments and sudden right-angled bends, usually as the result of intersecting fault or fracture systems in the underlying bedrock.

recumbent folds Folds that have been compressed so tightly that the fold structures have closed in on one another and collapsed, resulting in a nearly horizontal axial plane.

reef A ridge-like accumulation of the exoskeletons of large colonies of coral polyps or other reef-building marine organisms. The reef itself may be entirely beneath the water or may protrude above the water surface to form a coastal shelf or island. A fringing reef is attached to the mainland, while a barrier reef is separated from the mainland by a lagoon.

reg A stony desert from which the finer materials have been removed by the wind.

regelation A process of glacial motion resulting from the melting of ice as it is pressed into the up-ice side of a bedrock obstruction, followed by the refreezing of the water after it has flowed around the obstruction to the down-ice side.

region A portion of the Earth that displays relative similarity in one or more attributes.

regional metamorphism Large-scale metamorphism that occurs in conjunction with tectonic activity; especially found in the root zones of mountain systems.

regolith Weathered rock material, commonly located between unweathered bedrock (below) and soil (above).

relative glacier motion The advance or retreat of the margin of a glacier relative to the underlying surface.

relative humidity The percentage ratio of the amount of water vapor actually in the air to the maximum amount of water vapor the air could hold at that temperature and pressure.

relict A landform produced by geomorphic processes no longer active in the area.

relief (topographic) Differences in elevation within an area. The topographic relief of a mapped area is computed by taking the elevational difference between the highest and lowest point.

remote sensing The collection of information at a distance from a location by means of mechanical devices that receive emitted or reflected electromagnetic radiation.

representative fraction (or RF) scale A map scale stating a direct ratio between map distances and corresponding Earth distances.

respiration The release, by organisms, or energy from carbohydrate molecules within cells. The energy is then used in metabolism. For plants, it is the opposite of photosynthesis.

reversal of topography Occurs when intense folding or faulting causes rock strata to overturn, so the older strata overlie younger strata.

reverse fault A fault in which the vertical displacement exceeds the horizontal and a compressionary component is present, so that some overlapping of the two sides occurs.

revolution The yearly orbital motion of the Earth around the Sun.

rhumb line A line of constant compass bearing.

ria coast A highly irregular erosional coast dominated by numerous bays or estuaries that were produced by the gradual submergence of river valleys. The Middle Atlantic coast of the U.S. is a classic example.

Richter scale The most commonly used scale for measuring earthquake magnitudes. Each number on the scale has ten times the earthquake wave amplitude of the preceding number.

rift valley A deep, linear, steep-sided trough produced by the subsidence of a strip of land between two faults. It is sometimes formed during an early stage in the separation of two continental plate segments.

rills A series of nearly parallel small shoestring channels that result from the early stages of fluvial erosion of an unprotected slope.

riparian Riverbank vegetation that often flourishes in the moist soils along streams in arid and semiarid regions.

rip current A narrow zone of water moving rapidly seaward along a shoreline.

river A stream of water, usually of large size.

roche moutonée A bedrock outcrop in glaciated terrain that exhibits a smoothly rounded up-ice side and an irregularly plucked down-ice side.

rock A solid aggregate of minerals. Different rocks are defined on the basis of their proportional mineral constituency. Rocks may also consist of organic materials.

rock cycle The continuous cycle of rock formation, erosion, deposition, and reconstitution.

rockfall The falling of an individual rock or small rock mass down a nearly vertical cliff.

rockslide The sliding of a mass of detached rock down a moderate to steep slope.

rock glacier A glacierlike tongue of angular rock debris usually beginning in a cirque or other steep-walled area.

Rossby wave A large-scale north/south-oriented wave in the jet stream. These waves help determine the direction of movement of surface weather systems.

rotation The spinning of the Earth on its axis. This motion is the basis of our calendar day.

S waves Earthquake shear waves that are characterized by a side-to-side motion.

saline water intrusion The intrusion of salt water into a fresh water aquifer, often as the result of the excessive withdrawal of well water.

salinization The increase in soil salinity, usually in poorly drained arid areas, that occurs as a consequence of irrigation for a prolonged period. The salt is derived either from the deeper soil layers or from the irrigation water itself.

saltation An asymmetrical bouncing motion exhibited by relatively coarse textured sediments being transported either by a stream or by the wind.

salt marsh A marsh produced by the colonization of a coastal mudflat by marsh grasses.

sand bar A largely or entirely submerged ridge composed of sand that is formed in the nearshore oceanic waters by currents and larger waves.

sand shadow A fixed sand dune that forms in the lee of a stationary obstruction such as a rock.

sand sheet A flat sandy surface that displays little or no tendency toward dunal formation.

sandstone A clastic sedimentary rock comprised largely of sand-sized clasts that are usually composed of silica.

sandstorm A stinging mass of wind-blown sand grains extending only a meter or two above the ground.

Santa Ana wind A hot, dry, desert-derived wind that periodically affects the lowlands of coastal southern California. It generally produces a high fire danger.

savanna A tropical grassland that contains scattered trees or shrubs. Savannas are dominant in forest/grassland transitional areas in regions with alternating wet and dry seasons.

sclerophyll A plant with small, hard, thick, leathery leaves that serve to minimize transpirational losses.

scrub A dense growth of shrubs and small trees, typical of many tropical and subtropical areas with alternating wet and dry seasons.

sea cliff A cliff formed along or just behind a shoreline by wave undercutting in areas of competent (strong) materials.

seafloor spreading The process of formation of new oceanic crust by the extrusion of lava along divergent lithospheric plate boundaries.

sea ice Ice that forms on the ocean surface from the freezing of sea water. Most exists in the Arctic Ocean and near the coast of Antarctica.

seamount A circular or elliptical submarine mountain rising more than 900 meters (3,000 ft) from the deep sea floor. Seamounts are produced by undersea volcanic eruptions.

seawall A rock or wooden barrier built parallel to the coast to absorb and dissipate wave energy.

second An angular distance of 1/60th of a minute, or 1/3,600th of a degree.

secondary succession Vegetation succession that occurs following a disturbance to an area in which a soil cover and perhaps some seeds and vegetation remain. It typically proceeds much more rapidly than does primary succession.

sedimentary rock Rock that has formed from the lithification of sediments and typically accumulates in layers (strata).

sediments Solid materials of either organic or inorganic origin that were transported to their depositional sites by water, ice, wind, or gravity. The lithification of sediments forms sedimentary rocks.

seismograph An instrument used to measure earthquake shock waves.

semiarid climate A climate with a potential evaporation rate that slightly exceeds its mean precipitation total, so that it is moderately dry.

semideciduous A vegetation assemblage containing a mixture of deciduous and evergreen species.

shale The most abundant sedimentary rock type, composed of clay-sized particles, and often containing fossils. Most shales formed from muds deposited on the deep ocean floor.

shear strength The measure of the ability of surficial materials to resist downslope movements.

shield An ancient block of continental crust that has been tectonically stable for a lengthy geologic timespan.

shield volcano A broad, gently sloping volcano constructed primarily of basalt derived from fluid lava.

shortwave radiation Electromagnetic radiation of relatively short wavelength. Usually used to describe solar radiation (as opposed to longwave terrestrial radiation).

shrub See *bush*.

sial Derived from the words "silicon" and "aluminum," this term refers to the granitic rocks that comprise most of the continental crust.

silicate minerals A mineral family consisting primarily of silicon and oxygen. It is the dominant mineral group of the crust and mantle.

sill A relatively thin sheet of magma injected more or less horizontally between near-surface sedimentary rock strata.

siltstone A clastic sedimentary rock composed predominantly of silt-sized particles.

sima Derived from the words "silicon" and "magnesium," this term refers to the basaltic rocks that comprise the oceanic crust.

sinkhole A rounded depression formed by subsurface solution, especially in regions underlain by limestone. Sinkholes can develop either by gradual subsidence or by the sudden collapse of the surface. Also called a doline or a "sink."

sinking creek A stream that enters a swallow hole in a karst region and continues its course beneath the surface.

sleet A precipitation form consisting of small ice pellets produced by the freezing of descending raindrops.

slump The intermittent and relatively slow movement of a mass of earth or rock along a curved slip-plane.

small circle A circle, drawn on a sphere, that does not bisect the sphere and does not form a full circumference of the sphere.

snow A precipitation form consisting of crystalline ice formed by sublimation. Snow generally falls in flat hexagonal flakes.

soil A loose mixture of weathered rock material, organic matter, water, and air, that can support plant growth. Most undisturbed soils tend to develop distinctive layers (horizons) that parallel the surface and differ in their physical, chemical, and biological characteristics.

soil horizons Soil layers, roughly paralleling the surface, that differ in their physical, chemical, and biological characteristics as a result of their formation under differing environmental conditions.

soil pH The acidity or alkalinity of the soil, determined by its supply of exchangeable H + ions.

soil profile A vertical slice through the soil in which the various horizons are exposed for analysis.

Soil Taxonomy A nickname for the U.S. Comprehensive Soil Classification System—the soil classification system currently in official use in the United States.

soil texture The size of the individual soil particles. Different texture categories are determined by the soil's proportions of sand, silt, and clay-sized particles.

soil water Water that adheres in a molecular film to the soil particle surfaces, and partially occupies the pore spaces between the soil particles.

solar day The period of time needed for the Sun to make one apparent 360° circuit of the sky.

solar noon The exact time that the Sun reaches its zenith (high point in the sky) at any given locality.

solar system The Sun and the other astronomical bodies, notably the nine planets, that revolve around the Sun because of its gravitational attraction.

solifluction A form of earthflow that occurs on hillsides in periglacial environments as the surface layers thaw.

solstice A time when the axial tilt of the Earth is exactly parallel to the direction of the Sun's vertical rays. This occurs only twice each year—on the first days of summer and of winter.

solum The upper part of the soil profile, in which soil-forming processes are actively occurring. In mature soils this includes the A, E, and B horizons.

solution The change of matter from the solid or gaseous state into the liquid state by its combination with a liquid. *Also,* the result of such a change; a liquid combination of a liquid and a nonliquid substance.

source region A large land or water body over which air masses form and from which they derive their temperature and moisture characteristics.

South Pole The southern end of the Earth's axis, located at 90° S.

southern hemisphere The half of the Earth south of the equator.

specific heat The quantity of heat energy that must be absorbed or released by a particular substance in order to undergo a given temperature change.

specific humidity The ratio of the mass of water vapor in a parcel of air to the total mass of the air, including the water vapor. Usually stated in grams per kilogram.

spit A typically hook-shaped barrier beach that is attached to the mainland at one end and often extends partially across a bay on the other end.

spodosol A soil order containing acidic soils with a light colored eluvial E horizon and an illuvial B horizon. They are typically associated with a subarctic climate and with taiga vegetation.

spring The site of an outflow of subsurface water.

spring equinox See *vernal equinox.*

spring tides Exceptionally high and low tides, occurring approximately twice a month, caused by the parallel alignment of the Moon and Sun with the Earth.

spot elevation The stated elevation on a map of a specific reference point, such as the summit of a mountain.

squall line A line of heavy showers or thunderstorms associated with strong, gusty winds. Squall lines frequently develop in advance of strong cold fronts, in which case they are known as pre-frontal squall lines.

stable air Air with a vertical temperature profile that causes it to resist any lifting influences.

stack A small, steep-sided island located near the shoreline that is an erosional remnant of a retreating sea cliff.

stage The elevation of a water surface above a chosen datum plane.

stalactite An icicle-like travertine feature that hangs downward from the roof of a cave.

stalagmite The upward growing counterpart of a stalactite, formed of travertine, and found on the floor of a cave in a limestone region.

standard points or standard lines Points or lines of no distortion on a map projection.

standard time system The currently used time system for most of the world. Time is based on the east-west position of the Sun at meridians with longitudes divisible by 15, and changes by whole hour increments as time zones are crossed.

star dune A sand dune consisting of a large pyramidal sand accumulation with a number of radiating ridges.

stationary front A relatively stationary frontal boundary between two differing air masses. Such a front eventually will either dissipate, or will begin to move as a warm front or cold front.

steppe Short grasslands generally associated with semiarid climates in the low and middle latitudes. Grass heights are usually less than 0.6 meters (2 ft).

storm surge The very high ocean water levels that accompany the landfall of a hurricane.

strata Horizontal layers, usually of sedimentary rocks.

stratocumulus A low cloud type that typically forms a whitish or grayish layer with distinct cloud masses, usually in the form of rolls, between which patches of clear sky often appears.

stratopause The boundary between the stratosphere and the mesosphere. At this level, temperatures stop rising with increasing altitude.

stratosphere The thermal layer of the atmosphere above the troposphere, in which temperatures increase with altitude.

stratus From the Latin word meaning "layer," stratus clouds are low, dense, gray clouds that usually cover the entire sky.

stream A channeled body of water that flows downhill in response to gravity.

stream order A numerical system for classifying the approximate size of a stream, based on the number and nature of its tributaries. Two streams of a given order must join to form a stream of the next higher order.

stream stage The measured height of the water surface in a stream.

stream terrace A terrace consisting of the remnant of a former floodplain surface that was developed at a topographic level higher than the present stream valley level.

striations Sets of parallel bedrock scratches caused by the passage of glacial ice containing embedded rock fragments.

strike The compass orientation of a horizontal line in the plane of an inclined rock stratum or outcrop.

strike-slip fault See *transcurrent fault*.

stripped plain A plain formed when the erosional stripping away of weak rock strata leads to the uncovering of a horizontally bedded resistant rock layer.

structure (of soil) The characteristic shape, size, and organization of the soil aggregates or peds.

subduction The angular descent of the margin of an oceanic lithospheric plate into the asthenosphere, where it is eventually destroyed. Subduction results from a collision between two plates, with the denser plate being subducted.

sublimation The direct phase change of water vapor to ice, or vice-versa. Approximately 665 calories of energy are involved for each gram of H_2O that sublimes.

submarine canyon A deep, V-shaped canyon that is entrenched into the continental shelf. Such canyons are believed to have been formed initially by rivers when sea levels were lower, and to have been subsequently enlarged by the action of turbidity currents.

submarine trench A linear trench on the sea floor formed along the line where an oceanic lithospheric plate is being subducted. The world's greatest ocean depths occur in these trenches.

subpolar jet A jet stream that is produced by temperature contrasts between the low and high latitudes. Its position is closely related to the surface position of the polar front.

subpolar lows Zones of rising air and low pressure centered in the subpolar latitudes of both hemispheres, caused primarily by the Earth's rotation.

subsidence The vertical settling of the surface as a result of the removal of subsurface support by mining, solution, or other processes.

subsoil The soil horizon that underlies the topsoil. It usually contains little organic matter and is frequently an illuvial horizon.

subsolar point The point on the Earth's surface that has the Sun directly overhead.

subtropical belts of variable winds and calms The narrow belts of light and variable winds associated with the centers of the northern and southern hemisphere subtropical highs; also referred to as the "horse latitudes."

subtropical highs Zones of subsiding air and high pressure centered in the subtropics of both hemispheres, and caused primarily by the Earth's rotation.

subtropical jet A jet stream that is centered above the surface positions of the subtropical highs.

summer monsoon An onshore windflow of subcontinental proportions, produced by the higher air temperatures that develop over land than over water during the summer season.

summer solstice The day that one's hemisphere reaches its maximum inclination toward the Sun. The Sun is directly overhead at a latitude of $23^1/_2°$. Marks the official beginning of summer.

surface creep The slow transport of wind-blown surface materials by rolling or sliding movements. This process accounts for about 20 percent of all eolian surface sediment transport.

suspended load The portion of the sediment load of a stream that consists of fine textured solid particles transported within the flow, and kept from settling by turbulence.

suturing The fusion of two lithospheric plates into a single plate following their collision.

swamp A poorly drained area, frequently covered with hygrophytic vegetation.

swash The forward (up-beach) motion of water from a breaking wave.

symbols Devices employed on maps to represent specific mapped features.

syncline A downward fold in one or more rock strata. A ridge consisting of downfolded strata is termed a *synclinal ridge;* a valley consisting of downfolded strata is termed a *synclinal valley.*

system A set of interrelated components through which energy flows to produce orderly changes. A closed system has a limited supply of energy from an internal source. An open system has an unlimited supply of energy from an external source.

systematic approach An approach to study that organizes the things to be studied by type or category.

taiga The Russian name for the northern coniferous forest, the most poleward of the forest biomes.

talus cone A conical accumulation of angular rocks resting against a cliff or mountainside, derived largely from the process of frost wedging.

tarn A small lake occupying the enclosed basin of a cirque that no longer contains a glacier.

tectonic forces Landform producing forces originating inside the Earth, primarily as a result of the motions of tectonic plates.

temperature A measure of the thermal energy contained by an object or substance. This energy level is related to the rate of motion of its constituent molecules.

temperature inversion An increase in temperature with altitude in the troposphere. This situation represents a reversal of the normal decline in temperatures with altitude.

terminal moraine A linear accumulation of glacially deposited rock material that has formed along the line of maximum glacial advance. Its long axis is perpendicular to the direction of ice flow.

terrace A steplike surface that is relatively flat and linear. It is bounded on one side by a slope to a lower level and on the other by a slope to a higher elevation.

terranes Small pieces of continental crust, often of uncertain origin, that may be swept up by the movements of lithospheric plates and accreted onto the margin of a continent.

terrestrial radiation The long wave electromagnetic radiation emitted by the Earth system, which we sense as heat.

thermal energy A form of kinetic energy in which the temperature of the substance is related to the rate of motion of its molecules.

thermal high A high pressure system produced by the compaction of air as a result of cold surface temperatures.

thermal low A low pressure system produced by the expansion of air as a result of hot surface temperatures.

thermosphere The uppermost of the four thermal layers of the atmosphere, in which temperatures rise with increasing altitude.

throughflow The lateral flow of water near, but slightly below, the soil surface. It results from a temporary condition of saturation in the lower soil.

thrust fault A fault in which extreme compression causes one side to be thrust at a low angle over the other, often for a considerable distance.

thunderstorm A storm associated with cumulonimbus clouds, thunder and lightning. It is the world's most common storm type, and is often associated with heavy rains and strong local winds.

tidal bore A wave produced along the advancing front of the rising tide. It forms as the tide advances into a linear water body of restricted size, such as a river or bay.

tidal currents Currents in restricted bodies of water influenced by the tides that flow landward during the period of the rising tide and seaward during the period of the falling tide.

tidal inlet An oceanic inlet between barrier islands that connects the coastal lagoon to the open sea.

tides The rhythmic rise and fall of the ocean surface level produced by the gravitational influence of the Moon and Sun.

till Rock material deposited directly by glacial ice, and therefore displaying a lack of sorting.

till plain An extensive low relief surface covered by a thick accumulation of ground moraine. Much of the Midwest United States and the North European Plain are till plains.

topographic map A map that provides systematic information on surface elevations for the areas it depicts by means of colors, shading, contour lines, or some other method.

topography The combination of landform characteristics and distributions within a region.

topsoil The uppermost mineral horizon of the soil; often referred to as the A horizon.

tornado A tubular vortex of whirling air surrounding a central core of extremely low pressure, which extends downward from the base of a cumulonimbus cloud.

tower karst Steep-sided remnant limestone peaks of mountainous proportions.

trade winds The major wind belt of the tropics, blowing from the subtropical high pressure centers of each hemisphere toward the Intertropical Convergence Zone. Prevailing wind directions are northeasterly in the northern hemisphere and southeasterly in the southern hemisphere.

transcurrent fault A fault in which the two sides are predominantly horizontally offset. Also called a strike-slip fault.

translocation The transfer of dissolved materials and fine textured particles from one place to another in the soil as a result of soil water transport.

transpiration The evaporation from plant surfaces of water previously absorbed through the plant root systems.

transverse dune A linear sand dune with an undulating crestline that is elongated perpendicular to the prevailing wind direction.

travertine Calcium carbonate ($CaCO_3$) deposits generally found in caves. They assume a wide variety of forms, including stalactites, stalagmites, columns, and ribbons.

tree A large woody plant normally containing a single primary stem or trunk that develops many branches.

treeline The cold climate limit of tree growth at high latitudes and elevations. It normally coincides roughly with the 10° C (50° F) isotherm for the mean temperature of the warmest month.

trellis stream pattern Consists of a parallel pattern of major streams, which are joined by much smaller and shorter tributaries. This pattern typically develops in regions containing alternating bands of resistant and nonresistant sedimentary strata.

tributary A stream that flows into another larger stream.

Tropic of Cancer Located at 23¹/₂°N, it is the farthest north line of latitude that the Sun's vertical rays reach during the course of the year.

Tropic of Capricorn Located at 23¹/₂°S, it is the farthest south line of latitude that the Sun's vertical rays reach during the course of the year.

tropical depression A weak closed low pressure center that has formed over a tropical water body and has maximum sustained winds of less than 18 meters per second (40 mph). It may have the potential to develop into a tropical storm and, eventually, into a hurricane.

tropical storm A storm that has developed over a tropical water body and has maximum sustained winds of between 18 and 33.5 meters per second (40–74 mph). Tropical storms are given names, are capable of causing destruction if they strike land, and have the potential to strengthen into hurricanes.

tropics The low latitudes; often defined as those areas located between the Tropic of Cancer (23¹/₂°N) and the Tropic of Capricorn (23¹/₂°S).

tropopause The boundary between the troposphere and the stratosphere. At this level, temperatures stop falling with increasing altitude.

tropophyte A plant adapted to alternating favorable and unfavorable climatic conditions for growth. Most tropophytes grow in areas with alternating warm and cold or wet and dry seasons.

troposphere The lowermost of the four thermal layers of the atmosphere, in which temperatures normally decline with increasing altitude. Most weather phenomena exist in this layer.

tsunami A destructive ocean wave produced seismically by a sudden tectonically-induced movement of the sea floor.

tundra A cold-tolerant vegetation association consisting primarily of a mixture of small herbaceous perennials.

turbidity current A rapid submarine flow of dense, sediment-laden water down the continental slope at the outer edge of the continental shelf.

turbulence The chaotic eddying or swirling motions imparted to wind or flowing water by the effects of friction.

twilight The condition of semi-darkness caused by sunlight being scattered by air molecules and dust particles in the upper atmosphere when the Sun is less than 18° below the horizon.

typhoon The term used for a hurricane in the western North Pacific.

ultisol A soil order consisting of highly chemically weathered soils of limited natural fertility that have developed in warm, moist climates.

undertow The seaward movement of subsurface water near the shoreline. It occurs as a counterbalance to the landward flow of water carried by surface waves.

uniformitarianism The geologic principle stating that the Earth's surface features have evolved from the long continued application of currently operative geomorphic processes.

unloading A physical weathering process involving rock breakup resulting from the gradual decrease in gravitational pressure as erosion removes the overlying material.

unstable air Air with a vertical temperature profile that produces a tendency to rise, especially if the air is given an initial lift.

upwelling The flow of deep ocean water to the surface.

urban heat island The islandlike area of relatively warm temperatures associated with a city. The heat island results both from the nature of the city surface materials and from human activities that release heat.

valley A linear lowland usually occupied by a stream.

valley glacier An alpine glacier that advances down a preexisting valley.

valley train A linear train of glacio-fluvial materials deposited by a meltwater stream issuing from the margin of a glacier.

variable gas An atmospheric gas that varies appreciably in relative abundance over a short timespan.

vegetative association See *plant association.*

vegetative succession The gradual change in the composition of a plant community in response to changing environmental conditions following a disturbance such as a fire.

ventifact A desert rock with prominently wind polished, grooved, or faceted surfaces.

verbal map scale A map scale that is stated or written out fully, rather than indicated by means such as a graph or representative fraction.

vernal equinox The day that the Earth's axial inclination is exactly perpendicular with respect to the Sun, following a period in which one's hemisphere was inclined away from the Sun. Marks the official beginning of spring.

vertebrate An animal with a backbone or spinal column.

vertisol A soil order consisting of soils that contain large amounts of clay that swells when wetted and shrinks when dried. This results in a churning action that inhibits profile development.

volcanic neck A solid spire of hardened lava that was in the vent of a volcano at the end of its final eruption.

volcanism Processes involving the transfer of molten rock material either from one subsurface location to another, or its expulsion onto the surface.

volcano A hill or mountain constructed of materials from the interior of the Earth that have been ejected under pressure from a vent.

warm front A frontal boundary along which an advancing mass of warm air is overrunning and displacing relatively denser cold air.

warm ice glacier A wet-base glacier that has an underlying film of meltwater acting as a lubricant to aid slippage over its bedrock base.

water table The upper boundary of the zone of permanently saturated soil or rock; the boundary between soil water and ground water.

water vapor Water (H_2O) in the invisible gaseous state. It comprises an average of about 1.4% of the atmosphere at sea level and is the source of moisture for clouds and precipitation.

wash An often dry stream channel, usually in an arid climate. Also called a *wadi* or an *arroyo*.

wave A rhythmic vertical motion of the near-surface portion of any water body, usually generated by the wind. The water molecules move in circular or elliptical paths, with little or no net forward motion.

wave-built platform A platform of unconsolidated marine sediments that is generally deposited in the nearshore waters immediately seaward of a wave-cut platform.

wave-cut platform A rock platform that has been planed off at the base of erosionally effective wave action along an eroding coast.

wave refraction The bending of the angle of approach of incoming waves. Refraction occurs as the waves obliquely approach the shoreline and encounter a shallow bottom that slows and rotates them more directly toward shallow water.

weather The short-term condition of the atmosphere with respect to temperature, air pressure, wind, humidity, clouds, and precipitation.

weather front A boundary zone along which two or sometimes three air masses of differing temperature and moisture characteristics meet, often producing an extensive area of cloud cover and precipitation.

weather system An organized system such as an air mass, hurricane, or weather front. It consists of a combination of weather elements.

weathering The combined action of physical and chemical processes that disintegrate and decompose bedrock. An essential preliminary for soil formation and gradation.

weathering front The front (lowermost) margin of the weathering activities that are advancing downward from the surface into the underlying rock layers.

westerlies The major wind belts of the middle latitudes, blowing from the subtropical high pressure centers of each hemisphere toward the subpolar lows.

wet adiabatic rate The rate at which saturated air cools as it rises (or warms as it descends) through the troposphere. This rate averages approximately 3.2F°/1,000 feet, or 6.0C°/1,000 meters.

wilting point The soil moisture level at which plants begin to wilt. Wilting occurs because the remaining soil water is so tightly bound to the soil particles that the plants are unable to extract it.

wind The movement of air. Wind is caused by differences in air pressure.

wind shear Local variations in wind speed and direction that are sometimes dangerous to aircraft.

wind vane A mechanical device used to determine the direction of the wind.

windward coast A coast with a prevailing water to land windflow.

winter monsoon An offshore windflow of subcontinental proportions, produced by the colder temperatures that develop over land than over water during the winter season.

winter solstice The day that one's hemisphere reaches its maximum inclination away from the Sun. The Sun is directly overhead at a latitude of $23\frac{1}{2}°$ in the opposite hemisphere. Marks the official beginning of winter.

Wisonsinan stage The name used in North America for the most recent glacial period, or stage, of the Pleistocene epoch.

woodland A low density cover of woody vegetation. In contrast to a forest, it contains no continuous tree canopy.

xerophyte A plant especially adapted to grow in a region deficient in water.

zone of aeration The zone above the water table, where pore spaces or other openings are jointly occupied by soil water and gases.

zone of saturation The zone below the water table, where pore spaces or other openings are completely occupied by ground water.

INDEX

Page numbers in italics indicate illustrations.
Page numbers in bold indicate definitions.